PERIODIC TABLE OF THE ELEMENTS

1 IA	2 IIA	3 IIIB	4 IVB	5 VB	6 VIB	7 VIIB	8 VIIIB	9 VIIIB	10 VIIIB	11 IB	12 IIB	13 IIIA	14 IVA	15 VA	16 VIA	17 VIIA	18 VIIIA
1 **H** 1.0079																	2 **He** 4.00
3 **Li** 6.94	4 **Be** 9.01											5 **B** 10.81	6 **C** 12.01	7 **N** 14.01	8 **O** 16.00	9 **F** 19.00	10 **Ne** 20.18
11 **Na** 22.99	12 **Mg** 24.31											13 **Al** 26.98	14 **Si** 28.09	15 **P** 30.97	16 **S** 32.06	17 **Cl** 35.45	18 **Ar** 39.95
19 **K** 39.10	20 **Ca** 40.08	21 **Sc** 44.96	22 **Ti** 47.88	23 **V** 50.94	24 **Cr** 52.00	25 **Mn** 54.94	26 **Fe** 55.85	27 **Co** 58.93	28 **Ni** 58.71	29 **Cu** 63.54	30 **Zn** 65.37	31 **Ga** 69.72	32 **Ge** 72.59	33 **As** 74.92	34 **Se** 78.96	35 **Br** 79.91	36 **Kr** 83.80
37 **Rb** 85.47	38 **Sr** 87.62	39 **Y** 88.91	40 **Zr** 91.22	41 **Nb** 92.91	42 **Mo** 95.94	43 **Tc** 98.91	44 **Ru** 101.07	45 **Rh** 102.91	46 **Pd** 106.4	47 **Ag** 107.87	48 **Cd** 112.40	49 **In** 114.82	50 **Sn** 118.69	51 **Sb** 121.75	52 **Te** 127.60	53 **I** 126.90	54 **Xe** 131.30
55 **Cs** 132.91	56 **Ba** 137.34	71 **Lu** 174.97	72 **Hf** 178.49	73 **Ta** 180.95	74 **W** 183.85	75 **Re** 186.2	76 **Os** 190.2	77 **Ir** 192.2	78 **Pt** 195.09	79 **Au** 196.97	80 **Hg** 200.59	81 **Tl** 204.37	82 **Pb** 207.19	83 **Bi** 208.98	84 **Po** 210	85 **At** 210	86 **Rn** 222
87 **Fr** 223	88 **Ra** 226.03	103 **Lr** 262.1	104 **Rf**	105 **Db**	106 **Sg**	107 **Bh**	108 **Hs**	109 **Mt**	110 **Uun**	111 **Uuu**	112 **Uub**	113 **Uut**					

Lanthanides

57 **La** 138.91	58 **Ce** 140.12	59 **Pr** 140.91	60 **Nd** 144.24	61 **Pm** 146.92	62 **Sm** 150.35	63 **Eu** 151.96	64 **Gd** 157.25	65 **Tb** 158.92	66 **Dy** 162.50	67 **Ho** 164.93	68 **Er** 167.26	69 **Tm** 168.93	70 **Yb** 173.04

Actinides

89 **Ac** 227.03	90 **Th** 232.04	91 **Pa** 231.04	92 **U** 238.03	93 **Np** 237.05	94 **Pu** 239.05	95 **Am** 241.06	96 **Cm** 247.07	97 **Bk** 249.08	98 **Cf** 251.08	99 **Es** 254.09	100 **Fm** 257.10	101 **Md** 258.10	102 **No** 255

← Nonmetals

← Metalloids

Metals →

*Molar masses quoted to the number of significant figures given here can be regarded as typical of most naturally occurring samples.

Laboratory Techniques in Organic Chemistry

Supporting Inquiry-Driven Experiments

FOURTH EDITION

JERRY R. MOHRIG
Carleton College

DAVID G. ALBERG
Carleton College

GRETCHEN E. HOFMEISTER
Carleton College

PAUL F. SCHATZ
University of Wisconsin, Madison

CHRISTINA NORING HAMMOND
Vassar College

W. H. Freeman and Company
A Macmillan Higher Education Company

Publisher: Jessica Fiorillo
Acquisitions Editor: Bill Minick
Assistant Editor/Development Editor: Courtney Lyons
Associate Director of Marketing: Debbie Clare
Marketing Assistant: Samantha Zimbler
Project Editor: Georgia Lee Hadler
Copyeditor: Margaret Comaskey
Production Manager: Julia DeRosa
Art Director and Designer: Diana Blume
Photo Editors: Eileen Liang, Christine Buese
Project Management/Composition: Ed Dionne, MPS Ltd.
Printing and Binding: LSC Communications

Library of Congress Control Number: 2013955847

ISBN-13: 978-1-4641-3422-7
ISBN-10: 1-4641-3422-7

W. H. Freeman and Company
41 Madison Avenue, New York, NY 10010
Houndmills, Basingstoke, RG21 6XS, England
www.whfreeman.com

CONTENTS

Preface   xiii

PART 1   INTRODUCTION TO THE ORGANIC LABORATORY   1

ESSAY—The Role of the Laboratory   1

1   Safety in the Laboratory   3

    1.1   General Safety Information   4
    1.2   Preventing Chemical Exposure   5
    1.3   Preventing Cuts and Burns   8
    1.4   Preventing Fires and Explosions   9
    1.5   What to Do if an Accident Occurs   11
    1.6   Chemical Toxicology   13
    1.7   Identifying Chemicals and Understanding Chemical Hazards   14
    1.8   Handling Laboratory Waste   20
    Further Reading   21
    Questions   21

2   Green Chemistry   22

    2.1   The Principles of Green Chemistry   23
    2.2   Green Principles Applied to Industrial Processes   24
    2.3   Green Principles Applied to Academic Laboratories   28
    Further Reading   31
    Questions   32

3   Laboratory Notebooks and Prelab Information   32

    3.1   The Laboratory Notebook   33
    3.2   Calculation of the Percent Yield   35
    3.3   Sources of Prelaboratory Information   36
    Further Reading   39
    Questions   39

PART 2   CARRYING OUT CHEMICAL REACTIONS   41

ESSAY—Learning to Do Organic Chemistry   41

4   Laboratory Glassware   44

    4.1   Desk Equipment   45
    4.2   Miniscale Standard Taper Glassware   45
    4.3   Microscale Glassware   47
    4.4   Cleaning and Drying Laboratory Glassware   50
    Questions   51

5   Measurements and Transferring Reagents   **52**

  5.1   Using Electronic Balances   52
  5.2   Transferring Solids to a Reaction Vessel   54
  5.3   Measuring Volume and Transferring Liquids   55
  5.4   Measuring Temperature   62
  5.5   Measurement Uncertainty and Error Analysis   64
         Further Reading   72
         Questions   72

6   Heating and Cooling Methods   **73**

  6.1   Preventing Bumping of Liquids   73
  6.2   Conventional Heating Devices   74
  6.3   Heating with Laboratory Microwave Reactors   81
  6.4   Cooling Methods   85
  6.5   Laboratory Jacks   85
         Further Reading   86
         Questions   86

7   Setting Up Organic Reactions   **86**

  7.1   Refluxing a Reaction Mixture   87
  7.2   Addition of Reagents During a Reaction   89
  7.3   Anhydrous Reaction Conditions   90
  7.4   Inert Atmosphere Reaction Conditions   93
  7.5   Transfer of Liquids by Syringe Without Exposure to Air   101
  7.6   Removal of Noxious Vapors   103
         Further Reading   106
         Questions   106

8   Computational Chemistry   **107**

  8.1   Picturing Molecules on the Computer   107
  8.2   Molecular Mechanics Method   109
  8.3   Quantum Mechanics Methods: *Ab Initio,* Semiempirical, and DFT   115
  8.4   Which Computational Method Is Best?   121
  8.5   Sources of Confusion and Common Pitfalls   121
         Further Reading   124
         Questions   124

PART 3   BASIC METHODS FOR SEPARATION, PURIFICATION, AND ANALYSIS   **127**

ESSAY—Intermolecular Forces in Organic Chemistry   **127**

9   Filtration   **132**

  9.1   Filtering Media   132
  9.2   Gravity Filtration   134
  9.3   Small-Scale Gravity Filtration   135
  9.4   Vacuum Filtration   137
  9.5   Other Liquid-Solid and Liquid-Liquid Separation Techniques   140

9.6   Sources of Confusion and Common Pitfalls   140
       Questions   142

10   Extraction   142

10.1   Understanding How Extraction Works   143
10.2   Changing Solubility with Acid-Base Chemistry   147
10.3   Doing Extractions   149
10.4   Miniscale Extractions   152
10.5   Summary of the Miniscale Extraction Procedure   155
10.6   Microscale Extractions   155
       10.6a   Equipment and Techniques Common to Microscale Extractions   156
       10.6b   Microscale Extractions with an Organic Phase Less Dense Than Water   158
       10.6c   Microscale Extractions with an Organic Phase More Dense Than Water   160
10.7   Sources of Confusion and Common Pitfalls   161
       Questions   163

11   Drying Organic Liquids and Recovering Reaction Products   163

11.1   Drying Agents   163
11.2   Methods for Separating Drying Agents from Organic Liquids   166
11.3   Sources of Confusion and Common Pitfalls   168
11.4   Recovery of an Organic Product from a Dried Extraction Solution   169
       Questions   173

12   Boiling Points and Distillation   173

12.1   Determination of Boiling Points   174
12.2   Distillation and Separation of Mixtures   176
12.3   Simple Distillation   180
       12.3a   Miniscale Distillation   180
       12.3b   Miniscale Short-Path Distillation   183
       12.3c   Microscale Distillation Using Standard Taper 14/10 Apparatus   184
       12.3d   Microscale Distillation Using Williamson Apparatus   187
12.4   Fractional Distillation   188
12.5   Azeotropic Distillation   193
12.6   Steam Distillation   194
12.7   Vacuum Distillation   197
12.8   Sources of Confusion and Common Pitfalls   203
       Further Reading   205
       Questions   205

13   Refractometry   206

13.1   The Refractive Index   206
13.2   The Refractometer   208
13.3   Determining a Refractive Index   208
13.4   Sources of Confusion and Common Pitfalls   211
       Questions   211

14   Melting Points and Melting Ranges   211

14.1   Melting-Point Theory   212
14.2   Apparatus for Determining Melting Ranges   213

14.3   Determining Melting Ranges   215
14.4   Summary of Melting-Point Technique   217
14.5   Using Melting Points to Identify Compounds   218
14.6   Sources of Confusion and Common Pitfalls   219
         Further Reading   220
         Questions   220

15   Recrystallization   221

15.1   Introduction to Recrystallization   221
15.2   Summary of the Recrystallization Process   223
15.3   Carrying Out Successful Recrystallizations   224
15.4   How to Select a Recrystallization Solvent   225
15.5   Miniscale Procedure for Recrystallizing a Solid   228
15.6   Microscale Recrystallization   231
15.7   Microscale Recrystallization Using a Craig Tube   232
15.8   Sources of Confusion and Common Pitfalls   234
         Questions   235

16   Sublimation   236

16.1   Sublimation of Solids   236
16.2   Assembling the Apparatus for a Sublimation   237
16.3   Carrying Out a Microscale Sublimation   238
16.4   Sources of Confusion and Common Pitfalls   239
         Questions   239

17   Optical Activity and Enantiomeric Analysis   240

17.1   Mixtures of Optical Isomers: Separation/Resolution   240
17.2   Polarimetric Techniques   243
17.3   Analyzing Polarimetric Readings   247
17.4   Modern Methods of Enantiomeric Analysis   248
17.5   Sources of Confusion and Common Pitfalls   250
         Questions   251

PART 4   CHROMATOGRAPHY   253

ESSAY—Modern Chromatographic Separations   253

18   Thin-Layer Chromatography   255

18.1   Plates for Thin-Layer Chromatography   256
18.2   Sample Application   257
18.3   Development of a TLC Plate   260
18.4   Visualization Techniques   261
18.5   Analysis of a Thin-Layer Chromatogram   263
18.6   Summary of TLC Procedure   264
18.7   How to Choose a Developing Solvent When None Is Specified   265
18.8   Using TLC Analysis in Synthetic Organic Chemistry   267
18.9   Sources of Confusion and Common Pitfalls   267

Further Reading   269
Questions   269

19   Liquid Chromatography   **270**
19.1   Adsorbents   270
19.2   Elution Solvents   272
19.3   Determining the Column Size   273
19.4   Flash Chromatography   275
19.5   Microscale Liquid Chromatography   281
    19.5a   Preparation and Elution of a Microscale Column   281
    19.5b   Preparation and Elution of a Williamson Microscale Column   283
19.6   Summary of Liquid Chromatography Procedures   285
19.7   Sources of Confusion and Common Pitfalls   285
19.8   High-Performance Liquid Chromatography   287
Further Reading   291
Questions   291

20   Gas Chromatography   **291**
20.1   Instrumentation for GC   293
20.2   Types of Columns and Liquid Stationary Phases   294
20.3   Detectors   296
20.4   Recorders and Data Stations   297
20.5   GC Operating Procedures   299
20.6   Sources of Confusion and Common Pitfalls   303
20.7   Identification of Compounds Shown on a Chromatogram   304
20.8   Quantitative Analysis   305
Further Reading   308
Questions   308

PART 5   SPECTROMETRIC METHODS   **309**

ESSAY—The Spectrometric Revolution   **309**

21   Infrared Spectroscopy   **311**
21.1   IR Spectra   311
21.2   Molecular Vibrations   311
21.3   IR Instrumentation   316
21.4   Operating an FTIR Spectrometer   319
21.5   Sample Preparation for Transmission IR Spectra   319
21.6   Sample Preparation for Attenuated Total Reflectance (ATR) Spectra   323
21.7   Interpreting IR Spectra   325
21.8   IR Peaks of Major Functional Groups   330
21.9   Procedure for Interpreting an IR Spectrum   338
21.10   Case Study   339
21.11   Sources of Confusion and Common Pitfalls   341
Further Reading   344
Questions   344

22   Nuclear Magnetic Resonance Spectroscopy   348

22.1   NMR Instrumentation   350
22.2   Preparing Samples for NMR Analysis   353
22.3   Summary of Steps for Preparing an NMR Sample   357
22.4   Interpreting ^{1}H NMR Spectra   357
22.5   How Many Types of Protons Are Present?   357
22.6   Counting Protons (Integration)   358
22.7   Chemical Shift   359
22.8   Quantitative Estimation of Chemical Shifts   366
22.9   Spin-Spin Coupling (Splitting)   377
22.10   Sources of Confusion and Common Pitfalls   391
22.11   Two Case Studies   398
        Further Reading   405
        Questions   405

23 ^{13}C and Two-Dimensional NMR Spectroscopy   408

23.1 ^{13}C NMR Spectra   408
23.2 ^{13}C Chemical Shifts   412
23.3   Quantitative Estimation of ^{13}C Chemical Shifts   417
23.4   Determining Numbers of Protons on Carbon Atoms—APT and DEPT   427
23.5   Case Study   429
23.6   Two-Dimensional Correlated Spectroscopy (2D COSY)   431
        Further Reading   435
        Questions   435

24   Mass Spectrometry   441

24.1   Mass Spectrometers   442
24.2   Mass Spectra and the Molecular Ion   446
24.3   High-Resolution Mass Spectrometry   450
24.4   Mass Spectral Libraries   451
24.5   Fragment Ions   453
24.6   Case Study   459
24.7   Sources of Confusion   461
        Further Reading   462
        Questions   462

25   Ultraviolet and Visible Spectroscopy   465

25.1   UV-VIS Spectra and Electronic Excitation   466
25.2   UV-VIS Instrumentation   471
25.3   Preparing Samples and Operating the Spectrometer   472
25.4   Sources of Confusion and Common Pitfalls   474
        Further Reading   475
        Questions   475

26   Integrated Spectrometry Problems   476

PART 6   DESIGNING AND CARRYING OUT ORGANIC EXPERIMENTS   485

ESSAY—Inquiry-Driven Lab Experiments   485

27   Designing Chemical Reactions   488

27.1   Reading Between the Lines: Carrying Out Reactions Based on Literature Procedures   488
27.2   Modifying the Scale of a Reaction   494
27.3   Case Study: Synthesis of a Solvatochromic Dye   497
27.4   Case Study: Oxidation of a Secondary Alcohol to a Ketone   499
Further Reading   500

28   Using the Literature of Organic Chemistry   501

28.1   The Literature of Organic Chemistry   501
28.2   Searching the Literature of Organic Chemistry   504
28.3   Planning a Multistep Synthesis   506

Index   511

PREFACE

In preparing this Fourth Edition of *Laboratory Techniques in Organic Chemistry*, we have maintained our emphasis on the fundamental techniques that students encounter in the organic chemistry laboratory. We have also expanded our emphasis on the critical-thinking skills that students need to successfully carry out inquiry-driven experiments. The use of guided-inquiry and design-based experiments and projects is arguably the most important recent development in the teaching of the undergraduate organic chemistry lab, and it provides the most value added for our students.

Organic chemistry is an experimental science, and students learn its process in the laboratory. Our primary goal should be to teach students how to carry out well-designed experiments and draw reasonable conclusions from their results—a process at the heart of science. We should work to find opportunities that engage students in addressing questions whose answers come from their experiments, in an environment where they can succeed. These opportunities should be designed to catch students' interest, transforming them from passive spectators to active participants. A well-written and comprehensive textbook on the techniques of experimental organic chemistry is an important asset in reaching these goals.

Changes in the Fourth Edition

The Fourth Edition of *Laboratory Techniques in Organic Chemistry* builds on our strengths in basic lab techniques and spectroscopy, and includes a number of new features. To make it easier for students to locate the content relevant to their experiments, icons distinguish the techniques specific to each of the three common types of lab glassware—miniscale standard taper, microscale standard taper, and Williamson glassware—and also highlight safety concerns.

Sections on microwave reactors, flash chromatography, green chemistry, handling air-sensitive reagents, and measurement uncertainty and error analysis have been added or updated. The newly added Part 6 emphasizes the skills students need to carry out inquiry-driven experiments, especially designing and carrying out experiments based on literature sources. Many sections concerning basic techniques have been modified and reorganized to better meet the practical needs of students as they encounter laboratory work. Additional questions have also been added to a number of chapters to help solidify students' understanding of the techniques.

Short essays provide context for each of the six major parts of the Fourth Edition, on topics from the role of the laboratory to the spectrometric revolution. The essay "Intermolecular Forces in Organic Chemistry" provides the basis for subsequent discussions on organic separation and purification techniques, and the essay "Inquiry-Driven Lab Experiments" sets the stage for using guided-inquiry and design-based experiments. Rewritten sections on sources of confusion and common pitfalls help students avoid and solve technical problems that could easily discourage them if they did not have this practical support. We believe that these features provide an effective learning tool for students of organic chemistry.

Who Should Use This Book?

The book is intended to serve as a laboratory textbook of experimental techniques for all students of organic chemistry. It can be used in conjunction with any lab experiments to provide the background information necessary for developing and mastering the skills required for organic chemistry lab work. *Laboratory Techniques in Organic Chemistry* offers a great deal of flexibility. It can be used in any organic laboratory with any glassware. The basic techniques for using miniscale standard taper glassware as well as microscale 14/10 standard taper or Williamson glassware are all covered. The miniscale glassware that is described is appropriate with virtually any 14/20 or 19/22 standard taper glassware kit.

Modern Instrumentation

Instrumental methods play a crucial role in supporting modern experiments, which provide the active learning opportunities instructors seek for their students. We feature instrumental methods that offer quick, reliable, quantitative data. NMR spectroscopy and gas chromatography are particularly important. Our emphasis is on how to acquire good data and how to read spectra efficiently, with real understanding. Chapters on ^{1}H and ^{13}C NMR, IR, and mass spectrometry stress the practical interpretation of spectra and how they can be used to answer questions posed in an experimental context. They describe how to deal with real laboratory samples and include case studies of analyzed spectra.

Organization

The book is divided into six parts:
- Part 1 has chapters on safety, green chemistry, and the lab notebook.
- Part 2 discusses lab glassware, measurements, heating and cooling methods, setting up organic reactions, and computational chemistry.
- Part 3 introduces filtration, extraction, drying organic liquids and recovering products, distillation, refractometry, melting points, recrystallization, and the measurement of optical activity.
- Part 4 presents the three chromatographic techniques widely used in the organic laboratory—thin-layer, liquid, and gas chromatography.
- Part 5 discusses IR, ^{1}H and ^{13}C NMR, MS, and UV-VIS spectra in some detail.
- Part 6 introduces the design and workup of chemical reactions based on procedures in the literature of organic chemistry.

Traditional organic qualitative analysis is available on our Web site: www.whfreeman.com/mohrig4e.

Modern Projects and Experiments in Organic Chemistry

The accompanying laboratory manual, *Modern Projects and Experiments in Organic Chemistry*, comes in two complete versions:
- *Modern Projects and Experiments in Organic Chemistry: Miniscale and Standard Taper Microscale* (ISBN 0-7167-9779-8)
- *Modern Projects and Experiments in Organic Chemistry: Miniscale and Williamson Microscale* (ISBN 0-7167-3921-6)

Modern Projects and Experiments is a combination of inquiry-based and traditional experiments, plus multiweek inquiry-based projects. It is designed to provide quality content, student accessibility, and instructor flexibility. This laboratory manual introduces students to the way the contemporary organic lab actually functions and allows them to experience the process of science. All of its experiments and projects are also available through LabPartner Chemistry.

LabPartner Chemistry

W. H. Freeman's latest offering in custom lab manuals provides instructors with a diverse and extensive database of experiments published by W. H. Freeman and Hayden-McNeil Publishing—all in an easy-to-use, searchable online system. With the click of a button, instructors can choose from a variety of traditional and inquiry-based labs, including the experiments from *Modern Projects and Experiments in Organic Chemistry*. LabPartner Chemistry sorts labs in a number of ways, from topic, title, and author, to page count, estimated completion time, and prerequisite knowledge level. Add content on lab techniques and safety, reorder the labs to fit your syllabus, and include your original experiments with ease. Wrap it all up in an array of bindings, formats, and designs. It's the next step in custom lab publishing. Visit http://www.whfreeman.com/labpartner to learn more.

Acknowledgments

We have benefited greatly from the insights and thoughtful critiques of the reviewers for this edition:

Dan Blanchard, *Kutztown University of Pennsylvania*

Jackie Bortiatynski, *Pennsylvania State University*

Christine DiMeglio, *Yale University*

John Dolhun, *Massachusetts Institute of Technology*

Jane Greco, *Johns Hopkins University*

Rich Gurney, *Simmons College*

James E. Hanson, *Seton Hall University*

Paul R. Hanson, *University of Kansas*

Steven A. Kinsley, *Washington University in St. Louis*

Deborah Lieberman, *University of Cincinnati*

Joan Mutanyatta-Comar, *Georgia State University*

Owen P. Priest, *Northwestern University*

Nancy I. Totah, *Syracuse University*

Steven M. Wietstock, *University of Notre Dame*

Courtney Lyons, our editor at W. H. Freeman and Company, was great in so many ways throughout the project, from the beginning to its final stages; her skillful editing and thoughtful critiques have made this a better textbook and it has been a pleasure to work with her. We especially thank Jane Wissinger of the University of Minnesota and Steven Drew and Elisabeth Haase, our colleagues at Carleton College, who provided helpful insights regarding specific chapters for this edition. The entire team at Freeman, especially Georgia Lee Hadler and Julia DeRosa, have been effective in coordinating the copyediting and publication processes. We thank Diana Blume for her creative design elements. Finally, we express heartfelt thanks for the patience and support of our spouses, Adrienne Mohrig, Ellie Schatz, and Bill Hammond, during the several editions of *Laboratory Techniques in Organic Chemistry*.

We hope that teachers and students of organic chemistry find our approach to laboratory techniques effective, and we would be pleased to hear from those who use our book. Please write to us in care of the Chemistry Acquisitions Editor at W. H. Freeman and Company, 41 Madison Avenue, New York, NY 10010, or e-mail us at chemistry@whfreeman.com.

Introduction to the Organic Laboratory

Essay—The Role of the Laboratory

Organic chemistry provides us with a framework to understand ourselves and the world in which we live. Organic compounds are present everywhere in our lives—they comprise the food, fabrics, cosmetics, and medications that we use on a daily basis. By studying how the molecules of life interact with one another, we can understand the chemical processes that sustain life and discover new compounds that could potentially transform our lives. For example, organic chemistry was used to discover the cholesterol-lowering blockbuster drug, Lipitor®. Current research in organic semiconductors, which are more flexible, cheaper, and lighter in weight than silicon-based components, could lead to solar cells incorporated into clothing, backpacks, and virtually anything. The purpose of this textbook is to provide *you* with the skills and knowledge to make new discoveries like these, view the world from a new perspective, and ultimately harness the power of organic chemistry.

It is in the laboratory that we learn "how we know what we know." The lab deals with the processes of scientific inquiry that organic chemists use. Although the techniques may at first appear complicated and mysterious, they are essential tools for addressing the central questions of this experimental science, which include:

- What chemical compounds are present in this material?
- What is this compound and what are its properties?
- Is this compound pure?
- How could I make this compound?
- How does this reaction take place?
- How can I separate my product from other reaction side products?

Keep in mind that the skills you will be learning are very practical and there is a reason for each and every step. You should make it your business to understand **why** these steps are necessary and **how** they accomplish the desired result. If you can answer these questions for every lab session, you have fulfilled the most basic criterion for satisfactory lab work.

You may also have opportunities to test your own ideas by designing new experiments. Whenever you venture into the unknown, it becomes even more important to be well informed and organized *before you start any experiments.* Safety should be a primary concern, so you will need to recognize potential hazards, anticipate possible outcomes, and responsibly dispose of chemical waste. In order to make sense of your data and report your findings to others, you will need to keep careful records of your experiments. The first section of this textbook introduces you to reliable sources of information, safety procedures, ways to protect the environment, and standards for laboratory record-keeping. It is important to make these practices part of your normal laboratory routine. If you are ever unsure about your preparation for lab, ask your instructor.

There is no substitute for witnessing chemical transformations and performing separation processes in the laboratory. Lab work enlivens the chemistry that you are learning "on paper" and helps you understand how things work. Color changes, phase changes, and spectral data are fun to witness and fun to analyze and understand. Enjoy this opportunity to experiment in chemistry and come to lab prepared and with your brain engaged!

1

SAFETY IN THE LABORATORY

All of the stories in this chapter are based on the authors' experiences working and teaching in the lab.

Carrie used a graduated cylinder to measure a volume of concentrated acid solution at her lab bench. As she prepared to record data in her notebook later in the day, she picked up her pen from the bench-top and absent-mindedly started chewing on the cap. Suddenly, she felt a burning sensation in her mouth and yelled, "It's hot!" The lab instructor directed her to the sink to thoroughly rinse her mouth with water and she suffered no long-term injury.

This incident is like most laboratory accidents; it resulted from inappropriate lab practices and inattention, and it was preventable. Carrie should have handled the concentrated acid in a fume hood and, with advice from her instructor, immediately cleaned up the acid she must have spilled. She should never have introduced any object in the lab into her mouth. With appropriate knowledge, most accidents are easily remedied. In this case, the instructor knew from her shout what the exposure must have been and advocated a reasonable treatment.

Accidents in teaching laboratories are extremely rare; instructors with 20 years of teaching experience may witness fewer than five mishaps. Instructors and institutions continually implement changes to the curriculum and laboratory environment that improve safety. Experiments are now designed to use very small amounts of material, which minimizes the hazards associated with chemical exposure and fire. Laboratories provide greater access to fume hoods for performing reactions, and instructors choose the least hazardous materials for accomplishing transformations. Nevertheless, you play an important role in ensuring that the laboratory is as safe as possible.

You can rely on this textbook and your teacher for instruction in safe and proper laboratory procedures. **You are responsible for developing good laboratory habits**: Know and understand the laboratory procedure and associated hazards, practice good technique, and be aware of your actions and the actions of those around you. Habits like these are transferable to other situations and developing them will not only enable you to be effective in the laboratory but also help you to become a valuable employee and citizen.

The goal of safety training is to manage hazards in order to minimize the risk of accidental chemical exposure, personal injury, or damage to property or the environment.

- **Before you begin laboratory work,** familiarize yourself with the general laboratory safety rules (listed below) that govern work at any institution.
- **At the first meeting of your lab class,** learn institutional safety policies regarding personal protective equipment (PPE), the location and use of safety equipment, and procedures to be followed in emergency situations.
- **For each individual experiment,** note the safety considerations identified in the description of the procedure, the hazards

associated with the specific chemicals you will use, and the waste disposal instructions.

In addition to knowledge of basic laboratory safety, you need to learn how to work safely with organic chemicals. Many organic compounds are flammable or toxic. Many can be absorbed through the skin; others are volatile and can be ingested by inhalation. Become familiar with and use chemical hazard documentation, such as the Globally Harmonized System (GHS) of hazard information and Material Safety Data Sheets (MSDSs) or Safety Data Sheets (SDSs). Despite the hazards, organic compounds can be handled with a minimum of risk if you are adequately informed about the hazards and safe handling procedures, and if you use common sense while you are in the lab.

1.1   General Safety Information

General Safety Rules

1. **Do not work alone in the laboratory.** Being alone in a situation in which accidents can occur can be life threatening.
2. **Always perform an experiment as specified.** Do not modify the conditions or perform new experiments without authorization from your instructor.
3. **Wear clothing that covers and protects your body; use appropriate protective equipment, such as goggles and gloves; and tie back long hair at all times in the laboratory.** Shorts, tank tops, bare feet, sandals, or high heels are not suitable attire for the lab. Loose clothing and loose long hair are fire hazards or could become entangled in an apparatus. Wear safety glasses or chemical splash goggles at all times in the laboratory. Laboratory aprons or coats may be required by your instructor.
4. **Be aware of others working near you** and the hazards associated with their experiments. Often the person hurt worst in an accident is the one standing next to the place where the accident occurred. Communicate with others and make them aware of the hazards associated with your work.
5. **Never eat, drink, chew gum, apply makeup, or remove or insert contact lenses in the laboratory. Never directly inhale or taste any substance or introduce any laboratory equipment, such as a piece of glassware or a writing utensil, into your mouth.** Wash your hands with soap and water before you leave the laboratory to avoid accidentally contaminating the outside environment, including items that you may place into your mouth with your hands.
6. **Notify your instructor if you have chemical sensitivities or allergies or if you are pregnant.** Discuss these conditions and the advisability of working in the organic chemistry laboratory with appropriate medical professionals.
7. **Read and understand the hazard documentation regarding any chemicals you plan to use in an experiment.** This can be found in Material Safety Data Sheets (MSDSs) or Safety Data Sheets (SDSs).

8. **Know where to find and how to use safety equipment,** such as the eye wash station, safety shower, fire extinguisher, fire blanket, first aid kit, telephone, and fire alarm pulls.
9. **Report injuries, accidents, and other incidents to your instructor** and follow his or her instructions for treatment and documentation.
10. **Properly dispose of chemical waste**, including chemically contaminated disposable materials, such as syringes, pipets, gloves, and paper. **Do not dispose of any chemicals by pouring them down the drain or putting them in the trash can without approval from your instructor.**

Chemical Hygiene Plan

Your institution will have a chemical hygiene plan that outlines the safety regulations and procedures that apply in your laboratory. It will provide contact information and other information about local safety rules and processes for managing laboratory fires, injuries, chemical spills, and chemical waste. You can search the institutional web pages or ask your instructor for access to the chemical hygiene plan.

1.2 Preventing Chemical Exposure

Mary was wearing nitrile gloves while performing an extraction with dichloromethane. Although she spilled some solution on her gloves, she continued working until she felt her hands burning. She peeled off the gloves and washed her hands thoroughly, but a burning sensation under her ring persisted for 5 to 10 minutes thereafter. She realized that the dichloromethane solution easily passed through her gloves and she wondered whether her exposure to dichloromethane and the compounds dissolved in it would have an adverse effect on her health.

Personal Protective Equipment

This example illustrates the importance of understanding the level of protection provided by *personal protective equipment (PPE)* and other safety features in the laboratory.

- **Never assume that clothing, gloves, lab coats, or aprons will protect you from every kind of chemical exposure.** If chemicals are splashed onto your clothing or your gloves, remove the articles immediately and thoroughly wash the affected area of your body.
- If you spill a chemical directly on your skin, wash the affected area thoroughly with water for 10–15 min, and notify your instructor.

Eye protection. Safety glasses with side shields have impact-resistant lenses that protect your eyes from flying particles, but they provide little protection from chemicals. Chemical splash goggles fit snugly against your face and will guard against the impact from flying objects *and* protect your eyes from liquid splashes, chemical vapors, and particulate or corrosive chemicals. These are the best choice for the organic chemistry laboratory and your instructor will be able to recommend an appropriate style to purchase. If you wear prescription eyeglasses, you should wear chemical splash goggles over your corrective lenses. Contact lenses could be damaged from exposure to chemicals and therefore you should not wear them in the laboratory. Nevertheless, many organizations have removed restrictions on wearing contact lenses in the lab because concerns that they contribute to the likelihood or severity of eye damage seem to be unfounded. If you choose to wear contact lenses in the laboratory, you must also wear chemical splash goggles to protect your eyes. Because wearing chemical splash goggles is one of the most important steps you can take to safely work in the laboratory, **we will use a splash goggle icon (see margin figure) to identify important safety information throughout this textbook.**

Protective attire. Clothes should cover your body from your neck to at least your knees and shoes should completely cover your feet in the laboratory. Cotton clothing is best because synthetic fabrics could melt in a fire or undergo a reaction that causes the fabric to adhere to the skin and severely burn it. Wearing a lab coat or apron will help protect your body. For footwear, leather provides better protection than other fabrics against accidental chemical spills. Your institution may have more stringent requirements for covering your body.

Disposable gloves. Apart from goggles, gloves are the most common form of PPE used in the organic laboratory. Because disposable gloves are thin, many organic compounds permeate them quickly and they provide "splash protection" only. **This means that once you spill chemicals on your gloves, you should remove them, wash your hands thoroughly, and put on a fresh pair of gloves.** Ask your instructor how to best dispose of contaminated gloves.

Table 1.1 lists a few common chemicals and the chemical resistance to each one provided by three common types of gloves. A

TABLE 1.1	Chemical resistance of common types of gloves to various compounds		
		Glove type	
Compound	Neoprene	Nitrile	Latex
Acetone	Good	Fair	Good
Chloroform	Good	Poor	Poor
Dichloromethane	Fair	Poor	Poor
Diethyl ether	Very good	Good	Poor
Ethanol	Very good	Good	Good
Ethyl acetate	Good	Poor	Fair
Hexane	Excellent	Excellent	Poor
Hydrogen peroxide	Excellent	Good	Good
Methanol	Very good	Fair	Fair
Nitric acid (conc.)	Good	Poor	Poor
Sodium hydroxide	Very good	Good	Excellent
Sulfuric acid (conc.)	Good	Poor	Poor
Toluene	Fair	Fair	Poor

The information in this table was compiled from http://www.microflex.com, http://www.ansellpro.com, and "Chemical Resistance and Barrier Guide for Nitrile and Natural Rubber Latex Gloves," Safeskin Corporation, San Diego, CA, 1999.

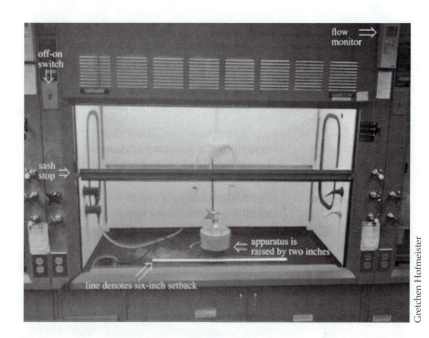

FIGURE 1.1 A typical chemical fume hood.

more extensive chemical resistance table for types of gloves may be posted in your laboratory. Additional information on disposable gloves and tables listing glove types and their chemical resistance are also available from many websites, for example:

http://www.microflex.com
http://www.ansellpro.com
http://chemistry.umeche.maine.edu/Safety.html

Chemical Fume Hoods

You can protect yourself from accidentally inhaling noxious chemical fumes, toxic vapors, or dust from finely powdered materials by handling chemicals inside a fume hood. A typical fume hood with a movable sash is depicted in Figure 1.1. The sash is constructed of laminated safety glass and can open and close either vertically or horizontally. When the hood is turned on, a continuous flow of air sweeps over the bench top and removes vapors or fumes from the area. The volume of air that flows through the sash opening is constant, so the rate of flow, or face velocity, is greater when the sash is closed than when it is open. Most hoods have stops or signs indicating the maximum open sash position that is safe for handling chemicals. If you are unsure what is a safe sash position for the hoods in your laboratory, ask your instructor.

Because many compounds used in the organic laboratory are at least potentially dangerous, the best practice is to run every experiment in a hood, if possible. Your instructor will tell you when an experiment *must* be carried out in a hood.

- **Make sure that the hood is turned on before you use it.**
- Never position your face near the sash opening or place your head inside a hood when chemicals are present. Keep the sash in front of your face so that you look through the sash to monitor what is inside the hood.
- Place chemicals and equipment at least six inches behind the sash opening.

- Elevate reaction flasks and other equipment at least two inches above the hood floor to ensure good airflow around the apparatus.
- When you are not actively manipulating equipment in the hood, adjust the sash so that it covers most of the hood opening and shields you from the materials inside.

A link to a YouTube video, created at Dartmouth College, which describes the function and use of fume hoods, can be found at: http://www.youtube.com/watch?v=nlAaEpWQdwA .

Chemical Hygiene

Poor housekeeping often leads to accidental chemical exposure. In addition to your own bench area, the balance and chemical dispensing and waste areas must be kept clean and orderly.

- If you spill anything while measuring out your chemicals, notify your instructor and clean it up immediately.
- After weighing a chemical, replace the cap on the container and dispose of the weighing paper in the appropriate receptacle.
- Clean glassware, spatulas, and other equipment as soon as possible after using them.
- Always remove gloves, lab coat, or apron before leaving the laboratory to prevent widespread chemical contamination.
- Dispose of chemical waste appropriately.

1.3   Preventing Cuts and Burns

As Harvey adjusted a pipet bulb over the end of a disposable glass pipet, the pipet broke and the broken end jammed into his thumb, cutting it badly. Harvey required hand surgery to repair a damaged nerve and he could not manipulate his thumb for several months afterward.

Cuts

While Harvey's accident was unusually severe, the most common laboratory injuries are cuts from broken glass or puncture wounds from syringe needles. For this reason, handle glassware and sharp objects with care.

- Check the rims of beakers, flasks, and other glassware for chips and discard any piece of glassware that is chipped.
- If you break a piece of glassware, use a dustpan and broom instead of your hands to pick up the broken pieces.
- Do not put broken glass or used syringe needles in the trash can. Dispose of them separately—broken glass in the broken glass container and syringe needles in the sharps receptacle.
- If a stopper, stopcock, or other glass item seems stuck, do not force it. Ask your instructor, who is more experienced with the equipment, for assistance in these cases.
- To safely insert thermometers or glass tubes into corks, rubber stoppers, and thermometer adapters, lubricate the end of the glass with a drop of water or glycerol, hold the tube near the lubricated end, and insert it slowly by gently rotating it.

- **Never push on the end of a glass tube or a thermometer to insert it into a stopper; it may break and the shattered end could be driven into your hand.**

Burns

Remember that glass and the tops of hot plates look the same when they are hot as when they are cold. Steam and hot liquids also cause severe burns. Liquid nitrogen and dry ice can quickly give you frostbite.

- Do not put hot glass on a bench where someone else might pick it up.
- Turn off the steam source before removing containers from the top of a steam bath.
- The screws or valve stems attached to the rounded handle that controls a steam line can become very hot; be careful not to touch them when you turn the steam on or off.
- Move containers of hot liquids only if necessary and use a clamp, tongs, rubber mitts, or oven gloves to hold them.
- Wear insulated gloves when handling dry ice and wear insulated gloves, a face shield, long pants, and long sleeves when dispensing liquid nitrogen.

1.4 Preventing Fires and Explosions

Michael was purifying a reaction product by distillation on the laboratory bench. The product mixture also contained diethyl ether. About halfway through the distillation, the distilled material caught fire. Michael's instructor used a fire extinguisher to put out the fire and assisted Michael in turning off the heating mantle and lifting the distillation system away from the heat source. As soon as possible, the entire apparatus was relocated to the fume hood and Michael was instructed to chill the receiving flask in an ice bath, to minimize the escape of flammable vapors from the flask.

Fires

Hydrocarbons and many of their derivatives are flammable and the potential for fire in the organic laboratory always exists. Fortunately, most modern lab procedures require only small amounts of material, minimizing the risk of fire. *Flammable* compounds do not spontaneously combust in air; they require a spark, a flame, or heat to catalyze the reaction. Vapors from low-boiling organic liquids, such as diethyl ether or pentane, can travel over long distances at bench or floor level (they are heavier than air) and thus they are susceptible to ignition by a source that is located up to 10 ft away. The best way to prevent a fire is to prevent ignition.

Four sources of ignition are present in the organic laboratory: **open flames**, **hot surfaces** such as hot plates or heating mantles (Figure 1.2), **faulty electrical equipment**, and **chemicals**. Flames, such as those produced by Bunsen burners, should be used rarely in the organic laboratory and only with the permission of your instructor. Hot plates and heating mantles, however, are used routinely. The thermostat on most hot plates is not sealed and can spark when it cycles on and off. The spark can ignite flammable vapors from

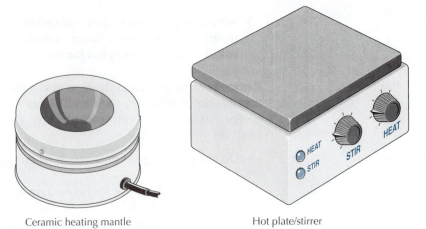

FIGURE 1.2 Heating devices.

Ceramic heating mantle   Hot plate/stirrer

an open container such as a beaker. An organic solvent spilled or heated recklessly on a hot plate surface can also burst into flames. Chemical reactions sometimes produce enough heat to cause a fire and explosion. For example, in the reaction of metallic sodium with water, the hydrogen gas that forms in the reaction can explode and ignite a volatile solvent that happens to be nearby.

- **Never bring a lighted Bunsen burner or match near a low-boiling-point flammable liquid.**
- Work in a fume hood, where flammable vapors are swept away from sources of ignition before they can catch fire.
- Flammable solvents with boiling points below 100°C—such as diethyl ether, methanol, pentane, hexane, and acetone—should be distilled, heated, or evaporated on a steam bath or heating mantle, never on a hot plate or with a Bunsen burner.
- Use an Erlenmeyer flask fitted with a stopper—**never an open beaker**—for temporarily storing flammable solvents at your work area.
- Before pouring a volatile organic liquid, remove any hot heating mantle or hot plate from the vicinity.
- Do not use appliances with frayed or damaged electrical cords; notify your instructor of faulty equipment so it can be removed and replaced.

Explosions

Explosive compounds combine a fuel and an oxidant in the same molecule and decompose to evolve gaseous products with enough energy for the hot, expanding gases to produce a shock wave. Ammonium nitrate, $NH_4^+NO_3^-$, explosively produces gaseous N_2O and H_2O when detonated. You will not handle explosive chemicals in the instructional laboratory, although some chemical reactions, when improperly performed, can rapidly generate hot gases and cause an explosion.

A more likely scenario that you could encounter is an explosion due to accidentally allowing pressure to build up inside a closed vessel. If the pressure gets high enough or if there is a weakness in the wall of the vessel, it can fail in an explosive manner.

- Unless your instructor specifies otherwise, **never heat a closed system!** Some glassware, however, is designed to sustain pressure when heated and it may be used in certain applications.
- Never completely close off an apparatus in which a gas is being evolved: always provide a vent in order to prevent an explosion.
- Routinely check flasks for defects, such as star cracks (Figure 1.3), which may lead to a catastrophic failure of the flask.
- Perform reactions in a hood and use the sash to cover the opening when you are not actively manipulating equipment. The hood sash is constructed of laminated safety glass, which is a blast shield.

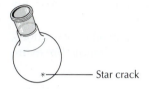

Star crack

FIGURE 1.3 Round-bottomed flask with a star crack.

Implosions

Implosions are the opposite of explosions. They occur when containers under vacuum cannot sustain the pressure exerted by the outside atmosphere and fail catastrophically. You will be handling evacuated flasks in the laboratory if you perform vacuum filtrations (Section 9.4), rotary evaporations (Section 11.4), or vacuum distillations (Section 12.7). Vacuum flasks are also used for holding very cold liquids (Section 6.4). Filter flasks and glassware used for rotary evaporation are heavy walled and designed to sustain pressure; therefore, the danger of implosion is small.

In order to prevent injuries from accidental implosions:

- Routinely check flasks for defects, such as star cracks (Figure 1.3), which may lead to a catastrophic failure of the flask.
- Perform vacuum-based procedures in a hood (with the hood sash serving as protection) or behind a safety shield, which is a heavy, portable buffer constructed of high-impact-resistant polycarbonate.
- Wrap containers that are routinely kept under vacuum with plastic mesh or electrical tape. Examples are filter flasks and Dewar flasks, which are vacuum-sealed thermos flasks for holding very cold liquids. Never use a Dewar flask that does not have a protective metal case on the outside. If a flask should implode, the metal case or tape or mesh will contain the broken glass and prevent flying shards from causing injury.

1.5   What to Do if an Accident Occurs

Always inform your instructor immediately of any safety incident or accident that happens to you or your neighbors. If a physician's attention is necessary, an injured person should always be accompanied to the medical facility; the injury may be more serious than it initially appears.

Fire

Colleges and universities all have standard policies regarding the handling of fires, which will be described in the Chemical Hygiene Plan and by your instructor. **Learn where the exits from your** laboratory are located. In case of a fire in the lab, get out of danger and notify your instructor as soon as possible.

Fire extinguishers. There are several types of fire extinguishers, and your instructor may demonstrate their use. Your lab is probably equipped with either class BC or class ABC dry chemical fire extinguishers suitable for solvent or electrical fires. At some institutions, instructors are the only people who are allowed to handle fire extinguishers in the laboratory.

- To use a fire extinguisher, aim low and direct the nozzle first toward the edge of the fire and then toward the middle.
- Do not use water to extinguish chemical fires.

Fire blankets. Fire blankets are used to smother a fire involving a person's clothing. Know where the fire blanket is located in your lab.

- If a person's clothing catches fire, ease the person to the floor and roll the person's body tightly in a fire blanket. When the blanket is wrapped around a person who is standing, it may direct the flames toward the person's face.
- If your clothing is on fire, do not run.

Safety shower. The typical safety shower dumps a huge volume of water in a short period of time and is effective when a person's clothing or hair is ablaze and speed is of the essence. **Do not use the safety shower routinely, but do not hesitate to use it in an emergency.**

Chemical Burns

The first thing to do if any chemical is spilled on your skin, unless you have been specifically told otherwise, is to wash the area well with water for 10–15 min. This will rinse away the excess chemical reagent. For acids, bases, and toxic chemicals, thorough washing with water will lessen pain later. Skin contact with a strong base usually does not produce immediate pain or irritation, but serious tissue damage (especially to the eyes) can occur if the affected area is not immediately washed with copious amounts of water. **Notify your instructor immediately if any chemical is spilled on your skin.** Seek immediate medical treatment for any serious chemical burn.

Safety shower. Safety showers are effective for acid burns and other spills of corrosive, irritating, or toxic chemicals on the skin or clothing. Remove clothing that has been contaminated by chemicals. Do this as quickly as possible while in the shower.

Eye wash station. Learn the location of the eye wash stations in your laboratory and examine the instructions on them during the first (check-in) lab session. If you accidentally splash something in your eyes, *immediately* use the eye wash station to rinse them with copious quantities of slightly warm water for 10–15 min.

- Do not use very cold water because it can damage the eyeballs.
- Position your head so that the stream of water from the eye wash fountain is directed at your eyes.

- **Hold your eyes open** to allow the water to flush the eyeballs for 10–15 min. Because this position can be difficult to maintain, assistance may be required. Do not hesitate to call for help.
- Move eyeballs up, down, and sideways while flushing with water to wash behind the eyelids.
- If you are wearing contact lenses, they must be removed for the use of an eye wash station to be effective, an operation that is extremely difficult if a chemical is causing severe discomfort to your eyes. Therefore, **it is prudent not to wear contact lenses in the laboratory.**
- Seek medical treatment immediately after using the eye wash for *any* chemical splash in the eyes.

Minor Cuts and Burns

Learn the location of the first aid kit and the materials it contains for the treatment of simple cuts and burns. All injuries, no matter how slight, should be reported to your instructor immediately. Seek immediate medical attention for anything except the most trivial cut or burn.

First aid kit. Your laboratory or a nearby stockroom may contain a basic first aid kit consisting of such items as adhesive bandages, sterile pads, and adhesive tape for treating a small cut or burn.

- Apply pressure to cuts to help slow the bleeding. Apply a bandage when the bleeding has stopped. If the cut is large or deep, seek immediate medical attention.
- When the cut is a result of broken glass, ensure that there is no glass remaining in the wound; if you are unsure, seek medical attention.
- For a heat burn, apply cold water for 10–15 min. Seek immediate medical attention for any extensive burn.
- For a cold burn, do not apply heat. Instead, treat the affected area with large volumes of tepid water and seek medical attention.

1.6   Chemical Toxicology

Most substances are toxic at some level, but the level varies over a wide range. A major concern in chemical toxicology is quantity or dosage. It is important that you understand how toxic compounds can be handled safely in the organic laboratory.

The toxicity of a compound refers to its ability to produce injury once it reaches a susceptible site in the body. A compound's toxicity is related to its probability of causing injury and is a species-dependent term. What is toxic for people may not be toxic for other animals and vice versa. A substance is *acutely toxic* if it causes a toxic effect in a short time; it is *chronically toxic* if it causes toxic effects with repeated exposures over a long time.

Fortunately, not all toxic substances that accidentally enter the body reach a site where they can be harmful. Even if a toxic substance is absorbed, it is often excreted rapidly. Our body protects us with various devices: the nose, scavenger cells, metabolism, and rapid exchange of good air for bad. Many foreign substances are detoxified and discharged from the body very quickly.

Action of Toxic Substances on the Body

Although many substances are toxic to the entire system (arsenic, for example), many others are site specific. Carbon monoxide, for example, forms a complex with the hemoglobin in our blood, diminishing the blood's ability to absorb and release oxygen; it also poisons the action of mitochondrial aerobic metabolism.

In some cases, the metabolites of a compound are more toxic than the original compound. An example is methanol poisoning. The formic acid that is formed by the body's metabolism of methanol affects the optic nerve, causing blindness. The metabolism of some relatively harmless polycyclic aromatic hydrocarbons produces potent carcinogenic compounds. As far as our health is concerned, it does not matter whether the toxicity is due to the original substance or a metabolic product of it.

Toxicity Testing and Reporting

Consumers are protected by a series of laws that define toxicity, the legal limits and dosages of toxic materials, and the procedures for measuring toxicities.

Acute oral toxicity is measured in terms of LD_{50}. (*LD* stands for lethal dose.) LD_{50} represents the dose, in milligrams per kilogram of body weight, that will be fatal to 50% of a certain population of animals. Other tests include dermal toxicity (skin sensitization) and inhalation toxicity. In the case of inhalation, LC_{50} (*LC* stands for lethal concentration) is used to standardize toxic properties. Toxicity information is included as part of the MSDS or SDS for chemicals that are commercially available. A wall chart of toxicities for many common organic compounds may be hanging in your laboratory or near your stockroom.

1.7 Identifying Chemicals and Understanding Chemical Hazards

A set of laboratory manual instructions read: "If a yellow/orange color persists in your reaction mixture, add $NaHSO_3$(aq) (sodium bisulfite solution) gradually by pipetfuls (with swirling to mix) until the color fades." Jody started this process and became concerned when the color did not disappear after adding five pipetfuls of solution. She approached the instructor, who asked to see the container of the solution she was using. This led to the discovery that Jody was, in fact, adding sodium bicarbonate ($NaHCO_3$(aq)) to the reaction mixture; she had read the label on the bottle as "sodium bi…" and assumed it was what she needed. Instead of adding a reducing agent, Jody was adding a base!

Although this laboratory mishap did not lead to an accident, it demonstrates a common and potentially dangerous oversight in the organic chemistry lab. Fortunately, Jody knew that something was wrong when she did not witness the expected color change. Safety in the laboratory critically depends on your knowledge of chemical names and structures, your understanding of chemical reactivity and potential hazards, the proper labeling of chemicals, and your careful attention.

Identifying the Chemical

Chemists have invested a great deal of energy in devising systematic names of chemicals for good reason, and you should never consider nomenclature to be "unimportant." The IUPAC (International Union of Pure and Applied Chemistry) naming system is fairly complex, however, and people are bound to make mistakes. In addition to IUPAC names, common names are still in regular use, and it can be confusing to work with compounds that are identified by multiple names. The American Chemical Society's Chemical Abstracts Service (CAS) has developed an identification system in which each chemical is given a unique number. By correlating the **CAS number** with structure, you can avoid the confusion associated with multiple names.

Commercial suppliers of chemicals, such as Sigma-Aldrich and Acros Organics, have electronic searchable databases of the names, structures, CAS numbers, properties, and hazard information associated with every chemical that they sell. Those databases are among the most convenient places to go for information:

http://www.sigmaaldrich.com
http://www.acros.com

A screenshot from a search for "acetyl chloride" on the Sigma-Aldrich website shows that, in addition to the name and structure, the CAS number, molecular weight, boiling point, and density are provided. These have been highlighted in blue boxes in Figure 1.4.

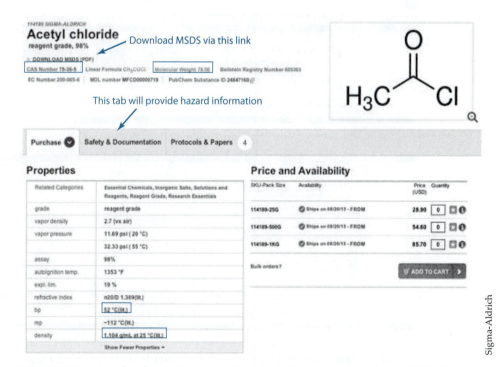

FIGURE 1.4 Screenshot from a Sigma-Aldrich search for the compound acetyl chloride, with some information highlighted in blue boxes. Screenshot captured from http://www.sigmaaldrich.com.

Global Harmonized System of Safety Information

A **Global Harmonized System (GHS)** of classifying and labeling chemicals has been developed by the United Nations for identifying the hazards associated with all chemicals that are manufactured and shipped around the world, often in large quantities. The United States Department of Labor Occupational Safety and Health Administration (OSHA) has revised its Hazard Communication Standard to align with the GHS. The GHS is the primary information system described in this textbook. GHS information is conveyed on chemical labels and chemical suppliers' websites, as shown in Figure 1.5, which is a screenshot from a search for acetyl chloride on the Acros website.

The safety information regarding acetyl chloride is shown in the third section of the screenshot, and the top half of this section provides GHS information:

• GHS pictograms are described in Figure 1.6, and some definitions of the hazard terms are provided in Table 1.2. Some hazards are represented by two different pictograms in Figure 1.6: Self-Reactives and Organic Peroxides are depicted by either the Flame or the Exploding Bomb, and Acute Toxicity is depicted by either the Exclamation Mark or the Skull and Crossbones. In these cases, the severity of the hazard dictates which pictogram is used. Greater hazards are labeled with the more serious pictogram (Exploding Bomb or Skull and

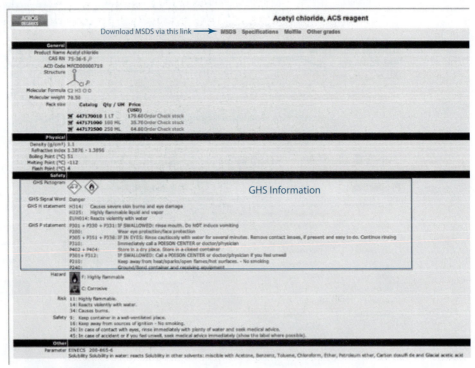

FIGURE 1.5 Screenshot from an Acros Organics search for the compound acetyl chloride, with GHS information and MSDS link highlighted in blue. Screenshot captured from http://www.acros.com.

Crossbones) and lesser hazards are labeled with the less serious pictogram (Flame or Exclamation Mark).

- Two signal words are used: "Warning" is less severe; "Danger" is more severe and is associated with the increased hazard categories 1 or 2. In the GHS system, smaller numbers indicate a greater hazard than bigger numbers.
- H (hazard) statements provide more specific information about the hazard.
- P (precautionary) statements explain how to minimize risks associated with handling the chemical and what to do in case of accidental exposure to the chemical.

The letter/number codes in front of the Hazard and Precautionary statements are for reference purposes. The Hazard, Risk, and Safety

TABLE 1.2	Definitions associated with chemical hazards
Term	**Definition**
Allergen	A chemical that causes an adverse immune response.
Aspiration	The entry of a liquid or solid chemical into the trachea and lower respiratory system through the oral or nasal cavity or from vomiting.
Carcinogen	A chemical that causes cancer or increases its incidence in humans or animals.
Corrosive	A chemical that causes destruction of living tissue at the site of contact.
Explosive	A chemical or mixture of chemicals that produces a damaging release of energy and gas when detonated by ignition, shock, or high temperature.
Flammable	A chemical that is easily ignited and burns rapidly. (Inflammable is a synonym; it means the same thing.)
Irritant	A chemical that causes reversible damage, such as redness, swelling, itching, or a rash at the site of contact.
Lachrymator	A chemical that causes tears.
LC_{50}	The concentration of a chemical in air or water that causes the death of 50% (one half) of a group of test animals.
LD_{50}	The amount of a chemical, given all at once, that causes the death of 50% (one half) of a group of test animals.
Mutagen	A chemical that increases the occurrence of mutations in populations of cells or organisms.
Organic peroxide	A chemical with structure R–O–O–R, which can explode upon concentration or sudden shock.
Oxidizer	A chemical that can rapidly bring about an oxidation reaction by supplying oxygen or receiving electrons.
Poison	An extremely toxic chemical.
Pyrophoric	A chemical that, within 5 min, spontaneously catches fire in air.
Self-reactive	A thermally unstable chemical that is liable to decompose in a strongly exothermic fashion in the absence of air. (In the GHS this excludes explosives, organic peroxides, and oxidizers.)
Sensitizer	A chemical that causes an adverse immune reaction upon repeated exposure.
Teratogen	A chemical that causes physical defects in a developing fetus.
Toxic	A chemical that causes adverse health effects in humans or animals.

The definitions in this table were compiled from Hill, Jr., R. H.; Finster, D. C. *Laboratory Safety for Chemistry Students;* Wiley: Hoboken, NJ, 2010; and *Globally Harmonized System of Classification and Labelling of Chemicals (GHS)*, 4th ed., United Nations: New York and Geneva, 2011.

FIGURE 1.6 Globally Harmonized System (GHS) pictograms indicating chemical hazards. The diamond surrounding a pictogram is normally shown in red.

entries in the bottom half of the safety section in Figure 1.5 are from the European system of coding hazards, which are self-explanatory based on analogy to the GHS.

Figure 1.5 shows that acetyl chloride has the GHS "Danger" signal word and the Corrosion and Flame pictograms, which are explained in more detail in the GHS Hazard and Precautionary Statements underneath the signal word. **Based on this information, you know to handle acetyl chloride with special care, to avoid skin and eye contact, and to avoid using it near ignition sources.** The simplest ways to achieve this are to work with acetyl chloride in a fume hood and to wear personal protection equipment for your eyes and hands. In addition, "reacts violently with water" is noted as a hazard. This indicates that you should avoid exposing this chemical to water and minimize its contact with water vapor present in the air. If you plan to work with acetyl chloride, ask your instructor what additional measures should be taken to prevent it from coming into contact with water.

The Four-Diamond Hazard Label

You may also encounter the color-coded four-diamond symbol, developed by the National Fire Protection Association (NFPA, Figure 1.7), on chemical labels. The four diamonds provide information on the hazards associated with handling specific compounds.

- **fire hazard** (top, red diamond)
- **reactivity hazard** (right, yellow diamond)

FIGURE 1.7 Four-diamond label for chemical containers indicating health, fire, reactivity, and specific hazards. The symbol in the specific hazard diamond indicates that the compound is reactive with water and should not come into contact with it.

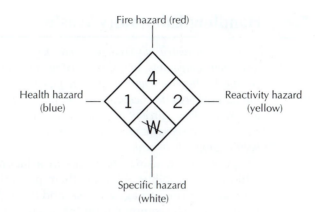

- **specific hazard** (bottom, white diamond)
- **health hazard** (left, blue diamond)

The numbers inside the diamonds indicate the level of hazard, with 1 being the least hazardous and 4 the most hazardous. Because this numbering system is opposite to the GHS, in which 1 indicates the greatest hazard, it can be confusing to work with the two systems. You should focus on learning and working with the GHS because it will eventually supplant the NFPA labeling system.

Safety Data Sheets

Currently, all laboratories must make available a **Material Safety Data Sheet (MSDS)** for every chemical used in the laboratory; under new OSHA regulations, these will be replaced by **Safety Data Sheets (SDSs)**. Every MSDS or SDS contains information on a list of topics required by law that describe the physical properties, hazards, safe handling and storage practices, and first aid information for a chemical. Manufacturers are required to prepare an MSDS or SDS for every chemical sold; the content is the same for a specific chemical, but the MSDS presentation format differs from one company to another. An MSDS from one company may be easy to read while that from another may be more difficult to understand. For this reason, a standardized format with 16 different sections has been prescribed for SDSs. You can access the MSDS (or SDS, as it is phased in) for every chemical you plan to work with in the laboratory from manufacturer's websites or from your institution. For example, the links to the MSDS for acetyl chloride from Sigma-Aldrich and Acros Organics are indicated in Figures 1.4 and 1.5.

The following websites have downloadable PDF files of MSDSs. The first requires you to register (for free) and the latter two require institutional subscriptions.

http://www.msds.com
http://www.MSDSonline.com
http://www.chemwatch.na.com

In addition to a complete MSDS, Chemwatch also provides mini MSDSs that briefly summarize the essential safety information for compounds in clear, concise language and pictograms.

1.8   Handling Laboratory Waste

Organic chemistry lab students were performing classification tests for unknown compounds, which required using small amounts of a variety of chemicals, dispensed with disposable pipets. As the lab period progressed, the odor of organic chemicals in the lab escalated to the point of being obnoxious. The source of odor was traced to a container for broken glass, where used disposable pipets that were contaminated with chemicals had been improperly discarded.

Any person using chemicals in a laboratory has a legal and ethical responsibility to handle them properly from the moment of purchase, during storage and use, and through appropriate disposal procedures. The common term for this mandate is "cradle to grave" responsibility. In the example above, the chemical residue in the pipets should have been removed and collected in an appropriate waste container before the pipets were discarded into the broken glass container.

At the end of every experiment, you may have a number of reaction by-products, such as aqueous solutions from extractions, filter paper and used drying agent coated with organic liquids, the filtrate from a reaction mixture or a recrystallization, and possibly a metal catalyst or other materials that need proper disposal. It is your legal obligation, as well as that of your instructor, the stockroom personnel, and your institution, to collect and handle all laboratory wastes in a manner consistent with federal and state requirements. Waste that cannot be reused or reclaimed must be disposed of by incineration or burial in a landfill. The method of disposal, which depends on local regulations and conditions, affects how waste is segregated and collected.

Satellite Accumulation Area

The waste containers in your lab will be located in a *satellite accumulation area*, which is a space for temporarily storing waste near where it is generated. Your instructor or laboratory personnel will assume responsibility for providing you with disposal instructions and for properly labeling and handling the waste. **It is your responsibility to check carefully—and then double-check—the label on a waste container BEFORE you place any waste in it.** If you are in doubt about what to do with something remaining from your experiment, consult your instructor. Placing waste in the wrong container may cause accidental emission of a toxic substance into the environment or may create an unsafe situation for workers managing the waste.

An organic laboratory will have several hazardous waste containers, labeled according to local regulations and protocols. In general, glass or polyethylene containers with tight-fitting caps are used for collecting chemical waste. These waste containers should be kept closed when not in use. Here are some ways that waste may be segregated in your laboratory:

- *Halogenated waste* is organic waste containing fluorine, chlorine, bromine, or iodine. It may be separated from other organic waste if incineration is an option for waste disposal; incineration of halogenated waste produces toxic HCl, for example.

- Organic waste is collected in *flammable waste* if it is not halogenated or *nonaqueous (without water) organic waste* containers.
- *Aqueous (water) waste* is collected separately from organic waste because it can react violently with some organic reagents and because it is treated differently upon storage and disposal. Often, aqueous waste is contaminated with organic compounds, which may be collected in an *aqueous (or water-containing) organic waste* container. Depending on local regulations, you may need to adjust the pH of aqueous waste.
- *Solid waste* consists of spent drying agents, filter paper coated with solvents, filter paper used in recrystallizations, and solid material remaining after a reaction.
- *Toxic metal waste* is waste containing heavy metals, such as chromium and mercury.

Sink or Trash Disposal

Except for a few materials that your instructor specifically deems to be harmless and acceptable under local regulations, you should NEVER dispose of any chemical or chemical-contaminated material in the sink or in a trash can.

Further Reading

Alaimo, R. J. (Ed.) *Handbook of Chemical Health and Safety;* American Chemical Society: Washington, D. C., and Oxford University Press: New York, 2001.

American Chemical Society. *Less Is Better: Guide to Minimizing Waste in Laboratories;* American Chemical Society: Washington, DC, 2002. Accessed electronically via http://www.acs.org

American Chemical Society. *Safety in Academic Chemistry Laboratories,* 7th ed.; American Chemical Society: Washington, DC, 2003. Accessed electronically via: http://www.acs.org

Armour, M. A. *Hazardous Laboratory Chemicals Disposal Guide,* 3rd ed.; CRC Press: Boca Raton, FL, 2003.

Furr, A. K. (Ed.) *CRC Handbook of Laboratory Safety,* 5th ed.; CRC Press: Boca Raton, FL, 2000.

Globally Harmonized System of Classification and Labelling of Chemicals (GHS), 4th ed., United Nations: New York and Geneva, 2011. Accessed electronically via: http://www.unece.org/

Hill, Jr., R. H.; Finster, D. C. *Laboratory Safety for Chemistry Students;* Wiley: Hoboken, NJ, 2010.

Lewis, Sr., R. J. *Rapid Guide to Hazardous Chemicals in the Workplace,* 4th ed.; Wiley: New York, 2000.

Lewis, Sr., R. J.; Sax, N. I. *Sax's Dangerous Properties of Industrial Materials,* 12th ed.; Wiley: Hoboken, NJ, 2012.

National Research Council of the National Academies. *Prudent Practices in the Laboratory: Handling and Management of Chemical Hazards;* National Academies Press: Washington, DC, 2011.

O'Neill, M. J. (Ed.) *The Merck Index: An Encyclopedia of Chemicals, Drugs, and Biologicals,* 15th ed.; Royal Society of Chemistry Publishing: Cambridge, UK, 2013.

School Chemistry Laboratory Safety Guide, U.S. Consumer Product Safety Commission and National Institute for Occupational Safety and Health: Bethesda, Maryland, 2006. Accessed electronically via: http://www.cpsc.gov or http://www.cdc.gov/niosh/

United States Department of Labor Occupational Safety and Health Administration Hazard Communication: https://www.osha.gov/dsg/hazcom/index.html

Questions

1. Name five important safety features that are found in your laboratory.
2. Locate the first aid kit in or near your laboratory. Based on your institution's Chemical Hygiene Plan, what is the procedure that should be followed if someone in the laboratory gets a minor cut to the skin?

3. A procedure calls for you to dissolve a compound in hot ethanol. Using one of the suggested online sources (such as the Sigma-Aldrich or Acros Organics websites), look up the boiling point and flammability of ethanol. What is the best method for heating ethanol?

4. Look up the list of Chemical Waste Policies in the Chemical Hygiene Plan at your institution. What is the policy for discarding broken glass?

5. Identify the type(s) of disposable gloves available in your organic chemistry lab. Would they provide good or excellent protection from the following chemicals: dichloromethane, ethyl ether, ethylene glycol, and hydrogen peroxide? (You may have to search the suggested websites, such as http://www.microflex.com, in order to fully answer this question.) For those chemicals against which your gloves do not provide good protection, what would you do if you spilled a small amount on your glove?

6. Suppose you plan to synthesize aspirin (acetylsalicylic acid) by reacting salicylic acid with acetic anhydride, using 85% phosphoric acid as a catalyst. In addition to the main product aspirin, acetic acid will be a side-product. Using one of the suggested online sources (such as the Sigma-Aldrich or Acros Organics websites), identify the CAS numbers for all of the reagents and products (five total) in this reaction.

7. (a) For which of the following compounds is it hazardous to breathe dust/vapor/fumes: acetylsalicylic acid, salicylic acid, acetic anhydride, acetic acid, phosphoric acid? (Use one of the suggested online sources, such as the Sigma-Aldrich or Acros Organics websites, or sources of MSDSs, to answer this question.)

   (b) Based on the boiling point or melting point of the compounds, which are you most likely to inhale accidentally?

   (c) Based on GHS hazard information, which would be most dangerous to inhale?

   (d) Of all these chemicals, which is most important to handle in a fume hood?

| Salicylic acid | Acetic anhydride | | Acetylsalicylic acid | Acetic acid |

CHAPTER

2

GREEN CHEMISTRY

You touch polycarbonate plastic every day; it is found in drinking bottles, food containers, eyeglass lenses, CDs and DVDs, and a variety of building materials. As with most plastics, the raw materials incorporated into traditional polycarbonates come from oil. Geoffrey Coates and his coworkers at Cornell University have recently developed a new family of catalysts that can effectively and economically use carbon dioxide (CO_2) in polycarbonate synthesis. This technology is being commercialized to prepare resins *for lining food and drink containers in order to replace resins derived from bisphenol A (BPA, a suspected endocrine disruptor). Fifty percent of the new resin (by weight) will be derived from sequestered CO_2 and it will require 50% less petroleum to produce than its traditional counterpart. If fully utilized, this technology could eliminate 180 million metric tons of CO_2 emissions annually.*

We are currently dependent on the petroleum industry for energy and raw materials to produce consumer goods. The carbon atoms that are incorporated into most fabrics, carpets, paints, disposable diapers, cleaning products, plastics, cosmetics, and medications come from oil, which continues to be one of our cheapest sources of raw material. Yet oil is not a renewable resource and its consumption contributes to climate change. *Green chemistry* is sustainable chemistry; its goal is to develop chemical processes that are safe, efficient, economical, and renewable. In 1998, Paul Anastas and John Warner articulated the principles that now govern the implementation of greener chemical practices in industry, government, and education. Although we cannot immediately replace all petroleum feedstocks with renewable ones, implementing green chemistry principles, as Coates and his colleagues have done, can make more efficient use of oil and reduce our dependence on it.

2.1   The Principles of Green Chemistry

Green chemistry has changed chemists' perspectives by requiring them to think beyond creating an innovative product and to evaluate the entire lifespan of the product, from "cradle to grave." It leads them to ask questions about the raw materials used to make a product, the process for making it, and its ultimate fate after use:

- How safe is the product for human health and the environment?
- What happens to the product once it is used or discarded?
- How safe and efficient is the process of making it?
- How much energy is consumed in the process of making it?
- How hazardous or renewable are the raw materials?
- How much waste is generated and how hazardous is it?

Chemists answer these questions for a proposed process in order to compare it to a competing process. They then evaluate the different strengths and weaknesses of both processes using the **Twelve Principles of Green Chemistry** (Table 2.1). A 100% green process is ideal, but it is almost impossible to accomplish all twelve principles in any one product or process. This analysis, however, provides a framework with which to prioritize different advantages and approach this ideal.

The challenge of identifying and developing viable renewable processes affects the goods we consume and the environment in which we live. By learning ways to meet this challenge, you will be better informed to make decisions about the work you do as a student, consumer, and future employee. The following sections should give you a sense that progress has been made in "greening"

T A B L E   2 . 1   The twelve principles of green chemistry

1. **Prevention:** avoid generating waste
2. **Atom Economy:** incorporate most atoms from the reagents into the product
3. **Less Hazardous Chemical Syntheses:** use and generate the least toxic materials
4. **Designing Safer Chemicals:** ensure that final products are nontoxic
5. **Safer Solvents and Auxiliaries:** use minimal and innocuous supporting materials
6. **Design for Energy Efficiency:** minimize energy requirements
7. **Use of Renewable Feedstocks:** whenever possible, use renewable raw materials
8. **Reduce Derivatives:** avoid introducing atoms that have to be removed later
9. **Catalysis:** use catalysts for efficient and less wasteful processes and re-use them when possible
10. **Design for Degradation:** plan for products to break down naturally into benign substances
11. **Real-Time Analysis for Pollution Prevention:** monitor the process to avoid accidental exposure to hazards
12. **Inherently Safer Chemistry for Accident Reduction:** avoid using chemicals that are highly reactive

---

The information in this table was compiled from http://www.epa.gov/sciencematters/june2011/principles.htm.

the chemistry performed in industrial and educational settings and will provide you with ideas and strategies of your own.

2.2   Green Principles Applied to Industrial Processes

The following examples are organized according to how they address the Twelve Principles of Green Chemistry. You will notice that although they do not meet all the criteria for a sustainable process, they represent real progress toward that goal.

Safer Solvents

One of the simplest ways to improve a process is to replace a hazardous solvent with a safer or more environmentally benign alternative. An example is the process of decaffeinating coffee. Caffeine has been extracted from green coffee beans using dichloromethane, which is a suspected carcinogen, or ethyl acetate, which is flammable but much safer for human health and the environment. Greener methods instead employ supercritical CO_2 or water for the extraction.

Supercritical CO_2. Carbon dioxide is a gas under normal conditions, but when it is subjected to conditions of temperature and pressure that exceed its critical point, 31.1°C and 73 atm pressure, it becomes a single fluid-like phase, called a *supercritical fluid.* Supercritical CO_2 is a very good solvent with properties similar to many common organic solvents. In addition to decaffeinating coffee, supercritical CO_2 can replace traditional and hazardous solvents in dry-cleaning clothing, cleaning electronic and industrial parts, and chemical reactions. At the end of these processes, the pressure is released and the escaping CO_2 gas can be easily recovered and recycled.

Water. In the quest for solvents that minimize health hazards and risks to the environment, water would appear to be ideal because it is readily available and nonhazardous. But a requirement for most reaction solvents is that they dissolve the reagents used in the reaction, and organic compounds are largely insoluble or only slightly soluble in water. Reactions in aqueous solution can be promoted with water-insoluble organic compounds, however, by using vigorous stirring, phase-transfer catalysts, or superheating by microwaves in sealed vessels (see Section 6.3).

Organic solvents. Some necessary reactions and separation processes require organic solvents; in these cases, the safest and environmentally most benign solvent that can accomplish the desired goal is the best choice. The American Chemical Society Green Chemistry Institute® (ACS GCI) has convened a body of representatives from pharmaceutical companies, called the Pharmaceutical Roundtable, to guide the chemical community in choosing greener organic solvents. This group has evaluated solvents in five categories—safety, health, environment (air), environment (water), and environment (waste)—and scored them from 1 (most benign) to 10 (least favorable) in each category. The scores for a selection of organic solvents, along with their boiling points and water solubilities, are listed in Table 2.2. These data indicate that solvents such as hexane, benzene, chloroform, dichloromethane, and ethyl ether should be avoided and replaced with lower-scoring alternatives where possible.

Catalysis

Catalysts are extremely important because they can make reactions more efficient and effective. They "activate" reagents by interacting

TABLE 2.2	ACS GCI Pharmaceutical Roundtable Solvent Selection Guide							
Solvent class	Solvent	Boiling point	Water solubility	Safety	Health	Env (air)	Env (water)	Env. (waste)
Hydrocarbon	Cyclohexane	81°C	0.05 g/L	6	5	4	7	2
Hydrocarbon	Heptane	98°C	Insoluble	6	4	4	7	2
Hydrocarbon	Hexane	69°C	Insoluble	6	7	5	8	1
Aromatic	Benzene	80°C	1.8 g/L	5	10	6	6	2
Aromatic	Toluene	110°C	0.5 g/L	5	7	6	6	2
Aromatic	Xylenes	~140°C	Insoluble	4	4	4	7	3
Halogenated	Chlorobenzene	131°C	0.5 g/L	3	5	5	8	6
Halogenated	Chloroform	61°C	8 g/L	2	9	7	7	6
Halogenated	Dichloromethane	40°C	13 g/L	2	7	9	6	7
Ester	Ethyl acetate	70°C	83 g/L	5	4	6	4	4
Ester	Isobutyl acetate	118°C	7 g/L	5	3	5	2	2
Ester	Isopropyl acetate	89°C	43 g/L	3	4	6	3	3
Ether	Ethyl ether	35°C	69 g/L	9	5	7	4	4
Ether	Tetrahydrofuran	66°C	Miscible	5	6	5	4	5
Ketone	Acetone	57°C	Miscible	4	4	7	1	5
Alcohol	Ethanol	79°C	Miscible	4	3	5	1	6
Alcohol	Ethylene glycol	197°C	Miscible	3	3	5	1	7
Alcohol	Methanol	65°C	Miscible	3	5	6	3	6

The information in this table was compiled from *The Merck Index,* 11th ed., and the ACS GCI Pharmaceutical Roundtable Solvent Selection Guide Version 2.0 Issued March 21, 2011: http://www.acs.org/content/acs/en/greenchemistry/industriainnovation/roundtable.html.

with them to lower the energy required to break and form new bonds, thus speeding up reactions and allowing lower temperatures to be used. Particularly useful are catalysts that can tolerate safer solvents, such as water or methanol, and that can be reused, saving resources and money. Chemists play important roles in developing greener processes by designing, synthesizing, and testing new catalysts.

Over the past 10 years, industry has embraced biological catalysts in the large-scale production of chemical feedstocks and fine chemicals. Selective enzyme-based catalysts are now used to produce pharmaceutical, cosmetic, and food products. These often function under mild conditions and in water solution, allowing for energy savings and reduced waste. Fermentation processes that capitalize on the enzymes in yeast are also widely used in industry, particularly for preparing biorenewable feedstocks from biomass.

Less Hazardous Chemical Syntheses

The Dow Chemical Company and BASF won a 2010 Presidential Green Chemistry Challenge Award for the production of propylene oxide by oxidation of propylene with hydrogen peroxide. Propylene oxide is a bulk commodity chemical used for making foam seat cushions and mattresses, detergents, and personal care products. The oxidation of propylene traditionally uses bleach or organoperoxides for this process. Bleach is made industrially from highly toxic and reactive chlorine gas. Organoperoxides are very reactive and the oxidation generates organic waste that needs to be recycled, which consumes extra energy (see reaction A below). The new process uses hydrogen peroxide, which is safer to handle, produces harmless water as waste, and requires less energy (see reaction B below). These benefits depend on the discovery of a catalyst that can activate hydrogen peroxide.

A. Organoperoxide process:

Propylene	Tert-butyl hydroperoxide		Propylene oxide	Tert-butyl alcohol
MW = 42	MW = 90	catalyst A	MW = 58	(waste)

B. Hydrogen peroxide process:

Propylene	Hydrogen peroxide		Propylene oxide	Water
MW = 42	MW = 34	catalyst B	MW = 58	(waste)

Atom Economy

Atom economy is a quantitative measure of how efficiently atoms of the starting materials and reagents are incorporated into the desired product. It represents the percentage of atomic mass of the starting materials that end up in the final product, assuming 100% yield in the reaction. The balanced equation for a reaction is used in the calculation of atom economy:

$$\text{atom economy} = \frac{\text{sum}(MW_{products})}{\text{sum}(MW_{reagents})} \times 100\%$$

It may be obvious that the Dow-BASF hydrogen peroxide process is more atom economical than the organoperoxide process, but the atom economy calculation enables this improvement to be quantified and compared with other alternatives. Here are the calculations for the organoperoxide (A) and hydrogen peroxide (B) processes:

$$\text{atom economy (A)} = \frac{MW_{\text{propylene oxide}}}{MW_{\text{propylene}} + MW_{\text{t-butyl hydroperoxide}}} \times 100\%$$

$$= \frac{58}{42 + 90} \times 100\% = \boxed{44\%}$$

$$\text{atom economy (B)} = \frac{MW_{\text{propylene oxide}}}{MW_{\text{propylene}} + MW_{\text{hydrogen peroxide}}} \times 100\%$$

$$= \frac{58}{42 + 34} \times 100\% = \boxed{76\%}$$

Reaction Efficiency

The concept of reaction efficiency was developed as a measure of the mass of reactant atoms actually contained in the final product. Suppose you had developed a catalyst that performs the hydrogen peroxide process but in only 55% yield, and the established organoperoxide process (A) occurs with 99% yield. Would it be worthwhile to switch to the hydrogen peroxide process? The reaction efficiency can help answer this question, as shown in the following equation:

$$\text{reaction efficiency} = \%\ \text{yield} \times \text{atom economy}$$

$$\text{reaction efficiency (hypothetical HOOH process)} = 55\% \times 0.76 = 42\%$$

$$\text{reaction efficiency (organoperoxide process)} = 99\% \times 0.44 = 44\%$$

The reaction efficiency indicates that only 42% of the mass of reactants would be recovered as product with your hypothetical catalyst, whereas 44% is transformed to product in the organoperoxide process. With your low-yielding process, it may not be worthwhile to switch. The Dow-BASF catalyst, however, is significantly higher yielding, making the hydrogen peroxide process superior.

Use of Renewable Feedstocks

The Dow-BASF synthesis described above is an important step toward greening an industrial process; however, the propylene feedstock originates from petroleum. Inventive chemists at the 2009 start-up company XL Terra, Inc. have developed a new plastic from biomass that has functional properties comparable to plastics made from petroleum. This plastic, Poly(Xylitan Levulinate Ketal) (PXLK), is made from a five-carbon carbohydrate (xylose) isolated from nonfood biomass, such as corn cobs or wood waste. After use, the plastic can be treated to recover and re-use the starting materials or left to biodegrade harmlessly. This is one example among many new chemical technologies that are being developed to replace petrochemical feedstocks with biorenewable feedstocks.

Synthetic Efficiency

The overall yield for a process is the product of the yields of each individual step; therefore, introducing additional steps in a process may diminish the yield. For example, if each step in a two-step synthesis occurred in 98% yield, the total overall yield would be $(0.98 \times 0.98) = 0.96$, or 96%. If the transformation involved six steps, each occurring in 98% yield, the overall yield would be only $(0.98)^6 = 0.88$, or 88%.

The effect of the number of steps on the efficiency of a process is shown by the synthesis of the common analgesic ibuprofen. The classic six-step route was developed at the Boots Pure Chemical Company, where ibuprofen was discovered. Several steps introduce carbon atoms that must be removed later and treated as waste. A greener route was developed by Boots-Hoechst-Celanese (BHC); it has only three steps, two of which have 100% atom economy. If each of these steps occurred in 98% yield, the overall yield would be $(0.98)^3 = 0.94$, or 94%. In reality, the yield for each step is higher. The only waste that is generated is acetic acid in the first step. Anhydrous hydrogen fluoride is both a catalyst and a solvent in the first step, but it is recovered and recycled with greater than 99.9% efficiency. Hydrogen fluoride is a very hazardous material and no doubt researchers are striving to develop safer alternatives. This Presidential Green Chemistry Challenge Award-winning process is commercialized and a Texas plant produces about 4000 tons of ibuprofen per year using this route.

Boots-Hoechst-Celanese Process:

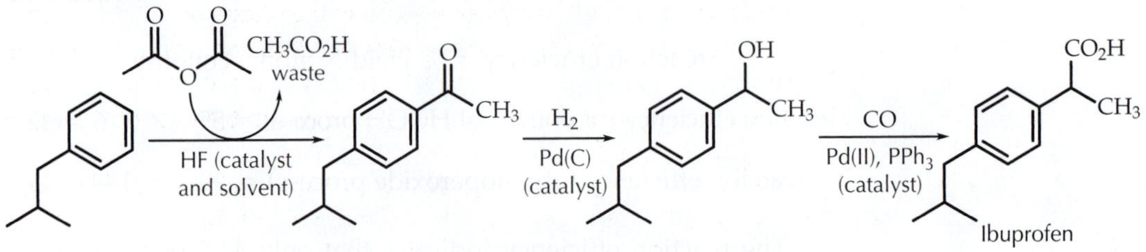

2.3   Green Principles Applied to Academic Laboratories

Apart from pedagogical or experimental value, the greatest concern in an academic laboratory is student safety. Experimental modifications that minimize student exposure to chemical hazards, reduce chemical waste, and decrease the need for safety equipment, such as fume hoods, have a high priority. Over the past 15 to 20 years, the single greatest change in instructional laboratories has been to reduce the scale of reactions performed by students. This lowers the potential for student exposure to chemical hazards and the amount of hazardous waste that is generated.

Safer Solvents

Just as in industry, academic laboratories are adapting to use solvents that pose fewer health and environmental hazards, such as water or ethanol, or that eliminate the need for solvents altogether. When a desired product can be isolated from mixtures by simple filtration, distillation, or sublimation, students do not need to handle flammable (ether) or possibly carcinogenic (dichloromethane) solvents in the laboratory. However, there are situations when it is impossible to separate an organic product without using an organic solvent. Below is a process for selecting the best solvent to use in these cases.

Extraction of an organic compound from an aqueous mixture. A common procedure for separating an organic compound from an aqueous mixture is extraction. For example, extraction is used to separate camphor from a water-based reaction mixture. Camphor is more soluble in an organic solvent than in water; when the reaction mixture is mixed with an organic solvent, the camphor will preferentially dissolve in the organic solvent. When the organic solvent is not very soluble in water, the water and organic phases separate from one another upon standing. A student can physically remove the organic phase and evaporate the organic solvent to recover the camphor. Often, dichloromethane is specified for this type of procedure. Would ethyl acetate be a "greener" alternative? To answer this question, look at the data for these two solvents presented in Table 2.2.

Dichloromethane   Ethyl acetate

According to the scores developed by the Pharmaceutical Roundtable, ethyl acetate is the better choice. It is less hazardous to human health and the environment, although it is also somewhat less safe because it is flammable. However, as long as ignition sources are not present, this should not be an issue.

There are two remaining concerns to address in making this switch: Ethyl acetate is significantly more soluble in water and has a higher boiling point than dichloromethane. You can decrease the solubility of water in ethyl acetate, however, by saturating the camphor-containing water solution with sodium chloride. The higher boiling point of ethyl acetate means that it requires more heat (energy) to remove the solvent and recover the camphor than does dichloromethane. Because the paramount consideration is student safety, ethyl acetate is the better choice.

Less Hazardous Chemical Syntheses

Oxidations that employ chromic acid (generated from CrO_3 and H_2SO_4) or molecular bromine (Br_2) are classic examples of reactions described in organic chemistry textbooks that have historically been performed in instructional laboratories. Unfortunately, chromium is a toxic heavy metal and bromine is corrosive and can be fatal if inhaled. Fortunately, the desired transformations can now be accomplished with safer reagents.

Oxidation of alcohols to ketones. Secondary alcohols *can be* oxidized to ketones with chromic acid, *but* aqueous sodium hypochlorite solution (NaOCl, household bleach) in the presence of acetic acid is a more benign oxidation.

$$\underset{\text{Alcohol}}{R-\underset{R}{\overset{\overset{\displaystyle OH}{|}}{\underset{|}{C}}}-H} \xrightarrow[\text{agent}]{\text{oxidizing}} \underset{\text{Ketone}}{R-\underset{R}{\overset{\displaystyle O}{\underset{|}{C}}}}$$

$$R-\underset{R}{\overset{\overset{\displaystyle OH}{|}}{\underset{|}{C}}}-H + NaOCl \xrightarrow[\substack{\text{acetic acid} \\ H_2O}]{} R-\underset{R}{\overset{\displaystyle O}{\underset{|}{C}}} + H_2O + NaCl$$

The green features of this experiment include the use of water rather than ether as the reaction solvent, the separation of organic product by steam distillation instead of extraction, the increased atom economy, and the lack of hazardous waste. (NaCl and water are harmless by-products.)

Another green laboratory oxidation uses the safer oxidant Oxone® in conjunction with catalytic amounts of sodium chloride. Oxone® is an inexpensive and nonhazardous triple salt with the formula: $2KHSO_5 \bullet KHSO_4 \bullet K_2SO_4$. The active reagent is $KHSO_5$, which oxidizes sodium chloride to generate sodium hypochlorite (NaOCl), the species that is thought to be responsible for oxidizing the alcohol. This experiment is not as atom economical as bleach oxidation because potassium bisulfate ($KHSO_4$) and potassium sulfate (K_2SO_4) are unused, although harmless, components of Oxone®. Nevertheless, Oxone® can be more reliable and produce higher yields than the corresponding oxidation with bleach.

$$R-\underset{R}{\overset{\overset{\displaystyle OH}{|}}{\underset{|}{C}}}-H \xrightarrow[\substack{\text{0.3 equiv NaCl} \\ H_2O/\text{ethyl acetate}}]{\substack{\text{0.6 equiv oxone} \\ (1.2 \text{ equiv } KHSO_5)}} R-\underset{R}{\overset{\displaystyle O}{\underset{|}{C}}}$$

Bromination of alkenes. Alkenes are traditionally converted into 1,2-dibromoalkanes with molecular bromine (Br_2). While this is a 100% atom-economical reaction, bromine is extremely hazardous to handle and the solvents employed in the reaction are halogenated, such as dichloromethane (see Table 2.2).

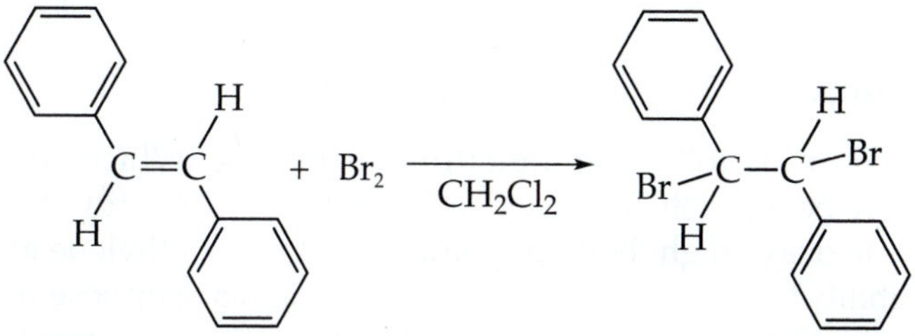

Trans-stilbene
MW = 180

MW = 340

A greener transformation uses hydrogen peroxide and HBr to generate Br_2 in an ethanol/water solvent. Although the reaction has less atom economy, the reagents are much safer to handle and the waste products are harmless. Ethanol is among the least harmful solvents listed in Table 2.2, and the desired product precipitates from solution and can be isolated by filtration.

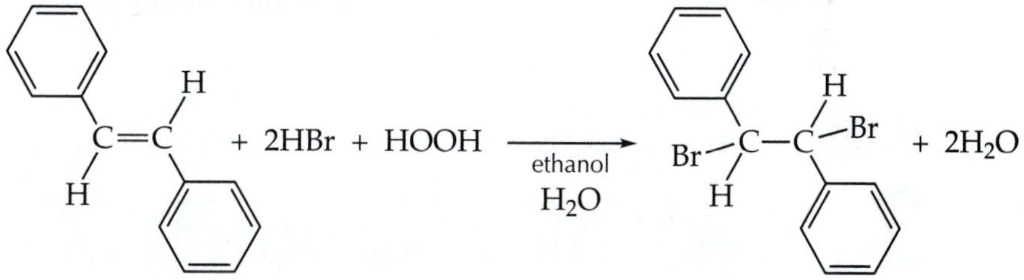

Trans-stilbene
MW = 180

MW = 340

New Developments   The examples presented here are only a few of many new developments in academic laboratories. For example, microwave heating of reactions is faster than conventional methods, leading to shorter reaction times and less energy usage (see Section 6.3). New, safer catalysts are continually being developed and implemented. The resources at the end of this chapter will direct you to more examples of the many improvements that have been made in green chemistry.

Further Reading

Anastas, P. T.; Warner, J. C. *Green Chemistry: Theory and Practice*; Oxford University Press: Oxford, 1998.

Doxsee, K. M.; Hutchinson, J. E. *Green Organic Chemistry Strategies, Tools, and Laboratory Experiments*; Brooks/Cole: Belmont, CA, 2004.

Dunn, P. J.; Wells, A. S.; Williams, M. T. (Eds.) *Green Chemistry in the Pharmaceutical Industry*; Wiley-VCH: Weinheim, 2010.

Lancaster, M. *Green Chemistry. An Introductory Text*, 2nd ed.; Royal Society of Chemistry: Cambridge, UK, 2010.

Roesky, H. W.; Kennepohl, D. K. (Eds.) *Experiments in Green and Sustainable Chemistry*; Wiley-VCH: Weinheim, 2009.

Websites

American Chemical Society Green Chemistry Institute® website: http://www.acs.org/content/acs/en/greenchemistry.html

Green Chemistry Journal: http://pubs.rsc.org/en/journals/journalissues/gc#!recentarticles&all

The Center for Green Chemistry and Green Engineering at Yale: http://greenchemistry.yale.edu/

The Journal of Chemical Education: http://pubs
   .acs.org/journal/jceda8

United States Environmental Protection Agency
   Green Chemistry website: http://www2.epa.
   gov/green-chemistry

Questions

1. Which of the following are attributes of a more benign solvent: low flammability, low toxicity, high boiling point, water solubility?

2. Which of the following strategies for developing a more sustainable process is NOT an option in industry: use a catalyst to introduce a more atom-economical reaction, reduce the scale of a reaction, reduce the number of steps in a process, use renewable feedstocks?

3. What is the simplest way that you could make an instructional laboratory procedure more benign?

4. Which organic solvent requires less energy to evaporate from a product: ethylene glycol or ethanol?

5. Suppose that a reaction you want to perform needs to be heated to at least 110°C in an organic solvent that can later be separated from water-soluble side products. Using Table 2.2, choose the most benign solvent that meets these criteria.

6. (a) Calculate the atom economy of the HBr/HOOH method for adding bromine to *trans*-stilbene (equation on page 31). (b) Suppose the yield is 85%. What is the reaction efficiency?

CHAPTER

3

LABORATORY NOTEBOOKS AND PRELAB INFORMATION

The effective practice of science depends on authentic laboratory records that document *how* the work was done, *when* the work was performed, *what* the results were, and *who* did the work. The standards for maintaining a laboratory notebook derive from these requirements for accuracy.

This means that:

- Your notebook is bound, with the pages numbered sequentially.
- You neatly record what you do at the time you do it, in permanent ink.
- You date the pages as you do experimental work.
- You accurately cite sources of information.
- You describe measurements and include units.
- You include names of lab partners, if relevant.
- You correct mistakes by making a single strike through faulty statements and substitute the right information.
- Your instructor is able to read and understand your notebook.

This does NOT mean that:

- Your notebook is a work of art.
- Your notebook is perfect and free of watermarks, stains, or blotches.

- You write down measurements on scraps of paper that you plan to transfer to your notebook later.
- You write in pencil and erase mistakes.
- There are missing pages in your notebook.

The rest of this chapter provides guidelines for accomplishing these goals and for finding appropriate sources of information.

3.1   The Laboratory Notebook

You need to complete a lab record in three stages for each experiment: **prelab, in lab, and postlab.** Your instructor will provide additional specific guidelines for lab notebook procedures at your institution, but here are some general suggestions.

Prelaboratory Preparation

The prelab is like a recipe used in cooking. It describes the materials, amounts, equipment, and procedures for performing the experiment. Procedures are always based on published work, such as a laboratory manual or an article from the primary literature. Depending on your instructor's expectations, you will either outline the procedure and bring a copy of the original work to the laboratory or you will write out a detailed procedure so you do not need access to the published procedure in the laboratory. The following items are generally included in the prelab preparation.

Experiment title. Write a title that clearly identifies what you are doing in the experiment or project.

Statement of purpose. Write a brief statement of purpose for the experiment, which includes the synthesis objective, analytical method, and/or conceptual approach.

Citation. Properly cite the reference that is the basis for your planned procedure, which could be your laboratory manual, another secondary source, or an article from the primary literature.

Balanced chemical equations. Write balanced chemical equations that show the overall process. Reaction mechanisms may be required by your instructor for prelab preparation.

Table of reagents and solvents. Construct a table with molecular weights, number of moles, and quantities (by weight or volume) of reagents. Also include physical constants for the reagents, solvents, and product(s), such as the densities of liquid compounds, boiling points of compounds that are liquids at room temperature, and melting points of organic solids.

 Safety information. Briefly list the safety precautions for all reagents and solvents you will use in the experiment (see Section 1.7). These can be incorporated into the table of reagents and solvents, if you like.

Method of yield calculation. Outline the computations to be used in a synthesis experiment, including calculation of the theoretical yield (see Section 3.2).

Procedure outline. The goal of the outline is to provide you with the background to carry out the experiment effectively and with understanding. This is especially important in experiments where you participate in the experimental design. Include the techniques to be used, such as reflux, filtration, or distillation. You might want to list the page in your lab manual or techniques book where the illustration of a particular glassware setup is shown, particularly if this is the first time you will be using it. Note the operations that you plan to perform in a fume hood or for which you will wear gloves.

Waste disposal. If the procedure states how to dispose of the waste remaining from the experiment, briefly summarize the instructions in your notebook.

Prelab questions. Answer any assigned prelab questions. Your instructor may request that these be prepared and turned in as a separate document.

In the Laboratory

During the lab period, write a *running account* **of the steps you perform, your measurements, and your observations.** This is the story of what you actually did and observed in the laboratory as you carried out the experiment. If this is incomplete, you often cannot interpret the results of your experiments once you have left the laboratory. It is difficult, if not impossible, to reconstruct an accurate running account at a later time.

Be aware of the physical properties and safety information for the chemicals that you are handling in the laboratory. Follow proper safety precautions when performing lab manipulations.

Include the following items in your running account.

Date(s). Write the date you actually carried out the experiment on each page.

Lab Partners. Include the name(s) of your lab partners, if relevant.

Procedures. Completely describe your actions in the laboratory. They should be described in sufficient detail so that they could be reproduced by another organic chemistry student who hasn't done the experiment.

Measurements. Record the **actual quantities** of all reagents as they are used, as well as the amounts of crude and purified products you obtain. If you weigh a material by taking the difference between the mass of a flask containing the material and the mass when it is empty, include both measured masses along with the calculated mass. Include temperatures and times, as appropriate. Report boiling points and melting points over the entire range of distillation or melting (for example, 58–60°C). Always record measurements using the appropriate number of **significant figures** (see Section 5.5) and include **units**. Describe the instrument type and conditions used for obtaining data, if relevant.

Observations. Describe changes in color, solubility, and temperature, and note the evolution of gases as you perform the experiment.

Here are some examples of the type of information to include in your running account:

- A white solid appeared, which dissolved when sulfuric acid was added.
- The solution turned cloudy when it was cooled to 10°C.
- An additional 10 mL of solvent were required to completely dissolve the yellow solid.
- The reaction was heated at 50°C for 25 min on a water bath.
- A small puff of white smoke appeared when sodium hydroxide was added to the reaction mixture.
- The NMR sample was prepared using 20 mg of recrystallized product in 0.7 mL of $CDCl_3$.
- A capillary OV–101 GC column heated to 137°C was used.
- The infrared spectrum was obtained from a cast-film sample.

Postlaboratory Interpretation of Your Experimental Results

After your experiments are complete, summarize and interpret your experimental data. This will normally include a section on interpretation of observations, analysis of physical and spectral data, a summary of your conclusions, calculation of a percent yield, and answers to any assigned postlab questions. Your instructor may specify additional material for this section and may request that you submit your post-laboratory analysis as part of a separate report.

Conclusions and summary. In an inquiry-based project or experiment, return to the question being addressed and discuss the conclusions you can draw from analysis of your data. For both inquiry-driven experiments and those where you learn about laboratory techniques, discuss how your experimental results support your conclusions. Include a thorough interpretation of NMR and IR spectra and other analytical results, such as TLC and GC analyses. Staple or tape properly labeled spectra and chromatograms into your notebook. Cite any reference sources that you used.

Percent yield. Synthetic chemists care deeply about percent yield; it is the single most important measure of success. The purity of a product is also crucial, but if a synthetic method produces very small amounts of the needed product, it is not much good. Reactions described in textbooks are often far more difficult to carry out in good yield than many books suggest.

3.2   Calculation of the Percent Yield

Calculate the percent yield when you report the results of a synthesis reaction. The percent yield is the ratio of the actual yield of product obtained relative to the theoretical yield (maximum amount possible), multiplied by 100:

$$\% \text{ yield} = \frac{\text{actual yield of product}}{\text{theoretical yield}} \times 100$$

Chemists typically calculate percent yields using mass as the measured value, but moles can also be used. As long as the actual and theoretical yield values have the same units, such as grams or moles, the yield calculation will be correct. Calculate the theoretical yield from the balanced chemical equation and the amount of limiting reagent, assuming 100% conversion of the starting materials to product(s).

Example. Consider the synthesis of 1-ethoxybutane from 1-bromobutane and sodium ethoxide. Notice that in the balanced reaction one mole of product is produced from one mole of 1-bromobutane and one mole of sodium ethoxide.

$$CH_3(CH_2)_3{-}Br + CH_3CH_2{-}O^-Na^+ \xrightarrow{\text{ethanol}} CH_3(CH_2)_3{-}O{-}CH_2CH_3 + NaBr$$

1-Bromobutane	Sodium ethoxide	1-Ethoxybutane
MW 137	MW 68.1	MW 102

density 1.27 g/mL^{-1}

The procedure specifies 4.50 mL of 1-bromobutane, 3.70 g of sodium ethoxide, and 20 mL of anhydrous ethanol. To calculate the theoretical yield, it is necessary to ascertain whether 1-bromobutane or sodium ethoxide is the **limiting reagent** by calculating the moles of each reagent present in the reaction mixture:

$$\text{moles of 1-bromobutane} = \frac{4.50 \text{ mL} \times 1.27 \text{ g/mL}}{137 \text{ g/mL}} = 0.0417 \text{ mol}$$

$$\text{moles of sodium ethoxide} = \frac{3.70 \text{ g}}{68.1 \text{ g/mol}} = 0.0543 \text{ mol}$$

Therefore, 1-bromobutane is the limiting reagent.

According to the balanced equation, equimolar amounts of the two reactants are required. Thus the theoretical yield, the maximum amount of product that is possible from the reaction, assuming that it goes to completion and that no experimental losses occur, is 0.0417 mol or 4.25 g of 1-ethoxybutane:

$$\text{theoretical yield} = 0.0417 \text{ mol} \times 102 \text{ g/mol}$$

$$= 4.25 \text{ g of 1-ethoxybutane}$$

The percent yield for a synthesis that produced 2.70 g of 1-ethoxybutane is 63.5%. Yields are customarily rounded off to the nearest integer values:

$$\% \text{ yield} = \frac{2.70 \text{ g}}{4.25 \text{ g}} \times 100 = 64\%$$

3.3   Sources of Prelaboratory Information

Handbooks have been the traditional sources of prelaboratory information on physical constants and safety information. Today, however, there are many websites where you can access this type of information more conveniently.

Online Resources

The simplest and quickest way to identify chemicals, their molecular weights, physical properties, and safety information is to use websites of companies that sell the chemicals. For example, the Sigma-Aldrich and Acros Organics websites provide physical properties, MSDSs, and spectral data for chemicals that they sell (described in Chapter 1). Canadian and United States occupational safety websites also have free searchable databases for chemical and safety information. Wikipedia, chembiofinder, or chemspider are additional, generally reliable sources of physical properties, and they may also link to MSDSs. At the time of publication, the following sites provided useful information:

http://www.sigmaaldrich.com/united-states.html
http://www.acros.com
http://www.ccohs.ca
http://www.cdc.gov/niosh/npg/npgdcas.html
http://en.wikipedia.org/wiki/Wikipediachemfind
http://www.chemspider.com
http://www.chembiofinder.cambridgesoft.com

For example, a screenshot of a search for acetyl chloride on the Sigma-Aldrich website provides physical properties under the "Purchase" tab (Figure 3.1) and safety information and links to spectral data under the "Safety & Documentation" tab (Figure 3.2). The "Protocols & Papers" tab provides citations to references that describe the preparation, properties, or use of the compound. These come from a variety of sources, such as *The Merck Index* (see below),

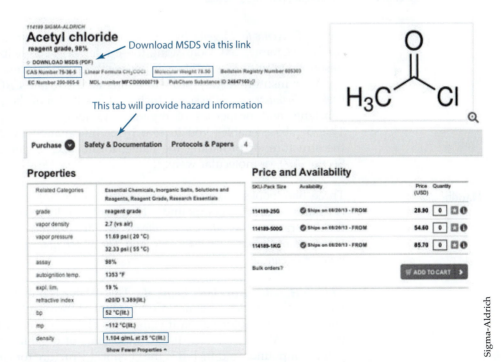

FIGURE 3.1 Screenshot of a search for the compound acetyl chloride, with some information and physical properties highlighted in blue boxes, posted on http://www.sigmaaldrich.com.

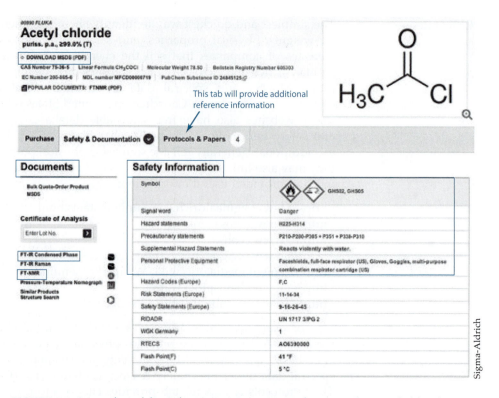

FIGURE 3.2 Screenshot of the Safety & Documentation information for acetyl chloride, posted on http://www.sigmaaldrich.com. The safety information and links to the MSDS and spectral data are highlighted in blue boxes.

Fiesers' Reagents for Organic Synthesis, or the primary chemical literature (see Chapter 28). Consulting these references can help you in designing your own laboratory experiments or procedures.

Your institution may provide online access to the *CRC Handbook of Chemistry and Physics,* which contains tables of the physical constants and properties of organic and inorganic compounds. This database enables you to search for information by drawing the structure, or providing the name, formula, CAS number (see Section 1.7), or molecular weight.

Handbooks

Three handbooks are particularly useful for finding the physical constants of organic compounds. They are often located in or near the laboratory: the *Aldrich Handbook of Fine Chemicals,* the *CRC Handbook of Chemistry and Physics,* and *The Merck Index: An Encyclopedia of Chemicals, Drugs, and Biologicals.*

Aldrich Handbook of Fine Chemicals. This catalog is published biennially by the Aldrich Chemical Company of Milwaukee, Wisconsin, and most of the information it contains is also available online via the Sigma-Aldrich website. It lists thousands of organic and inorganic compounds and includes the chemical structure for each one, a brief summary of its physical properties, references on IR, UV, and NMR spectra, plus safety and disposal information. There are also references to *Beilstein's Handbook of Organic Chemistry* and to *Fiesers'*

Reagents for Organic Synthesis (see Chapter 28). The entries are organized alphabetically by compound name, but you can also locate compounds using the formula, CAS number, or product number indexes at the back of the catalog.

CRC Handbook of Chemistry and Physics. This handbook is published annually and it contains a wealth of information, including extensive tables of physical properties and solubilities, as well as structural formulas, for more than 12,000 organic and 2,400 inorganic compounds.

To locate an organic compound successfully, you must pay close attention to the nomenclature used in the tables. In general, IUPAC nomenclature is followed, but a compound usually known by its common name may be listed under both names or even only under the common name. For example, the primary name of CH_3CO_2H is listed in the *CRC Handbook* as acetic acid, with ethanoic acid (its IUPAC name) given as the secondary name. No entry for ethanoic acid is listed. Conversely, the listing for $CH_3(CH_2)_5Br$ has 1-bromohexane as the primary name of the compound and *n*-hexyl bromide as the secondary name (synonym). In earlier editions of the *CRC Handbook,* substituted derivatives of compounds were listed under the heading of the parent compound rather than simply in alphabetical order by the first letter of the compound's name. For example, 1-bromohexane was listed under the parent alkane as "Hexane, 1-bromo-." A brief explanation of the nomenclature system, plus definitions of abbreviations and symbols, precedes the tables of organic compounds in all editions of the *CRC Handbook.*

The Merck Index. *The Merck Index: An Encyclopedia of Chemicals, Drugs, and Biologicals,* currently in its 15th edition, has over 10,000 organic compound entries that give physical properties and solubilities as well as references to syntheses, safety information, and uses. *The Merck Index* is particularly comprehensive for organic compounds of medical and pharmaceutical importance.

Further Reading

CRC Handbook of Chemistry and Physics; CRC Press; Taylor & Francis Group: Boca Raton, FL, electronic edition.

Haynes, W. M. (Ed.) *CRC Handbook of Chemistry and Physics,* 94th ed.; CRC Press; Taylor & Francis Group: Boca Raton, FL, 2013.

O'Neill, M. J. (Ed.) *The Merck Index: An Encyclopedia of Chemicals, Drugs, and Biologicals,* 15th ed.; Royal Society of Chemistry Publishing: Cambridge, UK, 2013.

Questions

1. True or false: After you have completed your experiments in the laboratory, your notebook should contain:
   (a) Sequentially numbered pages.
   (b) Entries written in pencil.
   (c) Blank spaces to fill in with details later.
   (d) No statements that were crossed out and rewritten.
   (e) Dates for all entries.
   (f) No detailed calculations—they should be done on the calculator or a separate sheet of paper and the answers copied into the book.

2. Suppose you plan to carry out the synthesis of aspirin (acetylsalicylic acid) by reacting salicylic acid with acetic anhydride, using 85% phosphoric acid as a catalyst. In addition to aspirin, acetic acid will be a side product. Using one of the suggested online sources (such as the Sigma-Aldrich or Acros Organics websites), look up the molar masses for all of the reagents and products in this reaction. Also look up the melting points of salicylic acid and acetylsalicylic acid and the boiling points and densities of acetic anhydride and acetic acid. Your answers should include both the molar masses and the physical properties for all four reagents and products.

3. Beginning with 1.0 g salicylic acid, 2.0 mL acetic anhydride, and five drops of 85% phosphoric acid, identify the limiting reagent and calculate the theoretical yield of acetylsalicylic acid. If you isolated 0.95 g of acetylsalicylic acid, what would be your percent yield?

Salicylic acid　　　　　Acetic anhydride　　　　　Acetylsalicylic acid　　　　　Acetic acid

2

Carrying Out Chemical Reactions

Essay— Learning to Do Organic Chemistry

Carrying out chemical reactions is, in many ways, much like being a master chef. Chefs, like chemists, use specialized equipment and master the techniques that allow them to reproduce detailed recipes (experimental reaction procedures). They also use their knowledge and creativity to design new dishes (new compounds) modeled from a palate of available foodstuffs (commercially available chemicals). And, of course, the cooking of food involves a series of chemical reactions.

Safely manipulating chemical substances and converting one compound into another—in short, doing chemistry—is highly creative and can be terrifically exciting. It can also be daunting due to the large variety of specialized equipment used in the lab. The names of the glassware used can be bewildering enough, let alone imagining how the pieces all fit together. But the information provided in the chapters that follow will guide you in acquiring the knowledge and skills you will need to do experimental organic chemistry.

Carrying out reactions in the lab is central to chemistry. Setting up and running a chemical reaction can be broken down into five parts:

(1) Preparation and setup
(2) Monitoring the reaction progress
(3) Post-reaction "workup," involving the quenching of excess reagents and rough preliminary purification of reaction products
(4) Final purification
(5) Analysis of the reaction products

In the beginning you will most likely be provided with explicit directions that specify the necessary glassware and equipment, and the required amounts of reagents and solvents to use. You will learn to perform the techniques of organic chemistry while gaining experience in properly documenting your work in your laboratory notebook. When you have the chance to adapt a procedure from the chemical literature or design your own experiments, you will have to make decisions about the scale of a reaction and about the appropriate glassware and equipment to use.

Glassware that is too large or too small for the scale of a reaction can be problematic. Liquids form a thin, almost invisible, coating on glass surfaces, and using glassware that is too large for the amounts of reagents used can result in the loss of chemicals and lower product yields, especially in the "microscale" reactions common in the modern laboratory. On the other hand, glassware that is too small for the scale of reagents used can be messy and even dangerous. Some commonly used organic solvents undergo a remarkable volume expansion when heated, which can result in overflow and potential fires. Developing a sense of scale in the laboratory is a skill all organic chemists must hone.

You will have at your disposal specialized equipment to handle both solid and liquid reagents. Reagents and solvents will have to be accurately measured and dispensed, requiring proper use of volumetric glassware, transfer pipets, and balances. The chapters in Part 2 will guide you in the proper use of laboratory glassware, as well as measuring and transferring equipment. It will also outline the methods used to safely heat and cool reaction mixtures, as reactions are often run at temperatures above or below room temperature.

Many reactions involve the use of highly reactive reagents, such as strong acids, bases, or oxidizing or reducing agents, that can react to give undesired side products if the reaction conditions are not carefully controlled. The reaction progress is often closely monitored and excess reagents are quenched as soon as the reaction is complete. Thin-layer chromatography and gas chromatography, two techniques commonly used to monitor reaction progress, are discussed in Part 4.

The fun really begins when the reaction is complete. The post-reaction processing of a chemical reaction mixture is called the "workup" and involves a series of steps which ensure that catalysts and excess reactive reagents are neutralized and separated from the desired reaction products. Workup methods are discussed in Part 3. For instance, if a strong acid, like HCl, is used in a reaction, the first step in the workup is invariably the neutralization of any excess acid using a mild base like aqueous sodium bicarbonate, producing the vigorous bubbling of CO_2 gas. NaCl from neutralizations and other water soluble by-products are generally removed from the usually water-insoluble organic reaction products by water extraction. After a crude product is isolated and dried to remove traces of water, a final purification is generally done. The purification method used will depend on the physical properties of the product and will involve techniques like recrystallization, distillation, or chromatography. These purification methods are discussed in Parts 3 and 4.

Finally, as with a chef's masterpiece, the final purified product is subject to a "taste test." Did you really make and isolate what you intended? Although it seems absurd today, at one time chemists routinely used taste as a method to characterize new chemicals. We know much more about chemical toxicology today. Tasting your products will not be an analytical method used in your laboratory! Instead, Part 5 of this text guides you through analytical methods like NMR and IR spectroscopy, which you will use to prove that your reaction products are what you think they are.

If you have not already done so, we urge you to carefully read Chapter 1 on laboratory safety before you begin your laboratory work. Doing organic chemistry safely should be a constant consideration while you are working in the laboratory.

LABORATORY GLASSWARE

You will find an assortment of glassware and equipment in your laboratory desk; some items will be familiar to you and other items may not. If your lab is equipped for miniscale experimentation, you will find specialized glassware called *standard taper glassware,* which has carefully constructed ground glass joints designed to fit together tightly and interchangeably. Standard taper glassware is available in a variety of sizes. If you will be carrying out microscale experimentation, you will use scaled-down glassware designed for the milligram quantities of reagents used in microscale work. There are two types of microscale glassware commonly used in the undergraduate organic laboratory—standard taper microscale glassware with threaded screw cap connectors and the Williamson microscale glassware that fastens together with flexible elastomeric connectors.

The fundamental chemistry is the same no matter which type of glassware you use, but there are some differences in how the various types of glassware are used and manipulated. Throughout this text, the discussions of many laboratory techniques include separate sections that describe procedures using each of the three types of glassware. **To quickly identify the section that pertains to your glassware, look for the appropriate icon in the margin. Figure 4.1 shows the icons for (a)** *miniscale standard taper glassware,* **(b)** *microscale standard taper glassware,* **and (c)** *microscale Williamson glassware.*

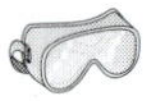

SAFETY PRECAUTION

Before you use any glassware in an experiment, check it carefully for cracks or chips. Glassware with spherical surfaces, such as round-bottomed flasks, can develop small, star-shaped cracks (Figure 4.2). Replace damaged glassware. When cracked glassware is heated, it can break and ruin your experiment and possibly cause a serious spill or fire.

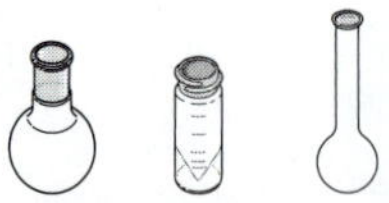

(a)   (b)   (c)

FIGURE 4.1 Icons used throughout this text to denote (a) miniscale standard taper glassware, (b) microscale standard taper glassware, and (c) microscale Williamson glassware.

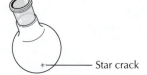

Star crack

FIGURE 4.2 Round-bottomed flask with a star crack.

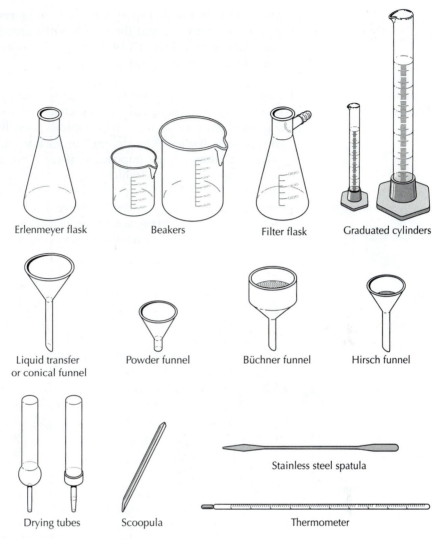

FIGURE 4.3 Typical equipment in a student desk.

4.1 Desk Equipment

A typical student lab desk contains an assortment of beakers, Erlenmeyer flasks, filter flasks, thermometers, graduated cylinders, test tubes, funnels, and other items. Your desk or drawer will probably have most, if not all, of the equipment items shown in Figure 4.3. Make sure that all glassware is clean and has no chips or cracks. Replace damaged glassware.

4.2 Miniscale Standard Taper Glassware

Standard taper glassware is designated by the symbol ⦵.

Standard taper glassware is designated by the symbol ⦵. All the joints in standard taper glassware have been carefully ground so that they are exactly the same size, and all the pieces fit together interchangeably. We recommend the use of ⦵ 19/22 or ⦵ 14/20

glassware for miniscale experiments. The numbers, in millimeters, represent the diameter and the length of the ground glass surfaces (Figure 4.4). A typical set of ℥ 19/22 glassware found in introductory organic laboratories is shown in Figure 4.5.

Greasing Ground Glass Joints

Because standard taper joints fit together tightly, they are often coated with a lubricating grease. The grease prevents interaction of the ground glass joints with the chemicals used in the experiment that can cause the joints to "freeze," or stick together. Taking apart stuck joints, although not impossible, is often not an easy task, and standard taper glassware (which is expensive) frequently is broken in the process. The use of grease does have its downside in that grease can contaminate a reaction mixture. Therefore, grease is sometimes omitted for reactions that do not involve strong bases, such as sodium hydroxide, which tend to promote the freezing of joints. In

FIGURE 4.4
Dimensions of ℥ 19/22 ground glass joints.

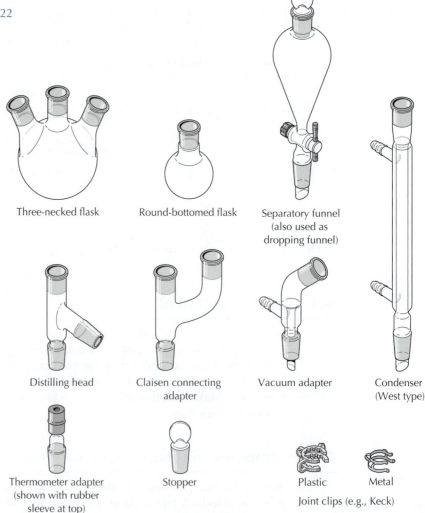

Three-necked flask   Round-bottomed flask   Separatory funnel
                                                     (also used as
                                                     dropping funnel)

Distilling head   Claisen connecting   Vacuum adapter   Condenser
                      adapter   (West type)

Thermometer adapter   Stopper   Plastic   Metal
(shown with rubber
sleeve at top)   Joint clips (e.g., Keck)

FIGURE 4.5 Standard taper glassware for miniscale experiments.

addition, contamination can be harder to avoid when working on a very small scale. A trace of grease in one gram of product is less of an issue than the same amount of grease in a few milligrams of product. Thus, microscale glassware with ground glass joints is never greased unless the reaction involves strong bases. Whether greased or not, ground glass joints are often held together using plastic or metal joint clips.

Types of grease for ⊤ joints. Several greases are commercially available. For general purposes in an undergraduate laboratory, a hydrocarbon grease, such as Lubriseal, is preferred because it can be removed easily. Silicone greases have a very low vapor pressure and are intended for sealing a system that will be under vacuum. Silicone greases are nearly impossible to remove completely because they do not dissolve in detergents or organic solvents.

Sealing a standard taper joint with grease. To seal a standard taper joint, apply **two thin strips of grease** almost the entire length of the inner joint about 180° apart, as shown in Figure 4.6. Gently insert the inner joint into the outer joint and rotate one of the pieces. The joint should rotate easily and the grease should become uniformly distributed so that the frosted surfaces appear clear.

Grease

FIGURE 4.6 Apply two *thin* strips of grease almost the entire length of the inner joint about 180° apart.

Using excess grease is bad practice. Not only is it messy, but it may contaminate the reaction mixture or coat the inside of reaction flasks, making them difficult to clean. Just enough grease to coat the entire ground surface thinly is sufficient. If grease oozes above the top or below the bottom of the joint, you have used much too much. Take the joint apart, wipe off the excess grease with a towel or tissue, and assemble the pieces again.

Removing grease from standard taper joints. When you have finished an experiment, use a tissue to wipe off as much grease from the joints as possible and then clean the glassware using a brush, detergent, and hot water. If this scrubbing does not remove all the grease, dry the joint and clean it with a tissue moistened with toluene or hexane.

SAFETY PRECAUTION

Toluene and hexane are irritants and pose a fire hazard. Wear gloves and work in a hood. Place the spent solvent in the appropriate waste container.

4.3   Microscale Glassware

When the amounts of reagents used for experiments are less than 300 mg or the total volume of a reaction mixture is less than 3 mL, microscale glassware is used. Recovering any product from an operation at this scale would be difficult if you were using 19/22 or 14/20 standard taper glassware; much of the material would be lost on the glass surfaces. Two types of microscale glassware are commonly used in undergraduate organic laboratories: standard taper glassware with threaded screw cap connectors and Williamson glassware that fastens together with flexible elastomeric connectors. Your instructor will tell you which type of microscale glassware is used in your laboratory.

 **Microscale Standard Taper Glassware**

The pieces of microscale standard taper glassware needed for typical experiments in the introductory organic laboratory are shown in Figure 4.7. The pieces fit together with 14/10 standard taper joints. The joints are held together by a threaded cap and O-ring, thus eliminating the use of clamps or joint clips. To assemble a joint, place the threaded cap over the inner joint, then slip the O-ring over the tapered portion. The O-ring holds the threaded cap in place. Fit the inner joint inside the outer joint and screw the threaded cap tightly onto the outer joint (Figure 4.8). A securely screwed connection effectively prevents the escape of vapors and is also vacuum tight. **Grease is not used with microscale glassware,** except when the reaction mixture contains a strong base, because its presence in reactions run on such a small scale could cause significant contamination of the reaction mixture.

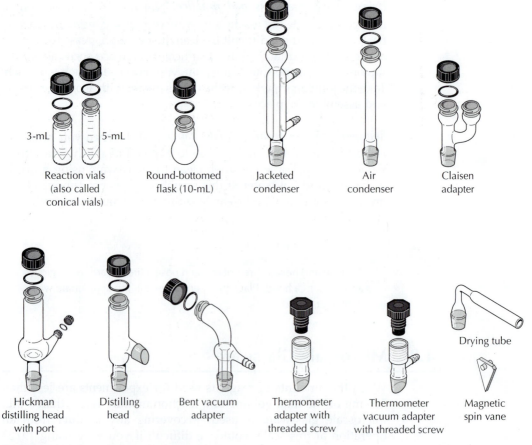

FIGURE 4.7 Microscale standard taper glassware.

FIGURE 4.8
Assembling a standard taper joint on microscale standard taper glassware.

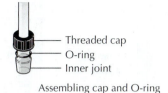

Threaded cap
O-ring
Inner joint

Assembling cap and O-ring

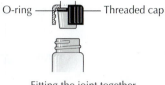

O-ring — Threaded cap

Fitting the joint together

The conical reaction vials and round-bottomed flasks can also be directly capped when additional glassware is not required or when used for sample storage (Figure 4.7). A Teflon septum fits inside the cap to cover the hole in it. One side of the septum is coated with silicone rubber. The harder, chemically inert Teflon side of the septum should face the inside of the vial. **When a cap and its septum are used to seal a conical vial or flask, an O-ring is not used.**

 Microscale Williamson Glassware

The various pieces of microscale Williamson glassware used in typical experiments in the organic laboratory are shown in Figure 4.9. This type of microscale glassware fits together with flexible elastomeric connectors that are heat and solvent resistant.

Grease is not used with a Williamson glassware connector. A flexible connector with an aluminum support rod fastens two pieces of

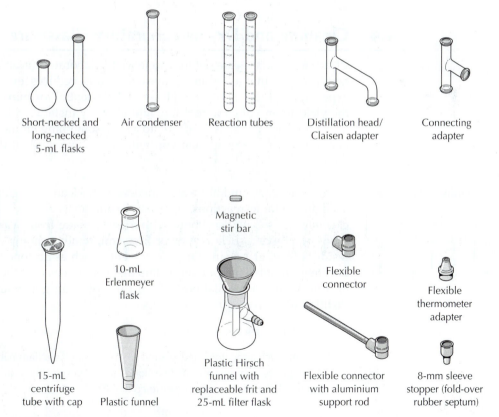

Short-necked and long-necked 5-mL flasks

Air condenser

Reaction tubes

Distillation head/ Claisen adapter

Connecting adapter

Magnetic stir bar

10-mL Erlenmeyer flask

Flexible connector

Flexible thermometer adapter

15-mL centrifuge tube with cap

Plastic funnel

Plastic Hirsch funnel with replaceable frit and 25-mL filter flask

Flexible connector with aluminium support rod

8-mm sleeve stopper (fold-over rubber septum)

FIGURE 4.9 Microscale Williamson glassware and other microscale apparatus. (Manufactured by Kontes Glass Co., Vineland, NJ.)

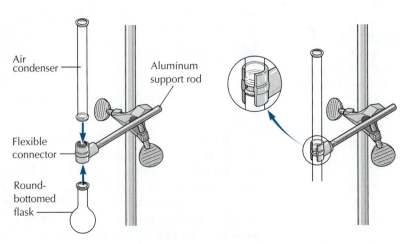

Air condenser

Aluminum support rod

Flexible connector

Round-bottomed flask

FIGURE 4.10
Assembling microscale Williamson glassware with a flexible connector.

Fitting the glassware into the flexible connector one piece at a time

Cutaway showing the two pieces of glassware fastened in connector

glassware together and provides attachment of the apparatus via a two-way clamp to a ring stand or vertical support rod. One piece of glassware is pushed into the flexible connector, and then the second piece is pushed into the other end of the connector, as shown in Figure 4.10. The flexible connector effectively seals the joint and prevents the escape of vapors.

4.4   Cleaning and Drying Laboratory Glassware

Cleaning the glassware before you leave the laboratory is an important practice that ultimately saves time and reduces everyone's exposure to chemicals. Clean glassware is essential for maximizing the yield in any organic reaction, and in many instances glassware also must be dry. It is best not to wash something immediately before using it, because then you will waste time while it dries in an oven.

Cleaning Glassware

Strong detergents and hot water are needed to clean most glassware used for organic reactions. You may want to wear gloves when cleaning glassware. Scrubbing with a paste made from water and scouring powder, such as Ajax or Bon Ami, removes many organic residues from glassware. Organic solvents, such as acetone or hexane, help dissolve the polymeric tars that sometimes coat the inside of a flask after a distillation. A final rinse of clean glassware with distilled water prevents water spots.

SAFETY PRECAUTION

Solvents such as acetone and hexane are irritants and are flammable. Wear gloves, use the solvents in a hood, and dispose of them in the flammable (nonhalogenated) waste container.

If soap and water prove ineffective, stubborn residues and traces of grease can be removed by soaking glassware in a *base bath*.

A base bath is a solution of alcoholic sodium hydroxide,* and a short soak will usually remove most residues. Strong bases can etch glassware, so soaking time should be kept to a minimum—typically 5 to 30 min—and the glassware must be thoroughly rinsed with water after this treatment. Because of possible deterioration through etching, volumetric glassware and items with glass frits should not be cleaned in a base bath.

SAFETY PRECAUTION

Strong bases, such as sodium hydroxide, cause severe burns and eye damage. Skin contact with alkali solutions starts as a slippery feel to the skin followed by irritation. Wash the affected area with copious amounts of water. Wear gloves, eye protection, and a rubber apron while cleaning glassware with alcoholic NaOH solution.

Drying Glassware

Dry glassware is needed for most organic reactions. The easiest way to ensure that glassware is dry is to leave it washed and clean at the end of each lab session. It will be dry and ready to use by the next laboratory period.

Oven drying of glassware. Wet glassware can be dried by heating it in an oven at 120°C for 20 min. Remove the dried glassware from the oven with tongs and allow it to cool to room temperature before using it for a reaction.

Drying wet glassware with acetone. Glassware that is wet from washing can be dried more quickly by rinsing it in a hood with a few milliliters of acetone. Acetone and water are completely miscible, so the water is removed from the glassware. The acetone is collected as flammable (nonhalogenated) waste; any residual acetone on the glassware is allowed to evaporate into the atmosphere. There is an environmental cost, as well as the initial purchase price and later waste disposal costs, in using acetone for drying glassware.

Questions

1. What type of glassware would be most appropriate for a reaction in which 75 mg of reactant A is treated with 95 mg of reagent B, in 2 mL of solvent? What if 0.80 g of reactant A is used in 20 mL of solvent?
2. What is the purpose of using grease on ground glass joints?
3. Why is grease generally not used with microscale standard taper glassware?

Under what circumstances might it be used with this glassware? Why?
4. Why is it important to make sure that glassware is dry before using it for a chemical reaction?
5. How can glassware for a reaction be dried quickly if it's found to be dirty and needs to be washed before beginning an experiment?

---

* Made by dissolving 120 g of NaOH in 120 mL of water and diluting to 1 L with 95% ethanol.

5

MEASUREMENTS AND TRANSFERRING REAGENTS

Whether you are carrying out miniscale or microscale experiments, you need to accurately measure both solid and liquid reagents as well as reaction temperatures and purification procedures. You will have to be aware of the uncertainties in the measurements you make and also use the appropriate number of significant figures for all of your measurements. You may also have the opportunity to evaluate data sets collected from experiments. Methods for weighing solids and liquids, measuring liquid volumes, transferring solids and liquids without loss, and measuring temperature are described in this chapter. In addition, the chapter offers a review of experimental uncertainties, significant figures, and basic error analysis.

5.1   Using Electronic Balances

Your laboratory is probably equipped with several types of electronic balances for weighing reagents. How do you decide which one to use to determine the mass of a reagent or product? As a general rule, a top-loading balance that weighs to the nearest centigram (0.01 g) is satisfactory for miniscale reactions using more than 2–3 g of a substance. However, in miniscale reactions using less than 2 g of reagent and in some microscale reactions, all reagent quantities should be determined on a balance that weighs at least to the nearest milligram (0.001 g). A top-loading milligram balance must have a draft shield to prevent air currents from disturbing the weighing pan while a sample is being weighed (Figure 5.1a).

When a quantity of less than 200 mg is required in a microscale reaction, its mass should be determined on an analytical balance (Figure 5.1b) that weighs to the nearest 0.1 mg (0.0001 g). Close the doors of the balance while weighing the sample.

Care of Electronic Balances

Electronic top-loading and analytical balances are expensive precision instruments that can be rendered inaccurate very easily by corrosion from spilled reagents. **If anything spills on the balance or the weighing pan, clean it up immediately.** Notify your instructor right away if the spill is extensive or the substance is corrosive.

Weighing Solids

No solid reagent should ever be weighed directly on a balance pan. Instead, weigh the solid in a glass container (vial or beaker), in an aluminum or plastic weighing boat, in a crinkle cup, or on glazed weighing paper. Then transfer it to the reaction vessel. Weighing directly into a round-bottomed flask or test tube, which are not stable on the balance pan, should be done only if the flask or test tube is held in a secure upright position inside a beaker or other stable container.

Tare mass. The mass of the container or weighing paper used to hold the sample is called the *tare mass* or just the *tare*. When weighing a specific quantity of reagent, the tare mass of the container or

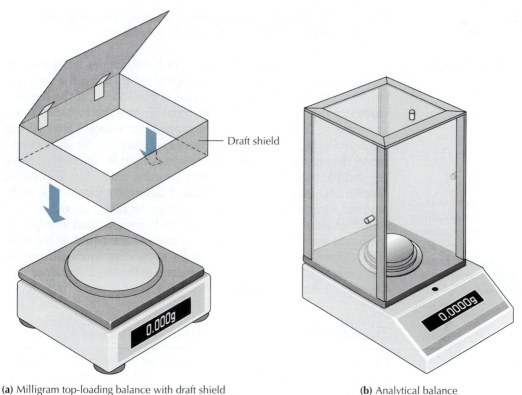

(a) Milligram top-loading balance with draft shield   (b) Analytical balance

FIGURE 5.1 Two types of balances.

weighing paper is simply subtracted by pressing the tare or zero button on the electronic balance before the sample is added. Then add the solid until the desired mass appears on the readout screen. For greatest accuracy, reweigh the sample container directly **after** the sample has been transferred from it. The difference between the combined mass and the mass of the container after sample transfer provides the mass of material actually delivered, accounting for traces of untransferred sample remaining in the container.

A vial or flask—with its label and cap or stopper—that will be used to hold a purified reaction product should be weighed **before the product is placed in it.** Be sure to record the tare mass of the container in your lab notebook. After the product has been collected, weigh the container holding the product again. Subtracting the mass of the empty container from the mass of the container with sample provides the mass of the sample.

How to weigh a solid. To weigh a specific quantity of a solid reagent, place a weighing boat, crinkle cup, or piece of diagonally folded glazed weighing paper on the balance pan and press the zero or tare button. Use a spatula to add small portions of the reagent until the desired mass (within 1–2%) is shown on the digital display. For example, the mass of a sample would not need to be exactly the 0.300 g specified, but normally it should be within ±0.005 g of that amount. **Record the actual amount you use** in your notebook. If the compound you are weighing is the limiting reagent, calculate the theoretical yield based on the actual amount you used, not on the amount specified in the experimental procedure.

Weighing Liquids

When you weigh a liquid, the mass of the container (tare) must be ascertained and recorded, or else subtracted by using the zero button on the balance, **before** the liquid is placed in it. If the liquid is volatile, a cap or stopper for the container must be included in the tare mass so the container can be capped immediately after adding the liquid to prevent losses due to evaporation. If a liquid is evaporating during the weighing process, its weight is constantly changing and the weight you record may be quite unreliable. To weigh a specific amount of a liquid compound, determine the volume of the required sample from its density and transfer that volume to a tared container. Ascertain the mass of the tared container and its cap, plus the liquid sample, to determine the mass of the liquid. If the mass of liquid needed is less than 1 g, an alternative to measuring the volume is to add the liquid drop by drop to the tared container until the desired mass is obtained.

Be very careful that liquid does not spill on the balance while you are weighing a liquid sample. Should a spill occur, clean it up immediately.

5.2   Transferring Solids to a Reaction Vessel

Once the mass of a solid reagent has been determined, the reagent must be transferred to the reaction vessel. If the sample is in a weighing boat, fold the boat diagonally before transferring the sample. If the sample is in a crinkle cup, pinch the edges of the cup together leaving a small opening so that the solid can slide out of it easily but not spill. If the sample is on a piece of glazed weighing paper with a diagonal fold (Figure 5.2a), overlap the two outside edges and firmly hold them between your thumb and index finger while transferring the solid (Figure 5.2b). A spatula can be used to aid in transferring the solid if it sticks to the weighing paper.

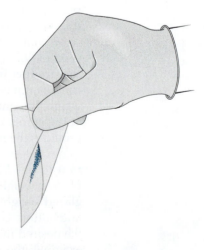

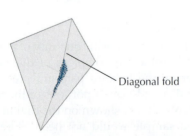

(a) Diagonally folded weighing paper with a solid sample

(b) Overlap opposite diagonal corners and hold firmly between thumb and index finger.

Diagonal fold

FIGURE 5.2 Preparing to transfer a solid sample from a weighing paper.

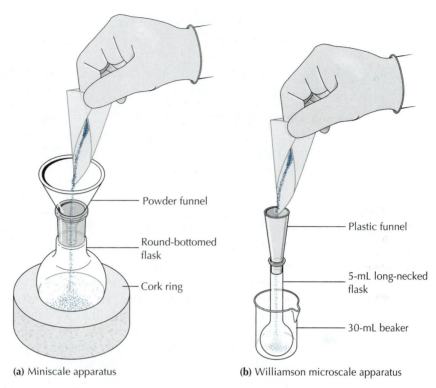

(a) Miniscale apparatus   (b) Williamson microscale apparatus

FIGURE 5.3 Transferring solids with a powder funnel.

Using a Powder Funnel

For reactions being run in miniscale round-bottomed flasks, transferring solids using a powder funnel serves to keep the solid from spilling and prevents any solid from sticking to the inside of the joint at the top of the flask (Figure 5.3a). The stem of a powder funnel has a larger diameter than that of a funnel used for liquid transfers so that solids will not clog it. Using a funnel is essential with Williamson microscale glassware because of the very small opening at the top of the round-bottomed flasks and reaction tubes (Figure 5.3b).

Transferring Solids to a Microscale Standard Taper Vial

Set the standard taper microscale vial in a small beaker so it will not tip. Pick up the weighing paper (see Figure 5.2a and b). Slide the overlapped edges further together to decrease the size of the opening at the bottom of the weighing paper (Figure 5.4a). Insert the tip of the paper into the conical vial and allow the solid to slide from the paper into the vial (Figure 5.4b).

5.3 Measuring Volume and Transferring Liquids

Several liquid volume measuring devices are used in the laboratory, including graduated cylinders, pipets, burets, dispensing pumps, syringes, and beakers and flasks with volume markings on them. The equipment used for measuring a specific volume of liquid depends on the accuracy with which the volume needs to be known. For example, the volume of a liquid reagent that is the limiting factor in a miniscale reaction may need to be measured with a graduated pipet and then weighed to determine the exact amount.

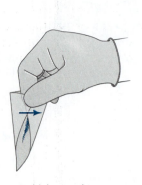

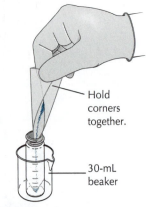

Hold
corners
together.

30-mL
beaker

FIGURE 5.4
Transferring solids with
a weighing paper into
a conical vial.

(a) Hold the weighing paper as shown,
and slide the overlapping edges
further together as shown in (b).

(b) Insert tip of paper
into conical vial.

If the liquid is a solvent or is present in excess of the limiting
reagent, volume measurement can be done with a graduated pipet
for microscale work and with either a graduated pipet or a graduated cylinder for miniscale work. The volume markings on beakers and flasks can be used only to estimate approximate volumes
and should never be used for measuring a reagent that will go into
a reaction.

**Graduated
Cylinders**

Graduated cylinders do not provide high accuracy in volume
measurement and should be used only to measure quantities of
liquids other than limiting reagents. The volume contained in a
graduated cylinder is read from the bottom of the *meniscus*, as
shown in Figure 5.5.

Graduated cylinders are not used to measure reagents for
microscale reactions; however, a 5- or 10-mL graduated cylinder
can be used for measuring volumes of extraction solvents greater
than 1 mL.

Dispensing Pumps

Dispensing pumps fitted to glass bottles come in a variety of sizes
designed to deliver a preset volume of liquid (>0.1 mL). **Dispensing
pumps should never be used for measuring the volumes of
limiting reagents.** Pumps in the 1-, 2-, and 5-mL range may be used
in miniscale reactions for other than limiting reagents and may
sometimes be used in microscale work for dispensing solvents.

Before you begin to measure a sample with a dispensing pump,
check that the spout of the pump is filled with liquid and contains

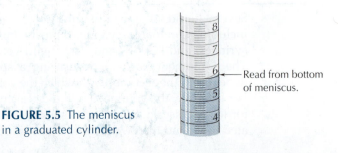

Read from bottom
of meniscus.

FIGURE 5.5 The meniscus
in a graduated cylinder.

no air bubbles that could cause a volume less than the preset one to be delivered. If air bubbles are present in the spout, pull up the plunger and discharge one or two samples into another container until the spout is completely filled with liquid. (Place the discarded samples in the appropriate waste container.) Dispense the sample *directly* into the container in which it will be used. If an accurate mass of the sample is necessary, dispense it into a preweighed container and then weigh the container and sample.

The effective operation of a dispensing pump demands that you slowly pull up the plunger until it reaches the preset volume stop (Figure 5.6). Hold the receiving container or reaction vessel under the spout and then gently push the plunger down as far as it will go to discharge the preset volume. Be sure that the last drop of liquid on the spout is transferred by touching the tip to the inside of the receiving vessel.

The exact amount of liquid delivered by a dispensing pump depends on the liquid's viscosity, density, surface tension, and vapor pressure. Normally, your instructor will adjust your pump to deliver the proper amount before your laboratory period. In the unlikely event that your pump has not been preset and checked for you, it may be calibrated by weighing a sample transferred with the pump. Dividing the mass of the liquid sample by the liquid's density provides the actual volume transferred, which can be compared to the nominal preset volume. If the actual and nominal volumes do not match, the preset volume position can be changed to accommodate the error and the pump recalibrated to verify the adjustment. Alternatively, the volume settings on many instruments can be adjusted from their factory settings to correct volume errors. Consult your instructor for the proper calibration and adjustment procedure for your pump.

In addition to analog pumps, like the one shown in Figure 5.6, a variety of digital pumps are also available. Your instructor will demonstrate the specific operating procedure of the bottle-top dispensing pumps used in your laboratory.

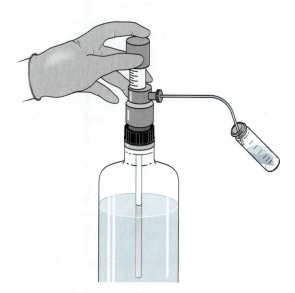

FIGURE 5.6
Dispensing pump.

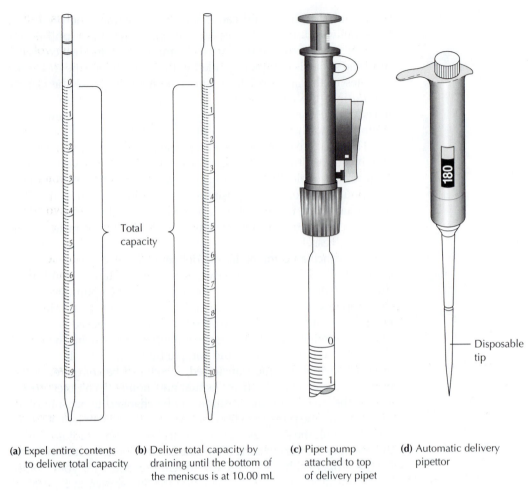

(a) Expel entire contents to deliver total capacity

(b) Deliver total capacity by draining until the bottom of the meniscus is at 10.00 mL

(c) Pipet pump attached to top of delivery pipet

(d) Automatic delivery pipettor

FIGURE 5.7 (a) and (b) Types of graduated pipets. (c) Pipet fitted with a pipet pump. (d) Automatic delivery pipettor.

Graduated Pipets

The small volumes used in microscale and many miniscale reactions are conveniently and accurately measured with graduated pipets of 1.00-, 2.00-, 5.00-, and 10.00-mL sizes (Figure 5.7). Two types of graduated pipets are available: one delivers its total capacity when the last drop is expelled (Figure 5.7a), and the other delivers its total capacity by stopping the delivery when the meniscus reaches the bottom graduation mark (Figure 5.7b). Both kinds of graduated pipets are frequently used to deliver a specific volume by stopping the delivery when the meniscus reaches the desired volume.

Pipets are usually filled and expelled using a pipet pump (Figure 5.7c). The pumps come in a variety of sizes to fit different sized pipets. To fill the pipet, turn the knurled wheel so that the piston moves up and liquid is drawn into the pipet. Turning the wheel in the opposite direction expels the liquid. Alternatively, a syringe attached to the pipet with a short piece of latex tubing can serve to fill the pipet and expel the requisite volume. The most accurate volumes are obtained by difference measurement—that is, filling the pipet to a convenient specific mark and then discharging the liquid until the required volume has been dispensed. The volume contained in a graduated pipet is read from the bottom of the meniscus. The excess liquid remaining in the pipet should be placed in the appropriate waste container.

Automatic Delivery Pipets

Small volumes of 10–1000 μL (0.010–1.000 mL) can be measured very accurately and reproducibly with automatic delivery pipets or pipettors. Automatic pipets have disposable plastic tips that hold the adjustable preset volume of liquid; no liquid actually enters the pipet itself, and the pipet should never be used without a disposable tip in place (Figure 5.7d). Automatic pipet plungers have two stops. To fill the pipet tip, **press the plunger to the first stop.** Holding the pipet vertically, dip the pipet tip in the liquid to be transferred and fill the tip by slowly releasing the plunger. To deliver the entire sample from the tip, **push the plunger fully to the second stop.** It is important to remember that the set volume of the pipet corresponds only to the volume achieved by filling the pipet from the first plunger stop. Automatic pipets are expensive, and your instructor will demonstrate the specific operating technique for the type in your laboratory.

Automatic pipets must be properly calibrated before use. Never assume that an automatic delivery pipet is calibrated accurately unless your instructor assures you that this is the case. As with dispensing pumps, the actual amount of liquid transferred depends on the liquid's density, viscosity, surface tension, and vapor pressure. You can calibrate an automatic pipet by delivering a preset amount of liquid from the pipet to a small, weighed flask or vial. Then weigh the flask to determine the exact mass of liquid. Dividing the mass of the sample by the density of the liquid provides the exact amount of liquid transferred. If the automatic pipet needs to be recalibrated, consult your instructor.

Syringes

A syringe with a needle attached works well for measuring and transferring the small amounts of reagents used in microscale reactions. Syringes are also utilized for measuring and transferring anhydrous reagents from a septum-sealed reagent bottle to a reaction vessel when inert atmospheric conditions are employed (see Section 7.4).

SAFETY PRECAUTION

A syringe needle can cause puncture wounds. Handle it carefully, keep the shield on it except when using it, and dispose of it **only** in a special "sharps" container.

Pasteur Pipets

Pasteur pipets are particularly useful for transferring liquids in microscale reactions and extractions. There are also times when it is helpful to know the approximate volume of liquid in a Pasteur pipet.

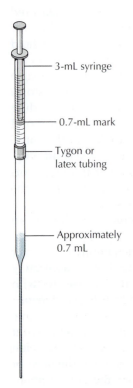

3-mL syringe

0.7-mL mark

Tygon or latex tubing

Approximately 0.7 mL

FIGURE 5.8 Using a syringe to estimate the volume of liquid drawn into a Pasteur pipet.

Approximating volumes with a Pasteur pipet. Pasteur pipets are suitable for measuring only approximate volumes because they do not have volume markings. **An approximate volume calibration of a Pasteur pipet is shown inside the front cover of this book.** Attaching a 1- or 3-mL syringe with a short piece of tubing to a Pasteur pipet also allows an approximate volume of the liquid to be estimated from the position of the plunger in the syringe as the liquid is drawn into the pipet (Figure 5.8).

Preventing dripping from Pasteur pipets. Volatile organic liquids tend to drip from a Pasteur pipet during transfers because the vapor pressure increases as your fingers warm the pipet and the rubber pipet bulb. Dripping can be minimized by either using a Pasteur filter-tip pipet or by pre-equilibrating the pipet and bulb with liquid vapors before drawing in the liquid.

Pasteur filter-tip pipets are prepared by using a piece of wire with a diameter slightly less than the inside diameter of the capillary portion of the pipet to push a tiny piece of cotton into the tip of the Pasteur pipet (Figure 5.9). A piece of cotton of the appropriate size should offer only slight resistance to being pushed by the wire. If there is so much resistance that the cotton cannot be pushed into the tip of the pipet, then the piece is too large. If this is the case, remove the wire and insert it through the tip to push the cotton back out of the upper part of the pipet, and tear a bit off the piece of cotton before putting it back into the pipet. The finished cotton plug in the tip of the pipet should be 2–3 mm long and should fit snugly but not too tightly. If the cotton is packed too tightly in the tip, liquid will not flow through it; if it fits too loosely, it may be expelled with the liquid. With a little practice, you should be able to prepare a filter-tip pipet easily.

An alternate method to prevent dripping involves pre-equilibrating the pipet and bulb with liquid vapors before filling

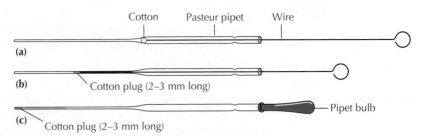

Cotton   Pasteur pipet   Wire

(a)

(b)   Cotton plug (2–3 mm long)

(c)   Cotton plug (2–3 mm long)   Pipet bulb

FIGURE 5.9 (a) and (b) Preparing a Pasteur filter-tip pipet. (c) The functional filter-tip pipet fitted with a rubber pipet bulb.

the pipet. Before drawing in a volatile liquid, hold the pipet tip just above the surface of the liquid and rapidly squeeze and release the pipet bulb five or six times. This process fills the pipet headspace with liquid vapors and the liquid subsequently drawn into the pipet tends not to drip. The technique is not quite as reliable as using a filter-tip pipet, but it works quite well for all but the most volatile solvents.

It may be tempting to invert a filled pipet to prevent dripping, but this should never be done. Inverting the pipet will cause some of the liquid to shoot out of the pipet and some to run into the pipet bulb where the liquid will become contaminated. Even if the bulb is new and clean, many organic liquids and solvents will dissolve or leach material from rubber pipet bulbs.

Plastic Transfer Pipets

Graduated plastic transfer pipets, available in 1- and 2-mL sizes, are suitable for measuring the volume of aqueous washing solutions used for microscale extractions and for estimating the volume of solvent in a microscale recrystallization (Figure 5.10). Most plastic transfer pipets are made of polyethylene and are chemically impervious to aqueous acidic or basic solutions, as well as alcohols, such as methanol or ethanol, and diethyl ether. They are not suitable for use with halogenated hydrocarbons because the plasticizer leaches from the polyethylene into the liquid being transferred.

Beakers, Erlenmeyer Flasks, Conical Vials, and Reaction Tubes

The volume markings found on beakers and Erlenmeyer flasks are only approximations and are not suitable for measuring any reagent that will be used in a reaction. However, the markings may be sufficient for measuring the amount of solvents in large-scale recrystallizations. The volume markings on conical vials and reaction tubes are also approximations and should be used only to estimate the volume of the contents, such as the final volume of a recrystallization solution, not for measuring the volume of a reagent used in a reaction.

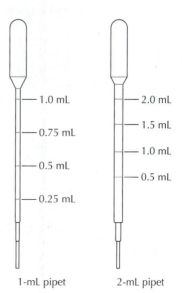

FIGURE 5.10 Graduated plastic transfer pipets.

1-mL pipet   2-mL pipet

5.4   Measuring Temperature

A number of temperature measurements must be made while carrying out chemical reactions and separations. For example, it may be necessary to maintain a constant temperature with a cooling or heating bath, to monitor the temperature of a reaction mixture, to determine the boiling point when carrying out a distillation, or to determine the melting point of a reaction product. There are numerous types of thermometers available, some suitable for a variety of tasks and others designed for specific purposes.

Types of Thermometers

Until recently, mercury thermometers were found in chemistry laboratories; however, concern for the environment, the toxicity of mercury, and the hazards of cleaning up a mercury spill from a broken thermometer have caused a number of states to ban the use of mercury thermometers in schools, colleges, and universities. They have been replaced by other types of temperature-measuring devices, such as nonmercury thermometers, metal probe thermometers, and digital thermometers that can be used with different types of temperature probes.

Nonmercury Thermometers

Nonmercury thermometers filled with alcohol or other organic liquids are now available; some of them can measure temperatures up to 300°C. Like mercury thermometers, nonmercury thermometers also need to be calibrated before using them for any temperature measurement where accuracy is essential, for example, when determining a melting or boiling point.

Digital Thermometers

Many types of temperature probes are available for use with digital thermometers. For example, the bead probe attached to the digital thermometer in Figure 5.11 can be used with an analog melting point apparatus (Figure 14.3). The use of a stainless steel or a Teflon-coated metal temperature probe with a digital thermometer can be used in a distillation. However, uncoated metal probes can react with hot organic vapors, particularly if they can be oxidized easily

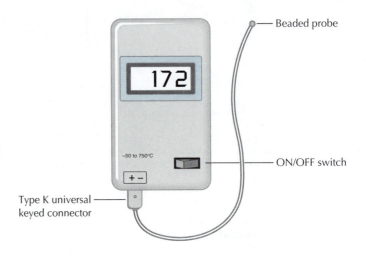

FIGURE 5.11 Digital thermometer.

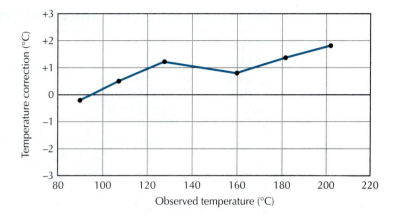

FIGURE 5.12
Thermometer
calibration graph.

or are acidic or corrosive; the use of an uncoated metal probe is not recommended for distillations of such compounds.

The length of a temperature probe that is positioned below the side arm of a distilling head needs to be determined experimentally by a series of distillations using pure compounds. Consult your instructor about the correct position within a distilling head for the type of temperature probe used in your laboratory.

Thermometer Calibration

The accuracy of a temperature determination is no better than the accuracy of the thermometer. You cannot assume that a thermometer has been accurately calibrated. Thermometers may give high or low temperature readings of 2°–3° or more.

A thermometer can be calibrated with a series of pure compounds whose melting points (see Chapter 14) are relatively easy to reproduce. The observed melting point corrections for the standard compounds in Table 5.1 can be plotted to determine the necessary temperature corrections (Figure 5.12). You can interpolate from your graph to ascertain the correction needed for any subsequent melting point determined with your calibrated thermometer.

TABLE 5.1	Compounds suitable for thermometer calibration
Compound	**Melting point, °C**
Benzophenone	48
Acetamide	81
Benzil	95
Benzoic acid	122
Phenacetin	135
Salicylic acid	160
Succinic acid	189
4-Fluorocinnamic acid	210
Anthraquinone	285

5.5 Measurement Uncertainty and Error Analysis

Suppose you are preparing for a reaction and you weigh 5.053 g of LiOH on a top-loading milligram balance. With your calculator you convert this amount into moles, using a molecular weight of 24 for LiOH, as shown in the equation below. You then record in your notebook "0.21054 mol LiOH."

$$5.053 \text{ g LiOH} \times \frac{1}{24 \text{ g/mol}} = 0.21054 \text{ mol LiOH}$$

In doing so, you have made two significant-figure mistakes. What are they?

The top loading balance weighs to ±1 milligram, so your mass is known to four significant figures; however, the value you used for the molecular weight of LiOH only had two significant figures. When you multiply or divide two numbers, the number of significant figures in the result is equal to the number of significant figures in the least precise figure—in this case, the molecular weight. Therefore, based on a molecular weight of 24, the number of moles should be rounded to two significant figures, or 0.21 mol LiOH. But, if you used this value, you would be short-changing your measurement. The accuracy of your mass determination is greater than the 0.21 mol value implies. The molecular weight of lithium hydroxide is known to five significant figures, so it makes sense to use a molecular weight that contains at least as many significant figures as the mass measurement. Thus, it would be appropriate to use 23.95 as the value for the molecular weight. Using this more exact value, the number of moles to record in your notebook would be 0.2110 mol—a value obtained from the calculation shown below and rounded to the appropriate number of significant figures (four).

$$5.053 \text{ g LiOH} = \frac{1}{23.95 \text{ g/mol}} = 0.21098 \text{ mol LiOH}$$

Would it really matter if you recorded 0.21 or 0.2110 mol in your notebook? Probably not; however, there are instances where this kind of precision does matter. It is best to get in the habit of using the proper number of significant figures in all laboratory situations, and your instructor will most likely demand it of you.

Types of error. You are probably familiar with basic error analysis from earlier science courses, but some of the major points are worth reviewing. There are three types of errors to be concerned with in the laboratory: ***human errors***, ***systematic errors***, and ***random errors***.

Human errors are mistakes due to inexperience, carelessness, or blunders on the part of the experimenter. These errors must be guarded against and eliminated, but when we talk about "error" in the laboratory, we are almost never referring to this type of error.

Systematic errors are those errors that are consistent from one measurement to another. For instance, a balance that consistently gives measurements that are 1 g too high exhibits a systematic error. Systematic errors are often due to faulty equipment or flaws in experimental design. These errors are not random and rooting them out requires calibration of instruments and careful analysis of experimental procedures. They cannot be estimated by statistical methods.

When we refer to "error" in the lab, most often we are talking about **random errors**. These errors cannot be avoided and are due to the physical limitations of measurements, such as the estimate of the last digit in a volume measurement using a graduated cylinder. Three different people may look at the meniscus of the liquid and reasonably propose a different value for that last digit in the measurement. Random errors, inherent in all measurements, are *not* mistakes. In this sense, **error** is synonymous with **uncertainty**. Uncertainties due to random errors can be estimated by statistical methods, and these are the types of errors that are the subject of error analyses.

Measurement Uncertainties and Significant Figures

Uncertainties in measurements are acknowledged through the proper use of significant figures. Electronic balances are generally accurate to ±1 unit in the last decimal place displayed on the readout. The uncertainty of volume measurements depends on the type of measuring device used. Knowing the uncertainty in a measurement will help you determine the number of significant figures to use in expressing the measured value. In the example of 5.053 g of LiOH, 5.05 are the *certain* digits, but the last digit in the value is *uncertain* by ±1 milligram. Together, the number of certain digits plus the first uncertain digit constitute the significant figures of a measurement. In other words, **the first uncertain figure should be the last significant figure.**

You should be particularly wary of measurements made with computer-controlled instruments. It is routine for computer outputs to list values to many, usually meaningless, decimal places. For instance, a computer-controlled FTIR spectrometer (see Chapter 21) is likely to provide infrared peak values out to two or more decimal places, for example, 3045.04 cm^{-1}. However, the digits to the right of the decimal point are beyond the accuracy of most spectrometers, except for the most expensive instruments, and should not be reported. Reporting those digits implies a greater accuracy than the measurement warrants, and routine IR data is customarily reported without decimals (e.g., 3045 cm^{-1}). When making any kind of measurement in the lab, whether by hand or using sophisticated instruments, it is important that you know the uncertainty of the measurement and record its value with the appropriate number of significant figures.

The rules for significant figures

1. Nonzero integers in a value always count as significant figures.
2. There are three types of zeroes that can appear in a value:
   (a) *Leading zeroes* precede all other nonzero integers; this type of zero never counts as a significant figure. For instance, the number 0.0035 contains only two significant figures. The three leading zeroes only serve to indicate the position of the decimal point.

(b) *Bounded zeroes* are zeroes found between nonzero digits in a value. Bounded zeroes always count as significant figures. For instance, the number 1.0035 contains five significant figures.

(c) *Trailing zeroes* are zeroes at the end of a number. Trailing zeroes only count as significant figures if the number contains a decimal point. For instance, the number 6500 contains only two significant figures. The zeroes do not count. On the other hand, 6500. has four significant figures. The decimal point in the second value implies that the number is certain out to all four digits that are shown. While technically correct, it is customary not to end a number with a decimal point—it is easy to confuse the decimal point with an end-of-sentence period, especially if the value happens to end a sentence. To avoid possible confusion it is better to express the four-significant-figure value as 6.500×10^3.

3. *Exact numbers* are numbers that have no uncertainty in them. For instance, numbers determined by counting something are exact numbers, such as five people, six experiments, and 10 determinations of a mass. These numbers are considered to have an infinite number of significant figures. Other examples include conversion factors, like "1000" when converting between grams and kilograms, or 2.54, the conversion between inches and centimeters (1 in = 2.54 cm).

Significant figures in calculations. Significant figures must be accounted for when performing calculations. Keeping proper account of significant figures, however, can become quite complex, depending on the calculations involved. Most of the calculations you will perform in the organic chemistry lab can be handled by a couple of simple rules.

1. When adding or subtracting numbers, the sum or difference will have the same number of decimal places as the least precise value used in the calculation. For example, consider the sum $3.590 + 4.3 + 0.0001 = 7.8901$. The proper value to record is 7.9, rounding the sum to the nearest tenth to match the number of decimal places in 4.3, the least precise figure.

2. When multiplying or dividing, the product or quotient has the same number of significant figures as the least precise value used in the calculation. For example, consider the product $24.5 \times 0.00021 \times 350.2 = 1.801779$. The proper result to report is 1.8, rounding the product to two significant figures to match the value with the fewest significant figures (0.00021).

Rounding

1. When carrying out a series of calculations, carry all digits through to the final result, then round the value to give the correct number of significant figures.

2. Consider only the digit directly to the right of the last significant figure when rounding. For instance, to round 3.849 to two

significant figures, the 4 is the digit to look at. Because 4 is less than 5, the number is rounded to 3.8. You should not round sequentially. That is, it would be incorrect to first round the 4 up to 5, based on the 9, and then round 3.85 up to 3.9 as the final result.

Accuracy and precision. **Accuracy** describes how close a measurement is to the true value. **Precision** refers to the reproducibility of a measurement. If you carried out a measurement many times and all of the measured values were very close to one another, you would be very precise in your measurements. However, precision does not imply accuracy, nor does accuracy imply precision. A balance that records masses of 2.001 g, 1.999 g, and 2.000 g in three separate determinations of the mass of a metal block known to weigh 1.000 g, would be a very precise balance, but not an accurate one. On the other hand, another balance that gave masses of 1.312 g, 1.006 g, and 0.723 g, in three determinations of the mass of the block, would be considerably more accurate, but much less precise.

Relative and absolute uncertainty. Suppose you weighed 25 g of a substance on a balance that had an uncertainty of ±1 g. In a separate measurement, you weighed a 2500-g sample on the same balance. Clearly, the second measurement is much better than the first—being within 1 g of 2500 g is pretty good, while being within 1 g of just 25 g is not so impressive, even though both measurements have the same *absolute uncertainty*, ±1 g. To compare the quality of the two measurements, look at their *relative uncertainty* (equation 1), which is expressed as a percentage.

$$\text{relative uncertainty} = \frac{|\text{absolute uncertainty}|}{\text{measured value}} \times 100 \qquad (1)$$

Applied to our examples, the relative uncertainty for the 25-g measurement is $1/25 \times 100 = 4\%$ (25 g ± 4%), whereas the relative uncertainty for the 2500-g measurement is $1/2500 \times 100 = 0.04\%$ (2500 g ± 0.04%). It would be best to use a more sensitive balance for weighing the 25 g sample! Be careful not to confuse absolute and relative uncertainties. Absolute uncertainties have the same units as the value of the measurement, whereas relative uncertainties are dimensionless quantities without units.

Propagation of Uncertainties

For most of the routine measurements and calculations in the organic chemistry lab, using proper significant figures will be sufficient for keeping track of uncertainties. When weighing reagents for a reaction, converting into moles, determining limiting reagents, and calculating reaction yields (Section 3.2), you need to record these values with the proper number of significant figures. It is unlikely, however, that you will need to explicitly denote, or propagate, the uncertainties in these values. For example, for the mass and moles of sodium hydroxide used in a reaction, typically you would record 0.302 g (7.55 mmol) in your notebook, and not 0.302 ± 0.001 g (7.55 ± 0.02 mmol). Indeed, it is customary to report percent yields only to the integer value, even if the uncertainties in your measurements would justify more significant figures (i.e., 83% yield, not 83.4% yield). Nevertheless, there may be instances in your lab work where you collect data that does require a more formal accounting for errors.

Propagating uncertainties through addition and subtraction. Suppose you need to add and subtract values for a series of measurements. Each measurement will have its associated absolute uncertainty, e_i, as shown.

$$3.03 \pm 0.04 \quad \leftarrow e_1$$
$$-4.88 \pm 0.05 \quad \leftarrow e_2$$
$$+1.02 \pm 0.04 \quad \leftarrow e_3$$
$$-0.15 \pm 0.02 \quad \leftarrow e_4$$

Sum $-0.98 \pm e_5$

To propagate the uncertainty through the arithmetic to give the final uncertainty associated with the sum, e_5, use *absolute uncertainties* of the measurements and apply the following formula (equation 2):

$$e_5 = \sqrt{e_1^2 + e_2^2 + e_3^2 + e_4^2} \tag{2}$$

Applying this equation to the uncertainties in our measurements gives

$$e_5 = \sqrt{0.04^2 + 0.05^2 + 0.04^2 + 0.02^2} = 0.078102$$

Rounded appropriately, the absolute uncertainty is 0.08 and the final result is expressed as -0.98 ± 0.08.

Propagating uncertainties through multiplication and division. When propagating uncertainty through multiplication and division operations, *relative uncertainties* are used. Consider the following expression, where each value is associated with its absolute uncertainty in parentheses. Note that all values will be rounded to the appropriate number of significant figures at the end of the calculation.

$$\frac{2.56(\pm 0.02) \times 4.01(\pm 0.04)}{0.95(\pm 0.03) \times 3.89(\pm 0.05)} = 2.77786 \pm e_5$$

To determine e_5, first convert all the absolute uncertainties into relative uncertainties.

$$\frac{2.56(\pm 0.78125\%) \times 4.01(\pm 0.99751\%)}{0.95(\pm 3.15789\%) \times 3.89(\pm 1.28534\%)} = 2.77786 \pm e_5$$

Equation 3, where $\%e_1$, $\%e_2$, $\%e_3$, and $\%e_4$ are the relative uncertainties in each measurement, is used to propagate the uncertainty in this multiplication and division calculation to give the relative uncertainty, $\%e_5$, of the answer:

$$\%e_5 = \sqrt{\%e_1^2 + \%e_2^2 + \%e_3^2 + \%e_4^2} \tag{3}$$

Applying this equation gives

$$\%e_5 = \sqrt{0.78125^2 + 0.99751^2 + 3.15789^2 + 1.28534^2} = 3.63727\%$$

The answer in terms of relative uncertainties and expressed with appropriate significant figures is $2.8 \pm 4\%$.

Finally, the relative uncertainty can be converted into the absolute uncertainty by calculating 4.0% of the answer (or 3.63727%, if you carry more figures through to the end before rounding):

$$2.77786 \times 0.0363727 = 0.101038$$

The answer and its absolute uncertainty, expressed with the appropriate number of significant figures, is 2.8 ± 0.1.

Standard Deviation as an Estimate of Uncertainty in an Average Value

In the course of your laboratory work you may need to interpret data collected from a number of repeated experiments and propose an estimate of the uncertainty in your results. There are a number of statistical methods that can be applied to estimate uncertainties. Here is an example illustrating the use of the standard deviation. You can refer to the sources listed at the end of the chapter for more information on the use of other, more rigorous methods, such as confidence intervals.

Example. In a lab project that sheds light on the effect of the nucleophile on the product ratio in an S_N2 reaction, a competition experiment was used in which 1-butanol was reacted with equimolar mixtures of HBr and HCl to form a mixture of 1-bromobutane and 1-chlorobutane (equation 4). After the reaction was complete, the relative amounts of the two products (expressed as percentages) were determined by gas chromatography (see Chapter 20). The same reaction was run and analyzed by each of the 20 students in the lab and their results are presented in Table 5.2.

$$CH_3CH_2CH_2CH_2OH + HBr + HCl \xrightarrow[\text{H}_2\text{O}]{\text{heat}} CH_3CH_2CH_2CH_2Br + CH_3CH_2CH_2CH_2Cl \quad (4)$$

The results look pretty good, but there is some variation in the product percentages from student to student. What is the "true" ratio of products in this reaction? It seems that taking the average would be a good start toward an answer. You can determine the average value $\bar{x}$, from the series of measurements, x_1, x_2, x_3, etc., as shown in equation 5, where n is the total number of measurements. In this example, you average the percentages for each of the two products in the example; the results are shown at the bottom of Table 5.2.

$$\bar{x} = \frac{x_1 + x_2 + x_3 + \cdots + x_n}{n} \quad (5)$$

The average of the product ratios in this example is 1-bromobutane:1-chlorobutane = 78.05:21.95, but how certain is this ratio? Is it reasonable to use four significant figures in the percentages? The *standard deviation, s*, can give an estimate of the uncertainty. It is a measure of how closely the data clusters around the average, and it bears the same units as the measurement in question.

TABLE 5.2 Student results from an S_N2 competition experiment

Student	% 1-Bromobutane	Deviation (d)[1] % 1-bromobutane	% 1-Chlorobutane	Deviation (d)[1] % 1-chlorobutane
1	81	2.95	19	−2.95
2	78	−0.05	22	0.05
3	84	5.95	16	−5.95
4	74	−4.05	26	4.05
5	78	−0.05	22	0.05
6	78	−0.05	22	0.05
7	75	−3.05	25	3.05
8	71	−7.05	29	7.05
9	80	1.95	20	−1.95
10	77	−1.05	23	1.05
11	76	−2.05	24	2.05
12	84	5.95	16	−5.95
13	81	2.95	19	−2.95
14	79	0.95	21	−0.95
15	72	−6.05	28	6.05
16	83	4.95	17	−4.95
17	86	7.95	14	−7.95
18	79	0.95	21	−0.95
19	68	−10.05	32	10.05
20	77	−1.05	23	1.05
Average	**78.05**		**21.95**	
Standard Deviation (s)		**4.59376**		**4.59376**

[1] d is the deviation of the value from the average: $d_n = x_n - \bar{x}$

The smaller the standard deviation, the less variation there is in the results from experiment to experiment.

To calculate the standard deviation (s) of a series of measurements, first determine the individual deviation, d, from the average for each measurement, where $d_n = x_n - \bar{x}$. The standard deviation can then be determined from the individual deviations by applying equation 6, where n is the number of individual measurements. Fortunately, your scientific calculator or spreadsheet program can do the calculation of a standard deviation for you with ease.

$$s = \sqrt{\frac{d_1^2 + d_2^2 + d_3^2 + \cdots + d_n^2}{n-1}} \tag{6}$$

The deviations for each measurement in this example are provided in Table 5.2 and you can apply these values in equation 6 to give the standard deviations shown at the bottom of the table. It should not be surprising that the standard deviations for the bromide and chloride products are identical. Because the percentages of the two products must add up to 100%, when the percentage of 1-bromobutane is high relative to the class average, the percentage of 1-chlorobutane will be equally low. You can see that the absolute values of the individual deviations in the bromide and chloride product for any one student are (must be) identical. Really, only the average and standard deviation for one of the products were needed.

The standard deviation and the average should end at the same decimal place, and generally the standard deviation is rounded to one significant figure. In this example, the standard deviation would be 5. Thus the final result, expressed with the appropriate significant figures, is 1-bromobutane:1-chlorobutane = 78:22 ± 5 (n = 20). The term in parentheses, n = 20, indicates how many individual determinations the final result was based upon.

It is important that you state your uncertainty interval; in this case, ±s is the standard deviation. Other measures of uncertainty, with different meanings than standard deviation, would give different uncertainty intervals.

What does the standard deviation really mean? It can be shown that about 68% of a large number of measurements ($n \to \infty$) fall within one standard deviation of the average value. In our example, you can see that 14 of the 20 (70%) student results fall within ±5 of 78:22. If you were to make a single determination of the bromide-to-chloride product ratio, the probability is 68% that the result would fall within ±5 of 78:22. Of course, if you were to increase the uncertainty interval, there would be a greater probability that the result of any one measurement would fall within the interval. For instance, if you reported the uncertainty to 2s (two standard deviations or ±10 in this example), then about 95% of the individual measurements would fall within the uncertainty of the measurement.

Rejection of data. The result from Student 19 in Table 5.2 looks a bit off compared to the others. Is this an *outlier* that should be rejected? Any apparently deviant datum should first be checked carefully. If the anomaly can be traced to some external factor, such as a spilled sample or an incorrect temperature reading, then the data point can be thrown out. Of course, all of the data points and experimental methods should be rigorously checked; if a mistake is detected in a measurement that does *not* appear to be anomalous compared to the rest of the data set, then it too must be rejected. If no external factor can be identified as the basis for rejecting an anomalous point, one can turn to a number of statistical tools to evaluate discordant results.

The Grubbs test is one commonly used test for outliers. To apply this test, the Grubbs statistic G is calculated as shown in equation 7.

$$G_{calculated} = \frac{|\,\text{discordant value} - \bar{x}\,|}{s} \qquad (7)$$

The $G_{calculated}$ value is then compared to the appropriate $G_{critical}$ value found in Table 5.3. If $G_{calculated}$ is greater than the $G_{critical}$ value, then the discordant data point can be rejected and a new average and standard deviation should be determined that excludes the questionable datum. It is not valid to apply the test a second time to another value, based on the new average and standard deviation.

We can apply the Grubbs test to the questionable data point in Table 5.2:

$$G_{calculated} = \frac{|68 - 78.05|}{5} = 2.01$$

Because 2.01 is less than 2.56, the value of $G_{critical}$ from Table 5.3 that corresponds to 20 measurements, Student 19's data should not be rejected.

TABLE 5.3	**Values of $G_{critical}$ for rejection of outliers**
Number of measurements	$G_{critical}$ (95% confidence)
3	1.15
4	1.46
5	1.67
6	1.82
7	1.94
8	2.03
9	2.11
10	2.18
11	2.23
12	2.29
13	2.33
14	2.37
15	2.41
16	2.44
17	2.47
18	2.50
19	2.53
20	2.56
21	2.58
22	2.60
23	2.62
24	2.64
25	2.66
30	2.75
40	2.87
50	2.96

Further Reading

Harris, D. C. *Quantitative Chemical Analysis*, 8th ed.; Freeman: New York, 2010, Chapters 3 and 4.

Taylor, J. R. *An Introduction to Error Analysis*, 2nd ed.; University Science Books: Sausalito, CA, 1997.

Questions

1. Why would a typical top-loading balance be a poor choice to weigh 75 mg of a reagent?
2. Why weigh a vial or flask that will be used to collect a reaction product before it is filled with the product?
3. Suggest a good way of measuring the transfer of 0.45 mL of a limiting reagent in a reaction.
4. How would you calibrate an automatic pipet?
5. How can a thermometer be calibrated?

6. (a) Solve for x in the following expression, using the appropriate number of significant figures in your answer: $x = 6.0246 + 2.01 - 5093.1$.
   (b) Solve for y in the following expression: $y = 0.0046 \times 3.9870 \times 456.9$.
7. Determine the absolute uncertainty, e, for the following expression (write the final answer and its absolute uncertainty with the appropriate number of significant figures):

$$\frac{1.324(\pm 0.004)}{1.25(\pm 0.07) \times 0.965(\pm 0.005)} = 1.0976165 \pm e$$

8. The following measurements were recorded in an experiment: 10.24, 9.39, 8.44, 11.34, 12.88, 10.04, 14.46, 8.18, 11.09, 11.11, 9.28, 10.78, 8.68. Calculate the average value for this data and its uncertainty, using standard deviation as your estimate of the uncertainty interval. Be careful to use appropriate significant figures.

9. Based on the Grubbs test, is any measurement in Problem 8 an outlier that should be discarded?

CHAPTER

6 HEATING AND COOLING METHODS

Many organic reactions do not occur spontaneously when the reactants are mixed together but require a period of heating to reach completion. Exothermic organic reactions, on the other hand, require removal of the heat generated during a reaction by using a cooling bath. Cooling baths are also used to ensure the maximum recovery of crystallized product from a solution or to cool the contents of a reaction flask. In addition, heating and cooling methods are utilized in other techniques of the organic lab, such as distillation (see Chapter 12) and recrystallization (see Chapter 15).

6.1 Preventing Bumping of Liquids

Liquids heated in laboratory glassware tend to boil by forming large bubbles of superheated vapor, a process called *bumping.* The inside surface of the glass is so smooth that no tiny crevices exist where air bubbles can be trapped, unlike the surfaces of metal pans used for cooking. Bumping can be prevented by the addition of inert porous material—a boiling stone or boiling stick—to the liquid or by mechanically stirring the liquid while it is heated. Without the use of boiling stones or stirring, *superheating* can occur, a phenomenon caused by a temperature gradient in the liquid—lower temperatures near the surface and higher temperatures at the bottom of the liquid near the heat source. Superheating can lead to loss of product and a potentially dangerous situation if the superheated liquid spatters out of the container and causes burns.

A heated liquid enters the vapor phase at the air–vapor interface of a pore in the boiling stone or stick. As the volume of vapor nucleating at the pore increases, a small bubble forms, is released, and continues to grow as it rises through the liquid. Because of the air trapped in the pores of a boiling stone or boiling stick, multiple small bubbles form instead of a few large ones. The sharp edges on boiling stones also catalyze bubble formation in complex ways not fully understood.

The *boiling stones* commonly used in the laboratory are small pieces of carborundum, a chemically inert compound of carbon and silicon. Their black color makes them easy to identify and remove from a solid product if they have not been removed earlier by filtration.

Boiling sticks are short pieces of wooden applicator sticks and can be used instead of boiling stones. Boiling sticks should not be used in reaction mixtures, with any solvent that might react with wood, or in a solution containing an acid.

Using Boiling Stones

One or two boiling stones suffice for smooth boiling of most liquids. **Boiling stones should always be added before heating the liquid.** Adding boiling stones to a hot liquid may cause the liquid to boil violently and erupt from the flask because superheated vapor trapped in the liquid is released all at once. If you forget to add boiling stones before heating, the liquid must be cooled well below the boiling point before putting boiling stones into it.

You should always add boiling stones or a boiling stick to any unstirred liquid before boiling it—unless instructed otherwise.

If a liquid you have boiled requires cooling and reheating, an additional boiling stone should be added before reheating commences. Once boiling stones cool, their pores fill with liquid. The liquid does not escape from the pores as readily as air does when the boiling stone is reheated, rendering it less effective in promoting smooth boiling.

Magnetic Stirring

Magnetic stirring is frequently used instead of boiling stones or boiling sticks. The agitation provided by stirring drives the vapor bubbles to the surface of the liquid before they grow large enough to cause bumping. Stirring is also a common method for preventing superheating.

6.2   Conventional Heating Devices

SAFETY PRECAUTION

These safety precautions pertain to all electrical heating devices.

1. The hot surface of a hot plate, the inside of a hot heating mantle, or the hot nozzle of a heat gun are fire hazards in the presence of volatile, flammable solvents. An organic solvent spilled on a hot surface can ignite if its flash point is exceeded. Remove any hot heating device from your work area before pouring a flammable liquid.

Flash point or autoignition temperature is the minimum temperature at which a substance mixed with air ignites in the absence of a flame or spark.

2. Never heat a flammable solvent in an open container on a hot plate; a buildup of flammable vapors around the hot plate could result. The thermostat on most laboratory hot plates is not sealed and it arcs each time it cycles on and off, providing an ignition source for flammable vapors. Steam baths, oil baths, and heating mantles are safer choices.

Heating Mantles

Many reactions are carried out in round-bottomed flasks heated with electric heating mantles shaped to fit the bottom of the flask.

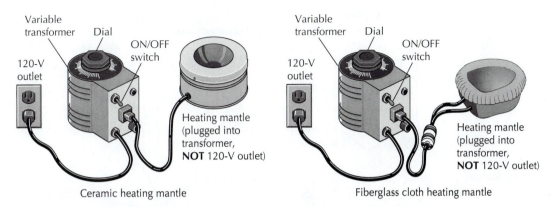

FIGURE 6.1 Heating mantle and variable transformer. (**Note**: The transformer dial is calibrated in percentage of line voltage, **not** in degrees.)

Several types of heating mantles may be available in your laboratory (Figure 6.1). One type of heating mantle has a metal housing and a ceramic well covering the heating element. These heating mantles can be used with flasks smaller than the designated size of the mantle because of radiant heating from the surface of the well. A different type of heating mantle consists of woven fiberglass with the heating element embedded between the layers of fabric. Fiberglass heating mantles come in a variety of sizes to fit specific sizes of round-bottomed flasks; a mantle sized for a 100-mL flask will not work well with a flask of another size.

Many types of heating mantles have no controls and must be plugged into a *variable transformer* (*or rheostat*) or other variable controller to adjust the rate of heating (see Figure 6.1). The variable transformer is then plugged into a wall outlet.

Heating mantles are supported underneath a round-bottomed flask by an iron ring or lab jack (see Section 6.5). Fiberglass heating mantles should not be used on wooden surfaces because the bottom of the heating mantle can become hot enough to char the wood.

Stir/Hot Plates

Hot plates work well for heating flat-bottomed containers such as beakers, Erlenmeyer flasks, and crystallizing dishes used as water baths or sand baths. Most laboratory hot plates incorporate magnetic stirring devices to allow the contents of the vessel to be stirred with a magnetic stir bar added to the flask.

Hot plates also serve to heat the aluminum blocks used with microscale glassware. Figure 6.2a shows a microscale setup for heating a standard-taper conical vial fitted with an air condenser; Figure 6.2b shows a microscale setup for heating a Williamson reaction tube and a round-bottomed flask fitted with an air condenser. Several types of aluminum heating blocks are available commercially. The blocks have holes sized to fit microscale reaction tubes or vials and a depression or hole for a 5- or 10-mL microscale round-bottomed flask. The blocks also have a hole designed to hold a metal probe thermometer so that the temperature of the block can be monitored.

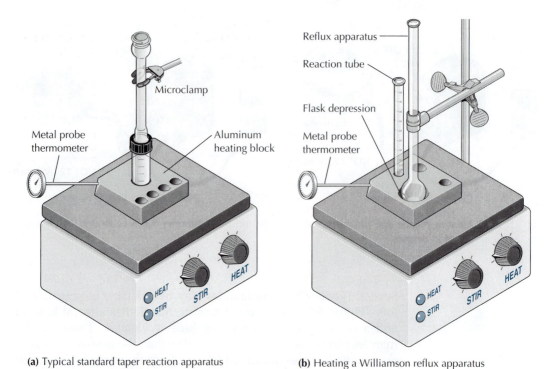

(a) Typical standard taper reaction apparatus with a conical vial and an air condenser

(b) Heating a Williamson reflux apparatus and a reaction tube

FIGURE 6.2 Aluminum blocks used for heating microscale glassware.

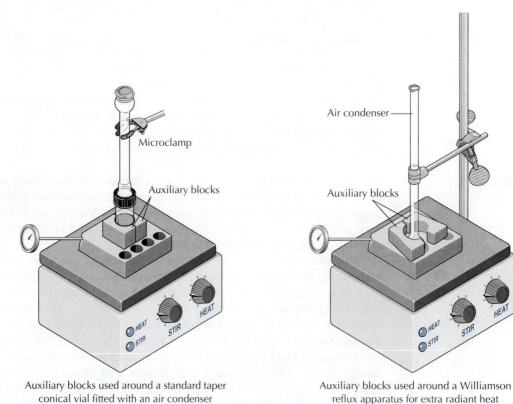

Auxiliary blocks used around a standard taper conical vial fitted with an air condenser

Auxiliary blocks used around a Williamson reflux apparatus for extra radiant heat

FIGURE 6.3 Using auxiliary aluminum blocks to provide extra radiant heat with microscale glassware.

Auxiliary aluminum blocks designed in two sections can be placed on top of the aluminum block around a vial or round-bottomed flask to provide extra radiant heat, as shown in Figure 6.3.

Sand Baths

A sand bath provides another method for heating microscale reactions. Sand is a poor conductor of heat, so a temperature gradient exists along the various depths of the sand, with the highest temperature occurring at the bottom of the sand and the lowest temperature near the top surface.

One method for preparing a sand bath uses a ceramic heating mantle about two-thirds full of washed sand (Figure 6.4a). A second method employs a crystallizing dish, heated on a hot plate, containing 1–1.5 cm of washed sand (Figure 6.4b); the sand in the dish should be level, not mounded. A thermometer is inserted in the sand so that its bulb or metal tip is completely submerged to the same depth as the contents of the reaction vessel. The heating of a reaction can be closely controlled by raising or lowering the vessel to a different depth in the sand as well as by changing the heat supplied by the heating mantle or hot plate.

SAFETY PRECAUTION

Sand in a crystallizing dish should not be heated above 200°C, nor should the hot plate be turned to high heat settings. Either situation could cause a crystallizing dish to crack. Also, it can take a half-hour or more for a sand bath to cool after use. Keep hot sand baths away from flammable items and allow abundant time for a bath to cool before handling it.

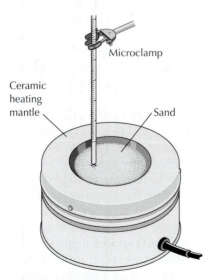

(a) Sand bath in ceramic heating mantle

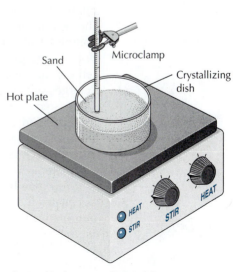

(b) Sand bath in crystallizing dish on hot plate

FIGURE 6.4 Sand baths.

Steam Baths

Steam baths or steam cones provide a safe and efficient way to heat low-boiling flammable organic liquids (Figure 6.5). Steam baths are used in the organic laboratory for heating liquids below 100°C and in situations where precise temperature control is not required. The concentric rings on the top of a steam bath can be removed to accommodate containers of various sizes. A round-bottomed flask should be positioned so that the rings cover the flask to the level of the liquid it contains. For an Erlenmeyer flask, remove only enough rings to create an opening that is slightly larger than one-half of the bottom diameter of the flask.

SAFETY PRECAUTION

Steam is nearly invisible and can cause severe burns. Turn off the steam before placing a flask on a steam bath or removing it. The metal screw on the steam valve handle may be hot enough to cause burns. Grasp the neck of a hot flask with flask tongs; **do not use** a test tube holder or a towel.

Steam baths operate at only one temperature, approximately 100°C. Increasing the rate of steam flow does not raise the temperature, but it does produce clouds of moisture within the laboratory or fume hood and in the sample you are heating. Adjust the steam valve for a **slow to moderate rate of steam flow** when using a steam bath.

A steam bath has two disadvantages. First, it cannot be used to boil any liquid with a boiling point above 100°C. Second, water vapor from the steam may contaminate your sample being heated on the steam bath unless special precautions are taken to exclude moisture.

Water Baths

When a temperature of less than 100°C is needed, a water bath allows for closer temperature control than can be achieved with the heating methods discussed previously. The water bath can be contained in a beaker or crystallizing dish and heated on a stir/hot plate (Figure 6.6). Magnetic stirring of the water bath prevents temperature gradients and maintains a uniform water temperature. A large paper clip can serve as an effective stir bar in the water bath. Compared to a regular stir bar, a paper clip takes up little space in the bath and it tends not to interfere as much with the stir bar in the reaction vessel. Once the desired temperature of the water bath is reached, it can be maintained by using a low heat setting on the hot plate.

You may be fortunate to have in your laboratory modern stir/hot plates that allow the bath temperature to be set and controlled automatically, via an attached temperature sensor, as shown in Figure 6.6b. The thermometer or temperature sensor should **always be held by a clamp or other support device** so it does not touch the wall or bottom of the vessel holding the water (Figure 6.6). It is very easy to bump a thermometer that is merely set in a beaker and propped against its lip, perhaps breaking it or upsetting the water bath. In addition, if a thermometer or temperature sensor is

Steam in

Drain   Steam bath

Steam in

Drain

Steam cone

FIGURE 6.5 Steam baths.

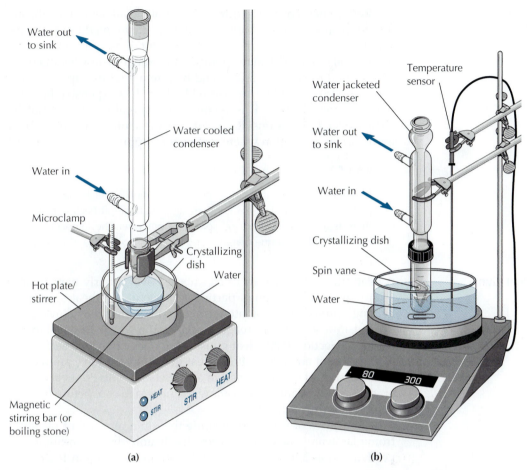

Water out to sink

Water cooled condenser

Water in

Microclamp

Hot plate/ stirrer

Magnetic stirring bar (or boiling stone)

Crystallizing dish

Water

HEAT
STIR
STIR
HEAT

(a)

Temperature sensor

Water jacketed condenser

Water out to sink

Water in

Crystallizing dish

Spin vane

Water

80   300

(b)

FIGURE 6.6 (a) Heating a miniscale reflux apparatus with a stir/hot plate and conventional thermometer. (b) Heating a microscale reflux apparatus using a digital stir/hot plate with integrated temperature sensor for automatic temperature control.

touching the bottom of the water bath, it may give a temperature reading that does not accurately reflect the temperature of the water surrounding the reaction vessel.

The reaction vessel should be positioned in the water bath with care. If magnetic stirring of a reaction mixture is needed, the reaction vessel should be centered on the stir/hot plate and clamped as close to the stirring motor as possible, without touching the bottom of the bath's container. Also, the reaction vessel should be submerged in the water bath no farther than the depth of the reaction mixture it contains. Proper reaction vessel placement may require adjusting the amount of water in the bath.

Oil Baths

Distillations of high-boiling liquids often need a heating bath of greater than 150°C (see Chapter 12). Water baths are limited to temperatures below 100°C, and a heating mantle may not offer fine enough temperature control for a successful distillation. In these cases magnetically stirred oil baths, heated on a hot plate, can provide the solution. The preceding discussion on using water baths also applies to using oil baths.

Both mineral oil (a mixture of high-boiling alkanes) and silicone oil are available for oil baths. Extremely stable, medium-viscosity silicone oil is ideal for heating baths, but it is quite expensive. Silicone oil is available in two temperature ranges—low temperature (designed for use up to 180°C) and high temperature (up to 230°C). Mineral oil that can be used for oil baths is less expensive but also poses a safety hazard: it is flammable. Mineral oil should not be heated over 175°C. Consult your instructor about using an oil bath if you are in a situation where one may be appropriate.

SAFETY PRECAUTION

Mineral oil is flammable. Care must be taken not to spill any on a hot plate. In addition, if any water gets into a mineral-oil heating bath, there is the danger of hot oil spattering out when the temperature exceeds 100°C and the denser water begins to boil.

Heat Guns

A heat gun allows hot air to be directed over a fairly narrow area (Figure 6.7). A heat gun is particularly useful as a heat source for heating thin-layer chromatographic plates after they have been dipped in a visualizing reagent that requires heat to develop the color (see Section 18.4). Heat guns usually have two heat settings as well as a cool air setting. If the heat gun does not have an integral stand, it should be suspended in a ring clamp with the heat setting on cool for a few minutes to allow the nozzle to cool before the gun is set onto the bench.

Other uses of heat guns include the rapid removal of moisture from glassware where dry but not strictly anhydrous conditions are needed and as a heat source in sublimations (see Section 16.2).

SAFETY PRECAUTION

Laboratory heat guns are not the same as domestic hair dryers. Heat guns can get extremely hot and can easily cause severe burns. Never point them at yourself or others.

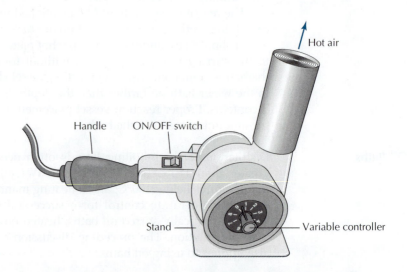

FIGURE 6.7 Heat gun.

Bunsen Burners

The use of Bunsen burners in the organic laboratory poses an extreme fire hazard because volatile vapors of organic compounds can ignite when mixed with air. Use of a Bunsen burner or other source of an open flame should be a rare event in an organic laboratory and should never be undertaken without your instructor's supervision.

6.3   Heating with Laboratory Microwave Reactors

The first commercial microwave ovens appeared in the late 1960s and today they are ubiquitous in kitchens. They offer a fast, convenient, and efficient way to heat food, and these same benefits can be realized using microwave reactors as heating devices in the laboratory. The first reports of microwave heating of organic reactions appeared in the 1980s, but the technique was slow to catch on. Early work involved heating reaction mixtures in sealed vessels using kitchen microwave ovens, but controlling reaction temperatures and pressures in kitchen ovens was very difficult. Without effective temperature and pressure control it was difficult to reproduce results and, worse, it was potentially dangerous.

In the 1990s microwave reactors designed specifically for laboratory use became commercially available and now microwave heating, or *microwave-assisted organic synthesis (MAOS)*, is routine in many laboratories. Compared to conventional heating methods, modern microwave devices typically allow reactions to proceed safely at much faster rates, using less energy, and often with higher yields and fewer side products.

SAFTEY PRECAUTION

Household microwave ovens must never be used to heat chemical reactions. These devices lack the ability to monitor and control reaction temperatures and pressures, and their use can result in fires and explosions. Only scientific microwave reactors should be used in the laboratory, and only in strict accordance with the manufacturers' instructions.

Physical Basis of Microwave Heating

The microwave frequency range, situated between radio waves and infrared radiation, is 0.3–300 gigahertz (GHz). Compared to visible and ultraviolet light, which is of sufficient energy to break chemical bonds, microwave radiation is very low in energy and cannot induce molecular fragmentation. Instead, absorption of microwaves increases the kinetic energy of molecules, resulting in heating. Nearly all microwave ovens operate at 2.45 GHz, corresponding to a wavelength of 12.25 cm. This is a dedicated frequency for microwave ovens, specified so as not to interfere with cellular phones, wireless networks, and other wireless technologies.

The most important mechanisms by which microwaves heat organic materials are *dipolar polarization* and *ionic conduction*. Like all electromagnetic radiation, microwaves consist of oscillating

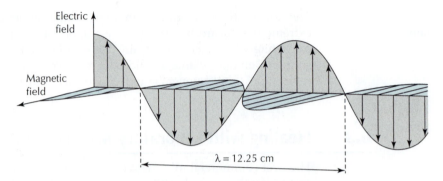

FIGURE 6.8 Electric field and magnetic field components of microwaves.

electric field and magnetic field components (Figure 6.8). Heating results from the interaction of the electric field component with polar molecules or ions.

Dipolar polarization operates with molecules that have dipoles, which respond to the electric field component by rotating in an attempt to align their dipole moments with the field as it oscillates. As the polar molecules rotate they collide with one another and their rotational energy is converted into translational kinetic energy, resulting in the observed heating. This mechanism explains why polar solvents, such as water, ethanol, and dimethyl sulfoxide (DMSO) are effectively heated by microwaves, whereas nonpolar solvents are not. Likewise, the mechanism also explains why gases are not heated by microwave irradiation. The distances between molecules in a gas are too great for an increase in the gas molecules' rotational motion to produce an increase in molecular collisions.

The ionic conduction mechanism operates on ions in solution. When irradiated by microwaves, dissolved ions oscillate back and forth under the influence of the oscillating electric field. The oscillation results in molecular collisions that heat the sample through the increase in kinetic energy. Ionic conduction is a stronger effect than dipolar polarization. Thus, salt water is heated more rapidly in a microwave oven than is pure water.

Microwave Reactors   Laboratory microwave reactors are either multimode or single-mode (also called monomode) units (Figure 6.9). Multimode reactors look

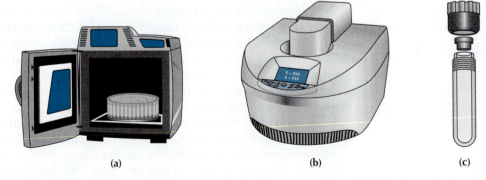

FIGURE 6.9   (a) A multimode reactor equipped with a 24-vessel rotor. (b) A single-mode reactor. (c) A 20-mL reaction vessel, sealing plug, and cap.

FIGURE 6.10
(a) Bird's-eye view of
the microwave
interference pattern
showing areas of high
and low energy in a
multimode reactor
chamber. (b) Bird's-eye
view of a single
microwave pattern in
the cavity of a single-
mode reactor.

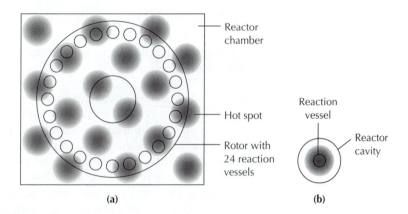

(a)

(b)

similar to kitchen microwave ovens. They have a large heating chamber where microwaves bounce around and interact with one another, creating an interference wave pattern with areas of high and low energy (called modes) that result in "hot" and "cold" spots (Figure 6.10a). Multimode reactors often have a turntable to average out the energy over the sample and can be configured with a rotor that accommodates multiple reaction vessels. The reactor shown in Figure 6.9a is equipped with a rotor that holds 24 20-mL reaction tubes.

The inhomogeneity of the microwave field in a multimode reactor can be a problem with small scale reactions (less than about 3 mL) because it can be difficult to get uniform heating of the sample, even with rotation on a turntable. Single-mode reactors with small reaction chambers that are the width of just one-half wave (mode) overcome this problem (Figure 6.10b). Samples as small as 0.2 mL can be heated evenly in single-mode reactors, with an upper limit typically of about 80 mL.

Most microwave reactions are performed in glass tubes sealed with caps designed to safely vent, and then reseal, in the event of overpressurization (Figure 6.9c). Depending on the particular reactor, the reaction vessels can be heated up to 300°C and sustain pressures of up to 300 psi. A powerful feature of sealed-tube systems is their ability to heat solvents to temperatures well above their normal boiling points, accessing very high temperatures in a short period of time. These conditions often enable reactions to be performed in minutes that might take hours or days to complete using conventional heating methods. Many microwave reactors can also accommodate reactions run in open round-bottomed flasks; however, this open-flask setup does not allow solvents to be heated above their boiling points.

Reaction temperature is typically monitored by an infrared sensor that measures the temperature of the walls of the reaction vessel, but some units are fitted with fiber-optic probes that can be inserted directly into the reaction mixture for more accurate temperature measurement. Some models are also equipped with sensors that monitor pressure in the reaction vessel. In the case of multimode units fitted with a rotor, one vessel is generally designated as the "control vessel" and is fitted with a fiber-optic temperature probe.

Because temperature is monitored and controlled by means of a single vessel, it is important that the contents of all of the vessels be the same or very similar because microwave energy is not absorbed equally by all materials.

Running a microwave-heated reaction. The detailed procedure for using a microwave reactor will depend on the particular reactor in your lab; nevertheless, the basic features and operating procedures are similar for most microwave units. To prepare for a reaction, a sealed-tube vessel is loaded with the reagents and solvents. Reaction mixtures are stirred by means of a magnetic stir bar added to the reaction tube before sealing. Once the reaction vessel or vessels are placed in the reaction chamber, the heating regime is programmed into the reactor.

There are three periods in a microwave-heated reaction, the *ramp time*, the *hold time*, and the *cooling time*. The ramp time is the period during which the sample is heated to the desired reaction temperature. Generally, the reactor power is adjusted to afford a ramp heating rate of 1–5°C/sec. The hold time is the period during which the temperature of the reaction mixture is held at the target temperature. During this time, the microwave power is automatically adjusted to keep the sample at a constant temperature. The cooling time begins at the end of the reaction. During this stage, heating is discontinued and the reactor cavity is cooled with a flow of cool air.

Microwave Reactor Safety

Used in accordance with manufacturers' instructions and recommendations, microwave heating is arguably safer than conventional laboratory heating methods. Nevertheless, the same safety precautions observed when using conventional methods should be applied when heating with microwaves. In addition, there are some microwave-specific precautions:

- Use only dedicated scientific microwave reactors in the laboratory.
- Think carefully about the type of reaction you are doing before applying microwave irradiation. Strongly exothermic reactions and reactions that generate gaseous products may not be appropriate. Consult your instructor before attempting a new reaction with a microwave reactor.
- Reaction vessels should always be checked for damage before use and should be filled according to the manufacturers' specifications. Over- or underfilling the vessels can be dangerous. Cap vessels only with the manufacturers' recommended caps.
- Magnetic stir bars used in a microwave reactor should not be exactly 3 cm in length. This length is ¼ wavelength of a 2.45 GHz microwave, making a 3-cm stir bar an effective antenna, resulting in dangerous arcing and overheating.
- At the end of a reaction cycle, be aware that the reaction vessel may be hot. Check its temperature before handling it. If the vessel is warm to the touch, it should be cooled in an ice bath before opening. Be sure to point the mouth of the vessel away from you and other persons before opening it because it is likely to retain some pressure. Open a microwave reaction vessel carefully in a fume hood.

6.4   Cooling Methods

Cooling baths are frequently needed in the organic laboratory to control exothermic reactions, to cool reaction mixtures before the next step in a procedure, and to promote recovery of the maximum amount of crystalline solid from a recrystallization. Most commonly, cold tap water or an ice/water mixture serves as the coolant. Effective cooling with ice requires the addition of just enough water to provide complete contact between the ice and the flask or vial being cooled. Crushed ice alone does not pack well enough against a flask for efficient cooling because the air in the spaces between the ice particles is a poor conductor of heat.

Temperatures as low as –20°C can be achieved by mixing solid sodium chloride with crushed ice in a one-to-three ratio, respectively. The amount of water added should be only enough to make good contact with the vessel being cooled.

Dewar Flasks

A cooling bath of 2-propanol and chunks of solid carbon dioxide (dry ice) can be used for temperatures from –30° to –70°C. **Caution:** Foaming occurs as solid carbon dioxide chunks are added to 2-propanol. The 2-propanol/dry ice mixture should be contained in a Dewar flask, a double-walled vacuum chamber that insulates the contents from ambient temperature (Figure 6.11).

SAFETY PRECAUTION

The inside silvered glass surface of a Dewar flask is very fragile and must be handled with care. There is a vacuum between the two glass walls of a Dewar flask. If the silvered glass is broken, an implosion occurs and shards of glass are released (see Section 1.4). Never use a Dewar flask that does not have a protective metal case on the outside. **Always use eye protection when using a Dewar flask.**

6.5   Laboratory Jacks

Laboratory jacks are adjustable platforms that are useful for holding heating mantles, magnetic stirrers, and cooling baths (Figure 6.12). A reaction apparatus is assembled with enough clearance between the bottom of the reaction or distillation flask and the bench top to position the heating or cooling device under the flask by raising the

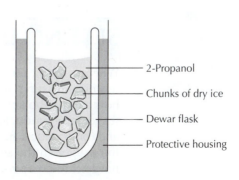

FIGURE 6.11 Dewar flask with a mixture of 2-propanol and dry ice.

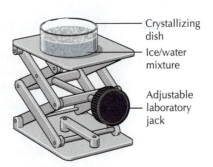

Crystallizing
dish

Ice/water
mixture

Adjustable
laboratory
jack

FIGURE 6.12 Laboratory
jack with ice bath.

platform of the lab jack. At the end of the operation, the heating or
cooling device can be removed easily by lowering the lab jack. Lab
jacks are also useful when carrying out reactions in a fume hood,
where reaction apparatuses should be elevated a few inches from
the hood floor to ensure good air flow around the equipment (see
Section 1.2).

Further Reading

De La Hoz, A., and Loupy, A. (Eds.) *Microwaves
in Organic Synthesis, Vol. 1, 3rd ed.;* Wiley-VCH:
Weinheim, 2012.

Kappe, C. O., Dallinger, D., and Murphree, S. S.
*Practical Microwave Synthesis for Organic Chem-
ists: Strategies, Instruments, and Protocols;* Wiley-
VCH: Weinheim, 2009.

Leadbeater, N. E., and McGowan, C. B. *Laboratory
Experiments Using Microwave Heating;* CRC
Press, Taylor & Francis Group: Boca Raton,
FL, 2013.

Questions

1. What is superheating of a liquid?
2. What is the purpose of adding boiling
   stones to a liquid that will be heated?
3. Why should boiling stones never be
   added to a hot liquid? What should you
   do if you begin to heat a liquid and then
   realize you forgot to add a boiling stone?

4. Specify one particular advantage that microwave heating has over conventional
   heating methods.
5. What is a Dewar flask used for?
6. When would you use a lab jack?

CHAPTER

7

SETTING UP ORGANIC REACTIONS

When carrying out organic reactions, it is often necessary to boil
reaction mixtures for extended periods, sometimes for days. But
how can you boil reaction mixtures without losing volatile solvents,
reagents, or products to evaporation? In other reactions, the reagents
and/or products may not be stable when exposed to water or to
the oxygen in the air, and some reactions involve compounds that
have noxious vapors. How do you handle all of these complications?

This chapter describes the methods and apparatus used to prevent loss of volatile compounds while maintaining a reaction mixture at the boiling point, to make additions of reagents to a reaction mixture during the course of a reaction, to keep atmospheric moisture or air from entering a reaction apparatus, and to prevent noxious vapors from entering the laboratory.

7.1   Refluxing a Reaction Mixture

Most organic reactions do not occur quickly at room temperature but require a period of heating. If the reaction were heated in an open container, the solvent and other liquids would soon evaporate; if the system were closed, pressure would build up and an explosion could occur. Chemists have developed a simple method of heating a reaction mixture for extended periods without loss of reagents. This process is called *refluxing,* which simply means boiling a solution while continually condensing the vapor by cooling it and returning the liquid to the reaction flask.

A condenser mounted vertically above the reaction flask provides the means of cooling the vapor so that it condenses and flows back into the reaction flask. Condensers are cooled by either water or air. When the boiling point of a reaction mixture is less than 150°C, a water-jacketed condenser is used to transfer heat from the vapor to the water running through the outer jacket of the condenser. For efficient heat transfer, water must be flowing through the outer jacket, but if the flow is too fast it is wasteful and the rubber hose may pop off the condenser's water inlet. For reaction mixtures with boiling points above 150°C, an air condenser is usually sufficient because the vapor loses heat rapidly enough to the surrounding atmosphere to condense before it can escape from the top of the condenser.

Extent of Heating

The amount of heat applied to a reflux apparatus should enable the liquid in the reaction mixture to boil at a moderate rate. With more heat, faster boiling occurs, but the temperature of the liquid in the flask cannot rise above the boiling point of the solvent or solution. If the system is boiling at too rapid a rate, the capacity of the condenser to cool the vapors may be exceeded and reagents (or product!) may be lost from the top of the condenser. In general, heat is applied so that the position at which the vapors condense is in the lower third of the condenser.

 Miniscale Reflux Apparatus

A funnel keeps the reagents from coating the inside of the ground glass joint.

Begin the assembly of a reflux apparatus by firmly clamping a round-bottomed flask to a ring stand or vertical support rod. Position the clamp holder far enough above the bench top so that a ring or a lab jack (see Section 6.5) can be placed underneath the flask to hold a heating mantle. Add the reagents to the reaction flask with the aid of a conical funnel for liquids and a powder funnel for solids. Add a boiling stone or magnetic stir bar to the flask.

If grease is being used on the standard taper joint, apply it to the lower joint of the condenser before fitting it into the top of the flask as shown in Chapter 4, Figure 4.6. Attach rubber tubing to the

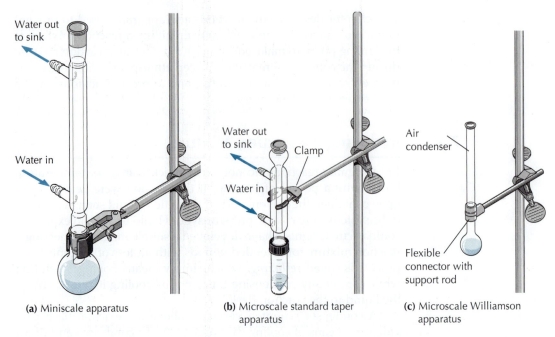

(a) Miniscale apparatus

(b) Microscale standard taper apparatus

(c) Microscale Williamson apparatus

FIGURE 7.1 Apparatus for simple reflux.

water jacket outlets as shown in Figure 7.1a. **Water must flow into the water jacket at the bottom inlet and out at the top outlet to the sink or drain**, to ensure that a column of water without any air bubbles surrounds the inside tube. Filling from the bottom allows the water jacket to completely fill with water even with a very low flow rate. Raise the heating mantle, supported on an iron ring or a lab jack, until it touches the bottom of the round-bottomed flask. At the end of the reflux period, lower the heating mantle away from the reaction flask.

 **Microscale Standard Taper Glassware**

Place the reagents for the reaction in a conical vial or 10-mL round-bottomed flask sitting in a small beaker so that it will not tip over. Put a boiling stone or a magnetic spin vane into the reaction vessel. Grease is not used on the joints of microscale glassware except when the reaction mixture contains a strong base such as sodium hydroxide (see Section 4.3). Fit the condenser to the top of the conical vial or round-bottomed flask with a screw cap and an O-ring as shown in Chapter 4, Figure 4.8. Fasten the apparatus to a vertical support rod or a ring stand with a microclamp attached to the condenser. Attach rubber tubing to the water jacket outlets (Figure 7.1b). **Water must flow into the water jacket at the bottom inlet and out at the top outlet to the sink or drain** to ensure that a column of water without any air bubbles surrounds the inside tube. Filling from the bottom allows the water jacket to completely fill with water even with a very low flow rate. Lower the apparatus into an aluminum heating block, sand bath, or water bath that is heated on a hot plate, or into a sand-filled ceramic heating mantle. At the end of the reflux period, raise the apparatus out of the heat source.

Microscale Williamson Glassware

Place a 5-mL round-bottomed flask in a 30-mL beaker and use the plastic funnel to add the reagents to the flask. Add a boiling stone or magnetic stir bar. Attach the air condenser to the flask using the flexible connector with the support rod. Clamp the apparatus to a vertical support rod or a ring stand as shown in Figure 7.1c. Wrap the air condenser with a wet paper towel or wet pipe cleaners to prevent loss of vapor when refluxing reaction mixtures containing solvents or reagents that boil under 120°C. Lower the apparatus into an aluminum heating block, sand bath, or water bath that is heated on a hot plate, or into a sand-filled ceramic heating mantle. At the end of the reflux period, raise the apparatus out of the heat source.

7.2 Addition of Reagents During a Reaction

Miniscale Glassware

When it is necessary to add reagents during the reflux period, a separatory funnel can be used as a dropping funnel. If the round-bottomed flask has only one neck, a Claisen adapter provides a second opening into the flask, as shown in Figure 7.2a. For a three-necked flask, the third neck is closed with a ground glass stopper, as shown in Figure 7.2b. If it is also necessary to maintain anhydrous conditions during the reflux period, both the condenser and the separatory funnel can be fitted with drying tubes filled with a suitable drying agent (see Section 7.3).

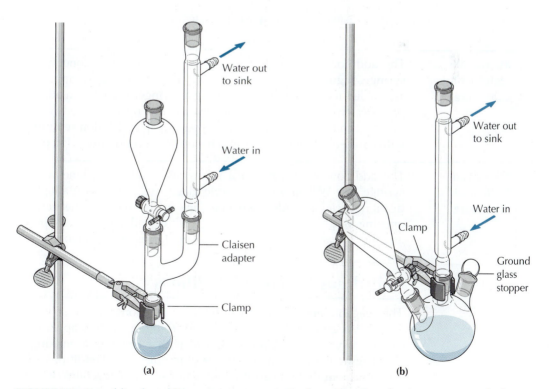

(a)   (b)

FIGURE 7.2 Assemblies for adding reagents to a reaction heated under reflux in (a) a one-necked reaction flask and (b) a three-necked flask.

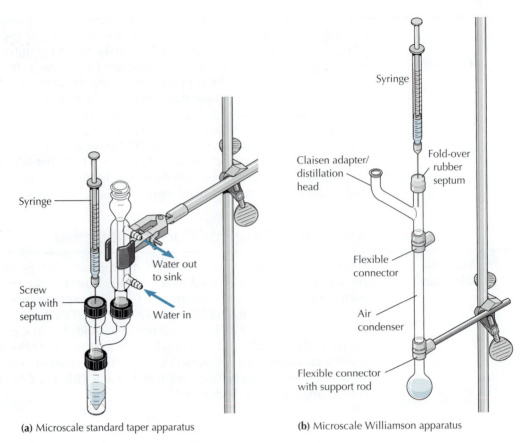

(a) Microscale standard taper apparatus   **(b)** Microscale Williamson apparatus

FIGURE 7.3 Using a syringe to add reagents to a microscale reaction.

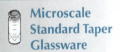

Microscale Standard Taper Glassware

The addition of reagents to a microscale reaction is done with a syringe. Figure 7.3a shows a microscale standard taper apparatus assembled for reagent addition using a syringe. The Claisen adapter provides two openings into the system. The opening used for the syringe can be capped either with a screw cap and Teflon septum or with a fold-over rubber septum. The top of the condenser is left open.

Microscale Williamson Glassware

The addition of reagents to a microscale reaction is done with a syringe. For Williamson microscale glassware, the Claisen adapter/distilling head provides two openings in the system. The vertical opening used for the syringe is capped with a fold-over rubber septum and the side-arm opening is left uncovered, as shown in Figure 7.3b.

7.3   Anhydrous Reaction Conditions

The reagents and/or products of some reactions can be sensitive to water, requiring that it be excluded from the reaction apparatus. The atmosphere, however, is replete with water vapor, making it challenging to maintain *anhydrous* (without water) conditions. The measures required for carrying out water-sensitive reactions depend on just how sensitive the reaction is. This section describes the use of drying tubes for excluding water from reactions that are only modestly sensitive to water. Drying tubes are effective for most of the water-sensitive reactions that you will carry out in your lab course. Section 7.4 describes more rigorous techniques for establishing and maintaining water- *and* air-free reaction conditions.

Preparing and using a drying tube. Keeping traces of water out of a reaction apparatus can be as simple as tightly capping a reaction flask; however, any reaction that requires heating or generates gaseous products cannot be closed to the atmosphere, or dangerous pressures would build up. In these cases, drying tubes can be used to maintain dry conditions while equalizing pressure with the atmosphere. Drying tubes are hollow plastic or glass tubes filled with a suitable drying agent, such as calcium chloride (see Section 11.1), and placed on openings in reaction apparatuses. Air can pass through the drying tube, keeping the contents of the reaction apparatus at atmospheric pressure; however, the drying agent will absorb moisture from any air passing through the tube into the reaction apparatus.

 Miniscale Glassware

For miniscale glassware, a thermometer adapter with a rubber sleeve serves to hold the plastic or glass drying tube (Figure 7.4a). A small piece of cotton is placed at the bottom of the drying tube to prevent drying agent particles from plugging the outlet of the tube; a piece of cotton is also placed over the drying agent at the top of the drying tube to keep the particles from spilling.

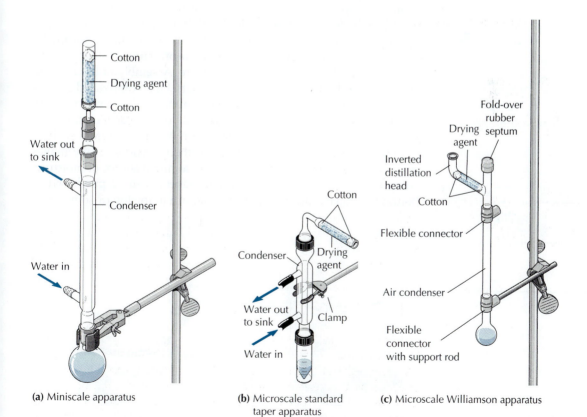

(a) Miniscale apparatus

(b) Microscale standard taper apparatus

(c) Microscale Williamson apparatus

FIGURE 7.4 Refluxing under anhydrous conditions.

Microscale Standard Taper Glassware

The L-shaped microscale standard taper drying tube has a ground glass inner joint that fits into the outer ground glass joint at the top of the condenser and is secured with an O-ring and screw cap (Figure 7.4b). A small piece of cotton is pushed into the drying tube to prevent the drying agent particles from falling into the reaction vessel; cotton is also placed near the open end of the drying tube to hold the drying agent in place.

 Microscale Williamson Glassware

Figure 7.4c shows how the Williamson microscale Claisen adapter/distilling head can be used as a drying tube. A small piece of cotton is pushed to the bottom of the side arm using the tip of a flexible plastic disposable pipet; a suitable drying agent, such as anhydrous calcium chloride, is added; and a second piece of cotton is placed at the top to keep the drying agent from spilling. The other opening is closed with a fold-over rubber septum. The Claisen adapter drying tube is fitted to the top of the air condenser with a flexible connector.

Handling Glassware and Solvents for More Rigorously Anhydrous Conditions

Establishing anhydrous reaction conditions begins with the glassware, reagents, and solvents used in the reaction. The surface of glass that has been in contact with the atmosphere is coated with a very thin layer of water, so glassware for anhydrous reactions is usually dried in an oven to remove this surface coating. If the glassware will not be immediately fitted to an apparatus with a drying tube, it should be placed in a desiccator while it cools (Figure 7.5a). An anhydrous atmosphere inside a closed desiccator is maintained by a drying agent, or desiccant (see Section 11.1). It will probably be necessary to slide the lid of the desiccator open slightly several times during the cooling process to relieve the increased air pressure inside the chamber caused by the heat from the glassware. Assembly of the reaction apparatus and addition of reagents should be accomplished as rapidly as possible to minimize their exposure to atmospheric moisture.

The reagents and solvents used for anhydrous reactions also need to be dry. Solid reagents can be stored in small desiccators such as the one shown in Figure 7.5b. Because the solvent generally makes up the bulk of a reaction mixture, it is a common source of moisture. Anhydrous solvents and other liquid reagents can be purchased in bottles that have a sealed cap with a septum in the top or some other

FIGURE 7.5
Desiccators.

(a) Large desiccator for storing oven-dried glassware

(b) Small desiccator for storing reagents

type of tight seal to exclude moisture. Anhydrous liquids should be transferred quickly to minimize contact with the atmosphere, and the bottles should be recapped immediately after transfer. Transferring liquids via syringe techniques, from bottles that are fitted with a septum, effectively minimizes contact with the atmosphere (see Section 7.5).

Molecular sieves. If a bottle containing a solvent or liquid reagent has been previously opened, or the presence of water is otherwise suspected, it may need to be dried using *molecular sieves* before use. Molecular sieves are extremely effective drying agents. They are porous aluminosilicate materials that have a three-dimensional structure riddled with a network of pores. They can be purchased with different pore sizes, denoted in Ångstroms (Å), which are determined by the particular composition of the material. For drying liquids, 3 Å molecular sieves are commonly used. This pore size is large enough to allow water to enter and become absorbed, but small enough to exclude molecules of most liquids being dried. The molecular sieves must be *activated* before use (or regenerated after use) by baking them in an oven at high temperature, usually under a vacuum, to drive out the water trapped in the pores. Molecular sieves can also be purchased in the activated form. To dry a liquid, about 10 g of 3 Å molecular sieves per 100 mL of liquid are added to the container and the liquid is simply allowed to sit overnight before use.* Storing a liquid over the sieves will keep it dry and ready for use.

Solvent purification systems. Your laboratory may have a commercial solvent purification system, providing ready access to dry and air-free solvents on demand. A modern solvent purification system, also called a Grubbs apparatus, uses a large stainless steel column filled with activated alumina (Al_2O_3) to dry the solvents. Activated alumina is a porous and highly water-absorbing form of aluminum oxide. A solvent is stored in a stainless steel reservoir (typically 18 L), which is connected to the alumina column. The solvent reservoir is pressurized with a dry, inert gas, such as argon. As the solvent passes through the column, the activated alumina absorbs traces of water. Before the solvent reservoir is connected to the solvent purification system, it is degassed by bubbling argon through the solvent. Thus the solvent that is delivered is free of both water and molecular oxygen.

7.4   Inert Atmosphere Reaction Conditions

Reagents or products that react with molecular oxygen and decompose must be handled in an *inert atmosphere*, which consists of an unreactive gas. Nitrogen is used most commonly, but argon is particularly effective because it is denser than air and will sink to the bottom of a reaction apparatus, pushing air away from the reaction mixture. If the compounds are also sensitive to moisture, the inert gas can be dried by passing it through a column packed with Drierite (see Section 11.1) or activated molecular sieves (Section 7.3).

*Williams, D. B. G.; Lawton, M. *J. Org. Chem.* **2010**, *75*, 8351–8354.

In this case, the glassware must also be oven dried. This section describes techniques for establishing and maintaining an inert atmosphere in a reaction or reagent vessel.

SAFETY PRECAUTIONS

1. Tanks of inert gases at high pressure must be handled with caution. Read the safety precautions for compressed gas cylinders and handle them only with the permission and supervision of your instructor.
2. The reaction assemblies described in the following techniques use syringe needles, which have sharp tips and can cause puncture wounds. Handle the needles with caution.
3. To avoid exposure to organic vapors, possible harm from accidental pressure build-up or violent reactions due to accidental exposure of reactive compounds to the air, perform air-sensitive reactions in a fume hood.

Inert Gas Sources

The most common sources of nitrogen and argon in the laboratory are steel cylinders of compressed gas. These are under very high pressure (up to 3000 psi, or pounds-per-square-inch) and they are usually equipped with a hand-wheel valve at the top of the cylinder. Gas cylinders must always be kept upright and individually secured with a strap, a chain, or a rack to a bench top or wall, to prevent the cylinder from falling over (Figure 7.6). If the valve were to be snapped off in a fall, the tank could become a damaging torpedo, propelled by the rapidly expanding gas (see the MythBusters video on YouTube that confirms this phenomenon: http://www.youtube.com/watch?v=ejEJGNLT084). Steel valve covers protect the valve

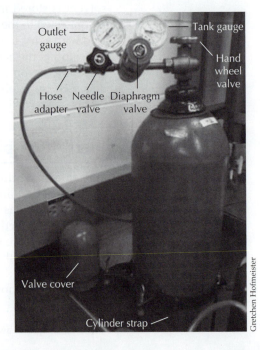

FIGURE 7.6 Compressed gas tank with two-stage regulator.

during transportation and should be screwed onto the cylinders when they are not being used.

The pressure and flow of gas from a cylinder is controlled with a two-stage regulator that has two adjustable valves (Figure 7.6). The diaphragm valve controls the pressure of the gas exiting the tank and the needle valve controls the flow of gas from the tank. The tank gauge measures the pressure inside the cylinder and the outlet gauge measures the pressure set by the diaphragm valve. When you use a gas from a steel cylinder, first open the wheel valve, followed by the diaphragm valve, and lastly the needle valve; when you are finished using the gas, close the valves in the reverse order.

For routine air-sensitive work, outlet pressures of approximately 3–5 psi are used. Figure 7.6 shows a hose adapter at the end of the needle valve, to which tubing is attached for filling balloons of inert gas or connecting with a manifold.

Sustaining an inert reaction atmosphere requires that vessel openings be sealed with septa or stoppers; however, introducing an inert gas at a pressure of 5 psi into a closed apparatus will cause stoppers and septa to pop out of the joints. This problem is circumvented by providing a means for relieving excess pressure: an exit port consisting of a septum pierced by a syringe needle, which vents to a fume hood directly or through a mineral oil bubbler (Figure 7.7). **Never connect a closed vessel to a tank of inert gas without providing a means for gas to escape the vessel, or dangerous pressures could develop.** Alternatively, the pressure of the inert gas source can be reduced to safe levels by using a balloon assembly or a gas manifold equipped with a bubbler.

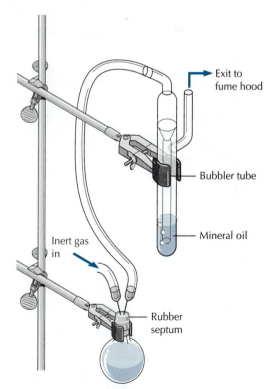

FIGURE 7.7 Mineral oil bubbler for maintaining safe positive pressure.

Gas Manifold with Bubbler

A gas manifold is particularly useful for handling highly air-sensitive reagents or for larger scale reactions. Gas manifolds can be simple or complex, and because they are constructed according to the needs of the user and the constraints of the institution, they vary considerably in design. The manifold consists of a piece of plastic, glass, or copper tubing, which is connected at one end to an inert gas source and at the other to a bubbler (Figure 7.8). One or more ports

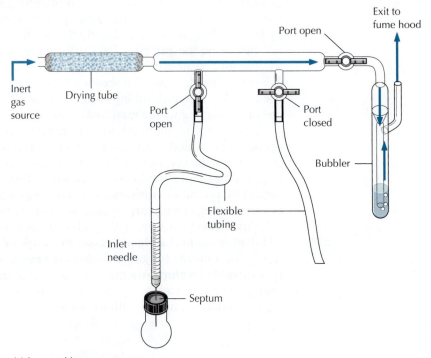

(a) Low positive pressure state

FIGURE 7.8 Gas manifold with bubbler. (a) The low positive pressure state. A closed reaction flask is connected to the manifold, which is under a positive pressure of inert gas controlled by the manifold bubbler. The primary flow of gas is indicated by the blue arrows. (b) The active purging state. The inert gas is diverted through the reaction flask by providing an exit port (a short needle) in the flask and isolating the manifold bubbler. The primary flow of gas is indicated by the blue arrows.

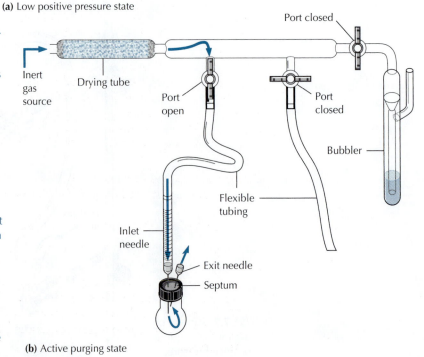

(b) Active purging state

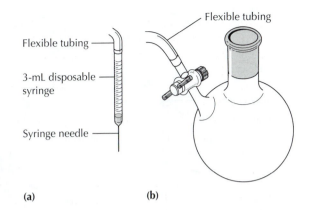

FIGURE 7.9
Connecting flasks with inert gas sources. (a) Tubing-needle adapter constructed from a 3-mL syringe, which has been modified by cutting off the top. (b) Flask with a stopcock-controlled inlet.

Flexible tubing

Flexible tubing

3-mL disposable syringe

Syringe needle

(a)   (b)

extend from the side of the tubing to connect to reaction flasks and reagent vessels. The ports have stopcocks or clamps to control access to the inert gas. In addition, a stopcock or clamp is located between the tubing and the bubbler to divert gas from the bubbler to an open port for specific operations.

Flexible tubing on a manifold port can be attached to a syringe needle, which then pierces a septum covering a flask opening. Adapters to connect tubing with syringe needles can be constructed from 3-mL disposable syringes, as shown in Figure 7.9a. Alternatively, specialty glassware that contains stopcock-controlled inlets can be connected directly to the flexible tubing on the manifold (Figure 7.9.b).

When the gas tank valve is opened, gas bubbles through the bubbler at the end of the manifold (the *Low Positive Pressure State*, Figure 7.8a), keeping the system pressure low. A flask connected to a port on the manifold will be exposed to safe pressures of inert gas when its port is opened. This is the configuration to use while waiting for a reaction that has been set up to go to completion. Because the inert gas is sweeping through the manifold toward the bubbler, however, it does not flush air out of a flask that is attached to a port. To displace air from a reaction flask, it is necessary to redirect the flow of gas from the bubbler to an exit port in the reaction flask. This is accomplished by closing access to the bubbler and providing a means for gas to escape from the flask, in a process called flushing or purging. This is the condition to use when preparing an apparatus and reagents for reaction (the *Active Purging State*, Figure 7.8b).

Balloon Assemblies

Many reagents are only moderately air-sensitive and can be handled using balloon assemblies (Figure 7.10) to establish and maintain an inert atmosphere.

Balloon assemblies are prepared as follows:

- Remove the plunger and cut the top off a 3-mL disposable plastic syringe.
- Fasten a balloon (latex, 12-mil wall thickness) to the top of the syringe with a rubber band that is doubled to make a tight seal.
- Fill the balloon with inert gas by carefully inserting the needle end of the syringe into a piece of tubing connected to the tank of inert gas.

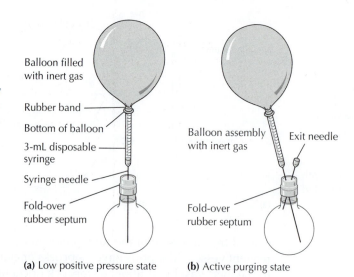

FIGURE 7.10 Balloon assemblies. (a) The low positive pressure state: a closed reaction flask is connected to the balloon assembly. (b) The active purging state: inert gas from the balloon assembly sweeps through the reaction flask by providing an exit port (a short needle) in the flask.

Balloon filled with inert gas

Rubber band

Bottom of balloon

3-mL disposable syringe

Syringe needle

Fold-over rubber septum

(a) Low positive pressure state

Balloon assembly with inert gas

Exit needle

Fold-over rubber septum

(b) Active purging state

- Remove residual air from the balloon assembly by repeating the following sequence **three times**: Fill the balloon with inert gas and pinch the neck of the balloon just above the top of the syringe barrel; disconnect the balloon assembly from the tank and allow the gas to be expelled from the balloon; quickly reconnect the balloon assembly to the tubing on the inert gas tank.
- Inflate the balloon a final time, tightly pinch its neck just above the top of the syringe barrel, and disconnect the assembly from the gas source. Allow a small amount of gas to escape the balloon to flush the syringe as you push the tip of the needle into a solid rubber stopper to stop the flow of gas.
- Use the balloon assembly as soon as possible after filling it with inert gas, because molecular oxygen from the atmosphere will slowly diffuse through the wall of the balloon and contaminate the inert gas.

Flushing an Apparatus

Flushing, or using an inert gas to sweep air out of a reaction apparatus, is routinely used in air-sensitive operations. This process, also called *purging*, establishes an inert atmosphere inside an apparatus that has been exposed to air. The inert gas source is introduced into one end of the apparatus and an exit needle is positioned so that the inert gas passes through the entire volume of the apparatus before exiting. Purging is easier to accomplish, especially for a large apparatus, with a continuous source of nitrogen from a manifold (Figure 7.8b). Nevertheless, balloons can be used to flush microscale flasks and vials (Figure 7.10b). The flushing process is carried out as follows:

- Seal the flask or apparatus openings with tightly fitted septa.
- Connect the inert gas source (manifold or balloon) to the lower part of the flask or apparatus.
- Create a gas outlet by piercing a septum with a disposable syringe needle that vents to a fume hood directly or through a bubbler. This should be positioned so that the inert gas is forced to sweep through the entire apparatus when traveling from inlet

to outlet. If using a manifold, close the access to the manifold bubbler to force the gas to travel through the apparatus.

- After flushing the system, remove the exit needle. **If using a manifold, immediately open the access to the manifold bubbler.** The manifold or balloon assembly will maintain a safe, positive pressure of inert gas throughout the course of the reaction. If using a balloon assembly for long reaction times, replace the balloon assembly periodically (at least every 24 hours).

Adding Reagents

Solid reagents can be added to a flask by briefly removing a stopper or septum and transferring the solid by funnel or weighing paper, and then purging the flask with inert gas. If the reaction is not very air-sensitive, the flask can be briefly opened to transfer liquids using graduated cylinders and pipets. The liquids can be degassed after the addition is complete. More sensitive reactions require the transfer of liquids or solutions without exposure to air using syringe techniques (see Section 7.5).

Degassing a Liquid or Solution

Dissolved oxygen gas can be removed from a liquid or solution by bubbling an inert gas through it in a process called *sparging*. The procedure for sparging is like flushing, except the gas inlet is submerged in the liquid (Figure 7.11). The flow of gas should cause active bubbling in the solvent. While this is most easily accomplished using a gas manifold, balloon assemblies may be used for small quantities of liquid, such as 5 mL; if no bubbles are observed,

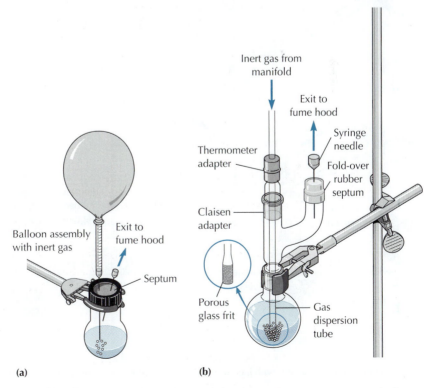

(a)

(b)

FIGURE 7.11 Degassing by sparging. (a) A microscale apparatus using a balloon assembly. (b) A miniscale apparatus using a gas manifold.

gently squeeze the balloon to increase the pressure of inert gas. Stirring the solvent while sparging can also make the process more effective. The time needed for degassing depends on the volume of solvent; ask your instructor for advice.

Refluxing Under an Inert Atmosphere

Flasks containing reaction mixtures that require heating should not be connected directly to the gas manifold or balloon assembly because the solvent vapors will migrate into the tubing and dissolve plastic or rubber components of the apparatus. When preparing a reaction mixture for reflux, use an adapter that enables the condenser to be flushed with inert gas (Figure 7.12a), and then move the

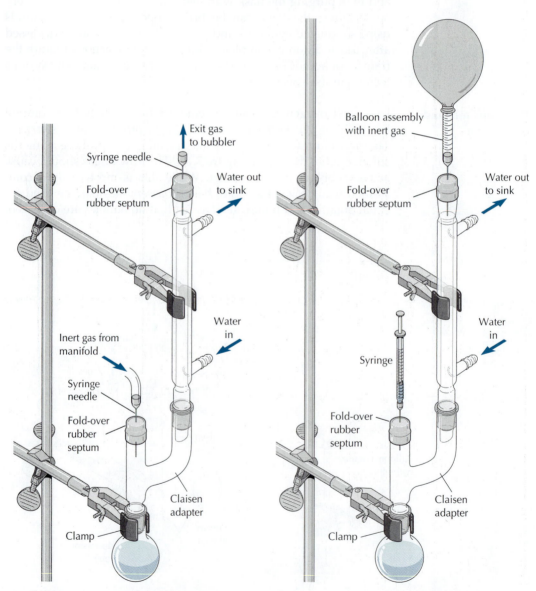

(a) Flushing reaction apparatus with inert gas   **(b)** Reaction apparatus with balloon assembly in place.

FIGURE 7.12 Miniscale apparatus for reflux under inert atmosphere. (a) Flushing the condenser. (b) Preparing for reflux; the syringe is removed before applying heat.

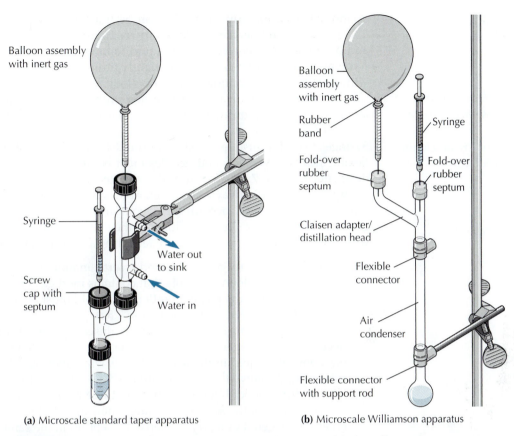

(a) Microscale standard taper apparatus

(b) Microscale Williamson apparatus

FIGURE 7.13 Microscale reaction apparatus with balloon assembly for reflux under inert atmosphere; the syringe is removed before applying heat.

inert gas inlet to the top of the condenser before heating commences (Figures 7.12b and Figure 7.13).

Adding Liquids to a Reaction in Progress

During the course of a reaction, small volumes of liquids can be introduced by syringe (Figures 7.12b and 7.13) but larger volumes require a pressure-equalizing funnel (Figure 7.14). This is a dropping funnel (Section 7.2) containing a side-arm tube, which allows gas in the funnel and the reaction flask to exchange in response to pressure changes during the addition. Liquids can be transferred to the pressure-equalizing funnel by removing the stopper, which will introduce air into the apparatus. After addition, the funnel can be flushed with inert gas and dissolved oxygen in the liquid can be removed by sparging. Alternatively, special transfer techniques (see Section 7.5) can be used to load the funnel without exposing the reagents to air.

7.5 Transfer of Liquids by Syringe Without Exposure to Air

FIGURE 7.14
Pressure-equalizing dropping funnel.

Air-sensitive reagents and dry solvents may require special techniques for transferring from reagent bottles to a reaction apparatus without exposure to atmospheric oxygen and moisture. Small quantities of liquids (up to 40 mL) can be transferred by syringe. The glassware and needles used for air-free transfer need to be dry, and both the vessel containing the liquid to be transferred and the receiving vessel must be sealed with tightly fitting septa and purged with inert gas. Two manifold ports or balloon assemblies are necessary for maintaining an inert atmosphere in both vessels. Do not attempt these operations without the supervision of your instructor.

Transferring Liquids by Syringe

Syringes used for air-free transfer are fitted with long (12–24 inch) flexible needles. Syringes and needles must be dry and stored in a desiccator (Section 7.3) before use. Wear protective gloves for these manipulations and always clean reusable syringes and needles immediately after use, according to your instructor's directions. Overall, the procedure involves flushing the syringe with inert gas, withdrawing the correct volume, and delivering a measured volume:

- Firmly clamp the reagent bottle and insert a manifold connection or balloon assembly that contains inert gas into the septum that seals the reagent bottle; a manifold must be in the low positive pressure state (see Figure 7.8a).
- Remove residual air from the syringe by inserting the syringe needle into the reagent bottle so that the tip accesses the inert gas above the liquid in the reagent bottle. Draw inert gas into the syringe, remove the syringe and needle, and expel the gas into the fume hood. Repeat this process **three times.** Because each puncture of the septum provides a new avenue for air to enter the reagent bottle, try to pierce the septum in the same place each time.
- Insert the needle of the syringe so that the tip extends below the surface of the liquid in the bottle (Figure 7.15a). By gently extending the plunger, slowly fill the syringe to a volume slightly larger than required. Allow the gas pressure in the reagent bottle to assist in filling the syringe. Do not pull on the plunger forcefully because this may cause leaks or generate gas bubbles.
- Withdraw the tip of the needle from the liquid but leave it in the bottle to access inert gas above the liquid. **Invert the syringe by gently bending the flexible needle into a U-shape** and allow any gas bubbles to migrate to the needle-end of the syringe. Push the plunger slowly to expel gas bubbles and adjust the volume of reagent to the desired amount (Figure 7.15b). **From this point until after delivering the reagent, keep the syringe inverted.**
- Introduce a small amount of inert gas from the space above the liquid into the needle by adjusting the plunger. This helps to avoid drips and exposure of the air-sensitive reagent to the atmosphere during transfer.
- Hold the inverted syringe with one hand. Use your other gloved hand to pull the needle out of the reagent bottle and quickly insert it through the rubber septum on the reaction apparatus.

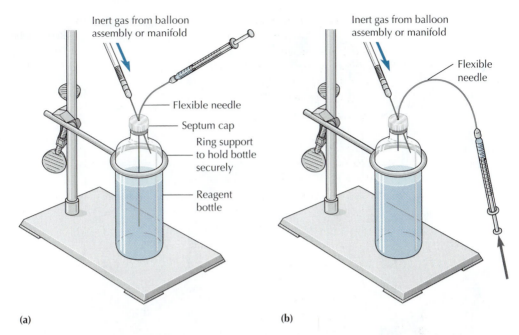

(a)   (b)

FIGURE 7.15 Filling a syringe with an air-sensitive reagent. Reprinted with permission from Aldrich Chemical Co., Inc., Milwaukee, WI.

- **Deliver the liquid into the reaction apparatus while the syringe is still inverted.** This will ensure that the small amount of inert gas at the tip of the needle is introduced before the liquid and that the quantity of liquid delivered is accurate.
- After delivery, the needle will still be filled with liquid. Pull back on the plunger to withdraw some gas from the space above the reaction solution and remove the syringe and needle from the flask.
- In the fume hood, clean the syringe and needle immediately, according to your instructor's directions.

7.6   Removal of Noxious Vapors

When a noxious acidic gas such as nitrogen dioxide, sulfur dioxide, or hydrogen chloride forms during a reaction, it must be prevented from escaping into the laboratory. Acidic or basic gases, such as HCl or NH_3, are readily soluble in water, so a gas trap containing either water or dilute aqueous sodium hydroxide for acidic vapors, or dilute hydrochloric acid solution for NH_3 vapors, effectively traps them. **Any reaction that emits noxious vapors should be performed in a hood.**

Miniscale Apparatus

Attach a U-shaped piece of glass tubing to the top of a reflux condenser by means of a one-hole rubber stopper or a thermometer adapter. Carefully fit the other end of the U-tube through a

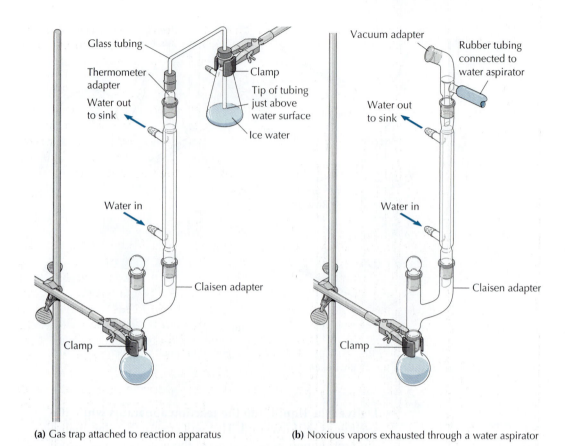

(a) Gas trap attached to reaction apparatus   (b) Noxious vapors exhausted through a water aspirator

FIGURE 7.16 Miniscale apparatus used to trap water-soluble noxious vapors.

one-hole rubber stopper sized for a 125-mL filter flask. Place about 50 mL of ice water, dilute sodium hydroxide solution, or dilute hydrochloric acid solution into the filter flask and position the open end of the U-tube **just above** the surface of the liquid, as shown in Figure 7.16a.

In laboratories equipped with water aspirators, a gas trap can be made by placing a vacuum adapter at the top of a condenser. The side arm of the vacuum adapter is connected by heavy-walled rubber tubing to the side arm of the water aspirator and the water turned on at a moderate flow rate. The noxious gases are pulled from the reaction apparatus and dissolved in the water passing through the aspirator (Figure 7.16b).

Microscale Standard Taper Apparatus

A gas trap for microscale reactions can be prepared with fold-over rubber septa, Teflon tubing (1/16 inch in diameter), and a 25-mL filter flask. To insert the Teflon tubing through a rubber septum, carefully punch a hole in the septum with a syringe needle and push a round toothpick through the hole. Fit the tubing over the point of the toothpick and pull the toothpick (with tubing attached) back through the septum, as shown in Figure 7.17. Repeat this process to also place a rubber septum on the other end of the tubing.

Half fill a 25-mL filter flask with ice water, or a dilute aqueous solution of acid or base if needed, and close the top with one

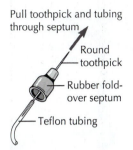

Pull toothpick and tubing through septum

Round toothpick

Rubber fold-over septum

Teflon tubing

FIGURE 7.17 Threading Teflon tubing through a rubber septum.

septum. Push the tubing down until the open end is **just above** the surface of the ice water or aqueous solution. Attach the other septum to the top of the condenser. The side arm of the filter flask serves as a vent (Figure 7.18a).

In laboratories equipped with water aspirators, a gas trap for microscale standard taper glassware can be made by placing a vacuum adapter at the top of a condenser. The side arm of the vacuum adapter is connected to the side arm of the water aspirator with heavy-walled rubber tubing and the water turned on at a moderate flow rate. The noxious gases are pulled from the reaction apparatus and dissolved in the water passing through the aspirator (Figure 7.18b).

 **Microscale Williamson Glassware**

A gas trap for microscale reactions using Williamson glassware can be prepared with three fold-over rubber septa, Teflon tubing (1/16 inch in diameter), and a 25-mL filter flask or a reaction tube (Figure 7.19). To insert the Teflon tubing through a rubber septum, carefully punch a hole in one septum with a syringe needle and push a round toothpick through the hole. Fit the tubing over the point of the toothpick and pull the toothpick (with tubing attached) back through the septum, as shown in Figure 7.17. Repeat this process to place a rubber septum on the other end of the tubing.

Half fill a 25-mL filter flask or a Williamson reaction tube with ice water or dilute acid or base solution if needed and close the top with one septum attached to the tubing. Push the tubing down until the open end is **just above** the surface of the water or aqueous solution. Attach the second septum to the top of the Claisen adapter. Close the other opening of the Claisen adapter with the third septum. If a filter flask serves as the trap, the side arm provides a vent

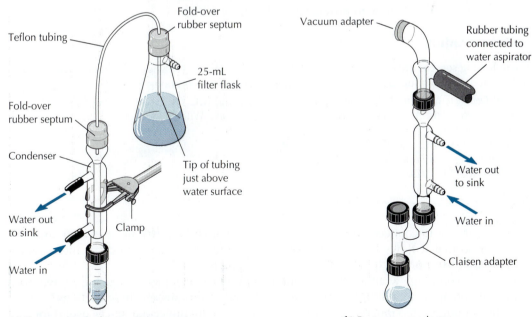

(a) Gas trap attached to reaction apparatus

(b) Bent vacuum adapter

FIGURE 7.18 Standard taper microscale apparatus used to trap water-soluble noxious vapors.

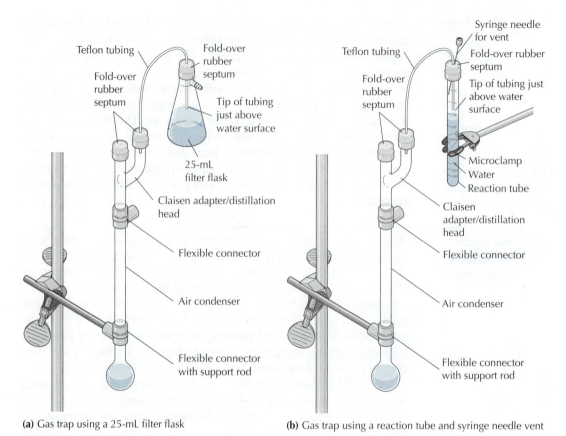

(a) Gas trap using a 25-mL filter flask

(b) Gas trap using a reaction tube and syringe needle vent

FIGURE 7.19 Williamson microscale apparatus used to trap water-soluble noxious vapors.

(Figure 7.19a); if the trap is a Williamson reaction tube, a syringe needle **must be inserted** into the septum attached to the reaction tube to provide a vent (Figure 7.19b).

Further Reading

The following sources provide additional details and information on a wide variety of reaction setups and methods for carrying out reactions under inert atmosphere conditions.

Aldrich Technical Bulletin AL-134, Handling Air-Sensitive Reagents; Aldrich Chemical Co., Inc., Milwaukee, WI.

Leonard, J., Lygo, B., and Procter, G. *Advanced Practical Organic Chemistry*, 3rd ed.; CRC Press: Boca Raton, FL, 2013.

Pirrung, M. C. *The Synthetic Organic Chemist's Companion*; Wiley: Hoboken, NJ, 2007.

Questions

1. Why is it best to connect water to a condenser so that it flows in at the bottom and out at the top?

2. How can you judge whether the amount of heat you apply to a reflux apparatus is sufficient but not too much?

3. Why do cylinders of compressed gases need to be kept upright and secured to a wall or bench top?

4. What is the difference between an anhydrous atmosphere and an inert atmosphere?

5. What is the purpose of using a mineral oil bubbler in air-sensitive reactions?

6. What are molecular sieves useful for?

8

COMPUTATIONAL CHEMISTRY

Using computational chemistry to study how proteins recognize potential drug candidates is a powerful tool in medicinal chemistry, and computer-assisted drug design has become standard in the pharmaceutical industry. *Computational chemistry* is the calculation of physical and chemical properties of compounds using mathematical relationships derived from theory and observation. Not only important for modeling the molecular docking of protein-ligand interactions, computational chemistry is often a starting point for doing chemical synthesis, exploring chemical reaction mechanisms, understanding the conformations of organic molecules, and predicting and understanding experimental results. Carrying out computations is often most useful before you go into the laboratory to perform your experiments.

Computational chemistry is often referred to as *molecular modeling.* However, we use the term *computational chemistry* to avoid confusion with molecular model sets, which you may have used to create three-dimensional structures of molecules.

8.1 Picturing Molecules on the Computer

Computational chemistry can be used to create three-dimensional images and two-dimensional projections of chemical structures. The computer images that result are completely interactive. These images are similar to a molecular model set, but computational chemistry is also much more. In molecular model sets, the bond lengths and bond angles are fixed at certain "standard values," such as 109.5° for the bond angle of a tetrahedral (sp^3) carbon atom. Anyone who has built a molecule containing a cyclopropane ring is well aware of the limitations of using a 109.5° bond angle for its "tetrahedral" carbon atoms. The structure of a molecule created on the computer can be optimized by changing bond lengths and angles until the structure represents the molecule's lowest energy conformation. Optimization means that the bond lengths and bond angles of the structure are allowed to deviate from their "standard values." Thus the molecule created on the computer is a more accurate picture of the actual molecule than can be obtained from using a molecular model set.

Computational Chemistry Programs

Most computational chemistry programs consist of interacting modules that carry out specialized tasks such as building a molecule, optimizing the molecular structure, and extracting physical properties from the calculation. The computer image of a molecule can be shown in a variety of ways—wire frame, ball-and-stick, and space-filling, to mention a few. Wire frame images are best to represent bond angles, lengths, and direction. A molecule's size and shape are probably best represented by a space-filling model. The rendering methods can be mixed to emphasize steric interactions in a specific portion of a molecule. The electron density surface can be displayed, providing a view of a molecule's overall shape and a map of the electrostatic potential on the molecular surface, highlighting regions of potential reactivity within the molecule. Molecular orbitals can also be superimposed onto a molecular structure.

Camphor
wire frame model

Camphor
ball-and-stick model

Camphor
space-filling model

Many physical and chemical properties can be extracted from an optimized molecular structure. These properties include bond lengths, bond angles, dihedral angles, interatomic distances, dipole moments, electron densities, heats of formation, and NMR chemical shifts. The computed properties are often very good approximations of the values determined by experiments.

Computational Methods

There are two major types of computational methods. The first, called *molecular mechanics*, is derived from a classical mechanical model, which treats atoms as balls and bonds as springs connecting the balls. In general, molecular mechanics methods pay attention to nuclei, while paying little attention to electrons. The second and more rigorous group of methods is based on *quantum mechanics*, which can be used to describe the physical behavior of matter on a very small scale. Quantum mechanics methods pay attention to both nuclei and electrons.

Following are some stand-alone computational chemistry packages available for modern microcomputers:

- Spartan from Wavefunction
- ChemBio3D from PerkinElmer
- HyperChem from HyperCube

As the field of computational chemistry continues to develop, new program suites become available with new methods, extended capabilities, and the ability to handle large molecules, including biopolymers. Most of these programs require a large server or computer cluster to operate efficiently, especially when employed by multiple users. Examples of this class of software are:

- Gaussian from Gaussian, Inc.
- MacroModel from Schrödinger
- GAMESS (General Atomic and Molecular Structure System) from the Gordon research group at Iowa State University

Inputting a molecular structure for calculation by one of these programs usually requires a graphical input module such as WebMO (available at http://www.webmo.net/). Software availability depends on an institution's computing facilities and licensing arrangements.

The following sections describe in general terms and provide examples of the types of calculations that are possible using these computational packages and also touch on their limitations. Because the operation of a program and its calculation modules differ from one package to another, the details of these packages are not discussed. Materials included with the packages provide comprehensive descriptions of the specific methods that the programs use.

8.2 Molecular Mechanics Method

The *molecular mechanics (MM)* method treats a molecule as an assemblage of classical balls (atoms) and springs (bonds, bond angles, and so on) connecting the balls. The *total energy* of a molecule, often called the *steric energy* or *strain energy,* is the sum of energy contributions from bond stretching, angle strain, strain resulting from improper torsion, steric or van der Waals interactions, and electronic charge interactions.

$$E_{\text{steric}} = E_{\text{bond stretching}} + E_{\text{angle bending}} + E_{\text{torsion}} + E_{\text{van der Waals}}$$
$$+ E_{\text{electrostatic interactions}}$$

The contributions are described by empirically derived equations. For example, the energy of bond stretching is approximated by the energy of a spring described by Hooke's Law from classical physics,

$$E_{\text{bond stretching}} = 1/2k \, (x - x_0)^2$$

in which k is a force constant related to bond strength and $(x - x_0)$ is the displacement of an atom from its equilibrium bond length (x_0). If a bond is stretched or compressed, its potential energy will increase, and there will be a restoring force that tries to restore the bond to its equilibrium bond length. The force constants for various types of bonds can be derived from experimental data and are incorporated into the molecular mechanics parameter set. The energy of the bond stretching in a molecule is the sum of the contributions from all of its bonds.

$$E_{\text{bond stretching}} = \sum_{i=1}^{i=n \text{ bonds}} 1/2k_i(x - x_0)_i^2$$

Other energy contributions are developed in a similar fashion. For example, if a bond has a force constant, k, which resists a change in the size of the bond angle, $E_{\text{angle bending}}$ must be systematically varied until it is minimized. Molecular mechanics calculations give good estimates for the bond lengths and angles in a molecule.

The collections of equations describing the various energies and their associated parameter sets are called *force fields.* Force fields differ because the parameters are optimized for specific types of molecules (e.g., proteins, organometallic compounds), for specific conditions (e.g., gas phase, condensed phase), or for specific properties such as NMR parameters (e.g., chemical shifts, coupling constants).

Following are some frequently used force fields:

- MM2, MM+, MM3, MM4 (**M**olecular **M**echanics) parameters originally developed for hydrocarbons with refinements to include other types of organic compounds.
- AMBER (**A**ssisted **M**odel **B**uilding and **E**nergy **R**efinement) parameters optimized for calculation of molecular dynamics of biomolecules.
- MMFF (**M**erck **M**olecular **F**orce **F**ield) parameters well suited for a wide range of molecules, from small organic compounds to biopolymers.
- CHARMM (**C**hemistry at **HAR**vard **M**olecular **M**echanics) parameters for a wide range of molecules, from small organics to proteins.
- BIO+, a version of the CHARMM incorporated into HyperChem.
- SYBYL parameters for most elements in the Periodic Table.

The kinds of energy outputs from a molecular mechanics calculation are listed here. These data come from using the ChemBio3D computational package with an MM2 force field, and they involved 19 iterations.

Stretch:	0.3406 kcal/mol
Bend:	0.3720
Stretch-Bend:	0.0893
Torsion:	2.1529
Non-1,4 VDW	-1.0609
1,4 VDW	4.6632
Total (steric energy):	**6.5571** kcal/mol

The absolute value of the steric energy of a molecule has no meaning by itself. Its calculated value can vary greatly from one force field parameter set to another. Steric energies are useful only for comparison purposes. The comparisons are most useful for conformers, such as chair and twist-boat cyclohexane, and diastereoisomers, such as *cis*- and *trans*-1,3-dimethylcyclohexane.

In the calculation of the total energy, each atom type is associated with an unstrained heat of formation. The relative heat of formation of each isomer is the sum of the heats for the unstrained atom types plus the strain energy.

Energies of Cyclohexane Conformers

WORKED EXAMPLE

The axial and equatorial conformers of cyclohexanes can be interchanged by way of a ring flip. In the simplest example, the axial hydrogen atoms of cyclohexane become equatorial hydrogen atoms and the equatorial hydrogen atoms become axial hydrogen atoms. Construct an energy profile for converting one chair conformer into its flipped chair conformer.

To perform this feat of molecular gymnastics, the cyclohexane ring twists and bends into several conformers. Starting at the chair conformer, it proceeds through a half-chair, then a twist-boat, then a boat conformation, then through another twist-boat and half-chair to the flipped chair conformer.

To construct the energy profile, you need to calculate the steric energies of each of the conformers in the pathway for converting one chair conformer into its flipped chair conformer. The computational chemistry package actually used to obtain the desired energies was Spartan 06, using an MMFF force field parameter set.

Construction and optimization of chair cyclohexane

1. If the computational chemistry package has a fragment library, select the chair cyclohexane. Otherwise, construct a ring of six carbons that roughly approximates a chair conformation.
2. Optimize the geometry (or minimize the energy) using the molecular mechanics module of the program. If the optimized structure is not in the chair conformation, judicious editing of the structure and optimization will usually afford the chair conformation.
3. Record the steric energy (−14.9 kJ/mol).

Construction and optimization of boat cyclohexane

1. If the computational chemistry package has a fragment library, select cyclopentane. Otherwise, construct a ring of five carbons.
2. Attach sp^3 carbons to the 1 and 3 positions of the cyclopentane ring. The attached carbons must be on the same side of the ring.
3. Make a bond between the two methyl groups to form bicyclo [2.2.1]heptane.
4. Optimize the geometry (or minimize the energy) using the molecular mechanics module of the program.
5. Delete the carbon atom that forms the one-carbon bridge of bicyclo[2.2.1]heptane. Optimize the geometry (or minimize the energy) using the molecular mechanics module of the program. This structure should be the boat conformation of cyclohexane.

6. Record the steric energy (13.0 kJ/mol).

Construction and optimization of twist-boat cyclohexane

1. Construct a chair cyclohexane.
2. Attach an sp^3 carbon atom to an axial position of the cyclohexane ring to create *axial*-methylcyclohexane.
3. Delete the ring carbon atom that is directly adjacent to the ring carbon bearing the methyl group.

4. Make a bond between the terminal carbons of the resulting six-carbon atom chain.
5. Optimize the geometry (or minimize the energy) using the molecular mechanics module of the program. This structure should be twist-boat cyclohexane.

6. Record the steric energy (9.9 kJ/mol).

Construction and optimization of half-chair cyclohexane

To construct this conformer, it is necessary to force five carbons of the cyclohexane ring to lie in the same plane.

1. Build a chain of six sp^3 carbon atoms.
2. Define the dihedral angle described by C2, C3, C4, and C5 to be 0° and lock the angle to that value.
3. Define the dihedral angle described by C1, C2, C3, and C4 to be 0° and lock the angle to that value.
4. Connect the terminal carbons, C1 and C6, with a bond.

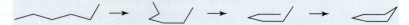

5. Optimize the geometry (or minimize the energy) using the molecular mechanics module of the program. Make sure the program respects the constraints. With Spartan 06 there is a Constraints box that must be checked.
6. Record the steric energy (28.2 kJ/mol).

Through the use of the calculated steric energies, an energy profile connecting each conformation of cyclohexane can now be constructed, as shown in Figure 8.1.

To recap, the steric energies are: chair, −14.9 kJ/mol; half-chair, 28.2 kJ/mol; twist-boat, 9.9 kJ/mol; boat, 13.0 kJ/mol.

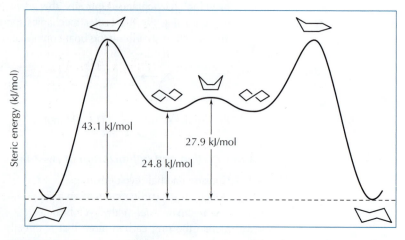

FIGURE 8.1 Energy profile for interconversion of the chair conformers of cyclohexane.

Equilibrium Constants for Axial and Equatorial Cyclohexane Conformers

Differences in steric energies can also be used to estimate equilibrium constants between interconverting conformers. At room temperature, methylcyclohexane is a mixture of *axial*-methylcyclohexane and *equatorial*-methylcyclohexane, and these are rapidly interconverting by way of a ring flip.

axial   *equatorial*

The relative amount of each conformer at equilibrium can be determined by the difference in energy between the two conformers, which is related to the equilibrium constant, K_{eq}, by the following relationships:

$$K_{eq} = \frac{\text{number of } eq\text{-methylcyclohexane molecules}}{\text{number of } ax\text{-methylcyclohexane molecules}}$$

$$\Delta G^0 = -RT \ln K_{eq} = -2.303 \, RT \log K_{eq}$$

where ΔG^0 is the change in Gibbs standard free energy in going from *axial*-methylcyclohexane to *equatorial*-methylcyclohexane, R is the gas constant (1.986 cal/deg mol), and T is the absolute temperature in degrees Kelvin (K).

Through the use of an MM2 force field, the steric energy of *axial*-methylcyclohexane is calculated to be 8.69 kcal/mol, and the steric energy of *equatorial*-methylcyclohexane is calculated to be 6.91 kcal/mol. If the difference in steric energy approximates the difference in free energy between the conformers, the free energy difference is −1.78 kcal/mol. The negative value for ΔG^0 signifies a release of energy in going from *ax*-methylcyclohexane to *eq*-methylcyclohexane. At room temperature (25°C, 298 K), the preceding equation becomes

$$-1.78 = -1.36 \log K_{eq}$$

$$\log K_{eq} = 1.31$$

$$K_{eq} = 20.4$$

At equilibrium, there would be approximately 20 molecules of *equatorial*-methylcyclohexane present for each molecule of *axial*-methylcyclohexane—close to the experimental value.

FOLLOW-UP ASSIGNMENT

Calculate the steric energies for *equatorial-tert*-butylcyclohexane and *axial-tert*-butylcyclohexane and use them to calculate the composition of their equilibrium mixture at 25°C. Construct the chair cyclohexane using the method on page 111 and then replace an equatorial hydrogen with a *tert*-butyl group. For the construction of *axial-tert*-butylcyclohexane, replace an axial hydrogen with a *tert*-butyl group. Optimize the geometries using the molecular mechanics module of your computational chemistry package. Record the two energies and calculate the equilibrium constant.

Energies of Butene Isomers: Limitations of Molecular Mechanics

WORKED EXAMPLE

Molecular mechanics methods work well for comparing the energies of conformers, but less well for isomeric compounds that are not conformers. Consider the case of the isomeric butenes: 1-butene, *cis*-2-butene, and *trans*-2-butene. The disubstituted 2-butenes are known to be more stable than 1-butene, and the *trans*-isomer of 2-butene is more stable than the *cis*-isomer. A quantitative experimental perspective comes from heats of formation (ΔH_f°) as well as heats of hydrogenation.

The hydrogenation of all three butenes produces butane.

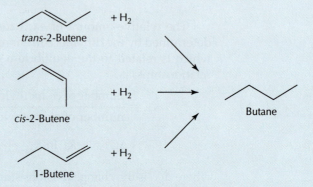

Thus the differences in the heats of hydrogenation are a measure of the relative energy levels of the alkenes (Figure 8.2).

The heats of hydrogenation and heats of formation follow:

	$\Delta H_{hydrogenation}$ (kJ/mol)	(ΔH°)	ΔH_f° (kJ/mol)
1-Butene	126.8	(0.0)	0.1
cis-2-Butene	119.7	(−7.1)	−9.2
trans-2-Butene	115.5	(−11.3)	−14.0

The data sets for both ΔH° and ΔH_f° indicate that *trans*-2-butene is more stable than the *cis* isomer by 4–5 kJ/mol and that 1-butene is the least stable of the three isomers.

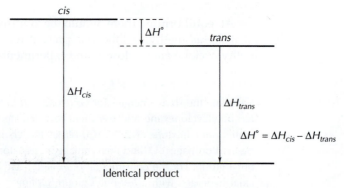

FIGURE 8.2 Energy diagram for the conversion of two isomers to a common product.

How well do the steric energies of these three butenes match the experimental data? With Spartan 06 and using the MMFF parameter set, the following results were obtained:

	Steric energy (kJ/mol)
1-Butene	22.7
cis-2-Butene	25.9
trans-2-Butene	20.3

The calculated steric energies indicate that the most stable isomer is *trans*-2-butene and the least stable isomer is *cis*-2-butene. This result does not agree with the experimental results. The molecular mechanics calculation is not reliable in comparing the energies of the butene isomers; however, calculations using the quantum mechanical methods described in Section 8.3 are far more reliable. Optimizing the geometry of the butenes using the AM1 parameter set (MOPAC) of the semiempirical quantum mechanical method in Spartan 06 gives the following results:

	ΔH_f° (kJ/mol)
1-Butene	0.7
cis-2-Butene	−7.1
trans-2-Butene	−11.4

Now the order of stability is correct and the differences in the calculated energies of the three isomers are close to the experimental results.

8.3 Quantum Mechanics Methods: *Ab Initio*, Semiempirical, and DFT

Quantum mechanical molecular orbital (MO) methods are based on solving the **Schrödinger wave equation**, $\hat{H}\Psi = E\Psi$, in which $\hat{H}$ is the Hamiltonian operator describing the kinetic energies and electrostatic interactions of the nuclei and electrons that make up a molecule, E is the energy of the system, and Ψ is the wavefunction of the system. Although simple in expression, the solution is exceedingly complex and requires extensive computational time. Even an organic molecule as simple as methane defies exact solution. The key to obtaining useful information from the Schrödinger relationship in a reasonable amount of time is choosing approximations that simplify the solution. When more approximations are used, however, the calculation is faster although the accuracy of the result may be degraded.

Ab Initio Quantum Mechanical Molecular Orbital (MO) Methods

Quantum mechanical MO models with the least degree of approximation are called **ab initio methods**. *Ab initio* is a Latin phrase that means "from the beginning" or "from first principles." Following are some common approximations that are used even in *ab initio* MO theory:

1. Nuclei are stationary relative to electrons, which are fully equilibrated to the molecular geometry (Born-Oppenheimer approximation).

2. Electrons move independently of each other, and the motion of any single electron is affected by the average electric field created by all the other electrons and nuclei in the molecule (Hartree-Fock approximation).
3. A molecular orbital is constructed as a linear combination of atomic orbitals (LCAO approximation).

Ab initio calculations use a collection of atomic orbitals called a **basis set** to describe the molecular orbitals of a molecule. There are numerous basis sets of varying complexity in use. The choice affects the accuracy of the calculation and the amount of time required for a solution. Normally, you should use the lowest degree of complexity that will answer your question or solve the problem.

The smallest basis set in common use is STO-3G, so called because it is a Slater-type orbital (STO) built from three Gaussian functions to describe each orbital. STOs have the same angular terms and overall shape as the hydrogen-like orbitals $1s$, $2s$, $2p$, and so on, but are different in that they have no radial nodes. The STO-3G basis set works reasonably well with first- and second-row elements that incorporate s- and p-orbitals. An *ab initio* calculation using an STO-3G basis set can often provide good equilibrium geometries.

The medium-sized 3-21G basis set is often a good starting point. The 3-21G symbolism signifies that three Gaussian functions are used for the wavefunction of each core electron, but the wavefunctions of the valence electrons are "split" two to one between inner and outer Gaussian functions, allowing the valence shell to expand or contract in size. The 6-31G* basis set, using more Gaussian functions and a polarization function on heavy atoms, provides better answers and is more flexible. However, it requires more calculation time, typically 10 to 20 times more than the same calculation using an STO-3G basis set.

Semiempirical Molecular Orbital (MO) Approach

The geometries and energies of organic molecules can be optimized by the *ab initio* MO method, using a 3-21G basis set with a desktop computer, but the calculation can take many minutes for the optimization of even a small organic molecule. For most practical purposes, a faster method of calculation is needed. The semiempirical **molecular orbital approach** introduces several more approximations that dramatically speed up the calculations. A geometry optimization using a semiempirical molecular orbital method is typically 300 or more times faster than one using an *ab initio* MO method with a 3-21G basis set.

The approximations generally used with semiempirical molecular orbital methods are as follows:

1. Only valence electrons are considered. Inner-shell electrons are not included in the calculation (this is also an option with *ab initio* MO calculations).
2. Only selected interactions involving at most two atoms are considered. This is called the *neglect of diatomic differential overlap*, or NDDO.
3. Parameter sets are used to calculate interactions between orbitals. The parameter sets are developed by fitting calculated results with experimental data.

Several popular versions of semiempirical methods follow:

- MNDO (**M**odified **N**eglect of **D**ifferential **O**verlap), an approximation for semiempirical calculations.
- AM1 (**A**ustin **M**ethod **1**), an improvement to MNDO introducing new atomic parameter sets.
- PM3, PM6 (**P**arameterized **M**odels **3** and **6**), a modification of the AM1 method especially useful for calculations on organometallic compounds.
- RM1 (**R**ecife **M**odel **1**), a reparameterization of AM1 that may give better results than AM1 or PM3 for organic molecules.

In many cases, AM1 is the method of choice for organic chemists; it should be used whenever possible before resorting to an *ab initio* calculation. For example, the optimization of 2-bromoacetanilide using an Apple Macintosh G-5 computer takes almost 18 min in an *ab initio* calculation using a 3-21G basis set; it takes 1.7 sec using the AM1 semiempirical method. The PM3 method is often used for inorganic molecules because it has been parameterized for more chemical elements. *MOPAC* or **M**olecular **O**rbital **Pa**ckage incorporates a number of semiempirical methods in a single program. As you become more familiar with computational chemistry, you will be able to experiment with various methods to find the one that works best for the molecules with which you are working.

The bromination of a benzene ring is an example of an electrophilic aromatic substitution reaction, which involves the reaction of Br_2 with the benzene ring to form a bromobenzenium cation in the rate-determining step. The bromobenzenium ion subsequently loses a proton to yield a bromobenzene product.

Bromobenzenium cation

In the case of a monosubstituted benzene, such as acetanilide, there are three possible monosubstituted products, the *ortho-*, *meta-*, and *para-*bromoacetanilides.

The reaction pathway with the lowest activation energy for the formation of the bromobenzenium ion will be favored. Because the formation of this cation is endothermic, the most stable bromobenzenium ion correlates with the rate-determining transition state. The energy profile for the formation of *para-*bromoacetanilide is shown in Figure 8.3.

You can use the semiempirical MOPAC molecular orbital method with the AM1 parameter set to calculate the heats of formation of the intermediate benzenium cations **1**–**3**, which would lead to the *ortho-*, *meta-*, and *para-*bromoacetanilides.

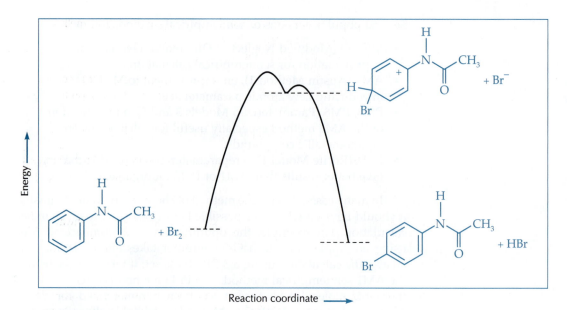

FIGURE 8.3 Energy profile for the bromination of acetanilide.

(1)   (2)   (3)

WORKED EXAMPLE

Construction and optimization of the bromobenzenium ion

1. Construct a 1, 4-cyclohexadiene molecule. Attach a bromine atom to one of the sp^3 carbon atoms of the molecule. At the other sp^3 carbon atom, delete one of the valences (or hydrogen atoms). Before optimizing the geometry, indicate that the molecule has a charge of +1 and is in the singlet state (all of its electron spins are paired).
2. Optimize the geometry using the semiempirical MOPAC method with the AM1 parameter set.
3. Record the heat of formation ($\Delta H_f = 923.2$ kJ/mol).

Construction and optimization of the intermediate bromobenzenium ions 1–3

1. Use a copy of the bromobenzenium ion that you have constructed to build the reactive intermediates **1–3**. For example, for the bromobenzenium ion **1**, attach an acetanilide group to the carbon *ortho* to the sp^3 carbon bearing the bromine atom.

2. Optimize the geometry using the semiempirical MOPAC method with the AM1 parameter set.
3. Record the heat of formation.

The intermediates leading to 3-bromoacetanilide and 4-bromoacetanilide can be created in a similar fashion. Record the heats of formation for these intermediates.

Using Spartan 06, the heats of formation are as follows:

ΔH_f (2-bromoacetamidobenzenium ion) = 695.6 kJ/mol
ΔH_f (3-bromoacetamidobenzenium ion) = 761.6 kJ/mol
ΔH_f (4-bromoacetamidobenzenium ion) = 681.6 kJ/mol

These results indicate that the lowest-energy, favored reaction pathway is the one that yields 4-bromoacetanilide.

Use of $\Delta\Delta H_f$ values to determine reactivity

We can also use MOPAC with the AM1 parameter set to gain insight into whether the acetamido group activates or deactivates the aromatic ring in the bromination reaction.

1. Build molecules of benzene and acetanilide.
2. Optimize the geometry of each molecule using the semiempirical MOPAC method with the AM1 parameter set.
3. Record the heat of formation for benzene and for acetanilide.

$$\Delta H_f \text{ (benzene)} = 92.1 \text{ kJ/mol}$$
$$\Delta H_f \text{ (acetanilide)} = -64.2 \text{ kJ/mol}$$

Now we can calculate the energy difference for the formation of the bromobenzenium ion intermediate in the bromination of benzene.

$$\Delta H_f \text{ (benzene to bromobenzenium ion)}$$
$$= 923.2 \text{ kJ/mol} -92.1 \text{ kJ/mol}$$
$$= 831.1 \text{ kJ/mol}$$

The bromobenzenium ion is 831.1 kJ/mol higher in energy than the starting material.

For the bromination of acetanilide, the reactive intermediate **3** is 745.7 kJ/mol higher in energy than the starting material.

$$\Delta H_f \text{ (acetanilide to 4-bromoacetamidobenzenium ion)}$$
$$= 681.6 \text{ kJ/mol} - (-64.2 \text{ kJ/mol})$$
$$= 745.8 \text{ kJ/mol}$$

The activation energy is lower for the bromination of acetanilide by over 60 kJ/mol. Thus the acetamido group activates the benzene ring toward electrophilic aromatic substitution.

You can also use the MOPAC package to calculate the positive charge distribution in the benzenium ions. The program can provide a color representation of the charge distribution. Because we do not have a palette of colors at our disposal, here are the electrostatic charge distributions at the carbon atoms of two relevant benzenium ions, as calculated by the AM1 parameter set of Spartan 06.

You can see that even in the benzenium ion itself, the positive charge is greater at the carbon atoms *ortho* and *para* to the *sp*³ carbon. The positive charge density is substantially greater at the para position of the bromoacetamidobenzenium ion, where the electron donating characteristics of the acetamido group stabilize this nearby positive charge.

Density Functional Theory (DFT)

In contrast to molecular orbital theory, the quantum mechanical *density functional theory (DFT)* optimizes an electron density rather than a wave function. Because the electron correlation energy as a function of the electron density can be included in the functional, DFT is more robust than MO theory with respect to calculating the electron-electron interaction term. DFT is very popular in the computational chemistry community and is a part of standard computational packages. The wave function approach has slightly broader utility, but DFT is often the method of choice to achieve a particular level of accuracy in the least amount of time for an average problem.

To determine a particular molecular property using DFT, such as the energy of a molecule, you need to know how the property depends on the electron density.

$$E[\rho(r)] = T_{ni}[\rho(r)] + V_{ne}[\rho(r)] + V_{ee}[\rho(r)] + E_{xc}[\rho(r)]$$

In this equation, $\rho(r)$ is the electron density at a specific position in space, and $E[\rho(r)]$ is called the *energy functional*. The electron density integrated over all space gives the total number of electrons. The equation allows the electrons to interact with one another and with an external potential—the attraction of the electrons to the nuclei.

$T_{ni}[\rho(r)]$ = the kinetic energy of the noninteracting electrons

$V_{ne}[\rho(r)]$ = the interaction of the nucleus and the electron

$V_{ee}[\rho(r)]$ = the classical electron-electron repulsion

$E_{xc}[\rho(r)]$ = the exchange-correlation energy, a combination of the correction to the kinetic energy deriving from the interacting nature of the electrons and all nonclassical corrections to the electron-electron repulsion energy

As with MO calculations, a basis set or sets for DFT is chosen to construct the density and a molecular geometry is selected. Then one guesses an initial electron density matrix and iteratively solves the basic DFT equation. After repeated iterations to minimize the ground state electronic energy and optimization of the molecular geometry, the desired molecular property can be calculated.

8.4   Which Computational Method Is Best?

The best computational method depends on the question you are asking and the resources at your disposal. Determination of molecular geometry is one of the easier aspects of computational chemistry. If you are simply trying to find the optimum (lowest energy) structures of organic molecules, molecular mechanics provides reasonable structures, and it is very fast. Accurate values for bond angles, bond lengths, dihedral angles, and interatomic distances can be determined from an optimized structure. In general, you are limited to typical organic compounds; for instance, there are few suitable parameter sets for carbon-metal bonds.

The energy differences between conformers determined by molecular mechanics are often very close to experimentally determined values, and they can be used to determine equilibrium ratios of the conformers. Because the calculations are fast, the energies of many conformers can be determined in a short time. This is especially useful when examining *rotamers,* conformations related by rotation about a single bond. As a classical mechanical model, however, molecular mechanics says nothing about electron densities and dipole moments. It also says nothing about molecular orbitals. However, the optimized structure from molecular mechanics can provide input data for other programs. Using a molecular mechanics calculation is often an efficient way to get an approximation that can be further refined with a quantum mechanical method, often saving computational time.

Semiempirical methods, which are significantly faster than *ab initio* calculations, provide reliable descriptions of structures, stabilities, and other properties of organic molecules. They are useful in calculating thermodynamic properties, such as heats of formation. The heats of formation can be used to compare energies of isomers, such as 2-methyl-1-butene and 2-methyl-2-butene, with greater accuracy than molecular mechanics may provide. The calculated heats of formation can also be used to approximate the energy changes in balanced chemical equations.

8.5   Sources of Confusion and Common Pitfalls

Computational chemistry is inherently complex, but most of the commercially available packages have been "human engineered," making it relatively easy to get started. When you get to a point in the process where you have a choice, a default option is usually provided. It is beneficial to acquaint yourself with the information provided with the package so you can make the best choices.

What Structure Should I Use to Start My Computation?

What to use for the correct structure is closely related to the method you use in building a molecule. In many packages, the user draws a two-dimensional projection, similar to the line formulas printed in a book, and the program translates it into a rough three-dimensional structure. However, if the projection is ambiguous, the program may create an unsuitable structure. For example, suppose you wanted to create *axial*-methylcyclohexane. The projection entered on the computer might look like this:

Viewing the structure created by this projection on the computer screen and then rotating it, you would probably observe a flat molecule, clearly unsuitable for optimizing the molecule's structure. To turn this projection into a three-dimensional structure usually requires some sort of "cleanup" or "beautifying" routine. This routine creates a three-dimensional structure using "normal" bond lengths and bond angles. In the case of methylcyclohexane, the structure typically becomes a cyclohexane in the chair conformation with a methyl group in an equatorial position.

Building a cyclohexane with a methyl group in the axial position usually requires the creation of the structure in stages. In this case, you need to create a chair cyclohexane and then replace one of the axial hydrogens with a methyl group. As you can see, the process involves building the framework first and then adding the necessary attachments at specific locations. Most computational chemistry packages contain templates or molecular fragments to assist in creating complex structures.

Can I Be Confident That My Calculated Structure Is the Global Minimum?

During the optimization of the geometry, the program tries to find the structural conformation with the lowest energy. At each point, it calculates the gradient or first derivative of the energy with respect to the motion of each atom in each Cartesian direction, and the geometry is perturbed in the direction of the resulting gradient vector. Each individual perturbation depends on the history of the energies and gradients from prior steps. This process is repeated until the gradient is computed to be zero, at which point a *local minimum* is likely to have been found (Figure 8.4).

FIGURE 8.4 Local and global minima resulting from energy minimization, showing that an energy profile is usually not a smooth curve with one minimum.

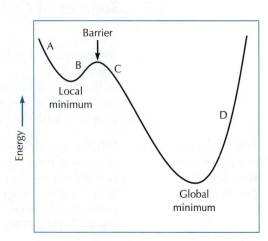

The energy surface is often uneven, with lumps, bumps, ridges, and several low spots. The low spot in which a minimization falls depends on where you start on the energy surface. In Figure 8.4, a start from point A or B will end up at the local minimum. A start at point C or D will end up at the desired *global minimum*. The calculation of *axial-* and *equatorial*-methylcyclohexane illustrates this point. The two structures are conformers that can be interconverted by way of a ring flip. *Axial*-methylcyclohexane is a local minimum and *equatorial*-methylcyclohexane is the global minimum. The barrier represents the strain energy required to flip the ring.

Systematic creation of starting structures. How do you know if a structure built with a computational chemistry package represents a local minimum or a global minimum? This question has led to many research projects. For our purposes, the answer is to create several different starting structures, carry out minimizations on each of them, and use the lowest energy as the global minimum. One of the several methods for systematically creating possible starting structures is *conformational searching*. Several conformations of a structure are created by rotating portions of the molecule connected by single bonds. Some modeling packages have routines called *sequential searching* which automate this process; in ChemBio3D this is called the *dihedral driver*. Other packages have methods such as Monte Carlo routines for generating random structures.

Molecular dynamics simulation. Yet another method of generating candidate structures for minimization is to use a molecular dynamics simulation program. This program simulates the motions of atoms within a structure. The molecule is given increased kinetic energy, the amount depending on the designated temperature. As the atoms move around, energy "snapshots" are taken at regular intervals. The structures with the lowest energies are used as starting structures for minimization. This method often propels molecules over energy barriers that are caused by steric interactions, bond strain, and torsional strain. The results of a molecular dynamics simulation can be plotted as the internal energy of a molecule versus time. In Figure 8.5, structures corresponding to low-energy conformers are designated with arrows. These conformers can be used as initial structures for energy minimizations by molecular mechanics or quantum mechanical calculations. Even using these methods, there is no guarantee that the global minimum will always be found with systems of fairly modest size. The situation is far more complex with a large molecule, such as a protein.

Does My Computational Result Represent Physical Reality?

Computational chemistry is based on theoretical models using approximations and parameter sets derived from theory and experiment. Thus it is important to keep a firm grip on reality at all times. You need to evaluate the result, especially a surprising result, and determine whether it makes sense chemically and physically, and not just accept the results of calculations as physical truth. In spite

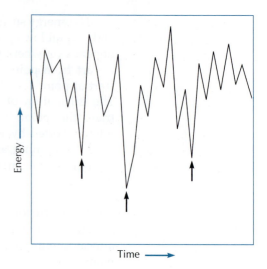

FIGURE 8.5 Output of a molecular dynamics simulation plotted as a graph of energy versus the conformation of the structure, changing with time.

of this caveat, computational chemistry is a highly valuable tool for gaining insights into organic chemistry.

Further Reading

Bachrach, S. M. *Computational Organic Chemistry,* Wiley: New York, 2007.

Cramer, C. J. *Essentials of Computational Chemistry: Theories and Models,* 2nd ed.; Wiley: New York, 2004.

Hehre, W. J. *A Guide to Molecular Mechanics and Quantum Chemical Calculations;* Wavefunction, Inc.: Irvine, CA, 2003.

Hehre, W. J.; Shusterman, A. J.; Huang, W. W. *A Laboratory Book of Computational Organic Chemistry,* Wavefunction, Inc.: Irvine, CA, 1998.

Questions

1. Which computational chemistry method, MM or QM, should be used to solve the problems posed in (a)–(e)? MM = molecular mechanics methods and QM = quantum mechanical methods.

   (a) Compare the heat of formation values for isomeric structures.

   (b) Predict the difference in energy between diaxial *cis*-1,3-dimethylcyclohexane and diequatorial *cis*-1,3-dimethylcyclohexane.

   (c) Predict the highest occupied molecular orbital of a compound.

   (d) Predict the charge distribution of a reaction intermediate.

   (e) Determine the lowest energy conformation of cyclodecane.

2. Which computational chemistry method would be most applicable for a study of kinetic versus thermodynamic control of product formation in a reaction? Explain why.

3. The reaction of (S)-(+)-carvone with $NaBH_4$ in methanol could produce compound **1** or compound **2** or a mixture of **1** and **2**.

(a) Add the missing C_3H_5 and OH substituents to the template provided to illustrate one possible chair-like conformation for each compound.

(b) If computational chemistry were used to determine if compound **1** or compound **2** is the most likely product (the lowest-energy conformation) in each reaction, would MM2 or MOPAC be the most appropriate method to use?

(c) How many chair-like conformations should be submitted for energy minimization? Describe their alcohol and alkene substituent orientations as axial or equatorial.

(d) What structure(s) would you input for compound **1**?

(e) What structure(s) would you input for compound **2**?

4. When two different students performed the computational chemistry exercise described in Question 3, they noticed that for several conformations their values were different. What problem is most likely occurring and how could the students make sure they obtained the lowest energy value?

5. Reduction of 3,3,5-trimethylcyclohexanone with sodium borohydride yields a mixture of *cis*-3,3,5-trimethylcyclohexanol and *trans*-3,3,5-trimethylcyclohexanol. Use molecular mechanics to determine the most stable conformer of each product.

6. Adamantane is a tetracyclic hydrocarbon, $C_{10}H_{16}$, incorporating four chair cyclohexane rings. Twistane is an isomeric tetracyclic hydrocarbon incorporating four twist-boat cyclohexane rings. Use semiempirical MOPAC calculations with the AM1 parameter set to optimize the geometries of adamantane and twistane. Record their heats of formation.

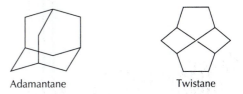

Adamantane   Twistane

Hints for Construction of the Molecules

Adamantane: Start with the chair conformation of cyclohexane. Attach carbon atoms to the three axial positions on the same side of the ring, attach a carbon atom to one of the three axial carbons atoms, and then make bonds between the newly attached carbon atom and the remaining two axial carbon atoms.

Twistane: Start with the twist-boat cyclohexane. Attach carbon atoms to the pseudo-axial positions at the 1,2,4,5 carbons of the ring, make a bond between the carbon atoms added at the 1 and 4 positions, and finally make a bond between the carbon atoms added at the 2 and 5 positions.

3

Basic Methods for Separation, Purification, and Analysis

Essay—Intermolecular Forces in Organic Chemistry

The structures of organic molecules and the making and breaking of covalent bonds in chemical reactions are the major focuses of classroom work in organic chemistry. After a discussion of intermolecular forces, mainly in the context of boiling points, the emphasis is on covalent bond chemistry. Except for hydrogen bonds, weak intermolecular forces may seem largely unimportant; however, many experimental techniques of organic chemistry—for example, the separation and purification of organic compounds—depend almost entirely on the weak forces between molecules.

Several categories of intermolecular interactions are listed here from strongest to weakest:

- Hydrogen bonding
- Dipole-dipole interactions
- Dipole-induced dipole interactions
- Induced dipole-induced dipole interactions

These electrostatic intermolecular forces are all concerned with favorable enthalpy changes that occur when molecules attract one another.

Hydrogen Bonding

Hydrogen bonding, often called H-bonding, occurs when hydrogen atoms are covalently attached to highly electronegative elements. Hydrogen atoms attached to atoms of these elements—most important are oxygen and nitrogen—can have reasonably strong electrostatic interactions, as well as weak orbital overlap, with electronegative atoms in nearby molecules. These interactions form intermolecular hydrogen bonds, whose energies are on the order of 15–20 kJ/mol (3.5–5 kcal/mol). This range of energies is only about

5% of the energy associated with covalent bonds, but it is enough to make hydrogen bonds the strongest of the weak intermolecular forces.

Perhaps the most dramatic example of intermolecular interactions by hydrogen bonding occurs between molecules of water. The high boiling point of water is an indication of the substantial intermolecular forces between water molecules. H_2O boils at 100°C whereas CH_4, whose molecules are approximately the same size, boils at −162°C. H_2O also boils over 160° higher than H_2S, which has a higher molecular weight and surface area. An intermolecular H-bonding network also gives ice an open tetrahedral structure, which makes ice a very unusual solid. Usually solids are more dense than their liquid phases, but ice floats because it is less dense than the liquid phase of water. Planet Earth would be a very different place without liquid water and floating ice.

Organic molecules that have hydrogen atoms covalently bonded to oxygen or nitrogen can also form H-bonds with water molecules or with other organic molecules that have oxygen or nitrogen atoms in them. Because C–C and C–H bonds are nonpolar, the majority of organic compounds are "greasy" and insoluble or only slightly soluble in water. Low-molecular-weight compounds (of about five or fewer carbons) that contain H-bonding functional groups are soluble in water, with the degree of solubility decreasing as the number of carbon atoms increases. Methanol (CH_3OH), ethanol (CH_3CH_2OH), and acetic acid (CH_3CO_2H) are completely *miscible* in water, which means they form a homogeneous solution with water when mixed in any proportion.

Dipole-Dipole and Dipole-Induced Dipole Interactions

Water is also distinguished by its polarity due to the relatively large charge separation in the polar O–H covalent bonds of water molecules. Just as bonds can be polar, entire molecules can be polar, depending on their shape and the nature of their bonds. Water has a large permanent dipole moment, which allows it to dissolve many inorganic and organic salts but not most organic molecules.

Molecules that have dipole moments can attract one another when their dipoles align, so there is an electrostatic attraction between them.

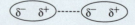

Some molecules, such as acetone (CH_3COCH_3), dimethyl sulfoxide (CH_3SOCH_3), and acetonitrile (CH_3CN), have significant dipoles—even though they have no hydrogen atoms that can H-bond with other molecules—which makes them polar solvents and miscible with water. Moreover, they can accept H-bonds from water, contributing to their water miscibility.

Molecules that have dipole moments can also induce dipoles in other nearby molecules that do not have dipole moments of their own. This process provides an attractive force, although it is usually not as great as the one provided by dipole-dipole interactions.

Induced Dipole-Induced Dipole Interactions

The weakest intermolecular interactions are induced dipole-induced dipole interactions, often called London dispersion forces. These intermolecular forces result from temporary charges on molecules due to fluctuations in the electron distribution within them. All covalent molecules exhibit this induced dipole-induced dipole polarization. The magnitude of these dispersion forces depends on how easily the electrons in a molecule can move in response to a temporary dipole in a nearby molecule, called *polarizibility*.

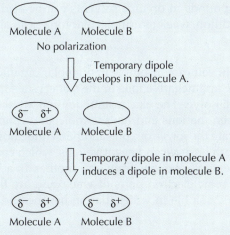

London dispersion forces are the only intermolecular interactions that attract alkane molecules to their neighbors. They play a major role in the structure of lipid bilayer membranes, where fatty acids having linear alkane chains of 11–19 CH_2 groups closely pack together to form the membrane.

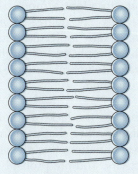

Diagram of a bilayer membrane. The fatty acid chains are attached as esters to molecules of glycerol, which also have ionic phosphates attached, shown as blue circles.

Van der Waals Forces

All weak intermolecular forces, with the exception of hydrogen bonds, are often referred to as van der Waals forces. The magnitude of van der Waals interactions depends on the surface areas of the interacting molecules. Thus, larger-molecular-weight compounds have higher boiling points, and isomers whose molecular shapes lead to larger surface areas also have higher boiling points.

When very large molecules interact, a combination of many hydrogen bonds and van der Waals electrostatic forces can produce a large cumulative effect with strong association between the molecules. These intermolecular forces can also occur between different portions within very large molecules. For example, they determine the three-dimensional shapes of proteins and nucleic acids (DNA and RNA).

Solubility

The solubility of one compound in another is governed both by intermolecular forces (enthalpy) and by entropy. Just as the distribution of 10 red M & Ms throughout a jar of blue M & Ms has greater entropy than a layer of 10 red M & Ms on top of the blue M & Ms, it is usually entropically favorable for solute molecules to be completely separated from one another and randomly distributed in a sea of solvent molecules. The solubility of most organic compounds in organic solvents can be attributed to the favorable entropy of mixing. In addition, weak intermolecular forces can also produce a favorable enthalpy of mixing.

Organic and inorganic salts are insoluble in nonpolar organic solvents, such as hexane, because the positive and negative ions in the salt crystals must be separated from each other in order to dissolve. The electrostatic ion-ion attraction is strong, and the weak interactions between the ions and hexane molecules cannot compensate for the energy required to separate the ions from one another. Water, however, has quite strong ion-dipole forces with both positive and negative ions, which can often compensate for the energy required to separate the ions from one another. In addition, cations such as Na^+ and K^+ attract the basic electron pairs of water and are thereby solvated. Thus, ionic salts are much more soluble in water than in hexane and in most other organic liquids.

Water and nonpolar organic solvents, such as hexane, do not dissolve in one another because nonpolar compounds cannot participate in hydrogen bonding and dipole-dipole interactions with water. Introducing a nonpolar material into water also disrupts the dynamic three-dimensional network of H-bonding interactions among water molecules and forces them to adopt a more structured two-dimensional cage surrounding the nonpolar molecules. The degree of order required for water to form solvent cages around individual hexane molecules reduces the freedom of motion of the water molecules far more than simply excluding the hexane molecules. Therefore, oily, nonpolar substances and water do not intermingle but instead interact to minimize the surface area of contact between them.

The *hydrophobic effect* describes the lack of solubility of greasy, nonpolar compounds in water, such as the linear alkyl chains in the lipid bilayers of membranes. Because water is the solvent used in nature, this phenomenon plays an important role in biological structure and reactivity. You are probably most familiar with the hydrophobic effect as the separation of oil and vinegar in salad dressing, in oil spills in the ocean, and as oil floating on top of water in a parking lot after it rains. The loss of entropy, which occurs when water forms ordered cages around nonpolar molecules, is the accepted thermodynamic rationale for the hydrophobic effect.

Acid-Base Effects on Solubility

Because acid-base reactions usually involve the transfer of protons between neutral and ionic species, pH affects the solubility of organic acids and bases in water. For example, at pH 9, solid benzoic acid is deprotonated to produce a water-soluble benzoate salt.

$$C_6H_5CO_2H(solid) + Na^+(aq) + OH^-(aq) \rightarrow C_6H_5CO_2^-(aq) + Na^+(aq) + H_2O$$

Benzoic acid   Benzoate anion

Acidification of the benzoate anion with sulfuric acid to pH 1 forms neutral benzoic acid, which is far less soluble in water and precipitates out of solution as a solid. Therefore, you can isolate benzoic acid from a basic aqueous solution by reducing the pH and filtering the resultant precipitate. Similar, but opposite, pH effects on solubility take place with amines, which are neutral molecules at high pH and ionic salts at low pH.

Intermolecular Forces in Separation and Purification

Part 3 is concerned mainly with the techniques that organic chemists use to separate liquids from other liquids by extraction and distillation and to separate solids from liquids by crystallization and filtration. Understanding the techniques of separation and purification of organic compounds depends on understanding the intermolecular interactions of liquids and solids.

9

FILTRATION

Filtration is an important technique for the physical separation of solids and liquids. Anyone who has brewed coffee has probably carried out a filtration to separate the liquid coffee solution from the coffee grounds. Filtration has several purposes in the organic laboratory:

- To separate a solid product from a reaction mixture or recrystallization solution
- To remove solid impurities from a solution
- To separate a product solution from a drying agent after an aqueous extraction

The miniscale filtrations commonly performed in the organic laboratory use conical funnels and Erlenmeyer flasks for gravity filtrations and either Buchner or Hirsch funnels and filter flasks for vacuum filtrations. All three types of funnels require the use of filter paper to separate the solid from the liquid in the mixture undergoing filtration, unless a funnel with a glass frit is used. The liquid that passes through the filter paper or glass frit is called the *filtrate*. The solid is called the *precipitate*. Microscale gravity filtrations are usually done with a Pasteur pipet packed with either cotton or glass wool. Microscale vacuum filtrations use smaller versions of the miniscale equipment. When and how to use each filtration method is explained in this chapter.

Although they are not strictly filtration techniques, decantation and centrifugation can also be used to separate solids from liquids in the organic laboratory.

9.1 Filtering Media

In any filtration, there needs to be a filtering medium that traps the solid being separated from its accompanying liquid. A variety of filtering media—filter paper, cotton, glass wool, micropore filters, and finely powdered solids called filter aids—are described in this section.

Filter Paper

Filter paper is used for both gravity and vacuum filtrations. For most filtrations performed in the introductory organic lab, a paper that provides medium filtering speed is satisfactory. Whatman is the major producer of filter paper for qualitative applications, and its various grades are listed in Table 9.1. Whatman No. 2 filter paper works well for both gravity and vacuum filtrations.

Special-purpose filter papers are also available. For example, when a filtrate contains the desired product and the solid being filtered is a by-product, a fast, hardened filter paper, such as Whatman 54, can be used. When an emulsion forms during an extraction, vacuum filtration through phase separator filter paper, such as Whatman 1PS, will usually break the emulsion.

TABLE 9.1	Some Whatman qualitative filter paper types with their approximate relative speed and retentivity	
Type number	Relative speed	Particle retention (μm)
Whatman 2	medium	> 8
Whatman 3[a]	medium-slow	> 6
Whatman 4	very fast	> 20–25
Whatman 5	slow	> 2.5
Whatman S & S 595	medium-fast	> 4–7

a. Thick—good for Buchner and Hirsch funnels.

Fluted Filter Paper

Fluted filter paper provides a larger surface for liquid-solid separations, which facilitates faster gravity filtration than does the usual filter paper cone. Faster filtration is especially important when filtering insoluble impurities from a hot recrystallization solution in order to prevent crystallization as the solution cools. Vacuum filtration does not work well for a hot solution because much of the solvent can be lost to evaporation and because the solution cools too rapidly, leading to premature crystallization.

To make a fluted filter, crease a regular filter paper in half four times (Figure 9.1a). Then fold each of the eight sections of the filter paper inward, so that it looks like an accordion (Figure 9.1b). Finally, open the paper to make a fluted cone, as illustrated in Figure 9.1c. Alternatively, commercially available filter paper already folded in this manner can be used.

Glass Fiber Filters

Glass fiber filter circles can be used instead of paper filters for vacuum filtration with Buchner or Hirsch funnels, which have a horizontal surface perforated with small holes. The filters are available in a wide range of sizes: 13–24-mm circles work well with Hirsch funnels; larger sizes can be used with Buchner funnels. Although glass fiber filters are more expensive than cellulose filter papers, they are particularly useful if the particles of the solid being filtered are very small.

Cotton and Glass Wool

Cotton or glass wool can be packed into a Pasteur pipet to make a useful filter in small-scale and microscale filtrations. The preparation and use of Pasteur filter pipets are described in Section 9.3.

Micropore Filters

Samples for instrumental analysis by NMR spectroscopy, polarimetry, or high-pressure liquid chromatography may contain very fine particles that would interfere with a measurement. A micropore filter will remove particles as small as 0.5 μm.

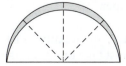 

FIGURE 9.1
Fluting filter paper.

(a) Crease filter paper.   (b) Fold each quarter inward.   (c) Fluted filter paper.

Draw a liquid sample into a syringe and attach a micropore filter to its end. Invert the syringe so the filter points upward, and push the syringe plunger just enough to force a few drops through the filter. Then reposition the filter pointing down and place it over a receiving vial. Press the plunger to force the solution through the filter into the vial. This filtered sample is then ready for analysis.

Use of a Filter Aid

Occasionally, you may encounter a mixture containing very fine particles of a by-product or other unwanted solid material that passes through filter paper or clogs filter paper pores and impedes filtration of the desired material. A filter aid such as Celite® facilitates the separation. Celite is a trade name for diatomaceous earth—a finely divided inert material derived from phytoplankton skeletons—which neither clogs the pores of filter paper nor passes through it. A filter aid should be used **only** for a mixture where the filtrate will contain the desired material and the solid adhering to the filter aid will be discarded.

In miniscale procedures, Celite may be added to a reaction mixture before vacuum filtration if the mixture contains a large quantity of unwanted fine particles that could clog the filter paper. In microscale procedures, the separation of fine particles of unwanted material from a liquid mixture is more easily carried out with a Pasteur pipet packed with silica gel or alumina as the filter aid.

9.2   Gravity Filtration

Gravity filtrations are used in the organic laboratory for several purposes: during a recrystallization where the desired product is completely dissolved in a hot solution but insoluble impurities remain, when activated charcoal is used to adsorb colored impurities present in a hot recrystallization solution, and to remove a drying agent from an organic solution.

Carrying Out a Gravity Filtration

The following procedure requires a minimum of 15 mL of liquid. Place a fluted filter paper (see Section 9.1) in a clean, short-stemmed funnel and put the funnel into the neck of a clean Erlenmeyer flask or, if the liquid will be distilled after filtration, into a round-bottomed flask. Wet the filter paper with a small amount of the solvent in the mixture so that the paper adheres to the conical funnel. When the liquid volume is less than 15 mL, the Pasteur filter pipet method described in Section 9.3 will prevent significant losses.

Filtering a room-temperature liquid. If the mixture is at room temperature, you can simply pour it into the filter paper and allow it to drain through the paper into an Erlenmeyer flask. Then add a few milliliters of the solvent to wash through any product that may have adhered to the filter paper.

Filtering a hot solution. If the mixture is a hot solution containing a dissolved solid, you must take precautions to prevent the solid from crystallizing during the filtration process. Add a small amount of the recrystallization solvent to the receiving flask (1–10 mL depending on the size of the flask). Then heat the flask, funnel, and solvent on

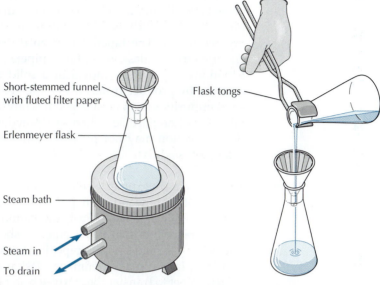

Short-stemmed funnel with fluted filter paper

Erlenmeyer flask

Steam bath

Flask tongs

FIGURE 9.2
Filtering solid impurities from a recrystallization solution.

Steam in

To drain

1. Heat receiving flask and funnel.

2. Pour hot solution through fluted filter paper.

a steam bath (Figure 9.2, Step 1) or clamp the flask in a water bath that is being heated on a hot plate in a hood. The hot solvent warms the funnel and helps prevent premature crystallization of the solute during filtration. If the steam bath is large enough, keep both flasks hot during the filtration process; if it is too small for both, keep the unfiltered solution hot and set the receiving flask on the bench top. Pour the hot recrystallization solution through the fluted filter paper (Figure 9.2, Step 2).

With recrystallizations from water or a water-ethanol mixture, the heating can be carried out directly on a hot plate. Do not use a hot plate with low-boiling flammable organic solvents, however, because the solvent may burst into flames if it comes into direct contact with a hot surface.

SAFETY PRECAUTION

Lift a hot Erlenmeyer flask with flask tongs.

9.3   Small-Scale Gravity Filtration

Pasteur pipets are excellent for filtration of small quantities of liquid when the tapered portion of the pipet is packed with glass wool or cotton.

SAFETY PRECAUTION

Glass Pasteur pipets are puncture hazards. They should be handled and stored carefully. Dispose of Pasteur pipets in a "sharps" box or in a manner that does not present a hazard to lab personnel or house-keeping staff. Check with your instructor about the proper disposal method in your laboratory.

Pasteur Filter Pipets

When a small amount of an organic liquid or solution needs to be separated from a solid that will be discarded, a Pasteur filter pipet provides the necessary filtration with minimal loss of the organic liquid. The tapered portion of the pipet is packed with either cotton or glass wool. Filter pipets are useful in a variety of situations—from the removal of a solid drying agent or a solid reaction by-product from an organic solution to the removal of solid impurities from liquid spectroscopy samples. If the solid contains very fine particles, such as a powdered catalyst, using glass wool or cotton may not provide sufficient filtration and a Celite filter pad is added.

Preparing and Using a Pasteur Filter Pipet

To prepare a filter pipet, use a pair of tweezers to pick up a small amount of cotton and then push it down into the pipet with a wooden applicator stick. Pack the cotton firmly into the bottom of the tapered portion of the pipet as shown in Figure 9.3. Use a microclamp to hold the filter pipet in a vertical position for the filtration and place a small Erlenmeyer flask underneath it. Use another Pasteur pipet to transfer the mixture being filtered to the filter pipet. The drying agent or solid impurities will adhere to the cotton. If the concentration of the liquid solution is not an important constraint, use a clean Pasteur pipet to add 1–2 mL of fresh solvent to the filter pipet to rinse all desired material from it and collect the rinse in the same Erlenmeyer flask.

S A F E T Y   P R E C A U T I O N

Wear gloves and use tweezers to handle glass wool.

Preparing a Celite Filter Pad in a Pasteur Pipet

Pick up a small amount of glass wool with tweezers and tightly pack it into the tapered portion of a Pasteur pipet using a wooden applicator stick as shown in Figure 9.3. Continue packing small portions until an approximately 2-cm depth is reached. Add approximately 1 cm of Celite on top of the glass wool to ensure efficient entrapment of very fine particles (Figure 9.4). Use a microclamp to hold the pipet and position the receiving container underneath it.

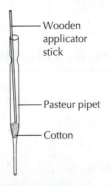

Wooden applicator stick

Pasteur pipet

Cotton

FIGURE 9.3
Pasteur filter pipet.

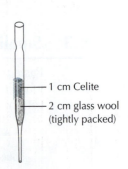

1 cm Celite

2 cm glass wool (tightly packed)

FIGURE 9.4
Pasteur filter pipet packed with a Celite filter pad.

9.4 Vacuum Filtration

Vacuum filtration is used to rapidly and completely separate a solid from the liquid with which it is mixed. The recovery of a crystallized product from a recrystallization procedure is a common application of vacuum filtration in the organic chemistry lab. Vacuum filtration is also employed when it is necessary to use a filter aid, such as Celite, to remove very finely divided insoluble solids from a solution; in this instance, it is the solution, not the solid, that is the desired product.

The vacuum source for a filtration can be either a water aspirator or a compressor-driven vacuum system. **Heavy-walled tubing must be used in vacuum filtration** so the tubing will not collapse from atmospheric pressure on the outside when the vacuum is applied. **If the tubing collapses, the vacuum filtration will not work.**

Funnels for Vacuum Filtration

Most funnels used for vacuum filtration have a flat, perforated or porous plate that holds filter paper to retain the solid from its accompanying liquid. They are made from porcelain, glass, or plastic. Figure 9.5 shows a porcelain Buchner funnel, a porcelain Hirsch funnel, and a plastic Hirsch microscale funnel with an integral adapter. Both Buchner and Hirsch funnels are available in a variety of sizes; select a size appropriate for the amount of material you will be collecting. For example, if you are filtering a mixture that contains 1–3 g of solid, use a 78- or 100-mm-diameter Buchner funnel. For filtering a mixture containing 0.2–1 g of solid, select a 43-mm-diameter Buchner funnel or a 16-mm Hirsch funnel. For microscale filtrations, use an 11-mm Hirsch funnel or a microscale plastic Hirsch funnel.

When using a Buchner or Hirsch funnel with perforations, **it is crucial to select the correct size of filter paper** for the funnel you are using (see Table 9.1). The paper must lie flat on the perforated plate and **just cover all the holes in the plate but not curl up the side.**

Miniscale Apparatus for Vacuum Filtration

The apparatus for a miniscale vacuum filtration consists of a Buchner funnel (or medium-size Hirsch funnel), neoprene adapter, filter flask, and trap flask or bottle (Figure 9.6). A trap flask is placed between the vacuum source and the filter flask to prevent back flow of water into the filter flask when a water aspirator is the vacuum source. With a compressor-driven vacuum system, the trap flask keeps any overflow from the filter flask out of the vacuum pump. **Both the filter flask and the trap flask must be firmly clamped to prevent the apparatus from tipping over.** The neoprene adapter ensures a tight seal between the filter flask and the Buchner funnel.

FIGURE 9.5 Funnels used for vacuum filtration.

Perforated plate

Perforated plate

Porous frit

Integral adapter

(a) Buchner funnel   **(b)** Hirsch funnel   **(c)** Plastic Hirsch funnel

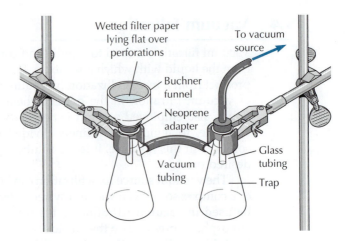

FIGURE 9.6
Apparatus for vacuum filtration.

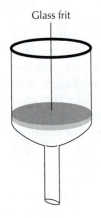

FIGURE 9.7
Buchner funnel with fritted glass disc.

Heavy-walled tubing is used to connect the vacuum line and filtration flask to prevent collapse of the tubing from atmospheric pressure when the vacuum is applied.

Place a piece of appropriately sized filter paper in the Buchner funnel and wet the paper with a small amount of the solvent present in the mixture being filtered. Turn on the vacuum source to pull the paper tightly over the holes in the funnel and then **immediately** pour the mixture into the funnel. At the end of the filtration, hold the filter flask firmly with one hand and **use the other hand to tip the Buchner (or Hirsch) funnel slightly to the side to break the seal before turning off the vacuum source.**

Unlike funnels with perforations and filter paper, funnels for vacuum filtration can also have a fritted glass disc, which need no filter paper (Figure 9.7). The same procedures are used with a fritted-glass funnel as described above for funnels with perforations except that no filter paper is used.

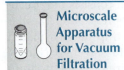

Microscale Apparatus for Vacuum Filtration

Microscale vacuum filtrations use a small porcelain Hirsch funnel with a 25-mL filter flask or an 18×150-mm side-arm test tube with a neoprene adapter assembled as shown in Figure 9.8a. When a plastic Hirsch funnel with an integral adapter is used, the funnel is simply inserted into a 25-mL filter flask; no neoprene adapter is used (Figure 9.8b).

Water aspirators and vacuum pumps sometimes provide a more powerful vacuum than is necessary for microscale vacuum filtrations and this can lead to the loss of valuable compounds. An alternative way to carry out a microscale vacuum filtration is to attach a 50-mL or 100-mL syringe to the side arm of the filter flask.* Withdrawing the syringe plunger can provide a gentle and effective vacuum (Figure 9.9).

A microscale filtration apparatus should always be firmly clamped at the neck of the filter flask or side-arm test tube; the

*Zhikin, D. M.; Kjonaas, R. A. *J. Chem Educ.*, **2013**, *90*, 142–143.

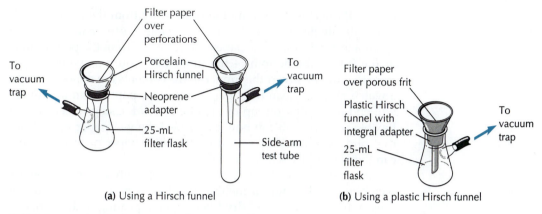

(a) Using a Hirsch funnel   (b) Using a plastic Hirsch funnel

FIGURE 9.8 Microscale apparatus for vacuum filtration.

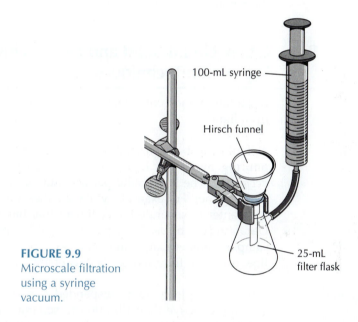

FIGURE 9.9
Microscale filtration
using a syringe
vacuum.

apparatus tips very easily when it is attached to the heavy-walled rubber tubing leading to the vacuum source. Place an appropriately sized filter paper or glass fiber filter in the Hirsch funnel so that it lies flat and just covers the holes in the funnel. Wet the filter paper with a small amount of the solvent present in the mixture being filtered. Use the vacuum source to pull the paper tightly over the holes in the funnel and then **immediately** pour the mixture into the funnel. At the end of the filtration, hold the filter flask firmly with one hand and **use the other hand to tip the Hirsch funnel slightly to the side to break the seal before turning off the vacuum source.**

Recovering Crystals After a Vacuum Filtration

Once the vacuum seal has been broken, the funnel can be separated from the filter flask or side-arm test tube. The easiest way to recover the solid is to invert the funnel over a clean watch glass or piece of weighing paper and allow the solid to pour onto the watch glass or paper. Not all of the solid will be released from the paper inside the funnel. Rather than scraping the solid away from the paper, which can contaminate the product with bits of paper, use a narrow flexible spatula to reach *behind* the paper and release it from the perforations or the frit on the bottom of the funnel. As you remove the paper from the funnel, invert the funnel so the contents are deposited onto a watch glass or clean filter paper. The filter paper from the funnel will end up resting on top of the pile of solid. Using forceps and a flexible spatula, you can now easily peel the wet paper away from the solid. Any remaining solid on the filter paper or in the funnel can then be gently scraped onto the pile.

If the solid is wet, allow it to dry (preferably overnight) before weighing or obtaining a melting point. If it is very wet, you can press it between pieces of dry filter paper, repeating with clean, dry paper until the bulk water or other solvent is removed. Then allow the remaining water or solvent to evaporate.

9.5   Other Liquid-Solid and Liquid-Liquid Separation Techniques

Decantation and centrifugation can also be used to separate solids from liquids.

Decantation

A liquid can be separated from a few large particles by **carefully pouring** away the liquid above the particles—a process called *decanting.* The large, solid particles stay in the bottom of the original container. For example, vintage wines are carefully poured into a decanter to separate the sediment that has formed in the bottle. Decanting can be used to separate a liquid from boiling stones. If the sample contains a large number of solid particles or the particles are fine, however, filtration is a better separation method.

Centrifugation

When a sample contains suspended particles, centrifugation may be more effective than filtration to separate the solid and the liquid. Centrifugation is also used to break liquid-liquid emulsions in microscale extractions. In fact, a microscale extraction is frequently carried out in a centrifuge tube to remove the lower layer with a Pasteur pipet, and if an emulsion forms, the tube can be spun in a centrifuge to separate the liquid phases.

During operation of a centrifuge, the sample tube **must be counterbalanced** by another centrifuge tube filled with an equal volume of liquid. A centrifuge containing unbalanced tubes vibrates excessively and noisily and may move around on the bench top. A balanced centrifuge makes a steady, uniform noise at full speed.

9.6   Sources of Confusion and Common Pitfalls

Much of the confusion regarding filtration arises in determining which method to select for a specific situation. As a general guide in miniscale experiments, if a solution contains unwanted solid material, use gravity filtration to separate the mixture; if the desired product is a solid, use vacuum filtration to recover it from the liquid. In microscale experiments, use a Pasteur filter pipet to remove unwanted solid; if the desired material is a solid, use vacuum filtration to recover it.

My Filtrate Looks Cloudy! What Should I Do?

Incomplete separation in a gravity filtration is probably caused by using the wrong type of filter paper. Tiny solid particles can go through filter paper designed for coarse solids. In vacuum filtrations, using filter paper of the wrong diameter can allow both the liquid and the solid particles to creep around the edges, which will lead to incomplete separation. There is a simple solution: Adjust the paper size or type and filter again.

If you used water to wet the filter paper for the vacuum filtration of a dry organic solution, water will contaminate the filtrate and probably result in a cloudy solution. It will also cause the solid in the funnel to be wet. The only remedy is to dry the filtrate or the filtered solid—whichever contains your desired compound. Think twice before you "wet" the paper and choose the solvent present in the mixture you plan to filter.

Sometimes people neglect to clean out a filter flask between different vacuum filtrations, resulting in mixtures of solutions in the filtrate that may cause solids to precipitate. It is always a good idea to start with a clean filter flask before filtering.

During a vacuum filtration, the filtered solution can evaporate and become cold. This will change the solubility of the dissolved solute and may cause it to crystallize or precipitate from the filtrate. If you are not collecting the filtrate, you usually don't need to worry about these solids.

My Funnel or Filter Paper Seems Clogged. What Should I Do?

Having liquid in the funnel that won't pass through the filter in a gravity filtration is perhaps the most frustrating part of any filtration. The pores in the filter paper can become clogged if paper of the wrong porosity is used. The answer to the problem usually is to interrupt the filtration and start over, using filter paper designed for the particle size the solid contains. If you are not collecting the solid, use a filter aid. In some cases, using a centrifuge for the separation may be more feasible.

Lack of suction in a vacuum filtration is usually caused by the collapse of thin-walled rubber tubing not designed for use with a vacuum. Replace the hoses with thick-walled vacuum tubing. Lack of suction could also be due to an inefficient vacuum system caused by insufficient power in the vacuum pump or water aspirator or by a leak in the system.

Sometimes during a vacuum filtration, the solvent evaporates so fast that dissolved solute precipitates in the paper or frit and clogs the funnel. If this happens, get a fresh funnel and paper and use a *passive vacuum* for the filtration. A passive vacuum is produced by briefly evacuating the filter flask and then isolating it from the vacuum source by pinching the vacuum tubing. This will retain a vacuum inside the filter flask but not continuously evacuate it. Periodically re-evacuate the filter flask and again isolate it from the vacuum source until the filtration is complete.

Questions

1. Why would a Hirsch funnel be more effective than a Buchner funnel for a small-scale vacuum filtration?
2. Explain the advantage that fluted filter paper has in a gravity filtration.
3. Why should a hot recrystallization solution be filtered by gravity rather than by vacuum filtration?
4. Explain why the filter flask can become quite cold to the touch during a vacuum filtration.

5. Why must the seal be broken in a vacuum filtration before the flow of water to a water aspirator is turned off?
6. To perform each of the following tasks, which type of filtration apparatus would you use?
   (a) Remove about 0.3 g of solid impurities from 5 mL of a liquid.
   (b) Collect crystals obtained from recrystallizing an organic solid from 20 mL of solvent.

CHAPTER

10   EXTRACTION

If Chapter 10 is your introduction to the separation and purification of organic compounds, read the Essay "Intermolecular Forces in Organic Chemistry" on pages 127–131 before you read Chapter 10.

Extraction is one of the most important techniques used in the separation and purification of organic compounds. Extractions are a major part of the workup procedure for isolating and purifying the products of organic chemical reactions. For example, water-insoluble organic compounds can be separated from aqueous mixtures by extracting them into organic solvents. Water-soluble salts and polar organic compounds are left behind in the water. Extractions are also used in food processing. An extraction process removes caffeine to produce decaffeinated coffee, and an extraction of vanilla beans with ethanol results in vanilla extract.

In an extraction a solution is mixed thoroughly with a second solvent that is *immiscible (insoluble)* with the first solvent, usually by shaking them together. When the two liquids are mixed together, a compound is transferred into the solvent in which it is more soluble. Carrying out two or three extractions of a water mixture with an organic solvent usually serves to separate and purify a desired organic compound.

The process of *liquid-liquid extraction* involves the distribution of a compound *(solute)* between two solvents that are immiscible in each other. Generally, although not always, one of the solvents in an extraction is water and the other is a much less polar organic solvent, such as diethyl ether, ethyl acetate, hexane, or dichloromethane. It is important to remember that immiscible liquids do not dissolve in one another. They form two liquid layers. You cannot extract a compound from water into ethanol because ethanol is miscible with water. If you add ethanol to water and shake up the mixture, you end up with only one layer! By taking advantage of the differing solubilities of a solute in a pair of immiscible solvents, however, compounds can be selectively transported from one liquid phase to the other during an extraction. The essay on intermolecular forces in organic chemistry that opens Part 3 describes the dipole-dipole forces between molecules and the structural factors that determine the solubility characteristics of organic compounds.

10.1 Understanding How Extraction Works

Aqueous Extractions

In a typical extraction procedure, an *aqueous phase (water)* and an immiscible organic solvent, often called the *organic phase,* are shaken in a separatory funnel (Figure 10.1). The solutes distribute themselves between the aqueous layer and the organic layer according to their relative solubilities. Inorganic salts generally prefer the aqueous phase, whereas most organics dissolve more readily in the organic phase. Two or three extractions of an aqueous mixture often suffice to almost quantitatively transfer a nonpolar organic compound to an organic solvent. Separation of low-molecular-weight alcohols or other polar organic compounds may require additional extractions or a different approach.

If at the end of an organic reaction you have an aqueous mixture containing the desired organic product and a number of inorganic by-products, extraction with an organic solvent immiscible with water can be used to separate the organic product from the by-products. The separatory funnel initially contains the aqueous reaction mixture (Figure 10.2a). When an organic solvent less dense than water is added to the separatory funnel and the funnel is stoppered and shaken to mix the two phases, the separated liquids would appear as shown in Figure 10.2b. Then the lower aqueous layer can be drained from the separatory funnel, leaving the organic layer containing the desired product in the funnel (Figure 10.2c). The separation of organic product and inorganic by-products normally is not entirely complete because the organic compound may have a slight solubility in water and the inorganic by-products may have a slight solubility in the organic solvent.

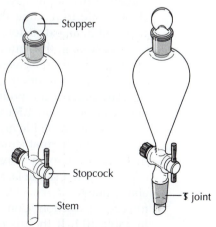

FIGURE 10.1
Funnels for extractions.

(a) Separatory funnel

(b) Dropping funnel, which can be used as a separatory funnel

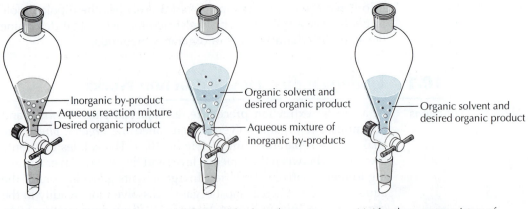

(a) Aqueous mixture of organic product and inorganic by-products
- Inorganic by-product
- Aqueous reaction mixture
- Desired organic product

(b) Most of the desired organic product has been transferred to the organic solvent.
- Organic solvent and desired organic product
- Aqueous mixture of inorganic by-products

(c) After the aqueous mixture of inorganic by-products has been drained from the separatory funnel, the organic solvent solution of the desired product remains in the separatory funnel.
- Organic solvent and desired organic product

FIGURE 10.2 Using extraction to separate an organic compound from an aqueous mixture.

Distribution Coefficient

When an organic compound is distributed or partitioned between an organic solvent and water, the ratio of solute concentration in the organic solvent, C_1, to its concentration in water, C_2, is equal to the ratio of its solubilities in the two solvents. The distribution of an organic solute, either liquid or solid, can be expressed by

$$K = \frac{C_1}{C_2} = \frac{\text{g compound per mL organic solvent}}{\text{g compound per mL water}} \tag{1}$$

K is defined as the *distribution coefficient,* or *partition coefficient.*

To reach equilibrium in the partition of organic compounds between an organic solvent and water, there has to be a thorough mixing process, which can be difficult to achieve. Vigorous shaking in a separatory funnel can lead to emulsions (see Section 10.3) which make the two liquid layers difficult to separate. Although it is not always safe to assume that an equilibrium distribution has been achieved in an extraction, there are important insights to be gained from discussing the equilibrium model. It is analogous to using the ideal gas model for gases, even though we know that it does not apply to every situation.

Any organic compound with an equilibrium distribution coefficient greater than 1.5 can be separated from water by extraction with a water-insoluble organic solvent. As you will soon see, working through the mathematics of the distribution coefficient shows that a series of extractions using small volumes of solvents is more efficient than a single large-volume extraction. A volume of solvent about one-third the volume of the aqueous phase is appropriate for each extraction. Commonly used extraction solvents are listed in Table 10.1. If the distribution coefficient K of a solute between

TABLE 10.1	Common extraction solvents				
Solvent	Boiling point, °C	Solubility in water, g/100 mL	Hazard	Density, g/mL	Fire hazard[a]
Diethyl ether	35	6	Inhalation, fire	0.71	++++
Pentane	36	0.04	Inhalation, fire	0.62	++++
Petroleum ether[b]	40–60	Low	Inhalation, fire	0.64	++++
Dichloromethane	40	2	$LD_{50}{}^{c} = 1.6$ mL/kg	1.32	+
Hexane	69	0.02	Inhalation, fire	0.66	++++
Ethyl acetate	77	9	Inhalation, fire	0.90	++

a. Scale: extreme fire hazard = ++++.
b. Mixture of hydrocarbons.
c. LD_{50}, lethal dose orally in young rats.

water and an organic solvent is large, a single extraction may suffice to extract the compound from water into the organic solvent. Most often, however, the distribution coefficient is less than 10, making multiple extractions necessary.

In general, the fraction of solute remaining in the original water solvent is given by equation 2.

$$\frac{(\text{final mass of solute})_{\text{water}}}{(\text{initial mass of solute})_{\text{water}}} = \left(\frac{V_2}{V_2 + V_1 K}\right)^n \qquad (2)$$

where

V_1 = volume of organic solvent in each extraction
V_2 = original volume of water
n = number of extractions
K = distribution coefficient

How Many Extractions Should Be Used?

Consider a simple case of extraction from water into ether, assuming a distribution coefficient of 5.0 for the organic compound being extracted. As an example, we will use 1.0 g of compound dissolved in 50 mL of water. Would the recovery of the desired compound be better if the water solution were extracted once with 45 mL of ether or two to three times with 15-mL portions of ether? The final mass of solute remaining in the water after extraction can be calculated using equation 2.

One extraction. Calculation of the amount of organic compound (solute) remaining in the water solution after one extraction using 45 mL of ether using equation 2 ($n = 1$):

$$(\text{final mass of solute})_{\text{water}} = x \text{ g}$$

$$(\text{initial mass of solute})_{\text{water}} = 1.0 \text{ g}$$

$$V_1 = 45 \text{ mL ether}$$

$$V_2 = 50 \text{ mL water}$$

$$\frac{\text{(final mass of solute)}_{\text{water}}}{\text{(initial mass of solute)}_{\text{water}}} = \frac{x \text{ g}}{1.0 \text{ g}} = \left(\frac{V_2}{V_2 + V_1 K}\right)^n = \left(\frac{50}{50 + (45 \times 5.0)}\right) \tag{3}$$

$x = 0.18$ g solute remaining in the water layer after the extraction

Thus, 0.82 g of solute was extracted into the ether layer and 0.18 g of solute remains in the water layer.

Two extractions. Calculation for two extractions, each using 15 mL of ether ($n = 2$):

$$\text{(final mass of solute)}_{\text{water}} = x \text{ g}$$

$$\text{(initial mass of solute)}_{\text{water}} = 1.0 \text{ g}$$

$$V_1 = 15 \text{ mL ether}$$

$$V_2 = 50 \text{ mL water}$$

$$\frac{\text{(final mass of solute)}_{\text{water}}}{\text{(initial mass of solute)}_{\text{water}}} = \frac{x \text{ g}}{1.0 \text{ g}} = \left(\frac{V_2}{V_2 + V_1 K}\right)^n = \left(\frac{50}{50 + (15 \times 5.0)}\right)^2 \tag{4}$$

$x = 0.16$ g solute remaining in water layer after second extraction

After two extractions with 15 mL of ether, a total of 0.84 g of solute has been extracted into the ether layers. The amount of solute separated by two extractions is comparable to that of the single extraction, but the process was carried out more effectively and used only 66% as much ether.

Three extractions. If a third extraction of the residual aqueous layer with 15 mL of ether were done, an additional 0.10 g of solute (10%) would be transferred from the aqueous layer to the ether layer, giving a total recovery of 0.94 g of solute. Only 6% of the organic compound would remain in the aqueous layer; most of it could be extracted with one more 15-mL portion of ether.

Drawing a flowchart of the extractions. It can be helpful to draw a flowchart that shows the steps in an extraction, particularly when multiple steps are involved. The flowchart shown here illustrates the three steps in separating 0.94 g of organic compound from a solution of 1.0 g of the compound in 50 mL of water, as described by the previous calculations. Recall that the distribution coefficient ($K = 5$) is relatively small; thus, three extractions are needed for satisfactory recovery of the organic compound.

At the end of the three extractions, the three ether solutions of the organic compound are combined before subsequent operations are used to purify and dry the combined ether solution and recover the purified organic compound. If more extractions are necessary, they can be illustrated by extending the flowchart below the point where the three ether solutions are combined into one product solution.

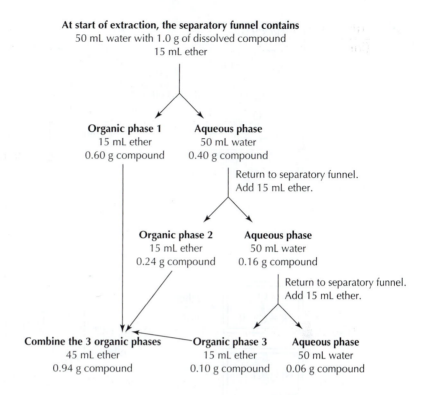

At start of extraction, the separatory funnel contains
50 mL water with 1.0 g of dissolved compound
15 mL ether

Organic phase 1
15 mL ether
0.60 g compound

Aqueous phase
50 mL water
0.40 g compound

Return to separatory funnel.
Add 15 mL ether.

Organic phase 2
15 mL ether
0.24 g compound

Aqueous phase
50 mL water
0.16 g compound

Return to separatory funnel.
Add 15 mL ether.

Combine the 3 organic phases
45 mL ether
0.94 g compound

Organic phase 3
15 mL ether
0.10 g compound

Aqueous phase
50 mL water
0.06 g compound

10.2 Changing Solubility with Acid-Base Chemistry

The catalysis of acids and bases is central to many organic chemical reactions, so it should come as no surprise that acids or bases are present in many organic reaction mixtures. Organic and inorganic acids tend to be soluble in water, but they are often soluble in organic solvents as well. Simple extraction is not very effective in separating these acids from the organic reaction products that you might want to separate and purify. Fortunately, there is a ready solution to the problem.

Whereas acetic acid, sulfuric acid, and hydrochloric acid are soluble in both water and diethyl ether, their conjugate bases (the $CH_3CO_2^-$, HSO_4^- or SO_4^{2-}, and Cl^- anions) are not soluble in organic solvents. Adding a water-soluble base, such as sodium bicarbonate ($Na^+HCO_3^-$), can change a distribution coefficient K from 1.0 to 0.001 simply by shaking an aqueous solution of 5% $NaHCO_3$ with the organic liquid containing an acid. Voila! The acid is extracted into the water as its ionic conjugate base, gaseous carbon dioxide is evolved, and the other organic compounds stay behind in the organic solvent. This constitutes a major separation.

Acid-base chemistry can also be used to extract water-insoluble amine bases into water by protonating the amines with dilute hydrochloric acid. The positively charged ammonium ions are usually quite soluble in water and not soluble in organic solvents.

Figure 10.3 shows an acid-base extraction flowchart for the separation of four organic compounds: a carboxylic acid, a phenol, an amine, and a ketone. It is important to understand the acid-base properties of these four classes of compounds: both the carboxylic

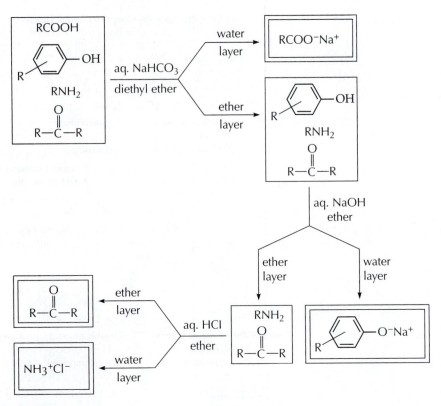

FIGURE 10.3 Separation of a four-component mixture by acid-base extractions.

acid and the phenol are acids, but they have different acid strengths. With a pK_a of ~5, the carboxylic acid is the stronger acid. The phenol has a pK_a of ~10 and requires a stronger base to be converted to the phenolate anion, its conjugate base.

$$RCOOH + Na^+HCO_3^- \longrightarrow RCOO^-Na^+ + H_2O + CO_2$$

pK_a ~ 5   Water-soluble
salt

phenol + $Na^+OH^- \longrightarrow$ phenolate $-O^-Na^+ + H_2O$

pK_a ~ 10   Water-soluble
salt

The amine requires a strong acid, such as hydrochloric acid, to convert it to a water-soluble salt. Under the normal pH range in water, the ketone is a neutral compound.

$$RNH_2 + HCl \longrightarrow RNH_3^+Cl^-$$

Water-soluble
salt

If you know the approximate amount of the organic compound present, you can calculate the amount of base or acid necessary to extract it into aqueous solution. The molar stoichiometry of acid-base reactions with HCl and NaOH is 1:1; it is best to use about a twofold excess of the calculated amount.

Using acid-base chemistry can also serve to separate and purify a carboxylic acid or an amine that is in an organic solvent, which also contains neutral organic by-products. Organic acids and bases that have been extracted into the aqueous layer can be recovered by neutralizing the extraction reagent. For example, RCOOH can be regenerated by acidifying the aqueous extract with 6M HCl until the solution becomes just acidic as indicated by pH paper. An amine can be recovered by making the aqueous solution alkaline with 6M NaOH until the solution reaches pH 12–13. If the carboxylic acid or amine is a solid, which will precipitate from the aqueous solution, it can be recovered by filtration. A liquid carboxylic acid or amine, or a solid carboxylic acid or amine that does not precipitate, can be extracted into an organic solvent, which can be dried (see Chapter 11) and the solvent evaporated.

10.3   Doing Extractions

A number of practical details need to be taken into account while carrying out an extraction:

- Density of the solvent used for the extraction
- Temperature of the extraction mixture
- Venting the separatory funnel
- Washing the organic phase
- Improving the efficiency of an extraction by salting out
- Preventing and dispersing emulsions
- Caring for the separatory funnel after an extraction

Density of the Solvent Used for the Extraction

Before you begin any extraction, look up the density of the organic solvent in Table 10.1 or use a handbook or online source to determine whether your extraction solvent is more dense or less dense than water. **The more dense layer is always on the bottom.** Organic solvents that are less dense than water form the upper layer in the separatory funnel (Figure 10.4a), whereas solvents that are more dense than water form the lower layer (Figure 10.4b). Occasionally, sufficient material is extracted from the aqueous phase to the organic phase or vice versa to change the relative densities of the two phases enough for them to exchange places in the separatory funnel. This can be confusing unless you are aware of what may be happening.

Temperature of the Extraction Mixture

Be sure that the aqueous extraction solution is at room temperature or slightly cooler before you add the organic extraction solvent. Most organic solvents used for extractions have low boiling points and may boil if added to a warm aqueous solution. A few pieces of ice can be added to cool the aqueous solution.

SAFETY PRECAUTION

Do not point a separatory funnel at yourself or your neighbor. Point the separatory funnel toward the back of the hood when venting it.

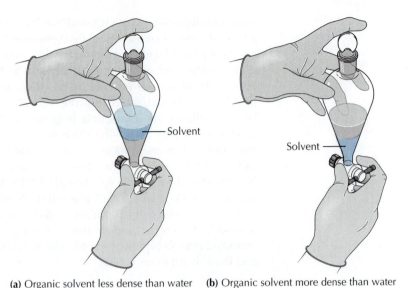

FIGURE 10.4
Solvent densities.

(a) Organic solvent less dense than water   **(b)** Organic solvent more dense than water

Venting the Separatory Funnel

Work in a hood while carrying out an extraction. Be sure that you vent an extraction mixture by carefully **inverting the stoppered separatory funnel** and immediately opening the stopcock before you begin the shaking process. If you do not do this, the stopper may pop out of the funnel and liquids and gases may be released (Figure 10.5). Pressure buildup in the separatory funnel is always a problem when using low-boiling extraction solvents such as diethyl ether, pentane, or dichloromethane.

Venting extraction mixtures is especially important when you use a dilute sodium carbonate or bicarbonate solution to extract an organic phase containing traces of an acid. Carbon dioxide gas is given off in the neutralization process. The CO_2 pressure buildup can easily force the stopper out of the funnel, cause losses of solutions, and possibly injure you or your neighbor. When using sodium carbonate or bicarbonate to extract or wash acidic contaminants from an organic solution, **vent the extraction mixture immediately after the first inversion and subsequently after every three or four inversions.**

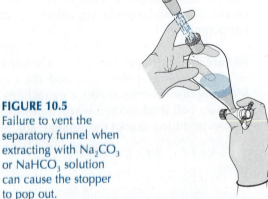

FIGURE 10.5
Failure to vent the separatory funnel when extracting with Na_2CO_3 or $NaHCO_3$ solution can cause the stopper to pop out.

Washing the Organic Phase

After an extraction is completed and the two immiscible liquids are separated, the organic layer is often extracted, with water or perhaps a dilute aqueous solution of an acid or a base. Chemists often use the term *washing* to describe this type of extraction. For example, a chemical reaction involving alkaline (basic) reagents often yields an organic extract that still contains some alkaline material. This alkaline material can be removed by washing the organic phase with a 5% solution of hydrochloric acid. Similarly, an organic extract obtained from an acidic solution should be washed with a 5% solution of sodium carbonate or sodium bicarbonate (see preceding section on venting). The salts formed in these extractions are very soluble in water but not in typical organic solvents, so they are easily transported into the aqueous phase. If acid or base washes are required, they are done in the same manner as any other extraction and are usually followed by a final water wash.

Another scenario in which you might want to wash an extraction solvent is when a solution of a nonpolar organic solvent contains a small amount of a polar organic solvent, such as ethanol or methanol. In this case using an acid or base is not helpful; water itself would be used for the washing extraction.

Improved Efficiency of Extraction by Salting Out

If the distribution coefficient for a substance to be extracted from water into an organic solvent is lower than 2.0, a simple extraction procedure may not be effective. In this case, a salting out procedure can help. *Salting out* is done by adding a saturated solution of NaCl (sometimes called *brine*), Na_2SO_4, or the salt crystals themselves, to the aqueous layer. The presence of a salt in the water layer decreases the solubility of the organic compound in the aqueous phase. Therefore, the distribution coefficient increases, allowing more of the organic compound to be transferred from the aqueous phase to the organic phase. Salting out can also help to separate a homogeneous solution of water and a water-soluble organic compound into two phases.

Preventing and Dispersing Emulsions

The formation of an *emulsion*—a suspension of insoluble droplets of one liquid in another liquid—is sometimes encountered while doing an extraction. When an emulsion forms, the entire mixture has a milky appearance, often with no clear separation between the immiscible layers, or there may be a third milky layer between the aqueous and the organic phases. Emulsions are not usually formed during diethyl ether extractions, but they frequently occur when aromatic or chlorinated organic solvents are used. An emulsion often disperses if the separatory funnel and its contents are allowed to sit in a ring stand for a few minutes.

Prevention of emulsions. Preventing emulsions is simpler than dealing with them. When using aromatic or chlorinated solvents to extract organic compounds from aqueous solutions, very gentle mixing of the two phases may reduce or eliminate emulsion formation. Instead of shaking the mixture vigorously, invert the separatory funnel and gently swirl the two layers together for 2–3 min. However, use of this swirling technique may mean that you need to extract an aqueous solution with an extra portion of organic solvent for maximum recovery of the product.

What to do if an emulsion forms. Should an emulsion occur, it can often be dispersed by vacuum filtration through a pad of the filter aid Celite. Prepare the Celite pad by pouring a slurry of Celite and water onto a filter paper in a Buchner funnel (see Sections 9.1 and 9.4). Remove the water from the filter flask before pouring the emulsion through the Buchner funnel. Return the filtrate to the separatory funnel and separate the two phases. Another method, useful when the organic phase is the lower layer, involves filtering the organic phase by vacuum filtration through a phase separator filter paper, such as Whatman 1PS. For microscale extractions, centrifugation of an emulsified mixture usually separates the two liquid phases (see Section 9.5).

Caring for the Separatory Funnel

When the entire extraction is complete, clean the funnel immediately and re-grease the glass stopcock to prevent a "frozen" stopcock later. Grease is not necessary with Teflon stopcocks, but they may also freeze if not loosened prior to storage.

10.4 Miniscale Extractions

Read Sections 10.1 and 10.2 before undertaking a miniscale extraction for the first time.

Before you begin an extraction, assemble and **label** a series of Erlenmeyer flasks for the aqueous phase and the organic phase for the number of extractions you will be doing. (Do **not** use beakers for the organic phase because the solvent can evaporate rapidly.) The solutions in an extraction tend to be colorless, so if the flasks are not clearly labeled it is easy to become confused about their contents by the end of the procedure.

SAFETY PRECAUTION

Wear gloves and work in a fume hood while doing extractions. Point the separatory funnel toward the back of the hood when venting it.

Extraction with an Organic Solvent Less Dense Than Water

Place a separatory funnel large enough to hold three to four times the total solution volume in a metal ring firmly clamped to a ring stand or upright support rod (Figure 10.6). **The stopcock must fit tightly and be closed.** If the separatory funnel has a glass stopcock, make sure the stopcock is adequately greased. If the separatory funnel has a Teflon stopcock, as shown in Figure 10.2, no grease is necessary. However, the nut on the threaded end of the stopcock must be tightened so that the stopcock fits snugly and yet can still be rotated with relative ease.

Pour the cooled aqueous solution to be extracted into the separatory funnel (Figure 10.6, Step 1). Add a volume of organic solvent equal to approximately one-third the total volume of the aqueous solution (Figure 10.6, Step 2), and put the stopper in place.

Remove the funnel from the ring and grasp its neck with one hand, holding the stopper down firmly with your index finger (Figure 10.6, Step 3). Invert the funnel and **open the stopcock**

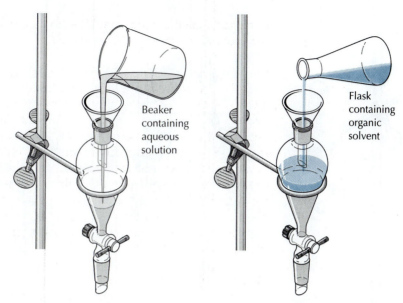

1. Add aqueous solution.

Beaker containing aqueous solution

2. Add organic solvent.

Flask containing organic solvent

3. Insert stopper and hold stopper with your finger.

4. Invert funnel and immediately open stopcock to release pressure, close the stopcock, and mix the layers by shaking the funnel.

FIGURE 10.6 Using a separatory funnel. *(Continued on next page)*

immediately to release the pressure from solvent vapors (Figure 10.6, Step 4). Close the stopcock and thoroughly mix the two liquid phases by shaking the mixture while inverting the separatory funnel four or five times. Then release the pressure by opening the stopcock. Repeat this shaking and venting process five or six times to ensure complete mixing of the two phases. Shaking too gently does not effectively mix the two phases; shaking too vigorously may lead to the formation of emulsions.

5. Use a ring stand to hold separatory funnel until layers separate.

6. Draw off bottom layer.

7. Pour off top layer.

FIGURE 10.6 *(Continued).*

Pour the top layer out of the top of the funnel so that it is not contaminated by the residual bottom layer adhering to the stopcock and tip.

Place the separatory funnel in the ring once more, remove the stopper, and wait until the layers have completely separated (Figure 10.6, Step 5). **Open the stopcock** to draw off the bottom layer into a labeled Erlenmeyer flask (Figure 10.6, Step 6). Pour the remaining organic layer out of the funnel through the top into a separate labeled Erlenmeyer flask (Figure 10.6, Step 7). Do this entire procedure each time you carry out an extraction.

If you are unsure which layer is the organic phase and which is the aqueous phase, you can check by adding a few drops of the layer in question to 1–2 mL of water in a test tube and observing whether it dissolves. **Do not discard any solution until you have completed the entire extraction procedure and are certain which flask contains the desired product.**

After the last extraction and separation of the lower aqueous phase, pour the remaining organic layer from the top of the separatory funnel into the Erlenmeyer flask that contains the other organic phases. The organic solution is now ready for a final aqueous washing, followed by the addition of an anhydrous drying agent (see Section 11.1).

SAFETY PRECAUTION

Wear gloves and work in a fume hood while doing extractions. Point the separatory funnel toward the back of the hood when venting it.

Extraction with an Organic Solvent More Dense Than Water

When extracting an aqueous solution several times with a solvent more dense than water, it is not necessary to pour the upper aqueous layer out of the separatory funnel after each extraction.

Simply drain the lower organic phase out of the separatory funnel into a labeled Erlenmeyer flask. Then add the next portion of organic solvent to the aqueous phase remaining in the funnel. At the end of the entire extraction procedure, drain the organic layer into the Erlenmeyer flask that contains the other organic phases. The organic solution is now ready for a final aqueous washing, followed by the addition of an anhydrous drying agent (see Section 11.1).

10.5   Summary of the Miniscale Extraction Procedure

1. After closing the stopcock, pour the aqueous mixture into a separatory funnel with a capacity three to four times the amount of the mixture.

2. Add a volume of immiscible organic solvent approximately one-third the volume of the aqueous phase and stopper the separatory funnel. **You must know the density of the organic solvent.**

3. Invert the funnel, grasping the neck with one hand and firmly holding down the stopper with your index finger. Open the stopcock to release any pressure buildup.

4. Close the stopcock, and shake the mixture while inverting the separatory funnel four or five times before releasing the pressure by opening the stopcock; repeat this shaking and venting process five or six times to ensure complete mixing of the two phases (see precautions about emulsions in Section 10.3).

5. Remove the stopper and allow the two phases to separate.

6. **For an organic solvent less dense than water,** draw off the lower aqueous phase into a labeled Erlenmeyer flask. Pour the organic phase from the top of the funnel into a second labeled Erlenmeyer flask. Return the aqueous phase to the separatory funnel. **For an organic solvent more dense than water,** draw off the lower organic phase into a labeled Erlenmeyer flask; the upper aqueous phase remains in the separatory funnel.

7. Extract the original aqueous mixture twice more with fresh organic solvent.

8. Combine the organic extracts in one Erlenmeyer flask and pour this solution into the separatory funnel. Extract the organic solution with dilute acid or base, if necessary, to neutralize any bases or acids remaining from the reaction.

9. Wash the organic phase with water or a saturated NaCl solution.

10. Dry the organic phase with an anhydrous drying agent (see Section 11.1).

10.6   Microscale Extractions

The small volumes of liquids used in microscale reactions should not be handled in a separatory funnel because much of the material would be lost on the surface of the glassware. Instead, use a conical vial or a centrifuge tube to hold the two-phase system and a Pasteur pipet to separate one phase from the other and transfer it to another container (Figure 10.7). The V-shaped bottom of a conical vial or a

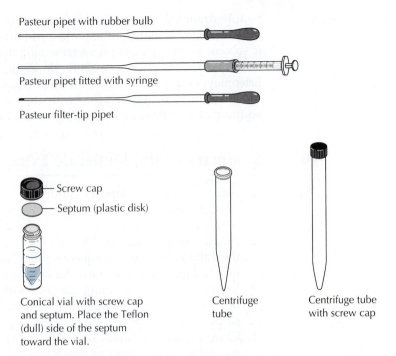

Pasteur pipet with rubber bulb

Pasteur pipet fitted with syringe

Pasteur filter-tip pipet

Screw cap

Septum (plastic disk)

Conical vial with screw cap
and septum. Place the Teflon
(dull) side of the septum
toward the vial.

Centrifuge
tube

Centrifuge tube
with screw cap

FIGURE 10.7
Equipment for
microscale extractions.

*Read Sections 10.1
and 10.2 before
undertaking a
microscale extraction
for the first time.*

centrifuge tube enhances the visibility of the interface between the two phases in the same way that the conical shape of a separatory funnel just above the stopcock enhances the visibility of the interface. Centrifuge tubes are particularly useful for extractions with combinations of organic and aqueous phases that form emulsions. The tubes can be spun in a centrifuge to produce a clean separation of the two phases.

10.6a Equipment and Techniques Common to Microscale Extractions

Before discussing specific types of extractions, we need to consider the equipment and techniques common to all microscale extractions.

Extractions involve the use of several containers. Before you begin an extraction, **carefully label all the conical vials and centrifuge tubes that will hold your aqueous and organic solutions.** The solutions in an extraction tend to be colorless, so if the containers are not clearly labeled, it is easy to become confused about their contents during the procedure. **Do not discard any solution until the entire extraction procedure is complete and you are certain which vessel contains the product.**

Conical Vials

Conical vials, with a capacity of 5 mL, work well for extractions in which the total volume of both phases does not exceed 4 mL. Conical vials tip over very easily, so **always place the vial in a small beaker.** The plastic septum used with the screw cap on a conical vial has a chemically inert coating of Teflon on one side. The Teflon looks dull and should be positioned toward the vial. (The shiny side of the septum is not inert to all organic solvents.)

Centrifuge Tubes

Centrifuge tubes with a 15-mL capacity and tight-fitting caps serve for extractions involving a total volume of up to 12 mL. Set centrifuge tubes in a test tube rack to keep them upright.

Mixing the Two Phases

Thorough mixing of the two phases is essential for complete transfer of the solute from one phase to the other. Mix the two phases by capping the conical vial or centrifuge tube and shaking it vigorously eight to 10 times. Slowly loosen the cap to vent the vial or centrifuge tube. Repeat the shaking and venting process four to six times.

Alternatively, or for a centrifuge tube without a screw cap, you can use the *squirt method*. Draw the two phases into a Pasteur pipet (with no cotton plug in the tip) and squirt the mixture back into the centrifuge tube five or six times to mix the two phases thoroughly. The use of a vortex mixer is another way to mix the two phases.

Separation of the Phases with a Filter-Tip Pipet

A filter-tip pipet (see Figures 5.9 and 10.7) provides better control for transferring volatile solvents such as dichloromethane or diethyl ether during a microscale extraction than does a Pasteur pipet without the cotton plug. The lower layer is more easily removed from a conical vial or centrifuge tube than the upper layer. **Expel the air from the rubber bulb before inserting the pipet to the bottom of the conical vial or centrifuge tube.** Slowly release the pressure on the bulb to draw liquid into the pipet. Maintain steady pressure on the rubber bulb while transferring the liquid to another nearby container. Hold the receiving container close to the extraction vial or centrifuge tube so the transfer can be accomplished without any loss of liquid (Figure 10.8).

Separation of the Phases with a Pasteur Pipet and Syringe

A pipet fitted with a small syringe can also be used to remove the lower layer (see Figures 5.8 and 10.7). Draw the liquid into the pipet with a steady pull on the syringe plunger until the interface between the layers reaches the bottom of the vial or tube. **Do not exceed the capacity of the Pasteur pipet (approximately 2 mL); no liquid should be drawn into the syringe.** Remove the pipet from the extraction vessel and transfer its contents to the receiving container. Hold the receiving container close to the extraction vessel so the transfer can be accomplished quickly without any loss of liquid (Figure 10.8). Depress the syringe plunger to empty the pipet.

FIGURE 10.8
Holding vials while transferring solutions.

What to Do If the Upper Phase Is Drawn into the Pasteur Pipet

The interface between the two phases in a conical vial or centrifuge tube can be difficult to see in some instances, and a small amount of the upper layer may be drawn into the Pasteur pipet. If this occurs, maintain steady pressure on the pipet with the rubber bulb or syringe and allow the two phases in the pipet to separate. Slowly expel the lower layer into the receiving container until the interface between the phases is at the bottom of the pipet. Then move the pipet to the original container and add the upper layer in the pipet to the remaining upper phase.

10.6b Microscale Extractions with an Organic Phase Less Dense Than Water

The microscale extraction of an aqueous solution with an organic solvent that is less dense than water and washing an ether solution with aqueous reagents are examples of this type of extraction.

SAFETY PRECAUTION

Wear gloves and work in a hood while doing extractions.

Two centrifuge tubes or conical vials and a test tube are needed for the extraction of an aqueous solution with a solvent less dense than water. Place the aqueous solution in the first centrifuge tube or conical vial, and add the organic solvent—diethyl ether in this example. Cap the tube or vial and shake it to mix the layers. Vent the tube (or vial) by slowly releasing the cap and allow the phases to separate. Repeat the shaking and venting four to six times. Alternatively, use the squirt method (five or six squirts) (see Section 10.6a) or a vortex mixer to mix the phases. Allow the layers to separate completely.

Put a filter-tip pipet or a Pasteur pipet fitted with a syringe (Section 10.6a) into the tube or vial with the tip touching the bottom of the cone (Figure 10.9, Step 1). Slowly draw the aqueous layer into the pipet until the interface between the ether and the aqueous solution is at the bottom of the V. Transfer the aqueous solution to the second centrifuge tube or conical vial (Figure 10.9, Step 2). The ether solution remains in the first tube.

In any extraction, no material should be discarded until you are certain which container holds the desired product.

Add a second portion of ether to the aqueous phase in the second tube, cap the tube, and shake it to mix the phases. Repeat the shaking and venting four to six times. After the phases separate, again remove the lower aqueous layer and place it in the test tube (Figure 10.9, Step 3). Transfer the ether solution in the first tube to the ether solution in the second tube with the Pasteur pipet (Figure 10.9, Step 4). Repeat the procedure if a third extraction is necessary.

Washing the Organic Liquid

If an experiment specifies washing an organic solution that is less dense than water with an aqueous solution, place the organic solution in a centrifuge tube or conical vial. Add the requisite amount of water or aqueous reagent solution, cap the tube (or vial), and shake it to mix the phases. Repeat the shaking and mixing four to six times. Open the cap to release any built-up vapor pressure and allow the layers

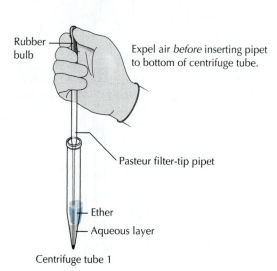

Rubber bulb

Expel air *before* inserting pipet to bottom of centrifuge tube.

Pasteur filter-tip pipet

Ether

Aqueous layer

Centrifuge tube 1

1. Remove lower aqueous phase with Pasteur pipet.

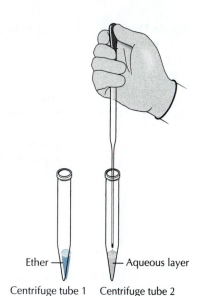

Ether

Aqueous layer

Centrifuge tube 1   Centrifuge tube 2

2. Transfer aqueous phase to tube 2.

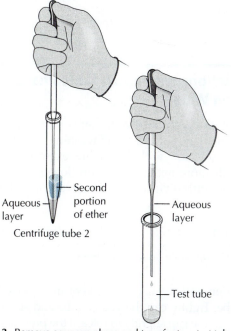

Aqueous layer

Second portion of ether

Centrifuge tube 2

Aqueous layer

Test tube

3. Remove aqueous phase and transfer to a test tube.

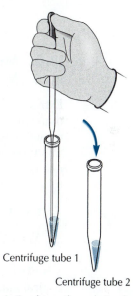

Centrifuge tube 1

Centrifuge tube 2

4. Combine ether solution from tube 1 with ether solution in tube 2.

FIGURE 10.9 Extracting an aqueous solution with an organic solvent less dense than water.

to separate. Transfer the lower aqueous layer to a test tube with a filter-tip pipet or a Pasteur pipet fitted with a syringe (see Figures 5.9 and 10.7). The upper organic phase remains in the extraction tube (or conical vial) ready for the next step (Figure 10.10), which may be washing with another aqueous reagent solution or, if the extractions are completed, drying with an anhydrous salt (see Section 11.1).

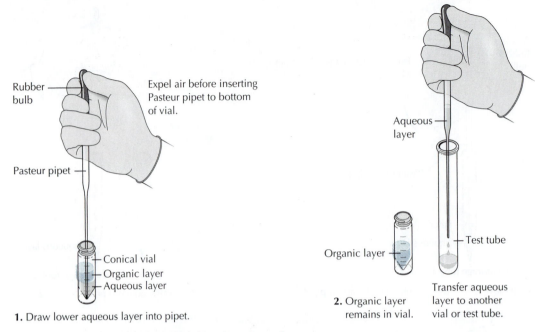

Rubber bulb

Pasteur pipet

Expel air before inserting Pasteur pipet to bottom of vial.

Conical vial
Organic layer
Aqueous layer

1. Draw lower aqueous layer into pipet.

Aqueous layer

Organic layer

Test tube

2. Organic layer remains in vial.

Transfer aqueous layer to another vial or test tube.

FIGURE 10.10 Washing an organic phase less dense than water.

10.6c Microscale Extractions with an Organic Phase More Dense Than Water

Extraction of an aqueous solution with a solvent that is more dense than water, such as dichloromethane (CH_2Cl_2), and washing a dichloromethane/organic product solution with water are examples of this type of extraction. The dichloromethane solution (lower phase) needs to be removed from the conical vial or centrifuge tube in order to separate the layers.

SAFETY PRECAUTION

Wear gloves and work in a hood while doing extractions.

In any extraction, no material should be discarded until you are certain which container holds the desired product.

Place the aqueous solution and the organic solvent in a labeled conical vial or centrifuge tube. Tightly cap the vial or tube and shake the mixture thoroughly. Loosen the cap slightly to release the pressure. Repeat the shaking and venting process four to six times. Alternatively, use the squirt method (five or six squirts) (see Section 10.6a) or a vortex mixer to mix the phases. Allow the layers to separate completely.

Put a filter-tip pipet or a Pasteur pipet fitted with a syringe (see Section 10.6a) into the conical vial or centrifuge tube with the tip touching the bottom of the cone (Figure 10.11, Step 1). Slowly draw the lower layer into the pipet until the interface between the two layers is exactly at the bottom of the V. Transfer the pipet to another centrifuge tube, conical vial, or test tube and expel the dichloromethane solution into the second container (tube 2 in Figure 10.11, Step 2). The aqueous layer remains in the extraction tube and can be extracted a second time with another portion of CH_2Cl_2. The second

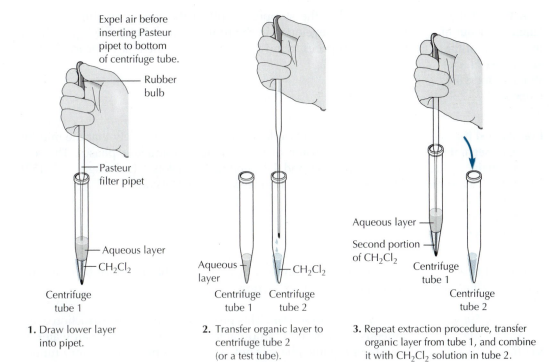

FIGURE 10.11 Extracting an aqueous solution with an organic solvent more dense than water.

dichloromethane solution is added to the second centrifuge tube after the separation (Figure 10.11, Step 3).

Washing the Organic Liquid

If the organic phase transferred to tube or vial 2 is being washed with an aqueous solution, the aqueous reagent is added to tube 2. Cap the tube or vial, shake it to mix the phases, and loosen the cap to release any pressure buildup. The lower organic phase is separated and transferred to another centrifuge tube (or conical vial) if more washings are necessary. Otherwise, the organic phase is transferred to a dry test tube for treatment with a drying agent (see Section 11.1).

10.7 Sources of Confusion and Common Pitfalls

How Do I Know Which Layer Is the Organic Phase in My Separatory Funnel (or Microscale Vial)?

Before beginning any extraction, ascertain the density of the organic solvent that you will be using. If the extraction involves dilute aqueous solutions of inorganic reagents, you can assume that their density is close to the density of water, 1.0 g/mL. If the density of the organic solvent is less than 1.0 g/mL, the organic phase will be the upper layer in the separatory funnel. If the density of the organic solvent is greater than 1.0 g/mL, the organic phase will be the lower layer. You can always verify which layer is which by adding a few drops of water to the top layer in the separatory funnel. If the top layer is the aqueous phase, the water drops will immediately dissolve. If the top layer is the organic phase, the water drops will fall through the organic phase to the lower aqueous phase.

Why Doesn't the Liquid Flow out of the Bottom of the Separatory Funnel When I Open the Stopcock?

It's likely that the stopper is still in the separatory funnel and a vacuum has developed in the funnel, which holds the liquid there. Close the stopcock, remove the stopper, reopen the stopcock, and all should be well.

I've Allowed the Layers to Separate in My Separatory Funnel (or Microscale Vial), but Three Layers Appear! Is That Possible?

If three, instead of two layers are visible, it is likely that the middle layer is an emulsion of the organic and aqueous phases. The section "Preventing and Dispersing Emulsions" (see pages 151–152) describes procedures for breaking up emulsions.

Why Can't I See Any Separation of Phases in the Separatory Funnel (or Microscale Vial)?

Several scenarios can lead to no discernible interface between the liquid phases in an extraction.

Solvent added to solvent. This problem occurs in the extraction of an aqueous solution with an organic solvent less dense than water. If the upper organic phase is not removed from the separatory funnel (or microscale vial) and the aqueous solution is not returned to the extraction vessel before the subsequent portion of organic solvent is added, no interface appears because the second portion of solvent is the same as the first solvent.

The upper layer is too small to be easily visible. Occasionally, the volume of the upper layer in a separatory funnel (or vial) is too small for the interface to be clearly visible. Draining some of the lower layer will increase the depth of the upper layer as the liquid moves toward the narrower conical portion of the funnel, and the interface will become visible. Another option is to add some additional solvent that will become part of the upper layer.

The refractive index of the two solutions is very similar. In rare instances, the refractive index of each solution is so similar that the interface is not visible. Usually adding more water to the aqueous phase will dilute the solution enough to change its refractive index and make the interface visible. Sometimes gently moving the funnel or vial will disturb the layers enough to make the interface noticeable.

I Forgot to Label My Containers. How Can I Tell Which Container Holds My Product?

It is **imperative** that all containers be clearly labeled to indicate their contents. If you are in doubt about the contents of any container, add a few drops of the solution in question to 1–2 mL of water in a small test tube and observe whether it dissolves. The organic phase will be insoluble.

A Prudent Practice

Never discard any solution during an extraction until you are certain that you know which container holds your product.

Questions

1. An extraction procedure specifies that an aqueous solution containing dissolved organic material be extracted twice with 10-mL portions of diethyl ether. A student removes the lower layer after the first extraction and adds the second 10-mL portion of ether to the upper layer remaining in the separatory funnel. After shaking the funnel, the student observes only one liquid phase with no interface. Explain.

2. A crude nonacidic product mixture dissolved in diethyl ether contains acetic acid. Describe an extraction procedure that could be used to remove the acetic acid from the ether.

3. What precautions should be observed when an aqueous sodium carbonate solution is used to extract an organic solution containing traces of acid?

4. When two layers form during a petroleum ether/water extraction, what would be an easy, convenient way to tell which layer is which if the densities were not available?

5. You have 100 mL of a solution of benzoic acid in water; the amount of benzoic acid in the solution is estimated to be 0.30 g. The distribution coefficient of benzoic acid in diethyl ether and water is approximately 10. Calculate the amount of benzoic acid that would be left in the water solution after four 20-mL extractions with ether. Do the same calculation using one 80-mL extraction with ether to determine which method is more efficient.

CHAPTER

11

DRYING ORGANIC LIQUIDS AND RECOVERING REACTION PRODUCTS

If Chapter 11 is your introduction to the separation and purification of organic compounds, read the Essay "Intermolecular Forces in Organic Chemistry" on pages 127–131 before you read Chapter 11.

Most organic separations involve extractions from an aqueous solution; no matter how careful you are, some water usually remains in the organic liquid. A small amount of water dissolves in most extraction solvents, and the physical separation of the layers in the extraction process may be incomplete. As a result, the organic layer usually needs to be dried with an anhydrous drying agent before recovering an organic product. Drying agents for organic liquids are usually anhydrous inorganic salts. After the drying procedure, the organic liquid needs to be separated from the drying agent and the solvent removed to recover the product. These operations are described in this chapter.

11.1 Drying Agents

The most common way to *dry* (remove the water from) an organic liquid is to add an *anhydrous* (deprived of water) drying agent that binds with water. Most anhydrous drying agents react with water to form crystalline *hydrates*, which are insoluble in the organic phase and can be removed by filtration:

$$n\text{H}_2\text{O} + \text{drying agent} \rightarrow \text{drying agent} \cdot n\text{H}_2\text{O}$$

Factors in Selecting a Drying Agent

Table 11.1 lists common drying agents used for organic liquids.

The factors that need to be considered when selecting a drying agent are:

- Capacity
- Efficiency
- Speed
- Chemical inertness

Capacity for removing water. The maximum number of moles of water bound in the hydrated form of the salt is called its *capacity.* It is the amount of water that can be taken up per unit weight of drying agent.

Efficiency. The *efficiency* expresses how much water the drying agent *leaves behind* in the organic liquid. The lower the efficiency value, the smaller the amount of water left in the organic liquid; thus, the drying agent is more efficient.

Speed of removing water. The *speed* with which the hydrate forms determines how long the drying agent needs to be in contact with the organic solution. A good general drying agent, such as $MgSO_4$, usually requires 5–10 min to remove water from an organic liquid. $CaCl_2$ and Na_2SO_4 usually require 15–30 min.

Chemical inertness. Drying agents must be *chemically inert* (unreactive) to both the organic solvent and any organic compound dissolved in the solvent. For example, bases such as K_2CO_3 and KOH are not suitable for drying acidic organic compounds because they undergo chemical reactions with these compounds. $MgSO_4$ is generally considered to be a neutral salt, but in the presence of water

TABLE 11.1   Common anhydrous chemical drying agents

Drying agent	Acid/base properties	Capacity	Efficiency[a]	Speed of drying	Comments
$MgSO_4$	Neutral	High	2.8	Fairly rapid	Good general drying agent
$CaCl_2$	Neutral	Medium to high	1.5	Fairly slow	Reacts with many organic compounds
Silica gel	Neutral	High	Low	Medium	Good general drying agent but somewhat expensive
Na_2SO_4	Neutral	Very high	25	Slow	Good for predrying; hydrate is unstable above 32°C
K_2CO_3	Basic	Low	Moderate	Fairly rapid	Reacts with acidic compounds
$CaSO_4$ (Drierite)	Neutral	Low	0.004	Fast	Fast and efficient but low capacity
KOH	Basic	Very high	0.1	Fast	Used to dry amines

a. Efficiency = measure of equilibrium residual water (mg/L of air) at 25°C

TABLE 11.2	Drying agents
Class of compounds	**Recommended drying agents**
Hydrocarbons, alkyl halides, and ethers	$MgSO_4$, $CaCl_2$, $CaSO_4$
Aldehydes, ketones, and esters	Na_2SO_4, $MgSO_4$, $CaSO_4$, K_2CO_3
Alcohols	$MgSO_4$, K_2CO_3, $CaSO_4$
Amines	KOH
Acidic compounds	Na_2SO_4, $MgSO_4$, $CaSO_4$

it is slightly acidic. Therefore, $MgSO_4$ is not suitable for drying solutions containing compounds that are especially acid sensitive. $CaCl_2$ reacts with alcohols and carbonyl compounds to form solid complexes and thus is not suitable for drying compounds with these functional groups.

Which Drying Agent Should I Use?

Table 11.2 lists suitable drying agents to use with various classes of organic compounds. Use it as a guide for selecting a drying agent if one is not specified in a procedure.

Some drying agents have a high capacity but leave quite a bit of water in the organic solution. Na_2SO_4 is a good example, as you can see from Table 11.1. It is particularly useful as a preliminary drying agent, but it is also widely used as a general-purpose drying agent because it is inexpensive and can be used with many types of compounds. However, the hydrate does not form quickly; it needs 15–30 min to form.

$MgSO_4$ is a good general-purpose drying agent, suitable for nearly all compounds. It has a high capacity for water and a reasonable efficiency, and it works fairly quickly. However, its exothermic reaction with water in the solution being dried sometimes causes the solvent to boil if the drying agent is added too rapidly. Slow addition of the drying agent prevents this.

$CaSO_4$ leaves little water behind, but it has a low capacity, which means that it works better after a preliminary drying of the liquid with Na_2SO_4 or $MgSO_4$.

Using a Drying Agent

To remove water from an organic liquid, add about 1 g of powdered or granular anhydrous drying agent per 25 mL of solution for a miniscale procedure. For microscale procedures, weigh the drying agent and use about 40 mg of drying agent per milliliter of solution.

Always place the organic liquid being treated with drying agent in an Erlenmeyer flask closed with a stopper to prevent evaporation losses.

Add the drying agent to the solution (Figure 11.1a). Swirl the flask to mix the drying agent with the liquid (Figure 11.1b). The first bit of drying agent you add will likely clump together (Figure 11.1c) as it binds water from the solution. Also, the surface of glass is polar and traces of water in the solution will often cling to the surface of the flask or vial. The added drying agent will tend to clump around these small water droplets and cling to the side of the flask. Clumping is particularly pronounced with $MgSO_4$ and Na_2SO_4. You have added enough when freshly added drying agent remains powdery and moves freely in the mixture while the flask is gently swirled. The solution may be stirred briskly with a

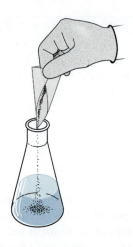

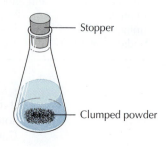

(a) Adding powdered drying agent to solution

(b) Swirling the mixture of solution and drying agent

(c) Drying agent clumped at bottom of flask

FIGURE 11.1 Adding drying agent to a solution.

magnetic stirring bar or a stirring rod or simply swirled occasionally by hand to ensure as much contact with the surface of the drying agent as possible.

If you are using granular $CaSO_4$ (Drierite) as your drying agent, it is useful to mix a small amount of indicating Drierite in it. The anhydrous form of indicating Drierite is blue, whereas the hydrated form is pink. If blue Drierite turns pink, you need to add more drying agent. The color is due to a small amount of $CuSO_4$ in the Drierite; the hydrated form of $CuSO_4$ has a pink color.

Often a preliminary drying period of 30–60 sec, followed by removal of the drying agent, is useful. Then allowing a second portion of drying agent to be in contact with the liquid for 10 min or more removes the water more completely than the use of a single portion.

11.2   Methods for Separating Drying Agents from Organic Liquids

After the drying agent has absorbed the water present in the organic liquid, it must be separated from the liquid by filtration (see Chapter 9). The container receiving the filtered liquid must be clean and dry and have a volume about two or three times the volume of the organic liquid.

Miniscale Separation of Drying Agents

Miniscale methods used to separate the drying agent from an organic liquid depend on whether the product is dissolved in a solvent.

The product is dissolved in a solvent. If the solvent will be evaporated to recover the product, place fluted filter paper in a small

FIGURE 11.2
Filtration of drying
agent from a solution
when the solvent will
be evaporated.

FIGURE 11.3 Filtration
of drying agent from
an organic liquid when
no solvent is present.

funnel and set the funnel in an Erlenmeyer flask (Figure 11.2). If the solvent will be distilled from the compound, use a round-bottomed flask as the receiving container and set it on a cork ring or in a beaker. Decant the solution slowly into the filter paper, leaving most of the drying agent in the flask. Rinse the drying agent with a few milliliters of dry solvent and also pour this rinse into the filter paper. The filtered organic liquid is then ready for the removal of the solvent.

A liquid product is not dissolved in a solvent. This method is not usually used for samples of less than 7–8 g because a significant amount of product can be lost on the surface of the glassware and drying agent. However in some extraction procedures, the organic liquid is *neat,* not dissolved in a solvent. In this situation, you must minimize the loss of liquid product during the removal of the drying agent. Instead of filter paper, tightly pack a small plug of cotton or glass wool about 5–6 mm in diameter into the outlet of the funnel. If the drying agent is powdery rather than granular, make sure the cotton plug is rolled very tightly. The plug traps the drying agent and absorbs only a small amount of the organic liquid (Figure 11.3). Slowly decant the liquid from the drying agent. The organic liquid is ready for the final distillation.

The drying agent is granular or chunky. If the drying agent is granular or chunky, for example $CaCl_2$ or Drierite, the cotton plug can be omitted and the liquid carefully decanted into the funnel, keeping all the drying agent in the original flask. The drying agent may or may not be rinsed with a few milliliters of solvent in this procedure. The organic liquid is ready for the final distillation or evaporation of the solvent.

Microscale Separation of Drying Agents

The separation methods that follow use Pasteur pipets in two different ways:

- Filter-tip pipets (see Chapter 5, Figure 5.9) fitted with a rubber bulb for the transfer of a liquid
- Pasteur filter pipets held by a clamp for the filtration

Method 1: Filtration of the organic liquid from the drying agent. After a microscale extraction, the organic liquid can be dried with a drying agent in a conical vial, a centrifuge tube, or a test tube. If the drying agent has large particles, such as calcium chloride, simply use a filter-tip pipet to remove the liquid from the drying agent and transfer it to a clean, dry container.

For granular or powdered drying agents, clamp a filter pipet in an upright position and use it as a filter funnel (see Chapter 9, Figure 9.3). Use a filter-tip pipet to transfer the liquid to the filtering funnel. Collect the filtered organic liquid in a clean, dry conical vial or small round-bottomed flask.

Method 2: Drying and filtration in one step. In this method, useful for a powdered drying agent such as magnesium sulfate, both

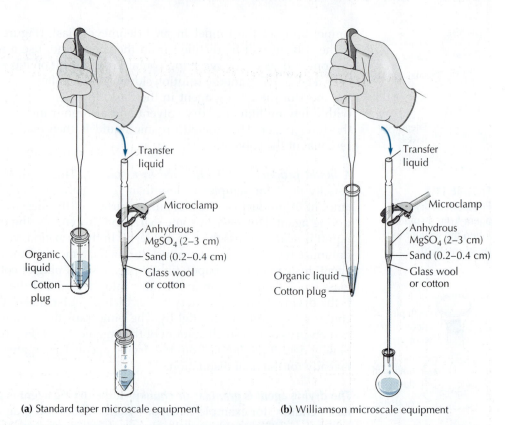

(a) Standard taper microscale equipment   **(b)** Williamson microscale equipment

FIGURE 11.4 Using microscale equipment and a Pasteur filter pipet containing anhydrous $MgSO_4$ to dry an organic liquid or solution.

drying and filtration are done simultaneously as the organic liquid passes through a Pasteur filter pipet containing anhydrous $MgSO_4$. A cotton or glass wool plug is packed into a Pasteur pipet and covered with a layer of sand (0.2–0.4 cm) and then with a layer of $MgSO_4$ (2–3 cm), as shown in Figure 11.4. The solution is transferred from its original container to the filtering pipet with a filter-tip pipet.

11.3   Sources of Confusion and Common Pitfalls

How Much Drying Agent Do I Need To Use?

The amount of drying agent necessary to remove residual water from the organic liquid cannot be specified exactly; it depends on how much water is present in the liquid. You need to learn to judge when enough drying agent has been added. If all the drying agent particles are clumped together and/or stuck to the sides of the flask, not enough has been used. Continue adding small amounts from the tip of a spatula until there is a thin layer of particles that look very similar to the original particles of the drying agent and that move freely in the flask. If indicating Drierite ($CaSO_4$) is the drying agent and it has turned a pink color, more anhydrous Drierite must be added.

Remember that using too much drying agent can cause a loss of product by its adsorption on the drying agent. If you have to add quite a bit of drying agent to reach the clumping point, you must have had a large amount of water present initially. In this case you may wish to add more organic solvent to minimize the loss of product.

What Is the White Liquid Around the Drying Agent?

When the drying agent is added to the organic liquid, a milky white liquid may appear around the drying agent particles, particularly when anhydrous calcium chloride in pellet form is being used. The pellets do not provide as much surface area for reaction with water as powder or granules do. The white liquid is a saturated water solution of calcium chloride. Continue adding pellets until the liquid is absorbed and some of the pellets move freely in the organic liquid. Allow at least 15–20 min for the drying agent to be effective.

How Do I Know When the Organic Liquid Is Dry?

Drying agents do not absorb water instantaneously and some drying agents act faster than others (Table 11.1). Allow a minimum of 10 min for the drying agent to become hydrated. When an organic liquid is dry, it will be clear and at least a portion of the drying agent will still have the particle size and appearance of the anhydrous form. If all the drying agent has become clumped or the organic liquid is still cloudy after 10 min, decant the organic liquid into a clean Erlenmeyer flask and add another portion of drying agent. Allow the mixture to be in contact with the fresh drying agent for another 10–15 min.

11.4 Recovery of an Organic Product from a Dried Extraction Solution

Once the extraction solution has been dried, the solvent must be removed to recover the desired organic compound. Evaporation of the solvent to the atmosphere has been a traditional method of recovering a product; however, concern for the environment and environmental laws now limit and sometimes prohibit this practice. Removing solvents by distillation or with a rotary evaporator are alternatives to evaporation; both methods allow the solvents to be recovered. Your instructor will advise you whether evaporation of solvents is allowed in your laboratory or if a method where the solvent is recovered must be used.

Evaporation Methods

For experiments in which the amount of solvent is small (less than 25 mL), it can be removed by evaporation on a steam bath in a hood or by blowing it off with a stream of nitrogen or dry air in a hood.

Boiling the solvent. Place a boiling stick or boiling stone in the Erlenmeyer flask containing the solution and heat the flask on a steam bath in a hood. The product will be the liquid or solid residue left in the flask when the boiling ceases. The last of the solvent can be blown off in a hood with a stream of nitrogen or dry air.

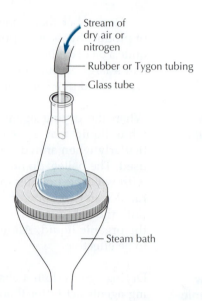

FIGURE 11.5
Using a stream of
nitrogen or dry air
to evaporate an
organic liquid.

Stream of
dry air or
nitrogen

Rubber or Tygon tubing

Glass tube

Steam bath

Evaporation with a stream of dry air or nitrogen. Evaporation is
a cooling process; therefore, gently heating the container holding
the solution will speed the process. However, the liquid should not
boil. The evaporation rate can be enhanced by directing a gentle
stream of dry air or nitrogen **above** the liquid in the container. **Note:**
If the end of the tube is close to or in the solution or the flow rate
of gas is too rapid, the liquid may spatter and some of the product
will be lost.

Figure 11.5 shows the apparatus for a miniscale evaporation
with a stream of nitrogen while heating with a steam bath adjusted
for a very slow rate of steam flow. A glass tube attached to rubber or
Tygon tubing that leads to the nitrogen source should be clamped so
the end is well above the liquid level.

In microscale evaporations, warm water suffices as the heat
source and the air or nitrogen flow is directed above the liquid
through a Pasteur pipet attached to rubber or Tygon tubing.
Figure 11.6a shows a standard-taper conical vial held by auxiliary
aluminum blocks set in a small beaker of warm water. Figure 11.6b
shows a Williamson reaction tube held by a microclamp in a small
beaker of warm water.

Distillation

Assemble the simple distillation apparatus shown in Chapter 12,
Figure 12.7. If the solvent is ether, pentane, or hexane, work in a
hood and use a steam bath or a water bath on a hot plate as a heat
source to eliminate the fire hazard an electric heating mantle poses
with the very flammable vapors from these solvents. Continue the
distillation until the solvent has completely distilled, an endpoint
indicated by a drop in the temperature reading on the thermometer,
when there is no longer any hot vapor surrounding the thermometer
bulb. The product and a small amount of solvent will remain in the
distilling flask. The remaining solvent can be removed by evaporation with a stream of dry air or nitrogen.

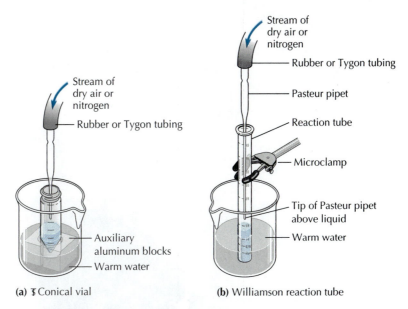

FIGURE 11.6 Microscale apparatus for using a stream of nitrogen or dry air to evaporate an organic liquid.

Using a Rotary Evaporator

A rotary evaporator is an apparatus for removing solvents rapidly in a vacuum (Figure 11.7). No boiling stones or sticks are necessary because the rotation of the flask minimizes bumping. Rotary evaporation is usually done in a round-bottomed flask that is no more than half filled with the solution being evaporated. A condenser allows the evaporated solvent to be condensed and recovered in a trap flask.

The following protocol is a generalized outline of the steps in using a rotary evaporator; consult your instructor about the exact operation of the rotary evaporators in your laboratory.

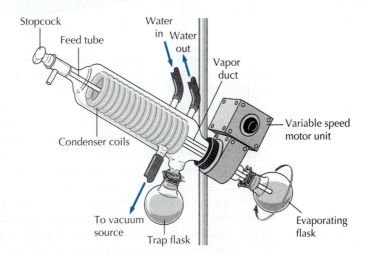

FIGURE 11.7 Diagrams for two models of rotary evaporators. *(Continued on next page)*

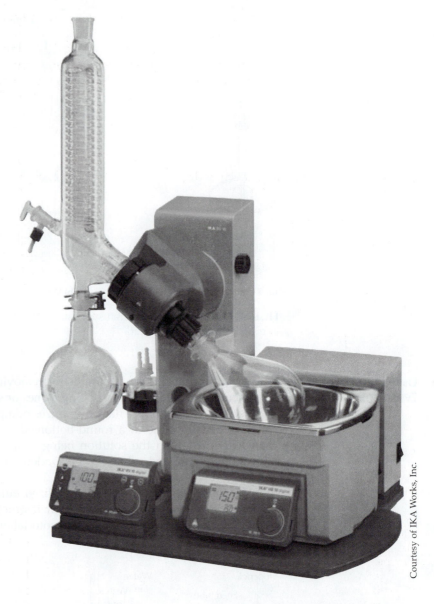

Courtesy of IKA Works, Inc.

FIGURE 11.7
(*Continued*).

Select a round-bottomed flask of a size that will be filled only halfway or less with the solution undergoing evaporation. Connect the flask to the rotary evaporator with a joint clip. Use an empty trap and be sure that it is also clipped tightly to the rotary evaporator housing. Position a room-temperature water bath under the flask containing the solution so that the flask is approximately one-third submerged in the water bath. Turn on the water to the condenser and then turn on the vacuum source. Make sure the stopcock is closed. As the vacuum develops, turn on the motor that rotates the evaporating flask. When the vacuum stabilizes at 20–30 torr or lower, begin to heat the water bath. A temperature of 50°–60°C will quickly evaporate solvents with boiling points under 100°C.

When the liquid volume in the round-bottomed flask no longer decreases, the evaporation is complete. Stop the rotation of the flask and remove the water bath. Open the stopcock slowly to release the vacuum and allow air to bleed slowly into the system. Hold the flask with one hand, take off the clip holding it to the evaporator, and remove the flask from the rotary evaporator. Turn off the vacuum source and the condenser water. Disconnect the trap from the rotary evaporator housing and empty the solvent in the trap into the appropriate waste or recovered solvent container.

Questions

1. Which would be a more effective drying agent, $CaCl_2$ or $CaCl_2 \cdot 6H_2O$? Explain.
2. (a) What are the disadvantages of using too little drying agent?
   (b) What are the disadvantages of using too much drying agent?
3. Which drying agent would you choose to dry a solution of 2-octanone (a ketone) in hexane? Explain your reasoning.
4. KOH is an excellent drying agent for some organic compounds. Would it be a better choice for a carboxylic acid (RCO_2H) or an amine (RNH_2)? Why?

CHAPTER

12

BOILING POINTS AND DISTILLATION

If Chapter 12 is your introduction to the separation and purification of organic compounds, read the Essay "Intermolecular Forces in Organic Chemistry" on pages 127–131 before you read Chapter 12.

Distillation is a method for separating two or more liquid compounds by taking advantage of their boiling-point differences. Unlike the liquid-liquid and liquid-solid separation techniques of extraction and crystallization, distillation is a liquid-gas separation in which vapor pressure differences are used to separate different compounds. For example, you may have seen the tall slender distillation columns in oil refineries where hydrocarbons are separated into various grades of gasoline. We will be discussing distillation on the miniscale and microscale level, where distillation is commonly used to separate liquid compounds from solvents following extractions and to purify liquid compounds. Recycling of solvents also often involves a distillation step.

A liquid at any temperature exerts a pressure on its environment. This *vapor pressure* results from molecules leaving the surface of the liquid to become vapor.

$$molecules_{liquid} \rightleftharpoons molecules_{vapor}$$

As a liquid is heated, the kinetic energy of its molecules increases. The equilibrium shifts to the right and more molecules move into the gaseous state, thereby increasing the vapor pressure. Figure 12.1 shows the relationship between vapor pressure and temperature for pentane, hexane, water, and octane.

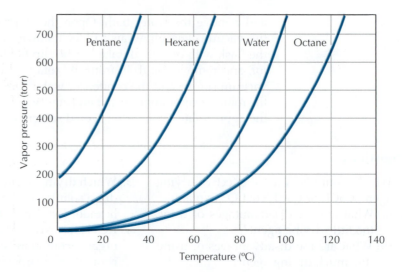

FIGURE 12.1 Examples of the dependence of vapor pressure on temperature.

12.1   Determination of Boiling Points

Boiling Point

The boiling point of a pure liquid is defined as the temperature at which the vapor pressure of the liquid exactly equals the pressure exerted on it by the atmosphere. At an external pressure of 1.0 atm (760 torr), the boiling point is reached when the vapor pressure equals 760 torr. However, at other pressures the boiling point of the liquid will be different. Table 12.1 gives boiling points of several liquids at different elevations. **When the boiling point of a substance is determined, both the atmospheric pressure and the experimental boiling point need to be recorded.**

Every pure and thermally stable organic compound has a characteristic boiling point at atmospheric pressure. The boiling point reflects its molecular structure, specifically the types of weak intermolecular interactions that bind the molecules together in the liquid state, which must be overcome for molecules to enter the vapor state. Intermolecular hydrogen bonding and dipole-dipole interactions always produce higher boiling points. Thus, polar compounds have higher boiling points than nonpolar compounds of similar molecular weight. In addition, increased molecular weight usually produces a larger molecular surface area and greater van der Waals interactions, which also leads to a higher boiling point.

TABLE 12.1	Boiling points (°C) of common compounds at different elevations (pressures)		
Compound	Death Valley, CA Elevation −285 ft P = 1.01 atm	New York City Elevation 0 ft P = 1.00 atm	Laramie, WY Elevation 7165 ft P = 0.75 atm
Water	100.3	100.0	92.2
Diethyl ether	35.0	34.6	26.7
Ethyl acetate	77.4	77.1	68.6
Acetic acid	118.2	117.9	108.7

Miniscale Determination of Boiling Points

The boiling point of 5 mL or more of a pure liquid compound can be determined by a simple distillation using miniscale standard taper glassware. The procedure for setting up a simple miniscale distillation is described in Section 12.3. When distillate is condensing steadily and the temperature stabilizes, the boiling point of the substance has been reached.

The microscale methods described here are an alternative for determining the boiling point of any pure liquid when only a very small sample of the liquid is available.

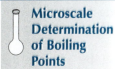

Microscale Determination of Boiling Points

Using a Williamson reaction tube. Place 0.3 mL of the liquid and a boiling stone in a reaction tube. Set the tube in the appropriately sized hole of an aluminum heating block (see Section 6.2). Alternatively, heat may be supplied by a sand bath (Section 6.2), in which case the tube and the thermometer need to be held by separate clamps. Clamp the thermometer so that the bottom of the bulb is 0.5–1.0 cm above the surface of the liquid; be sure that the thermometer does not touch the wall of the tube (Figure 12.2a).

Gradually heat the sample to boiling and continue to increase the rate of heating *slowly* until the ring of condensate is 1–2 cm above the top of the thermometer bulb. When the temperature reaches a maximum and stabilizes for at least 1 min, the boiling point of the liquid has been reached. Rapid or excessive heating of the tube can lead to superheating of the vapor and can also radiate heat from the tube to the thermometer bulb, causing the observed boiling point to be too high.

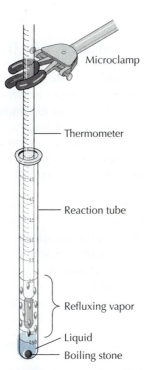

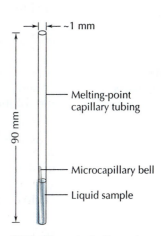

FIGURE 12.2
Apparatus for microscale boiling-point determinations.

(a) Williamson reaction tube or small test tube   (b) Capillary tube boiling point apparatus

Using a capillary tube. When only a few drops of a pure liquid are available, its boiling point can be determined with the same type of capillary tube that is used for melting points. A 10-μL syringe of the type used with a gas chromatograph works well for transferring a 4–5-μL sample into the capillary tube. If the liquid does not flow to the bottom of the tube, place the capillary tube in a centrifuge tube and spin it briefly in a centrifuge.

To prepare a microcapillary bell, obtain a 10-μL microcapillary tube that is about 40 mm long and cut the tube in half with a file or glass scorer. Hold the uncut end with tweezers and rotate the cut end in a small flame just long enough for the glass to melt and form a seal. Allow the tube to cool before inserting it **with the open end down** into the capillary tube containing the liquid sample (Figure 12.2b).

Determine the boiling point of the liquid by placing the capillary tube in a melting-point apparatus (see Section 14.2). Use the same heating procedure as for a melting-point determination (see Section 14.3). Increase the rate of heating fairly rapidly until the temperature is 15°–20°C below the expected boiling point of the compound; then decrease the rate of heating to about 2°C/min until a fine stream of bubbles emerges from the bottom of the micro-capillary bell. At this point, turn the heat controller down even lower to decrease the rate of heating. Carefully watch the stream of bubbles emerging from the bell and record the temperature when the last bubble emerges; this temperature is the boiling point of the compound. To verify it, **immediately repeat the determination** by increasing the rate of heating to 2°C/min to produce a second stream of bubbles.

12.2   Distillation and Separation of Mixtures

The boiling point of a liquid mixture depends on the vapor pressures of its components. Impurities can either raise or lower the observed boiling point. Consider, for example, the boiling charac-teristics of a mixture of pentane and hexane. The two compounds are mutually soluble, and their molecules interact with one another only by van der Waals forces. A solution composed of both pentane and hexane boils at temperatures between their two boiling points.

Raoult's and Dalton's Laws

If pentane alone were present, the vapor pressure above the liquid would be due only to pentane. When pentane is only a fraction of the solution, however, the *partial pressure* ($P_{pentane}$) exerted by pen-tane is equal to only a fraction of the vapor pressure of pure pentane ($P°_{pentane}$). The fraction is determined by $X_{pentane}$, the *mole fraction* of pentane, which is the ratio of moles of pentane to the total number of moles of pentane and hexane in the solution.

$$\text{Mole fraction of pentane:} \quad X_{pentane} = \frac{\text{moles}_{pentane}}{\text{moles}_{pentane} + \text{moles}_{hexane}}$$

$$\text{Partial pressure of pentane:} \quad P_{pentane} = P°_{pentane} \cdot X_{pentane} \tag{1}$$

The hexane present in the solution also exerts its own independent partial pressure.

$$\text{Mole fraction of hexane:} \quad X_{hexane} = \frac{\text{moles}_{hexane}}{\text{moles}_{pentane} + \text{moles}_{hexane}}$$

$$\text{Partial pressure of hexane:} \quad P_{hexane} = P^{\circ}_{hexane} \cdot X_{hexane} \tag{2}$$

The vapor pressure–mole fraction relationships expressed in equations 1 and 2 are valid only for ideal liquids in the same way that the ideal gas law strictly applies only to ideal gases. Equations 1 and 2 are applications of Raoult's law, named after the French chemist François Raoult, who studied the vapor pressures of solutions in the late nineteenth century.

Using Dalton's law of partial pressures, we can now calculate the total vapor pressure of the solution, which is the sum of the partial pressures of the individual components:

$$P_{total} = P_{pentane} + P_{hexane} \tag{3}$$

Figure 12.3 shows the partial pressure curves for pentane and hexane at 25°C using Raoult's law and the total vapor pressure of the solution using Dalton's law. The boiling point of a pentane/hexane mixture is the temperature at which the individual vapor pressures of both pentane and hexane add up to the total pressure exerted on the liquid by its surroundings.

Composition of the Vapor above the Solution

Being able to calculate the total vapor pressure of a solution can be extremely useful, but knowing the composition of the vapor above a solution is just as important. Qualitatively, it is not hard to see that the vapor above a 1:1 molar pentane/hexane solution will be richer in pentane as a result of its greater vapor pressure. Quantitatively, we can predict the composition of the vapor above a solution for which Raoult's law is valid simply by knowing the vapor pressures of its volatile components and the composition of the liquid solution.

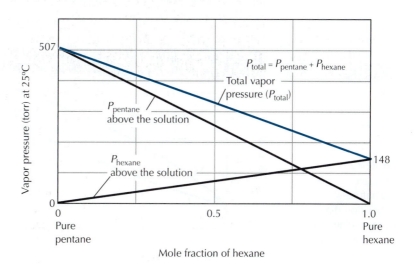

FIGURE 12.3 Vapor pressure-mole fraction diagram for pentane/hexane solutions at 25°C.

Here is an illustration of how it is done. A single expression for the total vapor pressure (equation 4) can be derived easily from equations 1, 2, and 3 because

$$X_{\text{hexane}} = 1.0 - X_{\text{pentane}}$$

$$P_{\text{total}} = X_{\text{pentane}}(P^\circ_{\text{pentane}} - P^\circ_{\text{hexane}}) + P^\circ_{\text{hexane}} \tag{4}$$

Applying the ideal gas law to the mixture of gases above a solution of pentane and hexane leads to equation 5. The quantity Y_{pentane} is the fraction of pentane molecules in the vapor above the solution.

$$Y_{\text{pentane}} = \frac{P_{\text{pentane}}}{P_{\text{total}}} \tag{5}$$

Finally, substituting equations 1 and 4 into equation 5 allows the calculation of the mole fraction of pentane in the vapor state (equation 6).

$$Y_{\text{pentane}} = \frac{P^\circ_{\text{pentane}} X_{\text{pentane}}}{X_{\text{pentane}}(P^\circ_{\text{pentane}} - P^\circ_{\text{hexane}}) + P^\circ_{\text{hexane}}} \tag{6}$$

Equation 6 allows us to construct a ***temperature-composition diagram*** (sometimes called a phase diagram) like the one shown in Figure 12.4. In addition to equation 6, all we need to know are the vapor pressures of pure pentane and pure hexane at various temperatures and the composition of the liquid. A similar diagram can also be constructed directly from experimental data.

It is useful to follow the dashed line in Figure 12.4, beginning at an initial liquid composition L_1, which has the molar composition of 75% hexane and 25% pentane. This mixture boils at 57°C, producing the vapor V_1, which has a molar composition of 52% hexane and 48% pentane. Notice that the mole fraction of the component with the lower boiling point is greater in the vapor than in the liquid. The new liquid that forms from the condensation of the vapor V_1 is L_2, which has the same composition as V_1. If liquid L_2 is vaporized, the new vapor will be even richer in pentane, shown by point V_2. Repeating the boiling and condensing processes a few more times allows you to obtain essentially pure pentane.

FIGURE 12.4
Calculated temperature-composition diagram for pentane/hexane solutions at 1.0 atm pressure.

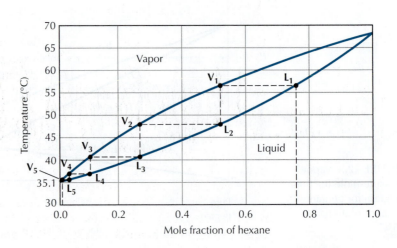

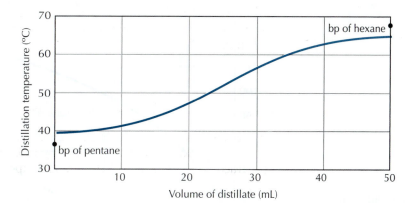

FIGURE 12.5

Distillation curve for simple distillation of a 1:1 molar solution of pentane and hexane.

Fractional and Simple Distillation

As pentane-enriched vapor is removed, the remaining liquid contains a decreasing proportion of pentane. The liquid, originally at L_1, now is richer in hexane (the component with the higher boiling point). As the mole fraction of hexane in the liquid increases, the boiling point of the liquid also increases until the boiling point of pure hexane, 69°C, is reached. In this way pure hexane can also be collected. The process of repeated vaporizations and condensations, called *fractional distillation,* allows you to separate liquid components of a mixture by exploiting the vapor pressure differences of the components (see Section 12.4) The condensed liquid is called the *distillate* or condensate.

In a *simple distillation,* perhaps only two or three vaporizations and condensations occur. Figure 12.4 shows that a simple distillation would not effectively separate a 1:3 molar solution of pentane and hexane because at least five vaporization-condensations steps are required. As the distillation proceeds, the remaining pentane/hexane solution does become increasingly concentrated in hexane and the boiling point of the solution increases, but the separation of pentane and hexane is not nearly complete. Figure 12.5 shows a distillation curve of vapor temperature versus volume of distillate for the simple distillation of a 1:1 pentane/hexane solution. The initial distillate is collected at a temperature above the boiling point of pure pentane and the final distillate never quite reaches the boiling point of pure hexane.

Now compare the temperature-composition diagram of the pentane/hexane system with that of a pentane/octane mixture. Whereas the boiling points of pentane (bp 36°C) and hexane (bp 69°C) differ by only 33°C, the boiling points of pentane and octane (bp 126°C) differ by 90°C, making it much easier to separate pentane from octane by distillation. Figure 12.6 shows that even with a 3:1 molar solution of octane and pentane, only two or three vaporizations and condensations are necessary to separate the two compounds, and thus a simple distillation would be reasonably successful in separating them. As the boiling point difference between two liquids becomes greater, simple distillation becomes increasingly more effective in their separation.

Even though simple distillation does not effectively separate a mixture of liquids whose boiling points differ by less than 60°–70°C, organic chemists use simple distillations in two commonly

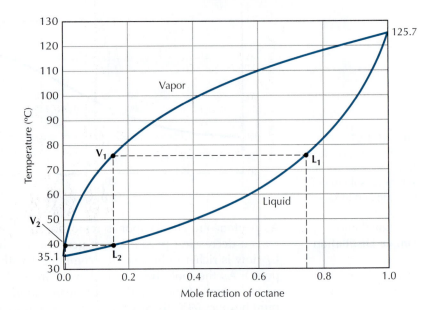

FIGURE 12.6
Calculated
temperature-
composition diagram
for pentane/octane
solutions at 1.0 atm
pressure.

encountered situations: (1) the last step in the purification of a liquid compound and (2) removal of a volatile solvent from an organic compound with a high boiling point.

12.3   Simple Distillation

In a simple distillation, **the distilling flask should be only one-third to one-half full of the liquid being distilled.** If the flask is too full, liquid can easily bump over into the condenser. If the flask is nearly empty, a substantial fraction of the material will be needed just to fill the flask and distilling head with vapor. When the desired liquid is dissolved in a large quantity of a solvent with a lower boiling point, the distillation should be interrupted after almost all of the solvent has been distilled and the higher-boiling liquids should be poured into a smaller distilling flask before continuing the distillation.

12.3a Miniscale Distillation

Figure 12.7 shows the miniscale apparatus for a simple distillation. The assembly of the apparatus is explained in detail in the following steps.

If you are using a digital thermometer, consult your instructor about the correct placement of the temperature probe in the distilling head. The digital thermometer probe can be inserted into a rubber septum stretched over the top standard taper joint of the distilling head.

**Steps in Assembling
a Miniscale
Apparatus for
Simple Distillation**

1. Select a round-bottomed flask of a size that will be one-third to one-half filled with the liquid being distilled. Place a clamp firmly on the neck of the flask and attach the clamp to a ring stand or support rod. Using a conical funnel, pour the liquid into the flask. Add one or two boiling stones.

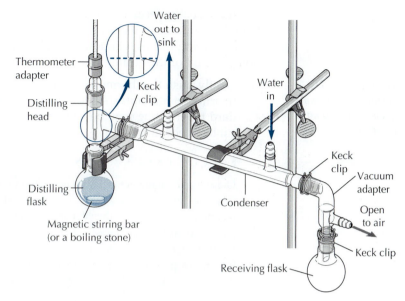

Water out to sink

Thermometer adapter

Keck clip

Water in

Distilling head

Keck clip

Vacuum adapter

Distilling flask

Condenser

Open to air

Magnetic stirring bar (or a boiling stone)

Keck clip

Receiving flask

FIGURE 12.7 Simple miniscale distillation apparatus. The enlargement shows the correct placement of a thermometer bulb for accurate measurement of the boiling point.

SAFETY PRECAUTION

Boiling stones should **never** be added to a hot liquid because they may cause a superheated liquid to boil violently.

2. *Lightly* grease the bottom joint and the side-arm joint (see Section 4.2) on the distilling head. Fit the distilling head to the round-bottomed flask and twist the joint to achieve a tight seal. Finish assembling the rest of the apparatus before inserting the thermometer adapter and thermometer. **Note: The distilling flask and distilling head need to be in a completely vertical position and the condenser positioned with a downward slant.**

A funnel keeps the ground glass joint from becoming coated with the liquid and prevents loss of product.

3. Attach rubber tubing to the outlets on the condenser jacket. Wire hose clamps are often used to prevent water hoses from being blown off the outlets by a surge in water pressure. Grease the inner joint at the bottom of the condenser, attach the vacuum adapter, and while the pieces are still lying on the desktop, place a Keck clip over the joint.

4. Clamp the condenser to another ring stand or upright support rod, as shown in Figure 12.7. If the clamp used to support the condenser has a stationary and a movable jaw, position it with the stationary jaw underneath the condenser and the movable jaw above. **The clamp attached to the condenser should have enough slack or play so that it does not cause tension in the connection between the condenser and distilling head that would make the connection loose.** Fit the upper joint of the condenser to the distilling head, twist to spread the grease, and place a Keck clip over the joint.

5. Lightly grease the inner joint at the bottom of the vacuum adapter and attach a round-bottomed flask to serve as the receiving vessel. Twist the joint to achieve a tight seal and immediately attach a Keck clip. **Although without a clip the receiver flask may stay attached to the vacuum adapter for a time, gravity will soon win out and the flask will fall and perhaps break.**

It is usually necessary to have at least two receiving vessels at hand; the first container is for collecting the initial distillate that consists of impurities with lower boiling points before the expected boiling point of the desired fraction is attained.

The use of Keck clips ensures that ground glass joints do not come apart.

6. Gently push the thermometer through the rubber sleeve on the thermometer adapter. Alternatively, a thermometer with a standard taper fitting may be used instead of the thermometer and rubber-sleeved adapter.

SAFETY PRECAUTION

Grasp the thermometer close to the bulb and push it gently 1–2 cm into the adapter. Move your hand several centimeters up the thermometer stem and repeat the pushing motion. Continue this process until the thermometer is properly positioned. Holding the thermometer by the upper part of the stem while inserting it through the rubber sleeve of the thermometer adapter could break the thermometer and force a piece of broken glass into your hand!

Proper positioning of the thermometer bulb is crucial.

7. Grease the joint on the thermometer adapter and fit it into the top joint of the distilling head. Adjust the position of the thermometer to **align the top of the thermometer bulb with the bottom of the side arm** on the distilling head (see detail in Figure 12.7).

A slow to moderate water flow rate suffices and lessens the chance of blowing the rubber tubing off the condenser.

8. Check to ensure that the rubber tubing is tightly attached to the condenser and that **water flows in at the bottom and out at the top. To prevent a flood, place the open end of the water outlet tubing into the sink. Slowly turn on the water.**

9. Place a heating mantle or other heat source under the distillation flask, using an iron ring or lab jack to support the mantle. If you are using a magnetic stirrer, turn it on (see Section 6.1).

10. Double-check the distillation apparatus, paying particular attention to the ground-glass joints to make certain that they are all fully seated in place, without any gaps. Then begin heating the flask.

If you are using an Erlenmeyer flask or graduated cylinder to collect the distillate, position the outlet of the vacuum adapter slightly inside the mouth of the receiving vessel. A **beaker should never be used as the receiving vessel** because its wide opening readily allows vapors to escape.

Carrying Out the Distillation

The expected boiling point of the liquid being distilled determines the heat input, which is often controlled by a variable transformer (see Section 6.2). Vaporization of a liquid with a high boiling point requires more heat than does that of a low-boiling liquid. Heat the liquid slowly to a gentle boil. A ring of condensate will begin to move up the inside of the flask and then up the distilling head. The temperature observed on the thermometer will not rise appreciably until the vapor reaches the thermometer bulb, which is measuring the vapor temperature, not the temperature of the boiling liquid. If the ring of condensate stops rising before it reaches the thermometer, increase the heat setting.

When the vapor reaches the thermometer, the temperature reading should increase rapidly. To achieve satisfactory separation of liquids that boil within 100°C of one another, adjust the heat input to maintain a collection rate of one drop every 1–2 sec. It may be necessary to increase the heat input during the distillation if the rate of collection slows.

Collect any liquid that condenses more than 5°C below the expected boiling point as the first fraction, or forerun, which is usually discarded; then change to a second receiving vessel to collect the desired liquid fraction when the temperature stabilizes at or slightly below the expected boiling point. Record the temperature at which you begin to collect the desired fraction.

As the end of a distillation approaches, it is essential to lower the heat source BEFORE the distillation flask reaches dryness (see Safety Precaution below). If the temperature begins to drop, this signifies that vapor is no longer reaching the thermometer bulb and that the distillation should be discontinued. Record the temperature at which the last drop of distillate is collected; the initial and final temperatures are the boiling range of a liquid fraction.

SAFETY PRECAUTION

> Be sure to leave a small residue of liquid in the boiling flask so that you will not overheat the flask and break it, nor will you char the last drops of residue, which causes cleaning difficulty. Moreover, some compounds, such as ethers, secondary alcohols, and alkenes form peroxides by air oxidation. If a distillation involving one of these compounds is carried to dryness, the peroxides could explode.

12.3b Miniscale Short-Path Distillation

When only 4–6 mL of liquid are distilled, a simple distillation apparatus can be modified to a short path by reducing the size of the glassware and shortening the condenser, as shown in Figure 12.8. The short path reduces the *holdup volume*, the volume of the distilling flask and fractionating column, which is filled with vapor during and after completion of a distillation. Short-path distillation also prevents distillate from being lost on the walls of a long condenser. A beaker or crystallizing dish of water surrounding the receiving flask replaces the condenser. If the liquid boils below 100°C, the beaker should contain an ice/water mixture. If the liquid boils above 100°C, a water bath provides sufficient cooling. For liquids that boil above 150°C, air cooling of the receiving flask suffices.

Figure 12.8b shows an even more efficient short-path distillation apparatus than the one shown in Figure 12.8a. In this apparatus the distilling head, a short condenser, and the vacuum adapter are combined in one piece of standard taper glassware. Using a pear-shaped distilling flask also leads to less loss of a valuable product. Despite the presence of a condenser, an ice/water bath is usually placed around the receiving flask for maximum cooling efficiency.

If you are doing a simple distillation as described in Section 12.3a, carry out the short-path distillation at a rate of less than one drop per second. While changing receiving flasks, it may be necessary to stop the distillation by removing the heat source.

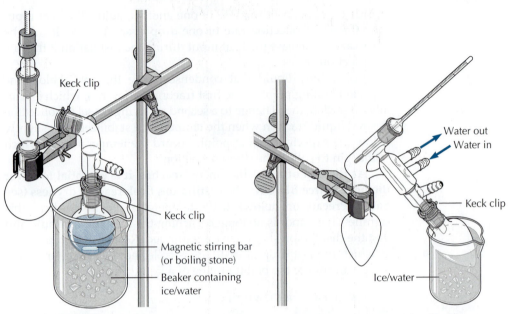

(a) ₮ 19/22 short-path apparatus   **(b)** ₮ 14/20 one-piece distilling head and condenser

FIGURE 12.8 Two types of short-path distillation apparatus.

12.3c Microscale Distillation Using Standard Taper 14/10 Apparatus

A microscale apparatus is required when the volume of a liquid to be distilled is less than 5 mL. For the distilling vessel, use a conical vial for 1–3 mL or a 10-mL round-bottomed flask for 4–5 mL of liquid. Set the vial or flask in a small beaker before putting the liquid into it. Add a magnetic spin vane to the vial or a magnetic stirring bar to the round-bottomed flask (see Section 6.1).

Assembly of a Short-Path Distillation Apparatus

Assemble standard taper microscale glassware into a short-path distillation apparatus with a 14/10 distillation head, a thermometer adapter (Figure 12.9), and a bent vacuum adapter, as shown in Figure 12.10. Begin by putting the thermometer through the threaded cap of the thermometer adapter, then push a small O-ring up the thermometer as shown in Figure 12.9. Fit a screw cap and a large O-ring over the ground glass joint at the bottom of the thermometer adapter. Place the thermometer adapter in the top of the distilling head and tighten the screw cap. Adjust the position of the thermometer in the adapter until **the top of the thermometer bulb is aligned with the bottom of the side arm of the distilling head** (see enlargement, Figure 12.10). Attach the bent vacuum adapter to the distilling head and the receiving vial to the open end of the bent adapter. Finally, attach the conical vial or round-bottomed flask holding the liquid to be distilled to the distilling head with a screw cap and O-ring; firmly clamp the apparatus to a ring stand or upright post.

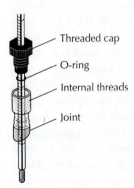

FIGURE 12.9
Thermometer adapter for 14/10 microscale glassware.

Place an aluminum heating block on a hot plate. Lower the distilling vessel into the heating block. The conical vial collecting the

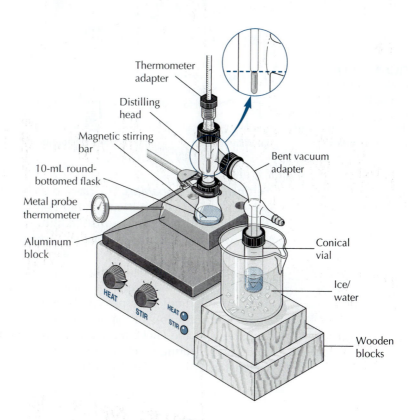

FIGURE 12.10
Short-path standard taper microscale distillation apparatus.

distillate should be half submerged in an ice/water bath for efficient condensation of the vapor.

For distillation of very volatile liquids, a water-jacketed condenser can be inserted between the distilling head and the vacuum adapter. Attach rubber tubing to the condenser so that water enters at the lower outlet and exits into the sink at the upper outlet.

Carrying Out the Distillation

The procedure for carrying out a microscale distillation is the same as that for a miniscale distillation. Follow the procedure described in Section 12.3a, pages 180–183. Have two conical vials available for the distillate: one for the forerun before the expected boiling point is reached, the other for the final product. Turn on the magnetic stirrer. Heat the aluminum block slowly to a temperature 20°–30°C above the boiling point of the liquid being distilled. Do the distillation at a rate of less than one drop per second. While changing the receiving vial, it may be necessary to stop the distillation by removing the heat source.

Using a Hickman Distilling Head

Another type of standard taper microscale distillation apparatus consists of a Hickman distilling head (Figure 12.11) and a 3–5-mL conical vial or a 10-mL round-bottomed flask. The Hickman distilling head also serves as the receiving vessel, which considerably reduces the holdup volume. Vapors condense on the upper portion of the Hickman still and drain into the bulbous collection well. One version of the Hickman still has a port at the side for easy removal of the condensate (Figure 12.11a).

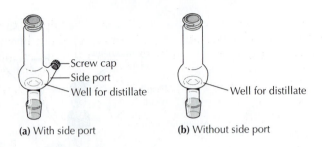

FIGURE 12.11
Hickman distilling heads. The condensate collects in the well at the bottom of the head in both versions.

—Screw cap
—Side port
—Well for distillate

—Well for distillate

(a) With side port   **(b)** Without side port

Setting up the apparatus. To carry out the microscale distillation, select a conical vial or 10-mL round-bottomed flask appropriate for the volume of liquid to be distilled; the vessel should be no more than two-thirds full. Use a Pasteur pipet to place the liquid in the vial and add a magnetic spin vane or a boiling stone. Attach the Hickman distilling head to the vial with a screw cap and O-ring. Usually an air condenser or a water-cooled condenser (for particularly volatile liquids) is placed above the Hickman distilling head to minimize the loss of vapor (Figure 12.12). Clamp the assembled apparatus at the Hickman distilling head and place the vial in an aluminum heating block. If you are using a spin vane, turn on the magnetic stirrer (see Section 6.1).

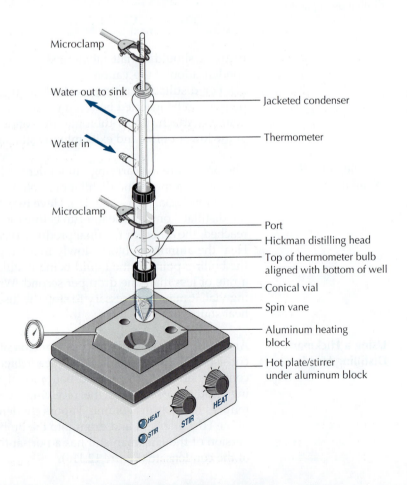

Microclamp

Water out to sink

Water in

Microclamp

Jacketed condenser

Thermometer

Port
Hickman distilling head
Top of thermometer bulb aligned with bottom of well
Conical vial
Spin vane
Aluminum heating block
Hot plate/stirrer under aluminum block

HEAT   STIR   HEAT
HEAT
STIR

FIGURE 12.12
Standard taper apparatus for a microscale distillation using a Hickman distilling head with a side port.

Grease is not used on ground glass joints of microscale glassware because its presence could contaminate the product.

It may be necessary to wrap the distillation vial loosely with glass wool to prevent rapid heat loss, but do not wrap the well of the Hickman distilling head.

Carrying out the distillation. Position a thermometer inside the condenser and the Hickman distilling head, with the top of the thermometer bulb aligned with the bottom of the head's collection well, as shown in Figure 12.12. Clamp the thermometer firmly above the condenser. Begin heating the aluminum block slowly to a temperature 20°–30°C above the boiling point of the liquid being distilled.

After the liquid in the vial boils, you should see a ring of condensate slowly moving up the vial and into the Hickman distilling head. The temperature observed on the thermometer rises as the vapor reaches the thermometer bulb. You may also see the upper neck of the Hickman distilling head become wet and shiny as the vapor condenses and begins to fill the well. The distillation must be done at a rate slow enough to allow the vapor to condense and not evaporate out of the condenser.

Removing the distillate. The collection well has a capacity of about 1 mL, so the distillate may need to be removed once or twice during a distillation. Open the port and quickly remove the distillate with a clean Pasteur pipet. Alternatively, withdraw the distillate using a syringe inserted through a septum in the screw cap of the port.

12.3d Microscale Distillation Using Williamson Apparatus

The Williamson microscale distillation apparatus is essentially a miniature version of the standard taper short-path distillation apparatus (see Section 12.3c). The apparatus consists of a 5-mL round-bottomed flask and a distillation head connected by a flexible connector with a support rod. The thermometer is held in place by the flexible thermometer adapter, as shown in Figure 12.13. The distillate is collected in a small vial that is at least three-fourths submerged in a 50-mL beaker of ice and water.

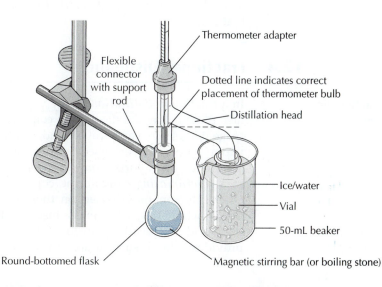

FIGURE 12.13
Williamson microscale distillation apparatus.

Thermometer adapter

Flexible connector with support rod

Dotted line indicates correct placement of thermometer bulb

Distillation head

Ice/water

Vial

50-mL beaker

Round-bottomed flask

Magnetic stirring bar (or boiling stone)

Assembling the Apparatus

Using a Pasteur pipet, transfer the liquid (no more than 3 mL) to the 5-mL round-bottomed flask and add a magnetic stirring bar or a boiling stone. Attach the flexible connector with a support rod to the flask and clamp the rod to a vertical support rod or ring stand. Fit the flexible thermometer adapter to the top of the distilling head and carefully push a thermometer through the adapter.

SAFETY PRECAUTION

Grasp the thermometer close to the bulb and push it gently 1–2 cm into the adapter. Move your hand several centimeters up the thermometer stem and repeat the pushing motion; continue this process until the thermometer is properly positioned. Holding the thermometer by the upper part of the stem while inserting it through the rubber sleeve of the thermometer adapter could break the thermometer and force a piece of broken glass into your hand!

Place the top of the thermometer bulb just below the side arm, as shown by the dashed line drawn across the distillation head in Figure 12.13. Fit the distillation head into the flexible connector holding the distillation flask. Place the receiving vial in a 50-mL beaker of ice and water, and position the vial under the outlet of the distillation head, as far as it will go. Put a sand bath or an aluminum heating block with a flask depression under the round-bottomed flask. The temperature of the sand bath or aluminum block needs to be 20°–50°C above the boiling point of the liquid being distilled. If you are using a magnetic stirrer, turn it on.

Carrying Out the Distillation

After the liquid in the flask boils, you should notice a ring of condensate slowly moving up the flask and into the distillation head. The temperature observed on the thermometer rises as the vapor reaches the thermometer bulb. The distillation should be done at a rate slow enough for the vapor to condense and not evaporate out of the system. It may be necessary to wrap a wet pipe cleaner or wet paper towel around the side arm of the distillation head to increase its cooling efficiency, particularly for the distillation of compounds that boil below 100°C.

12.4   Fractional Distillation

Fractionating Columns

In a *fractional distillation,* many vaporizations and condensations take place before the distillate is collected. As shown in Figure 12.4 (page 178), each vaporization and condensation cycle causes the vapor to become enriched in the more volatile compound. If a number of vaporization/condensation cycles are carried out in a *fractionating column,* the components of a mixture can be efficiently separated based on their vapor pressure differences. The fractionating column is inserted between the distillation flask and the distilling head of the distillation apparatus and provides a large surface area over which a number of separate liquid-vapor equilibria can occur. As vapor travels up a column, it cools, condenses into a liquid, then vaporizes again after it comes into contact with hotter vapor rising from below. The process can be repeated many times. If the fractionating column is efficient, the vapor that finally reaches the distilling head at the top of the column is composed entirely of the component with the lower boiling point.

Efficiency of a fractionating column. The efficiency of a fractionating column is expressed as its number of *theoretical plates*—a term best defined with the help of Figure 12.4 (page 178). Assume that the original solution being distilled has a molar composition of 75% hexane and 25% pentane. A fractionating column would have one theoretical plate if the liquid that is collected from the top of the column has the molar composition of 52% hexane and 48% pentane (L_2). In other words, a fractionating column has one theoretical plate if one complete vaporization of the original solution occurs in the column, followed by condensation of the vapor.

The column would have two theoretical plates if the liquid that distills has the molar composition L_3, which is 27% hexane and 73% pentane. Figure 12.4 indicates that a column with five theoretical plates would seem sufficient to obtain essentially pure pentane from the 1:3 pentane/hexane mixture present at the start of the distillation. However, as the distillation progresses, the residue in the boiling flask becomes richer in hexane, so a few more theoretical plates are required for complete separation of the two compounds.

Types of fractionating columns. Fractionating columns that can be used to separate two liquids boiling at least 25°C apart are shown in Figure 12.14. The larger the column surface area on which liquid-vapor equilibria can occur, the more efficient the column will be. The fractionating columns shown in Figure 12.14 have from six to eight theoretical plates. A fractionating column with eight theoretical plates can separate liquids boiling only 25°C apart.

A more efficient column can be made by packing a simple fractionating column with a wire spiral, glass helixes, metal sponge, or thin metal strips. These packings provide additional surface area on which liquid-vapor equilibria can occur. Care must be used in selecting packing materials, however, to ensure that the packing does not undergo chemical reactions with the hot liquids in the fractionating column.

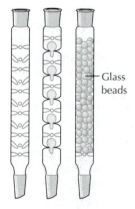

Glass beads

FIGURE 12.14
Examples of fractionating columns.

Effective Fractional Distillation

Figure 12.15 shows the separation of molecules of two compounds with different boiling points in a fractional distillation column. If the fractionating column has enough theoretical plates to completely separate a mixture of pentane and hexane, for example, the initial condensate will appear when the temperature is very close to 36°C, the boiling point of pentane. The observed boiling point will remain essentially constant at 36°C while all the pentane distills into the receiving vessel. Then the boiling point will rise

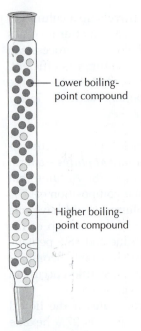

Lower boiling-
point compound

Higher boiling-
point compound

FIGURE 12.15
Separation of two
compounds with
different boiling points
in a fractional
distillation column.

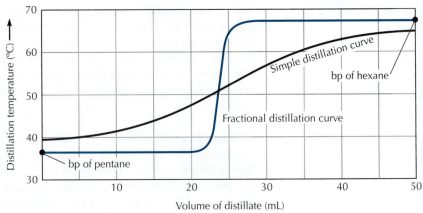

FIGURE 12.16 Distillation curve for the fractional distillation of a solution of pentane and hexane. The black line represents the distillation curve for a simple distillation of the same solution.

rapidly to 69°C, the boiling point of hexane. Figure 12.16 shows a distillation curve for the fractional distillation of pentane and hexane. The abrupt temperature increase in boiling point at approximately 22–24 mL of distillate demonstrates an efficient fractional distillation.

 Miniscale Fractional Distillation Apparatus

As in simple distillation, the distilling flask capacity should be about two times the volume of liquid being distilled. When the desired material is contained in a large quantity of a solvent with a lower boiling point, the distillation should be interrupted after the solvent has distilled, and the liquids with higher boiling points (the solution that remains in the boiling flask) should be transferred to a smaller flask before continuing the distillation.

Figure 12.17 shows the apparatus for a fractional distillation. Follow the steps listed in Section 12.3a for assembling a simple distillation apparatus, except for the addition of the fractionating column between the distillation flask and the distilling head. Be sure to add one or two boiling stones to the distilling flask, and be sure that the thermometer is placed correctly, as shown in the circled detail in Figure 12.7.

Carrying Out a Fractional Distillation

Rate of heating. Control of heating in a fractional distillation is extremely important; the heat needs to be increased gradually as the distillation proceeds. Applying too much heat causes the distillation to occur so quickly that the repeated liquid-vapor equilibria required to bring about maximum separation cannot occur. On the other hand, if too little heat is applied, the column may lose heat faster than it can be warmed by the vapor, thus preventing the vapor from reaching the top of the column. Thus, too little heat causes the thermometer reading to drop below the boiling point of the liquid, simply because vapor is no longer reaching the thermometer bulb.

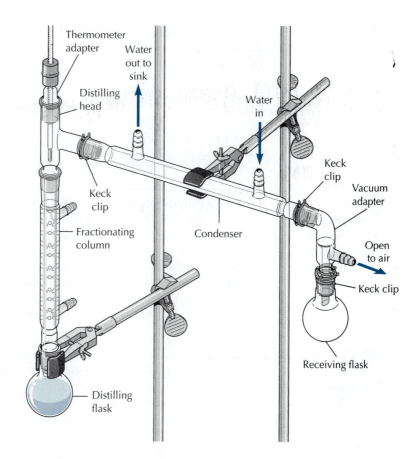

Thermometer adapter

Water out to sink

Distilling head

Water in

Keck clip

Vacuum adapter

Keck clip

Condenser

Fractionating column

Open to air

Keck clip

Receiving flask

Distilling flask

FIGURE 12.17
Miniscale fractional distillation apparatus. The fractionating column is inserted between the distilling flask and the distilling head.

Rate of distillation. The rate of distillation is always a compromise between the time the distillation takes and the efficiency of the fractionation. For an easy separation, one to two drops per second can be collected. Generally a slow, steady distillation where one drop is collected every 2–3 sec is a better rate. Difficult separations (when the boiling points of the distilling compounds are close together) require a slower distillation rate as well as a more efficient fractionating column—one with more theoretical plates. The distillation rate can be increased during collection of the last fraction, when all the lower boiling compounds have already been distilled.

Collecting the fractions. You will need a labeled receiving container (round-bottomed flask, vial, or Erlenmeyer flask) for each fraction you plan to collect. The cutoff points for the fractions are the boiling points (at atmospheric pressure) of the substances being separated. For example, in a fractional distillation of a solution of pentane (bp 36°C) and hexane (bp 69°C), the first fraction would be collected when the temperature at the distilling head reaches 34°–36°C. The temperature would stay at 36°C for a period of time while the pentane distills.

Eventually the temperature either rises or drops several degrees; a drop indicates that there is no longer enough pentane vapor to maintain the temperature at the thermometer bulb. At this point, increase the heat input and change to the second receiving flask.

Liquid then begins to distill again. Leave the second receiver in place until the temperature reaches 67°–69°C, near the boiling point of hexane; then change to the third receiving flask. The second receiver should contain only a small amount of distillate. Continue collecting fraction 3 (hexane) until only 1 mL of liquid remains in the distillation flask.

SAFETY PRECAUTION

A distillation flask should **never** be allowed to boil dry.

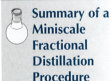

Summary of a Miniscale Fractional Distillation Procedure

1. Use a round-bottomed flask that has a capacity about two times the volume of the liquid mixture you wish to distill. Clamp the flask to a ring stand or upright support rod. Pour the liquid into the flask and add one or two boiling stones.
2. Set up the rest of the apparatus as shown in Figure 12.17.
3. Heat the mixture to boiling and collect the distillate in fractions based on the boiling points of the individual components in the mixture. Use a separate labeled receiving container for each fraction.

Microscale Fractional Distillation

Among the most efficient fractionating columns for microscale distillation are those with helical bands of Teflon mesh that spin at many rotations per minute. A microscale spinning-band distillation apparatus has a Teflon rod with a spiral molded along its axis, extending from the bottom of the column to the top (Figure 12.18). The spinning band wipes the condensate on the side of the column into a thin film and forces the rising vapors into contact with the descending condensate. The result is a large increase in the number of vaporization/condensation equilibria in the column. Spinning-band columns can have more than 100 theoretical plates and can be used to separate liquids that have a boiling-point difference of only a few degrees.

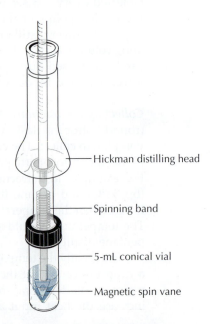

Hickman distilling head

Spinning band

5-mL conical vial

Magnetic spin vane

FIGURE 12.18
Hickman distilling head with spinning band apparatus.

12.5 Azeotropic Distillation

Up to this point, the mixtures to be separated have been solutions whose compounds interact only slightly with one another and thus approximate the behavior of ideal solutions. Most liquid solutions, however, deviate from ideality because of intermolecular interactions in the liquid state—hydrogen bonding, for example. In the distillation of some solutions, mixtures that boil at a constant temperature are produced. Such constant-boiling mixtures, called *azeotropes*, or *azeotropic mixtures*, cannot be further purified by distillation.

One of the best-known binary mixtures that forms an azeotrope during distillation is the ethanol/water system, shown in Figure 12.19. The azeotrope boils at 78.2°C and consists of 95.6% ethanol and 4.4% water by weight. The liquid that has this azeotropic composition will vaporize to a gas that has exactly the same composition because the liquid and vapor curves intersect at this point. No matter how many more liquid-vapor equilibria take place as the vapor travels up the column, no further separation will occur. Continued distillation never yields a liquid that contains more than 95.6% ethanol. Pure ethanol must be obtained by other means.

More detailed discussion about the formation of azeotropes from nonideal solutions can be found in the reference at the end of the chapter. Extensive tables of azeotropic data are available in references such as the *CRC Handbook of Chemistry and Physics*. Table 12.2 lists a few azeotropes formed by common solvents.

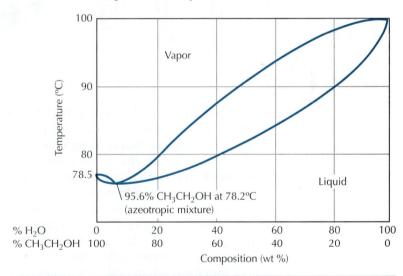

FIGURE 12.19
Temperature-composition diagram for ethanol/water mixtures.

TABLE 12.2		Azeotropes formed by common solvents		
Component X (bp)	% by wt	Component Y (bp)	% by wt	Azeotrope bp
Water (100)	13.5	Toluene (110.7)	86.5	84.1
Water (100)	1.4	Pentane (36.1)	98.6	34.6
Methanol (64.7)	12.1	Acetone (56.1)	87.9	55.5
Methanol (64.7)	72.5	Toluene (110.7)	27.5	63.5
Ethanol (78.3)	68.0	Toluene (110.7)	32.0	76.7
Water (100)	1.3	Diethyl ether (34.5)	98.7	34.2

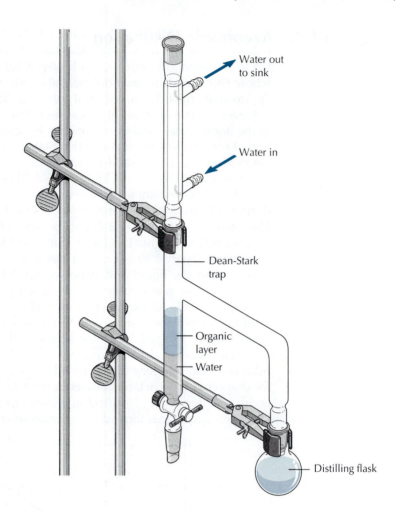

FIGURE 12.20 Dean-Stark apparatus for azeotropic removal of water from a reaction.

Azeotropic distillation is a useful way to remove a product, such as water, from a reaction mixture by codistillation with an immiscible organic liquid. Removing the water shifts the reaction equilibrium toward the product side. If the reaction were carried out in toluene, which is less dense than water, the vapor in the reflux condenser would contain an azeotropic mixture of toluene and water. When this mixture condenses, it falls into the Dean-Stark trap, shown in Figure 12.20, and separates into a layer of liquid toluene on top of the lower water layer. When the liquid level in the Dean-Stark trap reaches the level of the side arm, the toluene flows back into the reaction flask. The water can be removed through the stopcock at the bottom of the Dean-Stark trap.

12.6   Steam Distillation

Codistillation with water, called *steam distillation,* allows distillation of relatively nonvolatile organic compounds without using vacuum systems. Steam distillation can be thought of as a special kind of azeotropic distillation; it is especially useful for separating volatile organic compounds from nonvolatile inorganic salts or from the leaves and seeds of plants. Indeed, the process has found wide application in the flavor and fragrance industries as a means of separating essences or flavor oils from plant material. For example, limonene (oil of orange) can be separated from ground orange peels by steam distillation.

Mutual Insolubility and Vapor Pressure

Steam distillation depends on the mutual insolubility or immiscibility of many organic compounds with water. In such two-phase systems, at any given temperature each of the two components exerts its own full vapor pressure. The total vapor pressure above the two-phase mixture is equal to the sum of the vapor pressures of the pure components independent of their relative amounts.

Consider the codistillation of iodobenzene (bp 188°C) and water (bp 100°C). The vapor pressures ($P°$) of both substances increase with temperature, but the vapor pressure of water will always be higher than that of iodobenzene because water is more volatile. At 98°C,

$$P°_{iodobenzene} = 46 \text{ torr}$$

$$P°_{water} = 714 \text{ torr}$$

$$P°_{iodobenzene} + P°_{water} = 760 \text{ torr}$$

Therefore, a mixture of iodobenzene and water codistills at 98°C at 1.0 atm pressure.

An ideal gas law calculation shows that the mole fraction of iodobenzene in the vapor at the distilling head is 0.06 (46 torr/ 760 torr), and the mole fraction of water in the vapor is 0.94. However, because iodobenzene has a much higher molecular weight than water (204 g/mol versus 18 g/mol), its weight percentage in the vapor is much larger than 0.06, as the following calculation shows:

$$\frac{moles_{iodobenzene}}{moles_{water}} = \frac{P°_{iodobenzene}}{P°_{water}}$$

$$\frac{grams_{iodobenzene}/MW_{iodobenzene}}{grams_{water}/MW_{water}} = \frac{P°_{iodobenzene}}{P°_{water}}$$

Rearranging this expression and substituting for the molecular weights and vapor pressures allows you to calculate the weight ratio of iodobenzene to water in the distillate from the steam distillation:

$$\frac{grams_{iodobenzene}}{grams_{water}} = \frac{0.73}{1.00}$$

In other words, the distilling liquid contains 42% iodobenzene and 58% water by weight. In any steam distillation, a large excess of water is used in the distilling flask so that virtually all the organic compound can be distilled from the mixture at a temperature well below the boiling point of the pure compound.

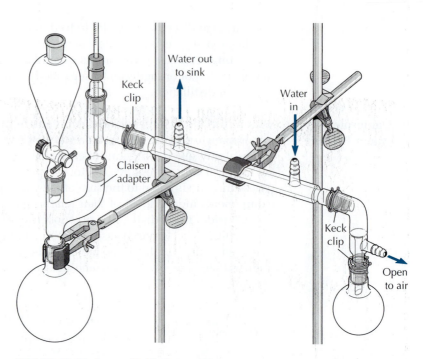

FIGURE 12.21 Steam distillation apparatus.

The steam distillation of most reasonably volatile organic compounds that are insoluble in water occurs between 80°C and 100°C. For example, at 1.0 atm, octane (bp 126°C) steam distills at 90°C, and 1-octanol (bp 195°C) steam distills at 99°C. The lower distillation temperature has the added advantage of preventing decomposition of the organic compounds during distillation.

Procedure for Steam Distillation

Use more water than the amount of organic mixture being distilled and select a distilling flask that will be no more than half filled with the organic/water mixture. Add one or two boiling stones to the flask. Modify a simple distillation apparatus by adding a Claisen connecting tube or adapter between the distillation flask and the distilling head. This adapter provides a second opening into the system to accommodate the addition of extra water without stopping the distillation (Figure 12.21).

Steps in a Steam Distillation

1. Set up the distillation apparatus using the procedure found in Section 12.3a but with the insertion of a Claisen adapter.
2. Pour the organic mixture and an excess of water into a firmly clamped distilling flask at least twice as large as the combined organic/water volume. Add one or two boiling stones.
3. Heat the mixture until the entire top organic layer has distilled into the receiving flask. The distillation rate can be as fast as you wish as long as the water-cooled condenser effectively cools the distillate. Sometimes it is worthwhile to collect some additional water after the organic material is no longer apparent in the distilling flask.
4. Separate the organic phase of the distillate from the aqueous phase in a separatory funnel (see Section 10.3).

12.7 Vacuum Distillation

Many organic compounds decompose at temperatures below their atmospheric boiling points. These compounds, however, can be distilled at temperatures lower than their atmospheric boiling points when a partial vacuum is applied to the distillation apparatus. Distillation at reduced pressure, called *vacuum distillation*, takes advantage of the fact that the boiling point of a liquid is a function of the pressure under which the liquid is contained (see Section 12.1). Although vacuum distillation is inherently less efficient for separating liquids than fractional distillation at atmospheric pressure, it is often the only feasible way to distill compounds with boiling points above 200°C.

A partial vacuum can be obtained in the laboratory with either a vacuum pump or a water aspirator. Vacuum pumps can easily produce pressures of less than 0.5 torr. The pressure obtained with a water aspirator can be no lower than the vapor pressure of water, which is approximately 13 torr at 15°C. In practice, an efficient water aspirator produces a partial vacuum of 15–25 torr.

The boiling point of a compound at any given pressure other than 760 torr is difficult to calculate exactly. As a rough estimate, a 50% drop in pressure lowers the boiling point of an organic liquid 15°–20°C. Below 25 torr, reducing the pressure by one-half lowers the boiling point approximately 10°C (Table 12.3).

The nomograph shown in Figure 12.22 provides a good way to estimate the boiling points of relatively nonpolar compounds at either reduced or atmospheric pressure. For example, if the boiling point of a compound at 760 torr is 200°C and the vacuum distillation is being done at 20 torr, the approximate boiling point is found by aligning a straightedge on 200°C in column B with 20 torr in column C; the straightedge intersects column A at 90°C, the approximate boiling point of the compound at 20 torr, as shown by the blue line on Figure 12.22. Similarly, the boiling point at atmospheric pressure can be estimated if the boiling point at a reduced pressure is known. When the boiling point in column A is aligned with the pressure in column C, a straightedge intersects column B at the approximate atmospheric boiling point. The nomograph gives a less accurate estimate of boiling points for polar compounds with stronger intermolecular forces in the liquid phase.

Monitoring the Pressure During a Vacuum Distillation

The pressure can be continuously monitored with a manometer (Figure 12.23) or read periodically with a McLeod gauge (Figure 12.24). If a water aspirator is used as the vacuum source, a trap bottle or flask must be used to prevent any back flow of water from entering the distillation apparatus.

TABLE 12.3	Boiling points (°C) at reduced pressures		
Pressure (torr)	Water	Benzaldehyde	Diphenyl ether
760	100	179	258
100	51	112	179
40	34	90	150
20	22	75	131

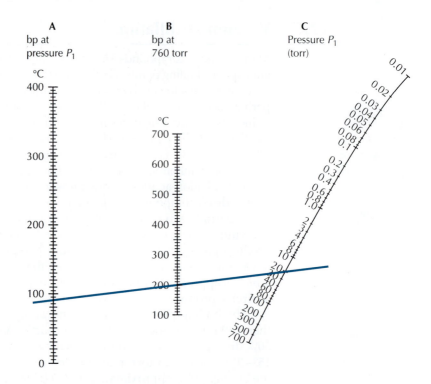

A
bp at
pressure P_1

B
bp at
760 torr

C
Pressure P_1
(torr)

FIGURE 12.22
Nomograph for
estimating boiling
points at different
pressures.

SAFETY PRECAUTION

Manometers and McLeod gauges normally contain mercury. Be careful not to spill any of the liquid mercury. If a mercury spill does occur, tell your instructor immediately.

When a vacuum pump is used as the vacuum source, a cold trap, kept at the temperature of isopropyl alcohol/dry ice (−77°C) or liquid nitrogen (−196°C), must be placed between the distillation

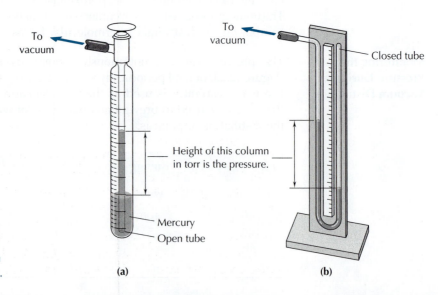

To
vacuum

To
vacuum

Closed tube

Height of this column
in torr is the pressure.

Mercury
Open tube

FIGURE 12.23 Two
types of closed-end
manometers used in
vacuum distillations.

(a)

(b)

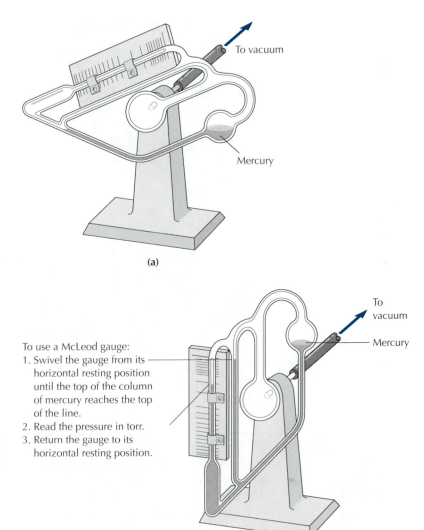

(a)

To vacuum

Mercury

To use a McLeod gauge:
1. Swivel the gauge from its horizontal resting position until the top of the column of mercury reaches the top of the line.
2. Read the pressure in torr.
3. Return the gauge to its horizontal resting position.

FIGURE 12.24
McLeod gauge used in vacuum distillations shown (a) in the horizontal resting position and (b) in the vertical measuring position.

To vacuum

Mercury

(b)

system and the pump. The trap collects any volatile materials that could otherwise get into the pump oil and cause a rise in the vapor pressure of the oil, which would decrease the efficiency of and possibly damage the pump. A pressure relief valve serves to close the system from the atmosphere and to release the vacuum after the system has cooled following the distillation. Consult your instructor before you do a distillation using a vacuum pump.

A McLeod gauge is often used to measure pressures below 5 torr. It works by compressing the gas inside the gauge into a closed capillary tube with a pressure great enough to be measured with a mercury column. Initially the gauge must be in the horizontal resting position with the mercury in the reservoir (Figure 12.24a). When the pressure inside the distillation apparatus has stabilized, the gauge is slowly rotated until the open-ended reference capillary tube is in the vertical position (Figure 12.24b). The pressure is indicated by the scale on the closed-end capillary tube when the

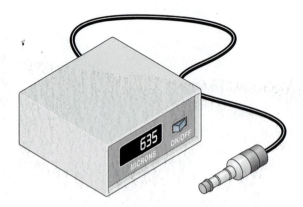

FIGURE 12.25 Digital, nonmercury vacuum gauge.

mercury level in the reference capillary tube reaches the calibration mark. After the pressure has been read, the gauge must be returned to the horizontal resting position.

When distillations are carried out at high vacuum, the distillation apparatus can be connected to a vacuum manifold, which has multiple ports equipped with stopcocks. To minimize leaks and for safety reasons, vacuum manifolds are mounted securely on metal racks. The pressure inside the vacuum system is often measured with a digital electronic gauge such as the one shown in Figure 12.25, which measures the pressure in microtorrs (10^{-3} torr). Consult your instructor before using a McLeod gauge or a vacuum manifold.

Apparatus for Miniscale Vacuum Distillation

The vacuum distillation apparatus shown in Figure 12.26 works adequately for most vacuum distillations, although a fractionating column may be needed to provide satisfactory separation of some mixtures. Because liquids often boil violently at reduced pressures, a Claisen connecting adapter is always used in a vacuum distillation to lessen the possibility of liquid bumping up into the condenser. If undistilled material jumps through the Claisen adapter into the condenser, you must begin the distillation again. Uncontrolled bumping during a vacuum distillation can be lessened by using a large distillation flask, by adding small pieces of wood splints in place of boiling stones, or by magnetic stirring.

S A F E T Y   P R E C A U T I O N

When you are using vacuum distillation, there is danger of an implosion that could scatter sharp shards of glass. Safety glasses must be worn at all times while carrying out a vacuum distillation.

If a satisfactory vacuum is to be maintained, **each connecting surface must be greased with high-vacuum silicone grease, and the rubber tubing to the aspirator or vacuum pump must be thick-walled** so that it does not collapse under vacuum. To prevent air leaks, rotate one part of the ground-glass joint against the other to make sure that the joint is evenly lubricated with the high-vacuum grease. Be sure to use a thin film of grease applied only at the top half of the inner joint. If the partial vacuum is not as low as expected, carefully check all connections for possible leaks.

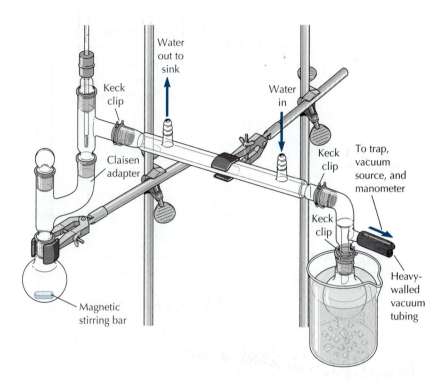

FIGURE 12.26
Vacuum distillation
apparatus.

To change the receiving flask using the apparatus shown in Figure 12.26, you must allow air into the distillation assembly to bring it back to atmospheric pressure. This often requires cooling down the distillation flask somewhat before allowing the air back in. Figure 12.27 shows a **"cow" receiver,** which allows the collection of four distillation fractions without breaking the vacuum. This apparatus is an efficient setup for vacuum distillations because the receiver can simply be rotated to change the receiver arm when beginning the collection of a new distillation fraction.

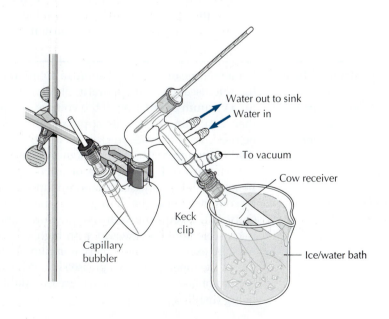

FIGURE 12.27 Short-path standard taper apparatus for vacuum distillation with capillary bubbler and cow receiver.

Figure 12.27 also shows how a very finely drawn-out capillary can provide a steady stream of very small bubbles to enhance the steadiness of a distillation. If you use a capillary bubbler, the bottom of the bubbler must be just above the bottom surface of the distilling flask and must always be below the liquid's surface. Do not use wood splints or boiling stones when you use a capillary bubbler; their violent motions may break the fragile tip of the bubbler, making it useless.

S A F E T Y   P R E C A U T I O N

Safety glasses must be worn at all times while carrying out a vacuum distillation because of the danger of an implosion, which can shatter the glassware.

 Steps in a Miniscale Vacuum Distillation

1. Add the liquid to be distilled to a round-bottomed flask sized so that it will be less than half filled. Add some wood splints or a magnetic stirring bar and set up the apparatus as shown in Figure 12.26, or use a capillary bubbler as shown in Figure 12.27.
2. Attach a trap and a manometer (see Figure 12.23), a McLeod gauge (see Figure 12.24), or a digital pressure gauge (see Figure 12.25) to the system and connect the apparatus to the vacuum source with thick-walled rubber tubing.
3. Double-check the distillation apparatus, paying particular attention to the ground-glass joints to make certain that they are all fully seated in place.
4. Close the pressure release valve and turn on the vacuum.
5. When the vacuum has reached an appropriate level, heat the distilling flask cautiously to obtain a moderate distillation rate. Periodically monitor the pressure during the distillation.
6. When the distillation is complete, remove the heat source and allow the apparatus to cool nearly to room temperature before allowing air into the apparatus. **Turn off the aspirator or vacuum pump only after the vacuum has been broken. If you have used a cold trap, empty its contents immediately.**

 Standard Taper Microscale Apparatus for Vacuum Distillation

The well in a Hickman distilling head has a capacity of only 1 mL.

For a volume of 2–5 mL of liquid, a 10-mL round-bottomed flask and the microscale 14/10 apparatus shown in Figure 12.28 can be used for a vacuum distillation. If the volume of liquid to be distilled is less than 2 mL, the microscale apparatus shown in Figure 12.29 should be used. In both cases thick-walled rubber tubing must connect the distillation apparatus to the vacuum source.

The ground glass joints of microscale glassware should not be greased. Usually clean standard taper joints are completely sealed by compression of the O-ring when the cap is screwed down tightly. Only if the requisite reduced pressure cannot be obtained should microscale joints be greased with high-vacuum silicone grease. Care must be exercised to use a very thin film of grease applied only at the top of the inner joints. No grease should be allowed to seep from the bottom of any joint because the grease might contaminate the liquid being distilled.

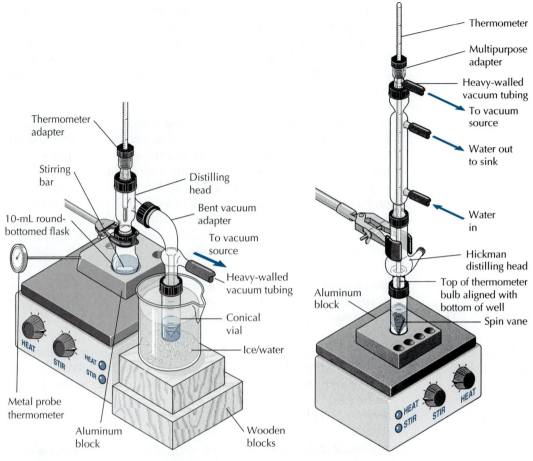

FIGURE 12.28 Short-path standard taper microscale apparatus for vacuum distillation.

FIGURE 12.29 Standard taper microscale apparatus for distillation with a Hickman distilling head.

SAFETY PRECAUTION

Safety glasses must be worn at all times while carrying out a vacuum distillation because of the danger of an implosion, which can shatter the glassware.

12.8   Sources of Confusion and Common Pitfalls

Distillation is an important method for separating and purifying organic liquids; however, carrying out successful distillations is a complex process that requires careful attention to a number of experimental details.

This Is My First Time Doing a Distillation. How Can I Avoid Common Problems?

Check your set-up before turning on the heat. After assembly of the glassware, it is good practice to double-check your distillation set-up before turning on the heat. Specifically, check to see that all ground glass joints are fully seated, with no gaps. With the multiple clamps used to secure different parts of the set-up in a miniscale simple (Figure 12.7) or fractional (Figure 12.17) distillation, it is easy for the glassware joints to separate if the various parts of the distillation apparatus are not perfectly aligned, especially those joints not secured with a Keck clip. Also make sure that the end of the rubber hose that carries water out of the condenser is placed securely in the sink.

Don't forget to add two or three boiling stones. Some form of agitation of the liquid in the distilling flask is needed to provide even, controlled boiling and to prevent bumping. Boiling stones work well for distillations done at atmospheric pressure, but be sure to add them *before* turning on the heat. Adding boiling stones to a superheated liquid can cause violent bumping. Boiling stones are not effective in vacuum distillations, but wood splints or stirring with a magnetic stirring bar serve as effective alternatives for both atmospheric and vacuum distillations.

Watch your distillation as it proceeds. Never leave a distillation unattended or let the distillation proceed to dryness. (See the safety precaution on page 183.)

What Type of Distillation Should I Use?

Simple distillation. Simple distillation is used in two commonly encountered situations: (1) to remove a low-boiling solvent from an organic compound with a high boiling point and (2) as the last step in the purification of a liquid compound to obtain a pure product and determine its boiling point.

Fractional distillation. Fractional distillation is used for the separation of a mixture of two or more liquid compounds whose boiling points differ by less than 60°–75°C.

Steam distillation. Steam distillation is used to separate volatile compounds from complex mixtures, often plant materials. It can also be used to separate an organic product from an aqueous reaction mixture containing inorganic salts.

Vacuum distillation. When the boiling point of a liquid compound is over 200°C, the compound may decompose thermally before its atmospheric boiling point is reached. The reduced atmospheric pressure of a vacuum distillation allows the compound to boil at a lower temperature and thus distill without decomposition.

Why Is the Thermometer Reading So Much Lower Than My Expected Boiling Point?

If the liquid in the distilling flask is boiling but the temperature recorded on the thermometer in the distilling head is still 25°–30°C, it is likely that the vapor has not yet reached the thermometer bulb. The space between the boiling liquid and the thermometer bulb in the distilling head must become filled with vapor before a temperature increase can be observed. Filling the space above the boiling liquid with vapor may require several minutes, depending on the rate of heating.

If the distillation is well under way and liquid is collecting in the receiving flask, yet the thermometer reading is still near room temperature, it is likely that the thermometer bulb is improperly positioned above the side arm (see Figure 12.7).

Why Did the Temperature Drop Suddenly in the Middle of My Distillation?

A sudden drop in temperature before all the liquid has distilled, especially in a fractional distillation, indicates a break between fractions. There is not enough vapor of the higher-boiling compound reaching the thermometer bulb to register on the thermometer. Increase the rate of heating until vapor again envelops the thermometer bulb.

When Do I Change Receiving Flasks?

Simple distillation. If you are conducting a simple distillation of a liquid that previously was dissolved in a low-boiling solvent, any liquid that distills at a temperature less than 5°C below the product's reported boiling point should be collected in a separate receiving flask. At 5°C or less from the expected boiling point of the liquid at the atmospheric pressure in your lab, change the receiving flask to the tared (weighed) receiving flask.

Fractional distillation. In a fractional distillation, the receiving flasks are changed soon after a sudden increase in temperature is noted, after a wait only long enough to allow the lower-boiling fraction to be washed out of the condenser. The sharp increase in temperature indicates that distillation of the lower-boiling component of the mixture is complete.

Further Reading

Haynes, W. M. (Ed.) *Handbook of Chemistry and Physics,* 94th ed. CRC Press: Boca Raton, FL, 2013.

Questions

1. Explain why the observed boiling point for the first drops of distillate collected in the simple distillation of a 1:1 molar solution of pentane and hexane, illustrated in Figure 12.5, will be above the boiling point of pentane.

2. The molar composition of a mixture is 80% hexane and 20% pentane. Use the phase diagram in Figure 12.4 to estimate the composition of the vapor over this liquid. This vapor is condensed and the resulting liquid is heated. What is the composition of the vapor above the second liquid?

3. A student carried out a simple distillation on a compound known to boil at 124°C and reported an observed boiling point of 116°–117°C. Gas chromatographic analysis of the product showed that the compound was pure, and a calibration of the thermometer indicated that it was accurate. What procedural error might the student have made in setting up the distillation apparatus?

4. The directions in an experiment specify that the solvent, diethyl ether, be removed from the product by using a simple distillation. Why should the heat source for this distillation be a steam bath, not an electrical heating mantle?

5. The boiling point of a compound is 300°C at atmospheric pressure. Use the nomograph (Figure 12.22) to determine the pressure at which the compound would boil at about 150°C.

6. Azeotropes can be used to shift chemical equilibria by removing products. Treatment of 1-butanol with acetic acid in the presence of sulfuric acid as a catalyst results in formation of butyl acetate and water. The mixture of 1-butanol/butyl acetate/water forms a ternary azeotrope that boils at 90.7°C. This azeotrope separates into two layers; the upper layer is largely butyl acetate, along with 11% 1-butanol, and the lower layer is largely water. Butyl acetate forms by an equilibrium reaction that does not especially favor product formation.

(a) Describe an apparatus by which azeotrope formation can be used to drive the equilibrium toward the products, thus maximizing the yield.

(b) How would you separate the 1-butanol/butyl acetate mixture that forms the upper azeotropic layer?

CHAPTER

13 REFRACTOMETRY

The *refractive index* is a useful physical property of liquids and solids that can be used to identify the composition of solutions. It is a measure of how much light is bent, or refracted, as it enters a liquid. The refractive index is widely used in the beverage and food processing industries, particularly in quality control. For example, you can determine the concentration of sucrose (table sugar) or ethanol in water by measuring the index of refraction of the aqueous solutions.

In the organic chemistry lab, refractometry is a simple, inexpensive technique that is used for liquid compounds. The refractive index can be measured quite accurately to four decimal places, and these measurements can be useful in assessing the purity of liquid compounds. The closer the experimental value is to the value reported in the literature, the purer the sample. You can confirm that a liquid is pure by comparing the experimentally measured refractive index with the published value. This is much like using a melting point to identify a solid compound, although the range of melting points is much greater than the range of refractive indexes. Extensive tables of organic compounds list refractive indexes.

13.1 The Refractive Index

A beam of light traveling from a gas into a liquid undergoes a decrease in its velocity because liquids are much denser than gases and the light beam is absorbed and then reemitted by the molecules as it travels through the liquid. If the light strikes the horizontal interface between gas and liquid at an angle other than 90°, the beam bends downward as it passes from the gas into the liquid.

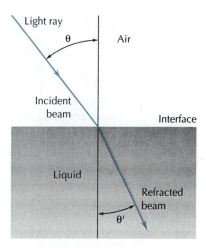

FIGURE 13.1
Refraction of light in a liquid.

Application of this phenomenon allows you to determine the refractive index, a measure of how much the light is bent, or *refracted*, as it enters the liquid (Figure 13.1):

$$n = \frac{V_{air}}{V_{liq}} = \frac{\sin \theta}{\sin \theta'}$$

The *refractive index, n,* represents the ratio of the velocity of light in a vacuum (or in air) to the velocity of light in the liquid being studied, where V_{air} is the velocity of light in air, V_{liq} is the velocity of light in the liquid, θ is the angle of the incident light, and θ' is the angle of the refracted light. Because the velocity of light in a liquid is always less than that of light in air, refractive index values are numerically greater than 1; for organic liquids the normal range is 1.3–1.7.

The density of organic liquids normally decreases with increasing temperature. This density change, in turn, affects the velocity of the light beam as it passes through the sample. Therefore, the temperature at which you measure the refractive index (20°C in the following example) is always specified by a superscript in the notation of *n*:

$$n_D^{20} = 1.3910$$

The wavelength of light used also affects the refractive index because light of differing wavelengths refracts at different angles. The *sodium D line*, which consists of two bright yellow closely spaced lines at 589 and 589.6 nm, usually serves as the standard wavelength for refractive index measurements and is indicated by the subscript D on the symbol *n*. If light of some other wavelength is used, the specific wavelength in nanometers appears as the subscript.

13.2 The Refractometer

The instrument used to measure the refractive index of a compound is called a *refractometer* (Figure 13.2). The instrument in Figure 13.2a includes a built-in thermometer for measuring the temperature of the refractive index reading, as well as a system for circulating water at a constant temperature around the sample holder. This type of refractometer uses a white light source instead of a sodium lamp and contains a series of compensating prisms that give a refractive index equal to that obtained with 589-nm light (the D line of sodium).

A sample is placed between two glass prisms, one of which is the illuminating prism and the other is a refracting prism. When the upper part of the hinged prism is lifted and tilted back, a few drops of sample can be placed on the lower prism. After the upper part of the hinged prism is set back on the lower prism, the light passes through the sample and is reflected by an adjustable mirror. When the mirror is properly aligned, the light is reflected through the compensating prisms and finally through a lens with crosshairs to the eyepiece.

Figure 13.2b shows a digital Abbé refractometer that uses a 589-nm integrated light source and has an LCD display that records the refractive index and the temperature; the readings are accurate to 0.0002 refractive index units. You need to adjust the view through the eyepiece in both types of refractometers.

13.3 Determining a Refractive Index

Do not use acetone or ethyl acetate to clean the refractometer prisms because they can dissolve the adhesive holding the prisms in place.

Four or five drops of liquid, which need to be free of water and other contaminants, are needed for a measurement. You must also record the temperature at which the refractive index is measured. Unless you are using a digital model that provides automatic temperature correction, you must apply a temperature correction to the experimental value before comparing it with a reported value.

Steps in Determining a Refractive Index

The following directions apply to the use of a refractometer such as the one shown in Figure 13.2a.

1. Check the surface of the prisms for residues from previous determinations. If the prisms need cleaning, place a few drops of ethanol on the surfaces and blot (do not rub) the surfaces with lens paper. Allow the residual ethanol to evaporate completely.

2. With a Pasteur pipet held 1–2 cm above the prism, place 4–5 drops of the sample on the measuring (lower) prism. Do not touch the prism with the tip of the pipet because the highly polished glass surface can scratch very easily, and scratches ruin the instrument. Lower the illuminating (upper) prism carefully so that the liquid spreads evenly between the prisms.

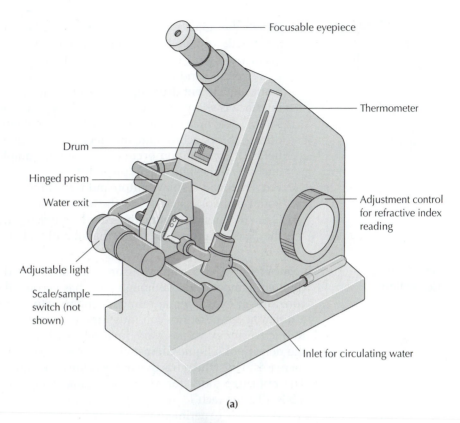

Focusable eyepiece

Thermometer

Drum

Hinged prism

Water exit

Adjustment control
for refractive index
reading

Adjustable light

Scale/sample
switch (not
shown)

Inlet for circulating water

(a)

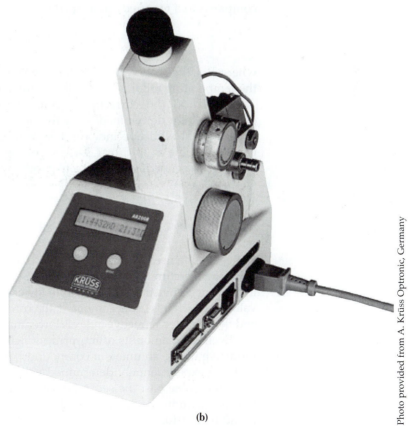

(b)

FIGURE 13.2
(a) Abbé-3L
refractometer.
(b) Digital Abbé
refractometer.

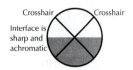

FIGURE 13.3 View through the eyepiece when the refractometer is adjusted correctly.

3. Rotate the adjustment control until the dark and light fields are exactly centered on the intersection of the crosshairs in the eyepiece (Figure 13.3). If color (usually red or blue) appears as a horizontal band at the interface of the fields, rotate the chromatic adjustment drum or dispersion correction wheel until the interface is sharp and uncolored (achromatic). Occasionally the sample evaporates from the prisms, making it impossible to produce a sharp, achromatic interface between the light and dark fields. If evaporation occurs, apply more sample to the prism and repeat the adjustment procedure.
4. Press the read display button and record the refractive index in your notebook. Then record the temperature.
5. Open the prisms, blot up the sample with lens paper, and follow the cleaning procedure with ethanol outlined in step 1.

Temperature Correction

Values reported in the literature are often determined at a number of different temperatures, although 20°C has become the standard. To compare an experimental refractive index with a value reported at a different temperature, you must first calculate a correction factor. Refractive index values vary inversely with temperature because the density of a liquid almost always decreases as the temperature increases. Thus, the refractive index is lower at higher temperatures. The refractive index for a typical organic compound decreases by 4.5×10^{-4} for each 1° increase in temperature.

When an experimental refractive index measured at 25°C is compared to a reported value at 20°C, the temperature correction is

$$\Delta n = 4.5 \times 10^{-4} \times (T_1 - T_2)$$

where T_1 is the observation temperature in degrees Celsius and T_2 is the temperature reported in the literature in degrees Celsius.

Then add the correction factor, including its sign, to the experimentally determined refractive index. For example, if your experimental refractive index is 1.3888 at 25°C, then you obtain a corrected value at 20°C of 1.3911 by adding the correction factor of 0.0023 to the experimental refractive index.

$$\Delta n = 4.5 \times 10^{-4} \times (25 - 20) = 0.00225 \text{ (round to 0.0023)}$$

$$n^{20} = n^{25} + 0.0023 = 1.3888 + 0.0023 = 1.3911$$

The correction needs to be applied before comparing the experimental value to a literature value reported at 20°C. If an experimental refractive index is determined at a temperature lower than that of the literature value to which it is being compared, the correction has a negative sign and the corrected refractive index is lower than the experimental value.

Even trace amounts of impurities (including water) change the refractive index, so unless a compound has been extensively purified, the experimentally determined value may not agree with the literature value past the second decimal place. It is not uncommon that a 1% impurity can change the refractive index of an organic liquid by 0.0010.

13.4   Sources of Confusion and Common Pitfalls

Why Is Everything Blurry When I Look Through the Eyepiece?

It's likely that the blurriness results from an incorrect focus for your eyes. Adjust the focus by moving the eyepiece up or down by rotating it.

Why Can't I Adjust the Chromatic Adjustment Drum to Produce a Sharp, Achromatic Interface?

Most likely you are attempting a measurement on a very volatile sample, which is evaporating from the prism as you are adjusting the refractometer. Add more sample and work quickly.

Why Can't I Reproduce the Literature Value for the Refractive Index of My Compound?

First, make sure you correctly applied the temperature correction factor to your observed refractive index value. If you did adjust your value with the temperature correction, impurities in the sample—either your sample or the sample used for the value reported in the literature—are likely the source of the problem. Keep in mind that the measured refractive index of a sample is very sensitive to even small amounts of impurities. Reproducing the last two decimal places of a literature value may not be possible without extensive purification of your sample.

Questions

1. A compound has a refractive index of 1.3191 at 20.1°C. Calculate its refractive index at 25.0°C.

2. To clean the glass surfaces of a refractometer, ethanol or isopropyl alcohol but not acetone or water is usually recommended. Why?

CHAPTER

14

MELTING POINTS AND MELTING RANGES

If Chapter 14 is your introduction to the purification and analysis of organic compounds, read the Essay "Intermolecular Forces in Organic Chemistry" on pages 127–131 before you read Chapter 14.

Molecules in a crystal are arranged in a regular pattern. Melting occurs when the fixed array of molecules in the crystalline solid rearranges to the more random, freely moving liquid state. The transition from solid to liquid requires energy in the form of heat to break down the crystal lattice. The temperature at which this transition occurs is the solid's ***melting point,*** an important physical property of any solid compound. The melting point of a compound is useful in establishing its identity and as a criterion of its purity. Until the advent of modern chromatography and spectroscopy, the melting point was the primary index of purity for an organic solid. Melting points are still used as a preliminary indication of purity.

14.1   Melting-Point Theory

The melting point, or more correctly the melting range, of a crystal-line organic compound is determined by the strength of the intermo-lecular forces between its molecules—hydrogen bonds, dipole-dipole interactions, and van der Waals interactions. These forces hold the molecules together in an orderly crystalline array and must be over-come for the molecules to enter the less orderly liquid phase. Polarity, large molecular surface area, and high molecular symmetry are asso-ciated with greater intermolecular forces and higher melting points.

Melting Behavior

The melting point is generally reproducible for a pure compound. Relatively pure compounds normally melt over a narrow tem-perature range of 0.5°–1.5°C, whereas impure substances often melt over a much larger range. The presence of even small amounts of impurities can depress the melting point a few degrees and cause melting to occur over a relatively wide temperature range. Adding greater amounts of an impurity generally causes a greater effect on the melting point.

Solid and liquid phases exist in equilibrium at their melting points, as shown by the solid blue curved line in Figure 14.1. This phase diagram plots the observed melting curve for mixtures of com-pounds A and B ranging from 100 mol % A (0 mol % B) to 0 mol % A (100 mol % B). A pure sample of compound A melts at temperature T_A, whereas pure compound B melts at temperature T_B. At T_A and T_B, pure samples of A and B melt sharply over a narrow temperature range.

When both compounds are present, melting begins at T_E, called the eutectic temperature. *Eutectic* refers to a mixture of elements or compounds that has a single chemical composition, which solidifies

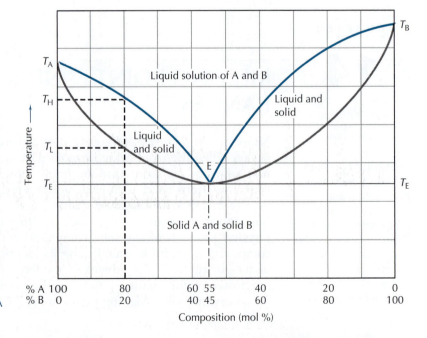

FIGURE 14.1 Melting-point composition diagram for the binary mixture A + B. T_A is the melting point of pure solid A, T_B of pure solid B, and T_E of eutectic mixture E. The temperature range T_L–T_H is the likely melting range that could be observed for a solid containing 80 mol % A and 20 mol % B.

Eutectics have important roles in the modern world. For example, silicon chips in semiconductors are bonded to gold-plated substrates through a silicon-gold eutectic.

at a lower temperature than any other composition. In Figure 14.1 the blue curved line has a minimum at T_E, called the eutectic point. This is the melting point of the eutectic composition, 55% A and 45% B. Now consider the melting behavior of a solid consisting of 80% of compound A and 20% of compound B. As this mixture of A and B is heated, the melting begins at T_E, but the observable melting, where a visible amount of liquid is present, occurs at T_L, the low end of the melting-point range. As heating continues, more liquid forms until T_H is reached and the entire sample becomes a liquid. The melting range is T_L–T_H. The lighter gray curves in Figure 14.1 are superimposed on the phase diagram to indicate where melting of the A-B mixture might occur.

Another way to look at this phenomenon is to compare freezing points with melting points. An impurity depresses the melting point of a solid just as it does the freezing point of a liquid. The freezing point and melting point are identical, although accurate freezing points are more challenging to obtain because liquids often supercool before they freeze. One practical application of depressed freezing points is salting roads to melt ice at a temperature lower than 0°C in northern climates. A eutectic mixture of NaCl and H_2O has a eutectic point of –21°C; it contains 23% salt by mass.

14.2   Apparatus for Determining Melting Ranges

The most common method for measuring melting points is to electrically heat a small amount of the solid in a capillary tube. Although analog methods can be used, most commercially available melting-point apparatuses now use digital methods. We will briefly discuss both. First, the things they have in common are: both digital and analog devices have a metal heating block, which normally has spaces for three melting-point tubes, allowing the simultaneous determination of more than one melting point. A thin-walled glass capillary tube holds each sample. A light illuminates the sample chambers, which are usually magnified for easy visualization.

The Digital Melting-Point Apparatus

The most significant advantage of a digital melting-point apparatus is that you can program the rate at which the temperature of the heating block increases. The temperature is measured with a platinum resistance sensor and recorded on a digital LCD (Figure 14.2). In digital measurements the start temperature, the temperature ramp, and the stop temperature can be programmed. For example, the starting temperature can be set to 10°C below the expected melting point and the temperature ramp set at 1°C/min or even 0.5°C/min, with the stop temperature set at 5°C above the expected melting point. Thus, a defined temperature program can be followed, which allows melting points to be reproduced easily. To measure an accurate melting point, it is important to be able to control the temperature increase near the melting point. If the temperature rises too quickly, the melting range will be greater and will likely be too high because it takes a finite amount of time for the solid sample to melt.

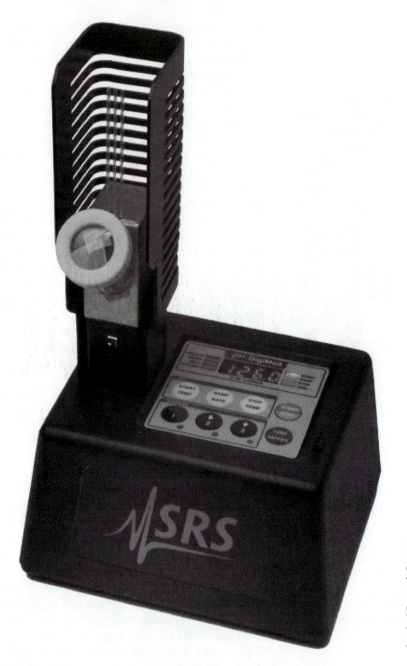

FIGURE 14.2 A modern digital melting-point apparatus.

Analog Melting-Point Apparatus

The traditional analog melting-point apparatus also uses electrical heating, but a rheostat controls the rate of heating by allowing continuous adjustment of the current (Figure 14.3). The higher the rheostat setting, the faster the rate of heating. However, the rate of heating at any particular setting increases more rapidly at the start and then slows as the temperature increases. The decreased rate of heating at higher temperatures allows for the slower heating needed as the melting point is approached. The temperature is measured with a traditional liquid-filled thermometer. Heat transfer

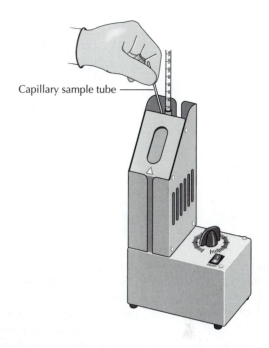

FIGURE 14.3 An analog melting-point apparatus.

Capillary sample tube

to the thermometer liquid is relatively slow; thus the response time is slower than with a digital thermometer imbedded in the metal heating block. It is more difficult to adjust the rate of heating in an analog melting-point apparatus, and typically analog melting points have a 1°–2° range at best.

14.3   Determining Melting Ranges

Sample Preparation

The melting range of an organic solid can be determined by introducing a small amount of the substance into a capillary tube with one sealed end. Such capillary tubes, which are approximately 1 mm in diameter, are commercially available.

Filling a capillary tube. Place a few milligrams of the dry solid on a piece of smooth-surfaced paper and crush the solid to a fine powder by rubbing a spatula over it while pressing down. Introduce the solid into the capillary tube by tapping the open end of the tube into the powder. A small amount of material will stick in the open end. Invert the capillary tube so that the sealed end is down; and holding it very near the sealed end, tap it lightly with quick motions against the bench top (Figure 14.4). Some melting-point apparatuses have a small built-in vibrator, which can replace the tapping technique for moving the solid sample to the bottom of the capillary tube.

SAFETY PRECAUTION

Care must be taken while tapping the capillary tube against the bench top; the tube could break and cause a cut.

The solid will fall to the bottom of the tube. Repeat this operation until the amount of solid in the tube is 2–3 mm in height. A small

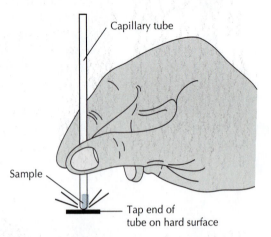

FIGURE 14.4 Tapping a melting-point tube to move all of the solid to the bottom of the capillary tube.

Capillary tube

Sample

Tap end of tube on hard surface

The ideal sample for a melting point is only 2–3 mm in height in the capillary tube.

sample is essential for accurate melting points. Melting-point determinations made with too much sample lead to a broad melting range because more time is required to melt the complete sample, as the temperature continues to rise while the sample melts.

An alternative method for getting the solid to the bottom of a capillary tube is to drop the tube down a piece of glass tubing about 1 m in length, with the bottom end resting on the lab bench. After a few trips down the glass tubing, the solid will usually fall to the bottom of the capillary tube.

Wet samples. If a solid is still wet from recrystallization, it will not fall to the bottom of a capillary tube but will stick to the capillary wall. This is probably a good thing because melting points of wet solids almost always have a wide range and thus are nearly worthless. If your sample is still wet, allow it to dry completely before continuing with the melting-range determination.

Thermometer Calibration

The accuracy of a melting-point determination can be no better than the accuracy of the temperature probe. You cannot assume that a thermometer has been accurately calibrated—although that may be the case, it is not always true. Thermometers can give high or low temperature readings of 1°–2°C or more. Section 5.4, page 63, describes a procedure for calibrating a thermometer. Many suppliers of melting-point apparatuses also sell sets of thermometer-calibration standard compounds. Modern digital melting-point instruments have the temperature probe calibrated to within 0.5°C.

Heating the Sample to the Melting Point

Digital melting-point apparatus. Place the melting-point tube in one of the illuminated sample lanes and program the apparatus for the start temperature, temperature ramp, and stop temperature. The sample can be heated rapidly until the temperature is 5°–10° below the expected melting point. The programmed temperature ramp of 0.5°–1.5° per minute will then take over. Carefully observe the sample until the first small drop of liquid appears, which is the onset of melting. Record the temperature that is displayed on the LCD. Continue your observations until the sample is completely melted and again record the temperature displayed on the LCD. The heating rate should be programmed to stop within a few degrees past the melting point. Remove the capillary tube and discard it in the container provided in your laboratory.

Analog melting-point apparatus. Set the rheostat dial for an appropriate rate of heating and turn on the electric current. The melting-point apparatus can be heated rapidly until the temperature is about 20°C below the expected melting point. Then decrease the rate of heating so the temperature rises only 1°–2° per minute and the sample has time to melt before the temperature rises above the true melting point. When you are taking successive melting points, remember that the apparatus needs to cool to at least 20° below the expected melting point before it can be used for the next determination.

Approximate melting point. If you do not know the melting point of a solid sample, you can make a quick preliminary determination of it by heating the sample rapidly and watching for the temperature at which melting begins. In a more accurate second determination, you can then carefully control the temperature rise.

Use a fresh sample for each determination. Always prepare a fresh sample for each melting-point determination; many organic compounds decompose at the melting point, making reuse of the solidified sample a poor idea. Moreover, many low-melting compounds (mp 30°–80°C) do not easily resolidify with cooling.

Reporting the Melting Range

Unless you have an extraordinarily pure compound in hand, you will always observe and report a melting range—from the temperature at which the first drop of liquid appears to the temperature at which the solid is completely melted and only a clear liquid is present. This melting range may be 1°–2° or slightly more. For example, salicylic acid often gives a melting range of 157°–159°C with an analog melting-point apparatus. An extremely pure sample of salicylic acid, however, melts over less than a 1° range (for example, 160.0°–160.5°C) and it may have 160°C listed as its melting point. Published melting points are usually the highest values obtained after several recrystallizations; the values you observe may be slightly lower.

14.4   Summary of Melting-Point Technique

1. Introduce the powdered, dry solid sample to a height of 2–3 mm into a capillary tube that is sealed at one end.
2. Place the capillary tube in the melting-point apparatus.
3. Adjust the rate of heating so the temperature rises at a moderate rate.

For a digital apparatus with temperature programming: set the start temperature to approximately 5°–10° below the expected melting point and the temperature ramp for a low but convenient temperature rise. The stop temperature should be approximately 5°C above the expected melting point.

For an analog apparatus using a liquid thermometer: decrease the rate of heating when the temperature reaches 10°–20° below the expected melting point so that the temperature rises only 1°–2° per minute. **Note:** There will be a time lag before the rate of heating changes.

4. If the temperature is rising more than 1°–2° per minute at the time of melting, determine the melting point again using a new sample.
5. Record the melting range as the range of temperatures between the onset of melting and the temperature at which only liquid remains in the tube. If you are using a digital apparatus, record the temperature profile that you programmed into the apparatus.

14.5   Using Melting Points to Identify Compounds

We have already discussed how impurities can lower the melting point of a compound. This behavior can be useful not only in evaluating a compound's purity but also in helping to identify the compound. Assume that two compounds have virtually identical melting ranges. Are the compounds identical? Possibly, but the identical melting ranges may be a coincidence. The use of a mixture melting point is one way to answer this question.

Mixture Melting Point

If roughly equal amounts of the two compounds are finely ground together with a spatula, the melting range of the resulting mixture can provide useful information. If there is a melting-point depression or if the melting range is expanded by a number of degrees, it is reasonably safe to conclude that the two compounds are not identical. One compound has acted as an impurity toward the other by lowering and broadening the melting range. If there is no change in the mixture's melting range relative to that for each separate compound, the two are probably the same compound.

Sometimes only a modest melting-point depression is observed. To know whether this change is significant, determine the mixture melting point and the melting point of one of the two compounds simultaneously in separate capillary tubes. This experiment allows simultaneous identity and purity checks. Infrequently, a eutectic point (point E in Figure 14.1) can be equal to the melting point of the pure compound of interest. In a case where you have accidentally used the eutectic mixture, a mixture melting point would not be a good indication of purity or identity. Errors of this type can be discerned by testing various mixtures other than a 1:1 composition. The subsequent use of 1:2 and 2:1 mixtures can avoid eutectic-point-induced misinterpretation.

Other Ways of Determining Identity

Spectroscopic methods (see Chapters 21–24) and thin-layer chromatography (see Chapter 18) are other ways to determine the identity of a solid organic compound.

14.6 Sources of Confusion and Common Pitfalls

When you heat a sample for a melting-point determination, you may see some strange and wonderful things happen before the first drop of liquid actually appears. The compound may soften and shrivel up as a result of changes in its crystal structure. It may "sweat out" some solvent of crystallization. It may decompose, changing color as it does so. None of these changes should be called melting. **Only the appearance of liquid indicates the onset of true melting.** Even so, it can be difficult to distinguish exactly when melting starts. In fact, even with careful heating, two people may disagree on a melting point by as much as 1°–2°C.

Why Aren't My Measured Melting Ranges Consistent from One Determination to the Next?

The rate of heating is the most important variable in determining melting ranges. In this regard, digital melting-point devices have a distinct advantage because you can record the exact temperature profile that you used in a previous melting-point determination. Heating faster than 1°–2° per minute may lead to an observed melting range that is higher than the correct one. If the rate of heating is extremely rapid (>10°C per minute), you may also observe thermometer lag with a liquid-filled thermometer, a condition caused by failure of the liquid's temperature to increase as rapidly as the temperature of the metal heating block. This error causes the observed melting range to be lower than it actually is. Determining accurate melting points requires patience.

In addition, a substantially greater amount of sample in the melting-point tube can result in a broad melting range. Fill the capillary to just 2–3 mm in height.

My Sample Disappeared While I Was Measuring the Melting Point! What's Going On?

Another possible complication in melting-point determinations occurs if the sample sublimes. Sublimation is the change that occurs when a solid is transformed directly to a gas, without passing through the liquid phase (see Chapter 16). If the sample in the capillary tube sublimes, it can simply disappear as it is heated. Many common substances sublime, for example, camphor and caffeine. You can determine their melting points by sealing the open end of the capillary tube in a burner flame before placing it in the melting-point apparatus (Figure 14.5a).

SAFETY PRECAUTION

Be sure no flammable solvents are in the vicinity when you are using a Bunsen burner.

My Sample Didn't Melt but Just Turned Brown. How Do I Interpret This?

Some compounds decompose instead of melting or as they melt, which is usually indicated by a change in color of the sample to dark red or brown. The melting point of such a compound is reported with the letter **d** after the temperature. For example, 186°C **d** means that the compound melts at 186°C with decomposition.

Sometimes decomposition occurs as a result of a reaction between the compound and oxygen in the air. If this is the case, when the air

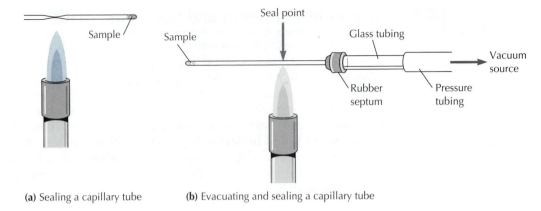

(a) Sealing a capillary tube   **(b)** Evacuating and sealing a capillary tube

FIGURE 14.5 Methods for sealing a capillary tube with a lab burner.

is evacuated from the capillary tube and the tube is sealed, the melting point can be determined without decomposition (Figure 14.5b).
Place the sample in the capillary tube as directed earlier. Punch a hole
in a rubber septum, insert the closed end of the capillary tube through
the inside of the septum, and then gently push most of the capillary
through the septum. Fit the septum over a piece of glass tubing that
is connected to a vacuum line. Turn on the vacuum source, and while
heating the upper portion of the capillary tube in a burner flame, hold
and pull on the sample end of the capillary tube until it seals.

Further Reading

Harding, K. E. *J. Chem. Educ.* **1999**, *76*, 224–225.
Skau, E. L.; Arthur, J. C. Jr. In *Physical Methods of
  Chemistry*, A. Weissberger and B. W. Rossiter
(Eds.); Wiley-Interscience: New York, 1971,
vol. 1, Part V.

Questions

1. A student performs two melting-point
   determinations on a crystalline product.
   In one determination, the capillary tube
   contains a sample about 1–2 mm in height
   and the melting range is found to be 141°–
   142°C. In the other determination, the
   sample height is 4–5 mm and the melting
   range is found to be 141°–145°C. Explain
   the broader melting-point range observed
   for the second sample. The reported melting point for the compound is 143°C.

2. Another student reports a melting range
   of 136°–138°C for the compound in
   Question 1 and mentions in her notebook
   that the rate of heating was about 12° per
   minute. NMR analysis of this student's
   product does not reveal any impurities.
   Explain the low melting point.

3. A compound melts at 120°–122°C on one
   apparatus and at 128°–129°C on another.

Unfortunately, neither apparatus is calibrated. How might you check the identity
of your sample without calibrating either
apparatus?

4. Why does sealing the open end of a
   melting-point capillary tube allow you to
   measure the melting point of a compound
   that sublimes?

5. A white crystalline compound melts at
   111°–112°C and the melting-point capillary
   is set aside to cool. Repeating the melting-
   point analysis with the same capillary
   reveals a much higher melting point of
   140°C. Yet repeated recrystallization of
   the original sample yields sharp melting
   points no higher than 114°C. Explain the
   behavior of the sample that was cooled
   and then remelted.

15

RECRYSTALLIZATION

A pure organic compound is one in which there are no detectable impurities. Because experimental work requires an immense number of molecules (Avogadro's number per mole), it is not true that 100% of the molecules in a "pure" compound are identical to one another. Seldom is a pure compound purer than 99.99%. Even if it were that pure, one mole would still contain more than 10^{19} molecules of other compounds. Nevertheless, we want to work with compounds that are as pure as possible. Along with chromatography, recrystallization is a major technique for purifying solid compounds.

If Chapter 15 is your introduction to the separation and purification of organic compounds, read the Essay "Intermolecular Forces in Organic Chemistry" on pages 127–131 before you read Chapter 15.

15.1 Introduction to Recrystallization

When a crystalline material (solute) dissolves in a hot solvent and then returns to a solid in a cooled solvent by crystallizing, the process is called *recrystallization.* Its success depends on the increasing solubility of the crystals in hot solvent and their decreased solubility when the solution cools, thereby causing the compound to recrystallize. Impurities in the original crystalline material are usually present at a lower concentration than in the substance being purified. Thus, as the mixture cools, the impurities tend to remain in solution while the highly concentrated product crystallizes.

Crystal Formation

Crystal formation of a solute from a solution is a selective process. When a solid crystallizes at the right speed under the appropriate conditions of concentration and solvent, an almost perfect crystalline material can result because only molecules of the right shape fit snugly into the crystal lattice. In recrystallization, dissolution of the impure solid in a suitable hot solvent destroys the impure crystal lattice, and recrystallization from the cold solvent selectively produces a new, more perfect (purer) crystal lattice. Slow cooling of the saturated solution promotes formation of pure crystals because the molecules of the impurities, which do not fit as well into the newly forming crystal lattice, have time to return to the solution. Therefore, crystals that form slowly are larger and purer than ones that form quickly. Indeed, rapid crystal formation (precipitation) traps the impurities because the lattice grows so quickly that the impurities are simply engulfed by the crystallizing solute as the crystals form.

Slow cooling of the hot solution often results in the formation of beautiful, pure crystals. Beautiful crystals are to an organic chemist what a home run is to a baseball player!

Solvent Properties

In general, a solvent with a structure similar to that of the solute will dissolve more solute than will solvents with dissimilar structures. Although the appropriate choice of a recrystallization solvent is a trial-and-error process, a relationship exists between the solvent's structure and the solubility of the solute—**like dissolves like.** In a recrystallization, the polarity of the solvent and that of the compound being recrystallized should be similar.

High-polarity solvents. Among the more polar organic solvents, both methanol and ethanol are commonly used for recrystallization because they dissolve a wide range of both polar and nonpolar compounds to the appropriate degree. Ethanol and methanol also evaporate easily and possess water solubility, which allows recrystallization from an alcohol/water mixture. Nonionic compounds generally dissolve in water only when they can associate with the water molecules through hydrogen bonding. Carboxylic acids, which readily form hydrogen bonds, are often recrystallized from water solution.

Low-polarity solvents. Organic solvents of low polarity also dissolve many nonionic organic compounds with ease. Even polar organic compounds can dissolve in solvents of low polarity if the ratio of polar functional groups per carbon atom is not too high and if hydrogen bonding can occur between the solute and the solvent. Among the low-polarity solvents, diethyl ether and ethyl acetate appear to provide the best solvent properties, although the low boiling point of diethyl ethyl (35°C) is a disadvantage and its extreme flammability requires careful attention to safety. Diethyl ether in combination with hexane or methanol has excellent solvent properties for recrystallizations.

Solvent boiling point. The boiling point of the solvent is another important property because the solvent needs to be volatile enough to evaporate fairly quickly from the crystals after they are recovered from the recrystallization solution. Therefore, most commonly used recrystallization solvents have boiling points at or below 100°C (Table 15.1).

TABLE 15.1   Common recrystallization solvents

Solvent	Boiling point, °C	Miscibility[a] in water	Solvent polarity	Comments
Diethyl ether	35	−	Low	Good solvent, but low bp limits its use
Acetone	56	+	Intermediate	Good general solvent
Petroleum ether[b]	60–80	−	Nonpolar	Good solvent for less polar compounds
Methanol	65	+	High	Good solvent for moderately polar compounds
Hexane	69	−	Nonpolar	Good solvent for less polar compounds
Ethyl acetate	77	−	Low	Good general solvent
Ethanol	78	+	High	Excellent general solvent
Cyclohexane[c]	81	−	Nonpolar	Good solvent for less polar compounds
Water	100		Very high	Solvent of choice for polar compounds
Toluene	111	−	Nonpolar	Good solvent for aromatic compounds, but slow to evaporate

a. Infinite solubility = +.
b. Petroleum ether (or ligroin) is a mixture of isomeric alkanes. The term "ether" refers to volatility, not the presence of an ether functional group.
c. May freeze if the cooling bath is less than 6°C.

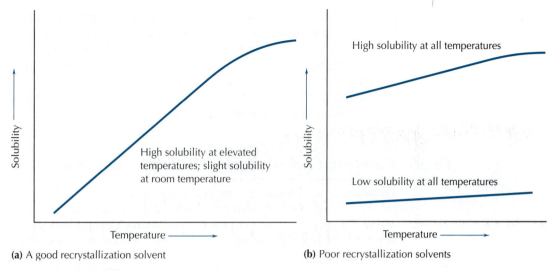

(a) A good recrystallization solvent

(b) Poor recrystallization solvents

FIGURE 15.1 Solubility graphs.

SAFETY PRECAUTION

Ether, hexane, and petroleum ether are very flammable and should be heated with a steam or hot-water bath. They should **never** be heated with a flame or on a hot plate. All organic solvents are toxic to some degree, so recrystallizations should be carried out in a fume hood.

Choice of a Recrystallization Solvent

The most crucial aspect of a recrystallization procedure is the choice of solvent because **the solute should have a maximum solubility in the hot solvent and a minimum solubility in the cold solvent.** There is always some loss of the compound being recrystallized because it has some solubility in the cold solvent. The goal is to attain maximum purification with minimum loss.

Figure 15.1a shows the solubility curve for a good recrystallization solvent with low solubility at lower temperatures and high solubility at higher temperatures. For recrystallization to work effectively, the solubility of the organic solid should not be too large or too small in the recrystallization solvent. If the solubility is too large, it is difficult to recover the compound, as illustrated by the upper curve in Figure 15.1b. If the solubility is too small, a very large volume of solvent will be needed to dissolve the compound or it simply may not dissolve sufficiently for recrystallization to be effective, as shown in the lower curve of Figure 15.1b.

15.2   Summary of the Recrystallization Process

Unless you are given explicit experimental directions, recrystallization can be challenging. It involves making decisions about which solvent or solvent mixture to use and how much to use. Before going into detail, here is an overview of the steps involved in a recrystallization.

1. Dissolve the solid sample in a minimum volume of hot solvent with a boiling stone or boiling stick present.

2. Cool the solution slowly to room temperature and then in an ice-water bath to induce crystallization.
3. Recover the crystals from the cooled recrystallization mixture by vacuum filtration.
4. Wash the crystals with a small amount of cold solvent.
5. Allow the crystals to air dry completely before weighing them and determining their melting point.

 15.3   Carrying Out Successful Recrystallizations

There are several important factors to consider in carrying out successful recrystallizations that apply to both miniscale and microscale recrystallizations. When you are recrystallizing a compound, attention to these details will make the process proceed more smoothly and successfully.

Erlenmeyer flasks are used for most miniscale and microscale recrystallizations because volatile solvents are more easily contained in them and potentially hazardous vapors are kept to a minimum. A beaker can serve as a recrystallization vessel when water, ethanol, or an ethanol-water mixture is used as the solvent.

Seed Crystals

Always set aside a small amount of the crude crystalline product to use as seeds for catalyzing the formation of crystals in the event that recrystallization does not occur. If no crystals appear in the cooled solution, it could mean that the solution is not saturated with your compound. It is also possible for a solution to be *supersaturated*. What enables a supersaturated solution to exist is a kinetic barrier to crystallization. This situation occurs more frequently than you might expect, especially if you concentrate solutions by boiling away some of the solvent. When a solution is supersaturated, crystals may not form until an appropriate surface is present on which crystal growth can occur. When you introduce a seed crystal or scratch the inside of the crystallization flask, crystals may appear to form instantly. The seed crystal or tiny particles of glass produced upon scratching provide surfaces that facilitate crystal growth. Deciding if a recrystallization solution is not saturated or is supersaturated can be difficult, but adding two or three small crystals of the compound to the solution will tell you.

Scale of the Recrystallization

Beakers are not normally used for recrystallizations because the solvent would evaporate too rapidly during heating.

The amount of solid to be recrystallized will determine the size of the vessel used for the recrystallization and the volume of solvent needed. For miniscale recrystallizations, you will probably never use an Erlenmeyer flask of smaller capacity than 50 mL. A 125- or 250-mL Erlenmeyer flask is usually appropriate for recrystallizations of 1–10 g. **A good rule of thumb is to use a flask two to three times larger than the amount of solvent you think you will need.** Microscale recrystallizations are usually done in 10- or 25-mL Erlenmeyer flasks or small test tubes.

The amount of solvent needed for the recrystallization will naturally differ if you are purifying 400 mg or 4.0 g of a compound. For example, you would not want to recrystallize 4.0 g of compound in

10 mL of solvent because it would be difficult to achieve much purification. **Ten to 40 times the amount of solvent as compound being recrystallized is usually used.**

Add a boiling stone or boiling stick to the recrystallization flask. **It is important to add the solvent incrementally and then allow the mixture to reach the boiling point before adding more solvent.** You want to use only the amount of solvent needed to dissolve all the solute in boiling solvent, thereby ensuring maximum recovery of the solute when the solution cools. If you are using approximately 20 mL of solvent, it works best to make incremental additions of solvent with a Pasteur pipet. If you are using a larger amount of solvent, pour small portions of warm solvent directly from the flask holding the solvent into the recrystallization flask.

Kinetics can also play a role in dissolution. Well-formed crystals are slower to dissolve than a powdered solid is. When dissolving a compound for recrystallization, you may add too much hot solvent if the rate of dissolution is slow. If the solution is too dilute, the compound will not crystallize upon cooling.

Insoluble Impurities

Consider a situation in which you have added 40 mL of warm solvent to your compound. When you heat the mixture to just under the boiling point of the solvent, most of the solid dissolves immediately. With the addition of another 5 mL of solvent, more of the solid dissolves. But after you have added another 10 mL of solvent and heated the mixture again to the boiling point, no more solid has gone into solution. In this situation you have to make accurate experimental observations and then act on them. For example, this may be the time to consider that your compound contains an insoluble impurity that needs to be removed by gravity filtration of the hot solution (see Section 9.2, Figure 9.2).

Maximum Recovery of Product

Many students recover a smaller amount of product from a recrystallization than they should because of mechanical losses on the walls of oversized flasks or during the filtration step. Losses also occur because (1) too much solvent is added, (2) premature crystallization occurs during a gravity filtration, or (3) the crystals are filtered before recrystallization is complete.

Ensuring Dry Crystals

When a higher-boiling-point solvent, such as ethyl alcohol, water, or toluene, is used as the recrystallization solvent, the recrystallized product dries slowly and should be allowed to dry at least overnight before its mass and melting point are determined. If water has been used as the recrystallization solvent, the drying procedure can be hastened by placing the crystals on a watch glass in a 50°C oven for 15–20 min. **Solids recrystallized from organic solvents should not be oven-dried because of the potential for a fire.**

15.4   How to Select a Recrystallization Solvent

Recrystallization is straightforward if you are told what solvent to use as well as the ratio of solvent to solute. It is much more challenging when you have to determine these factors yourself, especially if you use a mixture of two solvents. To be successful, you must consider the choices and then pay careful attention to your experimental observations and what they tell you. Use Table 15.1 and the essay on intermolecular forces at the beginning of Part 3 to decide on suitable candidates for the recrystallization solvent. Begin by carefully selecting what seems to be a good recrystallization solvent, using the following procedure.

Testing a Solvent

Careful measurements and observation are essential when testing potential solvents.

If you are working with less than 0.5 g of compound, the solid used for the tests can be recovered by evaporating the solvents.

Place a small sample (20–30 mg) of the compound to be recrystallized in a test tube and add 5–10 drops of a trial solvent. Shake the tube to mix the materials. If the compound dissolves immediately, your compound is probably too soluble in the solvent for recrystallization to be effective. If no solubility is observed, heat the solvent to its boiling point. If complete solubility is observed, cool the solution to induce crystallization. The formation of crystals in 10–20 min suggests that you have a good recrystallization solvent.

When you scale up a recrystallization from the test quantities, you need to be flexible enough to question your solvent choice if the recrystallization does not seem to be working. For example, if most of the crystals dissolve immediately in a small volume of solvent, you may have to boil away the solvent you are using and start again with a different solvent.

Two-Solvent Recrystallizations

Record the exact amount of each solvent used for the tests.

When no single solvent seems to work for a recrystallization, a pair of *miscible solvents*—solvents that are very soluble in one another—can often be used. Mixed-solvent pairs usually include one solvent in which a particular solute is very soluble and another in which its solubility is marginal to poor. Typical mixed-solvent pairs are listed in Table 15.2.

Mixed-solvent tests. To select a suitable mixed-solvent pair, place 20–30 mg of the solute in a test tube and add 5–10 drops of the solvent in which you expect it to be more soluble. Warm the solution nearly to its boiling point. When the solid dissolves completely, add the other solvent drop by drop until a slight cloudiness appears and persists as mixing continues, indicating that the hot solution is saturated with the solute. If no cloudiness appears, the compound is too soluble in this solvent pair for an effective recrystallization and another solvent pair should be tested.

TABLE 15.2   **Solvent pairs for mixed-solvent recrystallizations[a]**

Solvent 1	Solvent 2	Solvent 1	Solvent 2
Ethanol	Acetone	Ethyl acetate	Hexane
Ethanol	Petroleum ether	Methanol	Diethyl ether
Ethanol	Water	Methanol	Water
Acetone	Water	Diethyl ether	Hexane (or petroleum ether)
Ethyl acetate	Cyclohexane		

a. Properties of these solvents are given in Table 15.1.

If cloudiness appears, add the first solvent again in small portions until the cloudiness just disappears and then add a little more to ensure an excess. Let the solution cool slowly. The formation of crystals in 10–20 min suggests that you have found a good solvent pair.

Scaling up a mixed-solvent recrystallization. If one of the tests for a mixed solvent is more successful than those using a single solvent, scale it up for the recrystallization of your compound. Use approximately the same proportions of the two solvents in the scaled-up procedure as you used in the test.

In a mixed-solvent recrystallization, dissolve the solute in just enough of the hot solvent in which it is more soluble; then add a small excess of that solvent (about 10%) to prevent premature crystallization. Add the second solvent, in which the solute is sparingly soluble, in small portions until you have almost reached the saturation point of the solute. Detecting this point is challenging, but it helps to think of it in analogy to approaching the endpoint of an acid–base titration with phenolphthalein indicator. As you approach the endpoint in a titration, a slight pink cloudiness forms that dissipates when the solution is mixed. With each subsequent addition, the pink color persists a little longer, until the final addition causes the solution to turn permanently pink. In the same way, as you approach the saturation point of the solute a slight cloudiness will form, which persists a little bit longer with each subsequent addition. Your job is to stop adding the poor-solubility solvent *just before* it causes the entire solution to be cloudy. Then let the solution cool slowly before isolating the crystals. If crystals do not form, reheat the solution and add more of the less-soluble solvent and repeat this sequence until you obtain crystals.

If you reach the saturation point, the cloudiness will persist, indicating that small crystals of product have formed. At this point, you can add the first, more-soluble solvent, while heating, until the cloudiness completely disappears. Unfortunately, there often is a kinetic barrier to redissolving the crystals, and it is not unusual for redissolution to require the addition of too much of the first solvent in which the compound is more soluble. In this case, crystals will not form and you may need to evaporate most of the solvent and start over.

If the solute is very soluble in the first solvent, the volume of solvent compared to the amount of sample may be so small that the crystals will separate as a pasty mass that is difficult to filter. In this situation, you need to use more of the first solvent than will just dissolve the solute and then add a correspondingly larger amount of the second solvent. However, avoid using so much of the first solvent that no amount of the second solvent will produce crystal formation. Should this situation occur, the solvents need to be partially evaporated before cooling again; if crystallization still does not occur, remove all the solvents and test another solvent pair. If the solid is more soluble in the solvent with the lower boiling point, any excess solvent can simply be boiled away in the hood until cloudiness is reached.

15.5 Miniscale Procedure for Recrystallizing a Solid

The procedure for recrystallizing a solid involves three main steps:

- Dissolving the solid and removing insoluble impurities
- Cooling the solution to allow for crystal growth
- Collecting the recrystallized solid by vacuum filtration

S A F E T Y   P R E C A U T I O N S

1. Most organic solvents used for recrystallizations are volatile and flammable. Therefore, they should be heated on a steam bath or in a hot-water bath, not on a hot plate or with an open flame.
2. Lift a hot Erlenmeyer flask with flask tongs. **Note:** Test tube holders are not designed to hold an Erlenmeyer flask securely and the flask may fall onto the bench top and shatter.

If you are doing a mixed-solvent recrystallization, refer to the discussion on two-solvent recrystallizations in Section 15.4.

Step 1. Dissolving the Solid

Always set aside a small amount of the crude crystals to use as seeds in the event that recrystallization does not occur.

Boiling a mixed solvent (see Section 15.3) can preferentially remove the lower-boiling solvent, which will affect the solubility of the solute.

Place the solid to be recrystallized on a creased weighing paper and carefully pour it into an Erlenmeyer flask (Figure 15.2a). Alternatively, a plastic powder funnel may be set in the neck of the Erlenmeyer flask to prevent spillage (Figure 15.2b). Add one or two boiling stones or a boiling stick. Heat an appropriate volume of the solvent in another Erlenmeyer flask (see Figure 15.3). Then add small portions of hot (just below boiling) solvent to the solid being recrystallized. Begin heating the solid/solvent mixture, allowing it to boil briefly between additions, until the solid dissolves; then add about 10% excess solvent. Remember that some impurities may be completely insoluble, so do not add too much solvent in trying to dissolve the last bit of solid.

With particularly volatile organic solvents, such as ether or petroleum ether, it is often easier to add a small amount of cold

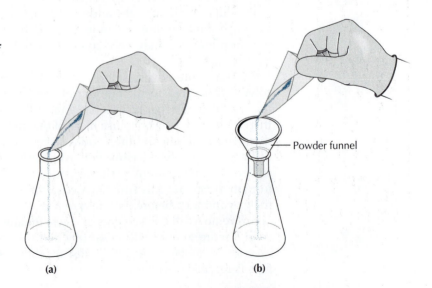

FIGURE 15.2 Two ways to add a solid to an Erlenmeyer flask for recrystallization.

(a)   (b)

Powder funnel

Boiling stick

Steam in

To drain

FIGURE 15.3 Heating a solution on a steam bath.

solvent and then heat the mixture nearly to boiling. Slowly add more cold solvent to the heated mixture until the solid just dissolves when the solution is boiling; then add about 10% excess solvent.

If you have no insoluble material or highly colored impurities in your hot recrystallization solution, cool the solution as described in step 2.

If you have **insoluble material** in the hot recrystallization solution, it needs to be removed before cooling the solution. Carry out the procedure described in Section 9.2 for filtering insoluble material.

If the compound you are recrystallizing is known to be colorless and if the recrystallization solution is deeply colored after the compound dissolves, treatment with activated charcoal (Norit, for example) may remove what is probably a small amount of intensely colored impurity. Activated charcoal has a large surface area and a strong affinity for highly conjugated colored compounds, allowing it to readily adsorb these impurities from the recrystallization solution. Using too much charcoal, however, may cause some of the compound you are purifying to be adsorbed by the charcoal and reduce your yield.

SAFETY PRECAUTION

Use gloves when handling activated charcoal pellets and **cool the hot solution briefly before adding the charcoal.** Adding charcoal to a boiling solution can cause the solution to foam out of the flask.

Add 40–50 mg of Norit activated-carbon decolorizing pellets to the **hot but not boiling** recrystallization solution. Then heat the mixture to just under boiling for a few minutes. (Boiling actually hinders decolorization, but heating to keep the compound in solution is necessary.) While the solution is still very hot, gravity filter it through a fluted filter paper (see Section 9.2).

Step 2. Cooling the Solution

The size and purity of the crystals obtained will depend on the rate at which the solution cools: the slower the cooling, the larger the crystals. Stopper the Erlenmeyer flask while the solution cools. Allow the hot solution to stand on the bench top until crystal formation begins and the flask reaches room temperature. The cooling process will take at least 20 min and occasionally may take 30 min or more before crystals appear. This slow cooling usually produces crystals of a reasonable purity and intermediate size. Once crystal growth appears to be complete, cool the solution for 10–15 min in an ice-water bath before recovering the crystals from the solution as described in step 3.

What to do if no crystals appear in the cooled solution. If no crystals appear in the solution after at least 15 min of cooling in an ice-water bath, add one or two seed crystals. If you do not have any seed crystals, scratch the bottom of the flask vigorously with a glass stirring rod. Tiny particles of glass scratched from the flask can initiate crystallization. If crystallization still does not occur, there is probably too much solvent. Boil off some of the solvent in the hood and cool the solution again.

Step 3. Collecting the Recrystallized Solid

To recover the recrystallized solid after crystallization appears to be complete, collect the solid by vacuum filtration (see Section 9.4), using a Buchner funnel, neoprene adapter, filter flask, heavy-walled rubber tubing, and trap bottle or flask (Figure 15.4). The trap flask avoids backflow of water from a water aspirator coming into contact with your remaining recrystallization solution; with a house vacuum system or vacuum pump, the trap flask keeps any overflow from the filter flask out of the vacuum line.

Choose the correct type and size of filter paper (see Section 9.1), one that will fit flat on the bottom of the Buchner funnel and just cover all the holes. Turn on the vacuum source and wet the paper with the recrystallizing solvent to pull it tightly over the holes in the funnel. Pour a slurry of crystals and solvent into the funnel.

After the filtration is complete, wash the crystals on the Buchner funnel with a small amount of cold recrystallization solvent (1–5 mL, depending on the amount of crystals) to remove any supernatant liquid adhering to them. To wash the crystals, allow air to enter the filtration system by removing the rubber tubing from the water aspirator nipple or vacuum system. Then turn off the water (to prevent backup of water into the system) or turn off the vacuum line. Loosen the neoprene adapter connecting the Buchner funnel to the filter flask. Cover the crystals with the **cold** solvent, reconnect the vacuum, and draw the liquid off the crystals. Initiate the crystal drying process by pulling air through the crystals for a few minutes. Again disconnect the vacuum by removing the rubber tubing from the vacuum source. Place the crystals on a *tared (preweighed)* watch glass. You will probably need to leave the crystals open to the air in your desk for a time to dry them completely. Remove any boiling stones or sticks before you weigh the crystals.

A Second Crop of Crystals

A second "crop" of crystals can sometimes be obtained by evaporating about half the solvent from the filtrate and again cooling the solution. This crop of crystals should be kept separate from the first crop of crystals until the melting points of both (see Section 14.3) have been determined. If the two melting points are the same, indicating that the purity is the same, the crops may be combined. Usually the

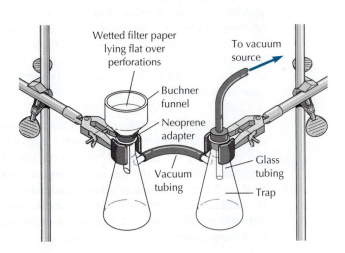

FIGURE 15.4
Apparatus for vacuum filtration. The second filter flask serves as a backflow trap.

Wetted filter paper lying flat over perforations

To vacuum source

Buchner funnel

Neoprene adapter

Vacuum tubing

Glass tubing

Trap

second crop has a slightly lower melting point and a larger melting range, indicating that some impurities crystallized with the desired product.

15.6 Microscale Recrystallization

Read Sections 15.1–15.3 before you undertake your first microscale recrystallization.

Microscale methods are used for recrystallizations of less than 300 mg of solid. If you are doing a mixed-solvent recrystallization, refer to the section on two-solvent recrystallizations in Section 15.4.

In many microscale recrystallizations, a 10- or 25-mL Erlenmeyer flask holds the recrystallization solution and a Hirsch funnel replaces the Buchner funnel for collecting the crystals. If the amount of solid being recrystallized is less than 150 mg, a 10-mL Erlenmeyer flask or a test tube can be used. The following steps outline the procedure for a microscale recrystallization.

Step 1. Dissolving the Solid and Removing Insoluble Impurities

Save a few crude crystals to use as seeds in the event that recrystallization does not occur.

Place the solid in a 25-mL or 10-mL Erlenmeyer flask or a test tube, depending on the mass of crude product to be recrystallized; add a boiling stick or boiling stone. With a Pasteur pipet, add only enough solvent to just cover the crystals. Use a hot-water or steam bath to heat the contents of the flask or test tube to the boiling point, then add additional solvent drop by drop, allowing the mixture to boil briefly after each addition. Continue this process until just enough solvent has been added to dissolve the solid. Be aware that some impurities may not dissolve; if this is the case, carefully transfer the solution into a clean flask or test tube using a filter-tip pipet (see Section 5.3).

Colored impurities. If colored impurities are present, cool the mixture slightly and add 10 mg of Norit activated-carbon decolorizing pellets (about 10 pellets). Keep the mixture heated to just under the boiling point. If the color is not removed after 1–2 min, add a few more Norit pellets and heat briefly. Prepare a Pasteur filter-tip pipet (see Section 5.3). Warm the Pasteur pipet by immersing it in a test tube of hot solvent and drawing the hot solvent into it several times. Then use the heated pipet to carefully separate the hot recrystallization solution from the Norit pellets and transfer it to a clean test tube or flask. If crystallization begins in the solution with the carbon pellets during this process, add a few more drops of solvent and warm the mixture to boiling to redissolve the crystals before completing the transfer.

Step 2. Cooling the Solution

The size of the crystals obtained will depend on the rate at which the solution cools: the slower the cooling, the larger and purer the crystals. Slow cooling usually produces crystals of a reasonable purity and intermediate size. To facilitate slow cooling, set the flask on a paper towel and cover the flask with a beaker; if the recrystallization was done in a test tube, place the test tube in an Erlenmeyer flask for the cooling period. Allow the solution to cool slowly to room temperature. The recrystallization process may take 20 min or more. Then chill the container containing the crystals in an ice-water bath to complete the crystallization process.

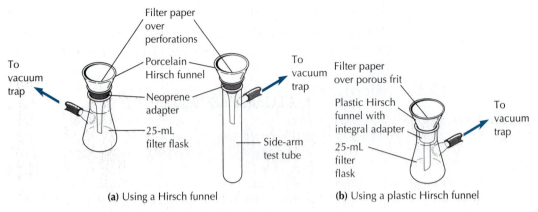

FIGURE 15.5 Vacuum filtration using a Hirsch funnel.

Step 3. Collecting the Recrystallized Solid

Assemble a filtration apparatus as shown in Figure 15.5a or b, using heavy-walled rubber tubing. Choose the correct size of filter paper—a size that fits flat on the Hirsch funnel and just covers the holes of the porcelain Hirsch funnel or the frit of the plastic Hirsch funnel. Clamp the filter flask firmly at the neck to a ring stand or apparatus rack.

Connect the filter flask to a vacuum trap as shown in Figure 15.4. Turn on the vacuum source and wet the paper with a few drops of the recrystallization solvent to pull it tightly to the funnel. Pour a slurry of crystals and solvent into the funnel. Alternatively, you can use a 50-mL syringe as the vacuum source (see Figure 9.9); withdrawing the syringe plunger can provide a gentle effective vacuum.

After the filtration is complete, wash the crystals on the Hirsch funnel with a few drops of cold recrystallization solvent to remove any supernatant liquid adhering to them. To wash the crystals, allow air to enter the filtration system by removing the rubber tubing from the water aspirator nipple or vacuum line. Turn off the water (to prevent backup of water into the system), or turn off the vacuum line and carefully loosen the neoprene adapter (or the plastic Hirsch funnel) from the filter flask. Add **cold** solvent one drop at a time to just cover the crystals, reconnect the vacuum, and draw the liquid off the crystals. Initiate the crystal drying process by pulling air through the crystals for a few minutes. Again disconnect the vacuum as described earlier.

Place the crystals on a *tared (preweighed)* watch glass. Allow the crystals to air dry completely on the watch glass before weighing them and determining their melting point. Remove any boiling stones before you weigh the crystals.

15.7   Microscale Recrystallization Using a Craig Tube

If you are working with less than 100 mg of a solid and 2 mL of solvent, the recrystallization can be carried out using a Craig tube, shown in Figure 15.6. The **Craig tube** is used to separate crystals from a solution with minimal loss of the solid product after a microscale recrystallization procedure. The outer part of the Craig tube is similar to a test tube, except that the diameter becomes greater partway up the tube, and the glass is ground at this point so that the inside surface is rough (Figure 15.6a). The rounded end of the inner part of the Craig tube (the plug) is also ground, so that when they are fit together they form a ground glass joint that allows liquid through but doesn't allow crystals to get through. The inner part can also be made of Teflon.

The outer part of the Craig tube is used as a test tube for a microscale recrystallization (see Section 15.6), using no more than 2 mL of solvent. After step 2 of the recrystallization has been completed in the outer Craig tube, insert the plug, to which you have attached a thin copper wire or strong thread around the narrow part of the inner plug (Figure 15.6b). Hold the Craig tube assembly upright and place a centrifuge tube over it with the copper wire or thread extending just below the lip of the centrifuge tube (Figure 15.6c).

Turn the entire assembly upside down so that the centrifuge tube is in an upright position and spin in a centrifuge, making sure that the centrifuge is balanced by placing another centrifuge tube filled with water on the opposite side. Centrifuge the sample for

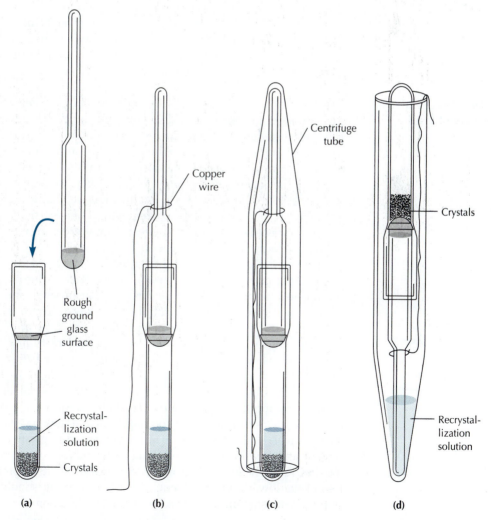

FIGURE 15.6 (a) A Craig tube with recrystallized solid and solution. (b) Craig tube assembly. (c) Centrifuge tube placed over the assembly. (d) Inverted Craig tube assembly after centrifugation.

several minutes until the solution from which the crystals grew goes to the bottom of the centrifuge tube. Use the copper wire or thread to pull the Craig tube out of the centrifuge tube. If the crystals have collected on the end of the inner plug (Figure 15.6d), it is a simple procedure to remove the plug and scrape the crystals onto a *tared (preweighed)* watch glass with a flat spatula.

If the crystals remain at the other end of the outer Craig tube, you can either continue to centrifuge the complete Craig tube assembly including the centrifuge tube or you can scrape the crystals from the inside surface of the Craig tube.

15.8   Sources of Confusion and Common Pitfalls

It is worthwhile to review Section 15.3, which discusses the importance of scale, volume of solvent, use of seed crystals, insoluble impurities, maximum recovery factors, and ensuring dry crystals. Probably the most confusing part of recrystallization is choosing the most effective recrystallization solvent by the methods of Section 15.4. This is the stage at which careful observations and thoughtful analysis of your experimental results can save a great deal of time in the long run. If loss of the crystals that you use for the solubility tests must be minimized, you can recover them by evaporation of the solvents.

How Much Solvent Should I Use?

The answer to this question depends on the solubilities of the compound in the hot and cold solvent and the amount of material being recrystallized. General recrystallization guidelines are always somewhat ambiguous because they cannot be applied in a straightforward manner for every one of the many thousands of organic compounds you might be recrystallizing. Choose a flask volume consistent with the scale of recrystallization (see Section 15.3). You do not want to use more solvent than will fit efficiently in the flask. If more solvent is required, it suggests that you might want to find a different recrystallization solvent.

A recrystallization is usually started with only enough solvent to cover the crystals in the recrystallization flask. After heating the solvent to boiling in a separate flask, add it in small increments, 1–5 mL for miniscale recrystallizations and a few drops for microscale recrystallizations. Reheat the recrystallization flask after adding **each** solvent increment. Add only enough solvent to just dissolve the crystals when the solvent is boiling, plus another 10% increment to provide a modest solvent excess.

Occasionally you may add large volumes of hot solvent to the recrystallization flask and see no effect on the solubility of your compound. If the rate of heating is too rapid, solvent may be evaporating from the recrystallization flask as fast as you are adding it. Evaporation is a particular problem when working with a mixed-solvent recrystallization. Rapid heating in this instance probably results in preferential loss of the lower-boiling solvent. The rate of heating should be at a setting that just maintains the solvent at its boiling point.

My Compound Is Not Crystallizing!

In many instances, recrystallization fails because too much solvent is used in the process. It is easy to add too much solvent when recrystallizing using mixed solvent systems because the system can see-saw between having too much of the more-soluble solvent and too much of the less-soluble solvent. Once there is too much solvent overall, your solid compound will not crystallize and you need to evaporate most of the solvent and start over.

Other solutions that fail to crystallize may be supersaturated and require overcoming the kinetic barrier to crystallization. The best approach is to add one or two seed crystals, but if you do not have a few seed crystals available, it may be possible to promote crystal formation by scratching the inside of the bottom of the flask vigorously with a glass stirring rod. Tiny particles of glass scratched from the flask can serve as centers for crystallization.

My Solution Turned Cloudy and the Separated Compound Looks Oily Instead of Crystalline

The formation of oils may be the most frustrating outcome of an attempted recrystallization. Instead of shining crystals, some solutions produce droplets of insoluble liquid as they cool, in a process called *oiling out*. This happens when a compound becomes insoluble at a temperature above its melting point. Oiling out frequently occurs if too little recrystallization solvent has been used, so that the compound becomes insoluble at too high a temperature. It is most common during recrystallization of a solute with a melting point near the boiling point of the solvent. Once a compound has oiled out, impurities dissolve in the oil and impede crystallization. As it cools, the oil often hardens into a viscous, glasslike substance.

If you have an oil rather than crystals, you can add more solvent so that the compound does not come out of solution at so high a temperature. It may also help to switch to a solvent with a lower boiling point (consult Table 15.1).

Some oils can be crystallized by dissolving them in a small amount of diethyl ether or hexane and allowing the solvent to evaporate slowly in a hood. Crystallization often occurs as the solution slowly becomes more concentrated. Once crystals form, seed crystals are available to assist further purification.

Questions

1. Describe the characteristics of a good recrystallization solvent.
2. The solubility of a compound is 59 g per 100 mL in boiling methanol and 30 g per 100 mL in cold methanol, whereas its solubility in water is 7.2 g per 100 mL at 95°C and 0.22 g per 100 mL at 2°C. Which solvent would be better for recrystallization of the compound? Explain.
3. Explain how the rate of crystal growth can affect the purity of a recrystallized compound.
4. In what circumstances is it necessary to filter a hot recrystallization solution?
5. Why should a hot recrystallization solution be filtered by gravity rather than by vacuum filtration?
6. Low-melting solids often "oil out" of a recrystallization solution rather than crystallizing. If this were to happen, how would you change the recrystallization procedure to ensure good crystals?
7. An organic compound is quite polar and is thus much more soluble in methanol than in pentane (bp 36°C). Why would methanol and pentane be an awkward solvent pair for recrystallization? Consult Table 15.1 to assist you in deciding how to change the solvent pair so that recrystallization would proceed smoothly.

CHAPTER

16 SUBLIMATION

Normally when you heat a solid, it melts; as you continue to heat the liquid phase, it boils, turning into the vapor phase. Some solids, however, don't melt at atmospheric pressure, but instead pass directly from the solid state to the gaseous state without going through a liquid state. This process is called *sublimation*. You may be familiar with solid CO_2. It sublimes, which explains its common name, *dry ice*. Carbon dioxide does not have a melting point at atmospheric pressure. The sublimation point for CO_2 at a pressure of one atmosphere is $-78°C$, well below its melting point of $-57°C$ at a pressure of 5.2 atmospheres.

Chapter 12 discussed how the vapor pressure of a liquid varies with temperature (see Figure 12.1), and how the boiling point of a liquid is the temperature at which its vapor pressure equals the pressure exerted on it by the atmosphere. The vapor pressure of a solid also varies with temperature and, when heated, the vapor pressure of some solids can increase above atmospheric pressure before the solid melts. Thus, the solid sublimes. Because the vapor can be resolidified, the vaporization-solidification cycle can be used as a purification method.

16.1 Sublimation of Solids

A number of substances exhibit appreciable vapor pressures below their melting points, such as iodine, camphor, and 1,4-dichlorobenzene (mothballs). You may already have seen iodine crystals evaporate to a purple gas during gentle heating and smelled the characteristic odor of camphor. These substances can change directly from the solid phase to the gas phase. Solid ice can also evaporate below its melting point. Freeze drying is a process used almost universally in biochemical and biomedical laboratories, as well as in food preservation. It efficiently removes water from solids through the evaporation of ice crystals. Although the vapor pressure of ice at its freezing point is only 4.6 mm Hg, when an effective vacuum is applied, the ice readily evaporates.

Even though many organic compounds have discrete melting points, they also have substantial vapor pressures below their melting points. Thus, they evaporate as they are heated in a melting point capillary (see Section 14.6). To determine the melting points of camphor and caffeine it is necessary to use a sealed capillary tube.

At reduced pressure, many organic compounds readily sublime. In the laboratory sublimation is used as a purification method for an organic compound if

1. it can vaporize without melting,
2. it is stable enough to vaporize without decomposition,
3. the vapor can be condensed back to the solid, and
4. the impurities present do not also sublime.

One advantage of using sublimation for purification of a solid is that no solvent is used; thus, none has to be removed later.

Sublimation is also a faster method of purification than recrystallization; however, it is not as selective because many organic compounds sublime. It is probably most effective in removing volatile organic compounds from a nonvolatile salt or other inorganic material. Use of reduced pressure, supplied by a vacuum source, makes decomposition and melting less likely to occur during the sublimation.

16.2 Assembling the Apparatus for a Sublimation

A miniscale sublimation apparatus under reduced pressure is shown in Figure 16.1. The apparatus consists of an outer tube and an inner tube. The outer vessel holds the sample being purified and is connected to a vacuum source. The inner tube, sometimes called a **cold finger**, usually contains an ice-water mixture that provides a cold surface on which the vaporized compound can recondense as a solid. The inner and outer tubes are sealed together by an O-ring seal.

The distance between the bottom surface of the inner tube and the outer tube is 0.5–1.0 cm. If the vapor has to travel a long distance, a higher temperature is needed to keep it in the gas phase, and decomposition of the solid sample may very well occur. If the surfaces are too close, impurities can spatter and contaminate the condensed solid on the surface of the inner tube.

Two simple microscale arrangements for sublimation under reduced pressure are shown in Figure 16.2. The inner test tube, which contains an ice-water mixture, serves as a condensation site for the sublimed solid. The outer vessel, a filter flask (Figure 16.2a) or a side-arm test tube (Figure 16.2b), holds the material being sublimed, and the side arm provides a connection to the vacuum source. The inner and outer vessels are sealed together by a neoprene filter adapter. The distance between the bottom surfaces of the inner tube and the outer tube or filter flask should be 0.5–1.0 cm, with the smaller distance used for smaller samples.

A side-arm test tube works well for 10–100 mg of material. The filter flask apparatus can be sized to suit the amount of material being purified. For example, 10–150 mg can be sublimed in a 25-mL

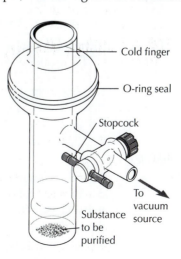

Cold finger

O-ring seal

Stopcock

To vacuum source

Substance to be purified

FIGURE 16.1 Miniscale sublimation apparatus.

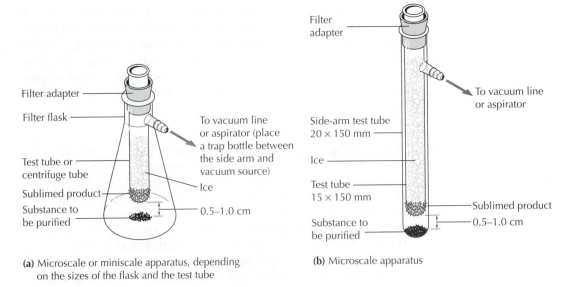

(a) Microscale or miniscale apparatus, depending on the sizes of the flask and the test tube

(b) Microscale apparatus

FIGURE 16.2 Two simple apparatuses for sublimation.

filter flask, whereas 100–300 mg or more material requires a 50- or 125-mL filter flask with a correspondingly larger test tube for the cold finger in the latter case.

A hot plate is a good way to heat the sublimation apparatus shown in Figures 16.1 and 16.2a. Because a test tube, however, does not have a flat bottom, a hot plate provides inefficient heat transfer, so either a sand bath or an aluminum block on a hot plate (see Section 6.2) can be used. Heat guns (see Figure 6.7), if available in your laboratory, have some advantages as a heat source in sublimations using the apparatus shown in Figure 16.2b. A heat gun can be moved up and down the sides of the test tube to vaporize any solid that has sublimed onto the sides of the outer side-arm test tube, driving it to the cold surface of the inner tube.

16.3   Carrying Out a Microscale Sublimation

Always wear safety glasses when using a vacuum. The pressure difference can cause glassware to implode, releasing glass shrapnel.

SAFETY PRECAUTION

The lip of the inner test tube must be large enough to prevent it from being pushed through the bottom of the filter adapter by the difference in pressure created by the vacuum. Slippage of the inner test tube could cause both vessels to shatter as the inner test tube hits the outer test tube or flask. Placing a microclamp on the inner test tube above the filter adapter helps keep the test tube from moving once it is positioned in the filter adapter.

Place the sublimation sample (10–100 mg) in a 25-mL filter flask or a side-arm test tube. Fit the inner test tube through the filter adapter and adjust its position so that the bottom of the inner test tube is 0.5–1.0 cm above the bottom of the flask or side-arm test tube; with smaller samples, aim for 0.5 cm. Turn on the water aspirator or vacuum line.

Ice-water is placed in the inner test tube after the vacuum is applied to prevent condensation of moisture from the air onto the tube before sublimation takes place.

After a good vacuum has been achieved, fill the inner test tube with ice, then gently heat the sublimation vessel. During the sublimation, you will notice material disappearing from the bottom of the outer vessel and reappearing on the cool surface of the inner test tube. If the sample begins to melt, briefly withdraw the heat source from the apparatus. If most of the ice melts, remove half the water from the inner test tube with a Pasteur pipet and add additional ice. Remember that it is not unlikely to have some nonvolatile impurities in your sample.

After the sublimation is complete, remove the heat source, allow the outer tube to cool, and **slowly let air back into the system** by gradually removing the rubber tubing from the water aspirator or other vacuum source. Then turn off the water flow in the aspirator or shut off the vacuum source and slowly disconnect the rubber tubing from the side arm of the filter flask or test tube. Carefully remove the inner test tube and immediately scrape the purified solid onto a tared weighing paper with a spatula. After weighing the sublimed solid, store it in a tightly closed vial.

16.4   Sources of Confusion and Common Pitfalls

As I Heated My Solid Sample, It Began to Turn Brown. Is This a Problem?

The color change is likely a sign of decomposition, which means the sample is getting too hot. A likely reason why it is not subliming faster at a lower temperature is that the pressure inside the sublimation tube is too high, probably caused by either a poor vacuum system or a leak in one of the joints in the apparatus. Stop heating and check for leaks!

How Do I Keep the Sublimed Solid from Falling into the Nonvolatile Impure Residue?

The most likely reason for this is a violent air current in the sublimation apparatus, produced by air rushing in as the vacuum is broken too quickly. The air must be let back in slowly by gently and carefully loosening the rubber hose from the inlet tube to the apparatus. Only after this is done should the vacuum be discontinued and the inner tube gently removed from the outer tube.

The Sublimed Solid Seems Wet; the Crystals Stick Together. What Happened?

The wetness is probably water that has condensed on the cold finger while it was exposed to the lab atmosphere. It's important to speedily scrape off the sublimed solid before humidity condenses on the cold surface. Once this occurs, you need to let the solid dry. If it has a melting point above 100°C and doesn't vaporize too easily, heat it in an oven on a watch glass at 50°C for 20 min. Remember that your compound sublimes, so don't heat it too hot or for too long.

Questions

1. Which of the following three compounds—polyethylene, menthol, or benzoic acid—is the most likely to be amenable to purification by sublimation? Give your reasoning.
2. A solid compound has a vapor pressure of 65 torr at its melting point of 112°C. Give a procedure for purifying this compound by sublimation.
3. Hexachloroethane has a vapor pressure of 780 torr at its melting point of 186°C. Describe how solid hexachloroethane would behave while carrying out a melting-point determination at atmospheric pressure (760 torr) in a capillary tube open at the top.

OPTICAL ACTIVITY AND ENANTIOMERIC ANALYSIS

Optical activity, the ability of substances to rotate plane-polarized light, played a crucial role in the development of organic chemistry because it linked the molecular structures that chemists write with the physical world. A major development in the structural theory of chemistry was the concept of the three-dimensional shape of molecules. When Jacobus van't Hoff and Joseph Le Bel noted the asymmetry possible in tetrasubstituted carbon compounds, they claimed that their chemical structures were identical to the physical structures of the molecules. Not only was the structural theory of organic chemists useful in explaining the facts of chemistry, it also happened to be true. Van't Hoff and Le Bel could make this claim because their theories of the tetrahedral carbon atom accounted not only for chemical properties but also for the physical property of optical activity.

17.1 Mixtures of Optical Isomers: Separation/Resolution

A molecule that possesses no internal mirror plane of symmetry and that is not superimposable on its mirror image is said to be *chiral,* or "handed." Chirality, a molecular property, is normally indicated by the presence of a *stereocenter*—a tetrahedral atom bearing four different substituents. A tetrahedral stereocenter is sometimes called a chiral or asymmetric center.

Chiral compounds possess the property of enantiomerism. *Enantiomers* are stereoisomers that have *nonsuperimposable mirror images.* Chiral compounds such as 2-butanol and the amino acid alanine, which contain only one stereocenter, provide simple examples of enantiomers.

2-Butanol   Alanine

Enantiomers and Racemic Mixtures

The enantiomers of 2-butanol have identical physical properties, including boiling points, IR spectra, NMR spectra, and refractive indexes, except for the direction in which they rotate plane-polarized light. Both enantiomers are optically active—one of them rotates polarized light in a clockwise direction and is called the *(+)-isomer.* The other enantiomer rotates polarized light counterclockwise and is called the *(−)-isomer.* The rotational power of (+)-2-butanol is exactly the same in the clockwise direction as that of (−)-2-butanol in the counterclockwise direction. Unfortunately, there is no simple theoretical way to predict the direction of the rotation of plane-polarized light on the basis of the configuration at a tetrahedral carbon stereocenter. Thus, it is not apparent which structure of 2-butanol or alanine is the (+)- or the (−)-enantiomer, although in some important instances, X-ray diffraction has been used successfully to make this kind of correlation.

Usually, simple compounds are optically inactive, even when their molecules are chiral. For example, you would normally find that a sample of 2-butanol is optically inactive. To understand this apparent paradox, consider the reduction of 2-butanone with sodium borohydride. This reaction can proceed in two ways. Hydride can react with 2-butanone from either the top side or the bottom side of the carbonyl double bond. The reaction occurs both ways at equal rates, giving rise to a 50:50 mixture of the enantiomers of 2-butanol—a product that is optically inactive:

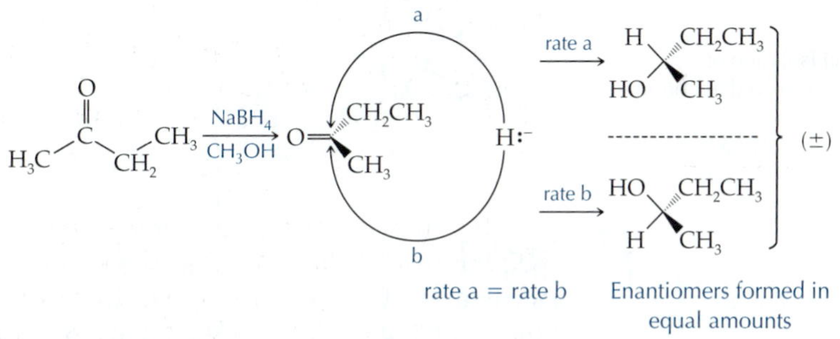

rate a = rate b   Enantiomers formed in equal amounts

An equal mixture of (+)- and (−)-enantiomers is called a racemic mixture. In the separation or resolution of a racemic mixture, the enantiomers are transformed into a pair of *diastereomers*— stereoisomers that have different physical and chemical properties. A mixture of two diastereomers is prepared from a racemic mixture by its reaction with an optically active substance. The diastereomers can then be separated by crystallization or chromatography because of their differential solubility or partitioning.

Resolution with Acids or Bases

The simplest reaction for preparing diastereomers from racemic mixtures is that of an acid with a base to form a salt. For *resolution* or separation of the two enantiomers in an acid–base reaction, the added reagent must be optically active. Reaction of a racemic amine, for example, with an optically pure carboxylic acid is a method for resolving the amine. Similarly, reaction of a carboxylic acid with an optically pure amine is a way of resolving the acid. Two different diastereomeric salts are produced in each of these reactions. These salts differ in their solubilities in various solvents and can be separated by fractional crystallization. The less soluble diastereomeric salt is the more easily obtained. The process for resolution of an amine with an optically pure carboxylic acid is represented in the following reactions:

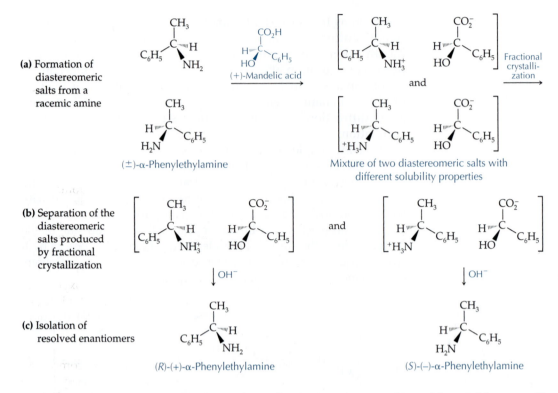

(a) Formation of diastereomeric salts from a racemic amine

(±)-α-Phenylethylamine

(+)-Mandelic acid

Mixture of two diastereomeric salts with different solubility properties

Fractional crystallization

(b) Separation of the diastereomeric salts produced by fractional crystallization

and

(c) Isolation of resolved enantiomers

(R)-(+)-α-Phenylethylamine

(S)-(–)-α-Phenylethylamine

If you examine the diastereomeric salts in (a) and (b), you will see that each salt has two stereocenters. When you compare their structures, you will find that the carbon stereocenters bearing the —OH group have the identical configuration in each salt, whereas the stereocenters bearing the —NH$_3^+$ group have opposite configurations. Thus the two salts are stereoisomers that are not mirror images; they are diastereomers.

Optically pure acids and bases, often isolated from plants, are frequently used for the resolution of racemic mixtures (Table 17.1); however, the diastereomers necessary for resolution do not need to be salts. For example, diastereomeric esters, formed by reaction of the enantiomers of an alcohol with an optically pure carboxylic acid, can also be used.

Enzymatic Resolution

An increasingly useful method for the resolution of racemic mixtures utilizes an enzyme that selectively catalyzes the reaction of one enantiomer. Because all enzymes are chiral molecules, the transition

TABLE 17.1	Optically active acids and bases used for resolutions
Bases	**Acids**
Brucine	Tartaric acid
Strychnine	Mandelic acid
Quinine	Malic acid
Cinchonine	Camphor-10-sulfonic acid
α-Phenylethylamine	

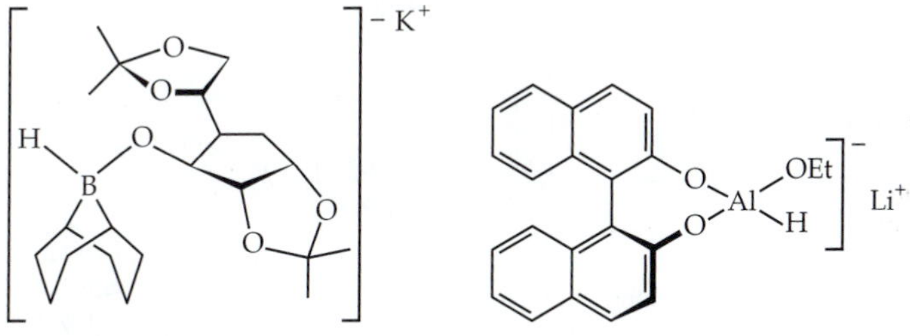

FIGURE 17.1 Chiral hydride reducing agents used in enantiospecific organic synthesis: (a) a chiral borohydride, (b) a chiral aluminohydride.

states for the reaction of an enzyme with two enantiomers are diastereomeric and the energies of these two transition states differ. Thus one of the enantiomers reacts faster than the other one. In many cases an enzyme reacts so much faster with one enantiomer that the specificity provides an excellent method for resolving a racemic mixture. For example, one enantiomer of an ester in a racemic mixture can be selectively hydrolyzed to a carboxylic acid by an esterase, whereas the other enantiomer is untouched. It is a straightforward matter to separate a pure enantiomer of the optically active carboxylic acid from the unreacted ester.

Enantiospecific Organic Synthesis

An alternative to separating the enantiomers of a chiral compound by a resolution process is the direct synthesis of optically active compounds using either chiral reagents or chiral catalysts. Return to the hydride reduction of 2-butanone on page 241 and notice that sodium borohydride produces a racemic mixture of the enantiomers of 2-butanol. The use of a borohydride reducing agent with a chiral group attached to boron, however, can produce an unequal mixture of (+)- and (−)-2-butanol. In some cases this approach gives a large excess of one enantiomer over the other. Figure 17.1 shows two different chiral hydride reducing agents.

Great progress has been made in enantiospecific organic synthesis in the past few years. The synthesis of pharmaceuticals, which are important to modern medicine, places great emphasis on the production of optically active drugs, which can be more effective and have fewer side effects than racemic drugs. The use of chiral catalysts in organic synthesis has been particularly effective in producing the optically active chiral precursors from which the drugs can be synthesized. Some of these catalysts are enzymes and others are chiral organic bases or Lewis acids. Optically active L-DOPA, the amino acid 3,4-dihydroxyphenylalanine, which is used in the treatment of Parkinson's disease, is synthesized commercially by using a chiral rhodium catalyst in the hydrogenation of an amino acid precursor.

17.2   Polarimetric Techniques

The traditional way to measure optical activity is with a polarimeter, a schematic description of which is shown in Figure 17.2. Polarimeters measure the degree of rotation of polarized light as it passes through an optically active material. All commercially available polarimeters have the same general features. The analyzer

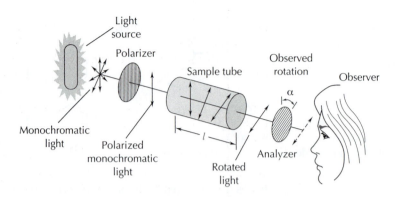

FIGURE 17.2
Schematic diagram of a polarimeter.

of a simple polarimeter is adjusted manually, whereas all the components of an automated polarimeter are housed in the instrument case and produce a digital readout of the observed rotation.

How a Polarimeter Works

The light beam approaching the polarizer in Figure 17.2 has wave oscillations in all planes perpendicular to the direction in which the beam is traveling. When the light beam hits the polarizer, which has ranks and files of molecules arranged in a highly ordered fashion, only the light whose oscillations are in one plane is transmitted through the polarizer. The light that gets through is called *plane-polarized light.* The remaining waves are refracted away or absorbed by the polarizer. In a rough analogy, the polarizer's molecules are ordered like the slats of wood in a picket fence. Only the light waves whose oscillations are parallel to the slats pass through the polarizer and into the sample tube.

The analyzer is a second polarizer whose ranks and files of molecules must also be lined up for the polarized light waves to be transmitted. If the polarized light has been rotated by an optically active substance in the sample tube, the analyzer must be rotated the same amount to let the light through. The rotation is measured in degrees, indicated by α in Figure 17.2. A modern automatic digital polarimeter is shown in Figure 17.3.

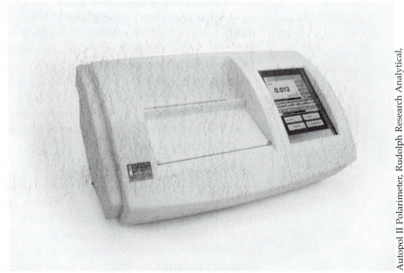

FIGURE 17.3 Modern digital polarimeter.

Autopol II Polarimeter, Rudolph Research Analytical, Hackettstown, NJ

**Use of
Monochromatic
Light**

Monochromatic light is used in polarimetric measurements because the optical activity or rotatory power of chiral compounds depends on the wavelength of the light used. For example, the rotation of 431-nm (blue) light is 2.8 times greater than the rotation of 687-nm (red) light. The traditional light source has been a sodium lamp, which has two very intense emission lines at 589 and 589.6 nm. This closely spaced doublet is called the *sodium D line.* Some polarimeters use a 589-nm LED. The intense 546.1-nm emission line of a mercury lamp is also used; the human eye is more sensitive to the mercury emission in the green region than to the sodium line in the yellow region of the spectrum. In modern digital polarimeters, the light source may be a xenon or tungsten/halogen lamp, which with appropriate filters allows the measurement of optical rotations at a number of wavelengths.

Polarimeter Tubes

Polarimeter tubes are expensive and must be handled carefully. They come in different lengths; 1-dm and 2-dm tubes are the most common for manual polarimeters in order to produce a large degree of rotation. Using longer tubes, however, also requires a larger amount of sample. The cell length is always given in decimeters (dm, 10^{-1} m) in the calculation of specific rotation (see Section 17.3). The periscope tube allows removal of any air bubbles from the light path that can be tedious to remedy when using a straight tube. The tubes shown in Figure 17.4a are closed with a glass plate and a rubber washer, both held in place by a one- or two-part screw cap. Be careful not to screw the cap too tightly because strain in the glass end plate can produce an apparent optical rotation. The tube shown in Figure 17.4b is designed so that samples can be introduced with a syringe.

Preparing polarimetric samples. The quantitative nature of optical rotation measurements requires careful technique as you prepare samples for polarimetric readings. Chiral samples should be weighed on analytical balances that read to the nearest 0.1 mg (see Section 5.1) and solutions for polarimetric measurements should be made in volumetric flasks. The three most common solvents for polarimetry are water, methanol, and ethanol. Aqueous hydrochloric acid is often used for optically active amines.

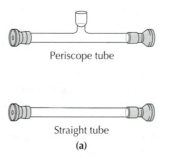

Periscope tube

Straight tube

(a)

Short-path tube
with syringe ports

(b)

FIGURE 17.4 Polarimeter tubes.

Cleaning a polarimeter tube. Before you fill a polarimeter tube, check to see that it is empty and dry by holding it up to the light and looking along its axis. If it is clean and dry, you can proceed to fill it with your optically active sample.

If the tube is not clean and dry, you should first clean it with some care. When you have cleaned the tube, it will have traces of water in it. Rinse the tube with the solvent you will be using for the solution of your optically active compound. After the tube has been well drained, rinse it with two or three small portions of your solution to ensure that the concentration of the solution in the polarimeter tube is the same as the concentration of the solution you have prepared. You may want to save these optically active rinses because your chiral compound can be recovered from them later.

Air bubbles and suspended particles. When you fill a polarimeter tube with a solution, remember that air bubbles in the tube are your enemy. Make sure that the tube has no air bubbles trapped in it; they will refract the light coming through. Also make sure that there are no suspended particles in a solution whose rotation you wish to measure, or you may get so little transmitted light that measurement of the rotation will be difficult. If you have a solution that you suspect may be too turbid for polarimetric measurements, filter it through a micropore filter using a syringe, or by gravity through a small plug of glass wool (see Section 9.1).

Making Polarimetric Measurements

Consult your instructor about the operation of the polarimeter in your laboratory.

Automatic digital polarimeters. If you use a clean tube and avoid air bubbles, making measurements using an automatic polarimeter is quite straightforward. Digital polarimeters produce a degree of rotation automatically to an accuracy of 0.02°–0.0003°. They usually have built-in temperature probes and allow you to use of a variety of light wavelengths.

Choose the temperature and the wavelength of light you wish to use. If necessary, obtain a blank polarimeter reading by filling the tube with distilled water. If a blank reading was necessary, drain the water from the tube, rinse it with the solvent you will use for your sample, and after it has been well drained, rinse it with two or three small portions of your sample solution to ensure that the concentration of the solution in the polarimeter tube is the same as the concentration of the solution you have prepared. After you have filled the polarimeter tube with your sample, enter its concentration and press set. The digital display will give you the optical rotation and will convert it to the specific rotation. After finishing your measurements, rinse the tube with distilled water. If you are using a tube that is filled with a syringe (Figure 17.4b), blow the cell dry with nitrogen or dry air.

Manual polarimeters. A manual polarimeter can be standardized by filling a tube with an optically inactive solvent such as distilled water or with the solvent being used for your sample. Adjust the

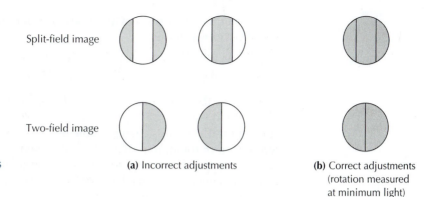

Split-field image

Two-field image

FIGURE 17.5
Representative images in the light field of a manual polarimeter.

(a) Incorrect adjustments

(b) Correct adjustments (rotation measured at minimum light)

instrument to the minimum-light position (see Figure 17.5). It can also be useful to check your ability to use the polarimeter properly by analyzing a 5.00% or 10.00% solution of sucrose in water. Determine the specific rotation of your sample based on the average of five to seven readings of the optical rotation.

Various optical devices allow easier measurement of optical rotation. They depend on a sudden change of contrast when the minimum amount of light is transmitted by the analyzer. Some manual polarimeters have a split-field image or two fields divided through the middle (Figure 17.5a). The analyzer is rotated in a clockwise or counterclockwise direction until a point is reached where every field is of equal minimum intensity and the divided fields are no longer visible (Figure 17.5b). The accuracy of manual-reading polarimeters is generally 0.05°–0.1° using a Vernier scale.

17.3   Analyzing Polarimetric Readings

Specific Rotation

The magnitude of the optical rotation depends on the concentration of the optically active compound in the solution, the length of the light path through the solution, the wavelength of the light, the nature of the solvent, and the temperature. A typical rotation of common table sugar (sucrose) is written in the following manner:

$$[\alpha]_D^{20} = +66.4° \, (H_2O)$$

The symbol $[\alpha]_\lambda^T$ is called the *specific rotation* and is an inherent property of a pure optically active compound. T signifies the temperature of the measurement in degrees Celsius, and λ is the wavelength of light used. In the sucrose example, the sodium D line was used. The specific rotation is calculated from the observed angle of rotation:

$$[\alpha]_\lambda^T = \frac{\alpha}{l \cdot c}$$

where α is the observed angle of rotation, l is the length of the light path through the sample in decimeters, and c is the concentration of the sample (g/mL). When a pure, optically active liquid is used as the sample, its concentration is simply the density of the liquid.

Sometimes a rotation of an optically active substance is given as a *molecular rotation:*

$$[M]_\lambda^T = \frac{M}{100}\,[\alpha]$$

where M is the molecular weight of the optically active compound.

The value of the specific rotation can change considerably from solvent to solvent. It is even possible for an enantiomer to have a different sign of rotation in two different solvents. Such solvent effects are due to specific solvent-solute interactions. The intrinsic specific rotation is generally considered to be a constant in dilute solutions at a particular temperature and wavelength. However, if you wish to compare the optical activity of a sample with that obtained by other workers, you should use the same concentration in the same solvent. Sucrose makes an excellent reference compound for polarimetry because its intrinsic specific rotation in water is essentially independent of concentration up to 5–10% solutions.

A change in the specific rotation due to temperature variation may be caused by a number of factors, including changes in molecular association, dipole-dipole interactions, conformation, and solvation. When nonpolar solutes are dissolved in nonpolar solvents, variation in the specific rotation with temperature may not be large. But for some polar compounds, the specific rotation varies markedly with temperature. Near room temperature, the specific rotation of tartaric acid may vary by more than 10% per degree Celsius.

17.4   Modern Methods of Enantiomeric Analysis

You can also use chiral chromatography or spectroscopy for measuring the enantiomeric composition of organic compounds. Three methods are especially powerful:

- High-performance liquid chromatograpy (HPLC)
- Chiral gas chromatography (GC)
- Nuclear magnetic resonance (^{1}H NMR) spectroscopy

These methods can be used to determine how successful a resolution of optical isomers or an enantiospecific organic synthesis has been. They have the advantage of needing much smaller samples to determine enantiomeric ratios than polarimetry usually requires.

Chiral Chromatography

Enantiomeric composition can be determined by chiral HPLC (see Section 19.8) or chiral GC (see Chapter 20). When a mixture of enantiomers passes through a chiral chromatographic column, each enantiomer has a different attraction for the chiral stationary phase—differences that lead to separation of the enantiomers. Typical stationary phases that produce this effect are chiral cyclodextrins, which are macrocyclic polyglucose compounds with six or more glucose units. These are often immobilized on the chromatographic columns by bonding to silica gel. The less tightly coordinated enantiomer passes through the column more rapidly than the enantiomer that is selectively retained by the chiral stationary phase.

NMR Analysis

If the enantiomers of a chiral carboxylic acid undergo reaction with an optically pure amine in an NMR tube, a mixture of diastereomeric salts is produced; these diastereomers can have subtly different ^1H NMR spectra (see Chapter 22). Neutralization of a mixture of enantiomers of a chiral amine by an optically pure carboxylic acid can serve the same purpose. The NMR spectra may be fairly complex, and the chiral enantiomers generally need to have a clean singlet for one of their NMR signals so that integration can be used reliably to determine the enantiomeric composition.

Chiral Shift Reagents for NMR Analysis

Chiral lanthanide shift reagents are sometimes used to produce a diastereomeric mixture for NMR analysis. Derivatives of camphor provide shift reagents that are rich in chiral character. $Eu(hfc)_3$, tris[3-heptafluoropropylhydroxymethylene)-(+)-camphorato]europium (III), is such a compound, which undergoes rapid and reversible coordination with Lewis bases, (B:), establishing the following equilibrium:

$$Eu(hfc)_3 + B: \rightleftharpoons B: Eu(hfc)_3$$

The complex $B:Eu(hfc)_3$ brings a paramagnetic ion, Eu^{3+}, into close proximity to the chiral organic base (B:), which induces changes in the ^1H NMR chemical shifts of the chiral base. The chemical shifts are different in each of the two coordinated enantiomers because the formation of the diastereomeric pair causes the protons of the two enantiomers to become nonequivalent.

Identification of the ^1H NMR signals of the α-protons of the organic base α-phenylethylamine and integration of their areas allows determination of the composition of the $B:Eu(hfc)_3$ complex, which equals the enantiomeric composition of the organic base

Eu(hfc)$_3$
(tris[3-heptafluoropropyl-
hydroxymethylene-(+)-
camphorato]europium III)

α-Proton

α-Phenylethylamine

Complex

Enantiomeric Ratio and Enantiomeric Excess

The purity of optically active compounds can be reported in terms of enantiomeric ratio or enantiomeric excess. *Enantiomeric ratio (er)* is simply the ratio of the relative amounts of the (+)- and (−)-enantiomers, which can be determined from the ratio of peak areas obtained using chiral HPLC, chiral GC (see Section 20.4), or ^1H NMR (see Sections 22.6 and 22.10).

When you are using optical activity to determine optical purity, *enantiomeric excess (% ee)* is a useful criterion. Enantimeric excess can be calculated from the expression

$$\% \text{ ee} = \left(\frac{[\alpha]_{\text{observed}}}{[\alpha]_{\text{pure}}} \right) \times 100\%$$

where $[\alpha]$ is the specific rotation of the chiral compound.

If we determine a specific rotation of 6.5° for 2-butanol, we can calculate the enantiomeric excess (% ee) of the sample if we know the specific rotation of pure (+)-2-butanol ($[\alpha] = +13.00$):

$$\% \text{ ee} = \left(\frac{6.5}{13.00} \right) \times 100\% = 50\%$$

It is instructive to examine the composition of 100 molecules of a mixture of (+)- and (−)-2-butanol with a % ee of 50%. We have an excess of 50 (+)-2-butanol molecules, which causes the optical activity. The remaining 50 molecules, because they produce no net optical activity, are composed of 25 (+)-2-butanol molecules and 25 (−)-2-butanol molecules. Thus, we have a total of 75 (+)-2-butanol molecules and 25 (−)-2-butanol molecules. When the % ee is 50%, the enantiomeric ratio is 75%/25% or 3:1. Many enantiospecific organic syntheses produce products with % ee values of 98% or greater.

17.5   Sources of Confusion and Common Pitfalls

I'm Getting Confused by All These New Words. How Can I Get Help?

There are a number of complex words that are used in discussions of stereochemistry and optical activity and the 3D structures of organic molecules. Understanding the concept of chirality itself involves spatial recognition that is easier for some than for others. The use of 3D molecular models and group learning are a great help to give substance to the ideas and terms used in the study of optical activity.

Enantiomer, diastereomer, and *racemic* are key terms used in Chapter 17. It is important to know their meanings. You may want to think of them as vocabulary words in a foreign language. It takes desire and effort to learn the vocabulary, but the conversation after you have learned these terms is far richer. Besides, you have probably noticed that they are used over and over again.

How Do I Choose Which Solvent and Wavelength of Light to Use When I Do Polarimetry?

Most of the time you will use experimental conditions that other scientists have already used and published. You can use chemistry handbooks and online sources to find this information.

My Polarimetric Results Don't Match the Data That I Expected to Find. What's Going On?

The quantitative nature of optical rotation measurements requires careful technique as you prepare samples for polarimetric readings using analytical balances and volumetric flasks. Beyond your accuracy in preparing the samples, there are factors to consider in the polarimetric measurements themselves that may affect your results. How pure is your solvent? Alcohols may contain water if they are from a bottle that has often been exposed to moist air. Does the concentration of your solution match the concentration used for the published value of the specific rotation? How accurate is the temperature setting? If all of these factors seem to be in order, it is useful to evaluate your assumptions in calculating the reading that you expected.

Questions

1. A sample of 2-butanol has a specific rotation of +3.25°. Determine the % ee and the molecular composition of this sample. The specific rotation of pure (+)-2-butanol is +13.0°.

2. A sample of 2-butanol (see question 1) has a specific rotation of −9.75°. Determine the % ee and the molecular composition of this sample.

3. The structures of strychnine (R = H) and brucine (R = CH₃O) are examples of alkaloid bases that can be used for optical resolutions and enantiospecific organic synthesis. These molecules are rich sources of chirality (respectively, $[\alpha]_D$ = −104° and −85° in absolute ethanol). Assume that nitrogen inversion is slow and identify the eight stereocenters in each of these two nitrogen heterocyclic compounds.

R = H, strychnine
R = CH₃O, brucine

4. Only one of the two nitrogens in strychnine and brucine acts as the basic site for the necessary acid-base reaction for a resolution. Which nitrogen, and why?

5. An optical rotation study gives α = +140° as the result. Suggest a dilution experiment to test whether the result is indeed +140°, not −220°.

4

Chromatography

Essay—Modern Chromatographic Separations

Few experimental techniques rival chromatography for purifying organic compounds and separating complex mixtures. Chromatography got its name from the fact that it was originally used to separate mixtures of colored substances—the pigments in green leaves. Once chemists realized that chromatography could be used to separate color-less substances as well, its development took off. Chromatography has revolutionized the practice of chemistry and many areas of modern biology because it can be used to separate almost anything, including proteins, nucleic acids, carbohydrates, and even viruses. This book describes partition chromatography for the analysis of mixtures and for the preparation of pure products.

Principles of Partition Chromatography

Partition chromatography is a physical method of separation in which the components of a mixture are *partitioned*, or distributed, between two phases: a stationary phase and a mobile phase. The *mobile phase*, which can be a liquid or a gas, moves in a definite direction and passes over the *stationary phase*. Components of a mixture are intro-duced at one end of the stationary phase and, as they interact with the mobile phase, different compounds migrate at different rates and separate from one another.

A chromatography system is like a highway (the stationary phase) between Los Angeles and San Francisco, where a continuous train of cars (the mobile phase) travels slowly enough for people to jump on and off at will. Individual people (the components to be separated) start in Los Angeles at the same time and travel in the cars, stopping along the way to shop or enjoy the view. Then—for the sake of this analogy—they jump on different cars. The time it takes them to reach San Francisco depends on how often and how long they stop. This is the same for molecules; their different rates of travel in chromatography directly depend on their different degrees of attraction to the stationary phase.

Molecular attraction is based on intermolecular forces (see the Essay, "Intermolecular Forces in Organic Chemistry," page 127–131). A substance that has stronger attractive forces with the stationary phase will migrate more slowly than a substance that has weaker attractive forces. At the same time, the ability of a substance to dissolve in and move with the mobile phase enables it to travel more easily over the stationary phase. Chromatography involves a series of equilibria between the stationary and mobile phases for attraction of a compound. These equilibrations take place at the interface between the two phases. Because the molecules interact with the stationary phase near its surface, they are not dissolved into the bulk material but instead they are *adsorbed* into an outer layer of the stationary phase. Adsorption is not an accidental misspelling of absorption. The difference between the two can be appreciated by imagining someone throwing a banana cream pie in your face. The pie that sticks to your face is adsorbed, whereas the pie you manage to ingest is absorbed.

As the mobile phase passes over compounds adsorbed on the stationary phase, some of the adsorbed molecules dissolve into the mobile phase and then a little later they get re-adsorbed back onto the stationary phase. This partitioning between the two phases occurs in the same way that compounds are distributed between the aqueous and organic phases in liquid-liquid extractions (see Chapter 10). Because each compound has a unique equilibrium constant (or distribution coefficient) between the two phases, each will travel at a unique rate. Repeating the equilibria many times during chromatography amplifies even small differences in distribution coefficients and enables compounds to be separated.

Chromatography in the Organic Lab

Four different types of chromatography are introduced in this book: thin-layer chromatography (TLC), liquid chromatography (LC), high-performance liquid chromatography (HPLC), and gas chromatography (GC). The method you choose will depend on your purpose (analytical or preparative), the resources available to you, and the properties of the compounds you are separating.

The most commonly used stationary phase in TLC and LC is silica gel, finely ground silica (SiO_2) particles that are coated with a thin layer of water molecules. Intermolecular hydrogen bonding and dipole-dipole interactions allow polar organic compounds to be attracted by the water-coated silica gel much more than nonpolar organic compounds. Therefore, polar compounds are carried more slowly by the mobile solvent phase and leave a chromatography column later than nonpolar compounds do. In the same manner, polar solvents move compounds faster on a TLC plate or through an LC column than nonpolar solvents do.

Regardless of the specific chromatographic method you use, it is important to understand how compounds interact with the stationary and mobile phases. The scale of your chromatography system should also fit the scale of your separation. Just as accommodating a larger crowd of people would require more cars and more access points for stopping, separating more material by chromatography requires larger amounts of the stationary and mobile phases. If there are too many cars for the size of the road, you produce a traffic jam.

THIN-LAYER CHROMATOGRAPHY

Thin-layer chromatography (TLC) has become a widely used analytical technique because it is simple, inexpensive, fast, and efficient, and it requires only milligram quantities of material. TLC is especially useful for determining the number of compounds in a mixture, for helping to establish whether two compounds are identical, and for following the course of a reaction.

In TLC, glass, metal, or plastic plates are coated with a thin layer of adsorbent, which serves as the *stationary phase*. The stationary phase is usually polar—silica gel is most widely used. The *mobile phase* is a pure solvent or a mixture of solvents; the appropriate composition of the mobile phase depends on the polarities of the compounds in the mixture being separated. Most nonvolatile solid organic compounds can be analyzed by thin-layer chromatography; however, TLC does not work well for many liquid compounds because their volatility can lead to loss of the sample by evaporation from the TLC plate.

If Chapter 18 is your introduction to chromatographic analysis, read the Essay "Modern Chromatographic Separations" on pages 253–254 before you read Chapter 18.

Overview of TLC Analysis

To carry out a TLC analysis, a small amount of the mixture being separated is dissolved in a suitable solvent and applied or spotted onto the adsorbent near one end of a TLC plate. Then the plate is placed in a closed chamber with the edge nearest the applied spot immersed in a shallow layer of the mobile phase, called the *developing solvent* (Figure 18.1). The solvent rises through the stationary phase by capillary action, a process called *developing the chromatogram.*

As the solvent ascends the plate, the compounds in the sample separate as a result of their many equilibria between the mobile and stationary phases. **The more tightly a compound binds to the stationary adsorbent, the more slowly it moves on the TLC plate** (Figure 18.2). When silica gel is the stationary phase, the developing solvent moves nonpolar substances up the plate most rapidly. As the chromatogram develops, polar substances travel up the plate slowly or not at all.

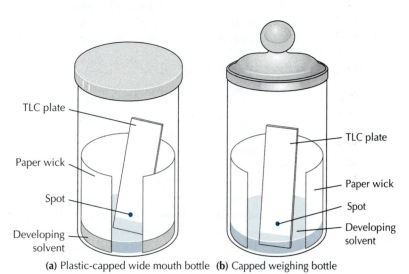

FIGURE 18.1
Developing chambers containing a thin-layer plate.

TLC plate

Paper wick

Spot

Developing solvent

TLC plate

Paper wick

Spot

Developing solvent

(a) Plastic-capped wide mouth bottle   **(b)** Capped weighing bottle

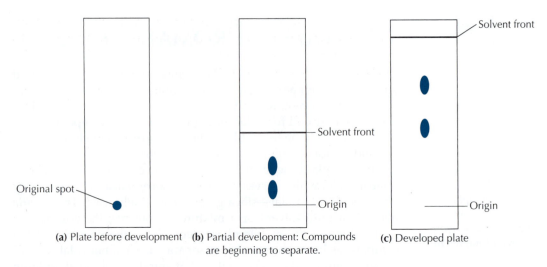

Original spot

(a) Plate before development   (b) Partial development: Compounds   (c) Developed plate
                                 are beginning to separate.

Solvent front

Solvent front

Origin

Origin

FIGURE 18.2 Steps in development of a TLC plate.

The TLC plate is removed from the developing chamber when the *solvent front* (leading edge of the solvent) is 1–1.5 cm from the top of the plate. The position of the solvent front is marked immediately, before the solvent evaporates, with a pencil line. The plate is then placed in a hood to dry.

You need to be able to observe the positions of colorless compounds on a plate in order to evaluate a chromatogram. Several methods are available for *visualizing* the compounds in the sample. Some compounds are colored and their spots can easily be seen. If the TLC plate has a fluorescent indicator, the plate can be visualized by exposure to ultraviolet light. Alternatively, the compounds can be visualized using a reagent that produces colored spots. The developed and visualized TLC plate is then ready for analysis.

18.1   Plates for Thin-Layer Chromatography

Thin-layer chromatographic plates consist of a solid support—such as glass, metal, or plastic—with a thin layer of an adsorbent that coats the solid surface and provides the stationary phase.

Adsorbents

Silica gel ($SiO_2 \cdot xH_2O$) is the most commonly used general-purpose adsorbent for partition chromatography of organic compounds. Aluminum oxide ($Al_2O_3 \cdot xH_2O$, also called alumina) can also be used as a polar adsorbent. Several intermolecular forces cause organic molecules to bind to these polar stationary phases. Only weak van der Waals forces bind nonpolar compounds to the adsorbent, but polar molecules can also adsorb by dipole-dipole interactions and hydrogen bonding. The strength of the interaction varies for different compounds, but generally **the more polar the compound, the more strongly it binds to silica gel or alumina.** Another type of silica gel adsorbent—used for reverse-phase chromatography—has a nonpolar surface that adsorbs less-polar compounds more strongly than polar compounds.

Silica gel and aluminum oxide. Silica gel and alumina adsorbents are prepared from activated, finely ground powders. Activation usually involves heating the powder to remove some of the adsorbed water. Silica gel is somewhat acidic; usually it effectively separates acidic and neutral compounds that are not too polar. Aluminum oxide is available in acidic, basic, and neutral formulations for the separation of relatively nonpolar compounds.

If the plastic seal on a package containing precoated silica gel or alumina TLC sheets has been broken for a long time, or the TLC sheets have not been purchased recently, they should be reactivated before use to remove some of the adsorbed water. To do this, simply heat the sheets in a clean oven for 20–30 minutes at the temperature recommended by the manufacturer.

Adsorbents for reverse-phase TLC. The adsorbents used on plates for reverse-phase thin-layer chromatography are based on silica gel modified by replacing the hydroxyl groups normally attached to silicon atoms with alkoxy groups and with long-chain alkyl groups, such as $-(CH_2)_{17}CH_3$. The alkyl chains provide a nonpolar liquid stationary phase. The solvents used in reverse-phase TLC are quite polar, for example, methanol or acetonitrile, and are often mixed with water. In reverse-phase TLC, the order of movement up the TLC plate is reversed; more polar compounds travel faster up the TLC plate than less polar compounds, which bind more tightly to the nonpolar adsorbent surface.

Backing for TLC Plates

A number of manufacturers sell TLC plates that are precoated with a layer of adsorbent.

We suggest using 2.5 × 6.7 cm or 2.9 × 6.7 cm TLC plates; 24 or 21 plates can be cut from a standard 20 × 20 cm sheet.

Plastic backing. Plastic-backed silica gel plates are usually the least expensive option. They can be cut to any desired size with a paper cutter or sharp scissors. The adsorbent surface is of uniform thickness, usually 0.20 mm. Results are quite reproducible, and sharp separation is normal. The plastic backing is generally a solvent-resistant polyester polymer. The adsorbent is bound to the plastic by solvent-resistant polyvinyl alcohol, which binds tightly to both the adsorbent and the plastic. Precoated plastic plates that have a fluorescent indicator are also available; these plates facilitate the visualization of many colorless compounds with a UV lamp (see Section 18.4).

Glass and aluminum backing. TLC plates with a glass or aluminum backing are also available in standard 20 × 20 cm sheets. Both types can be heated without melting the backing—an important property if the plate is to be visualized with a reagent that requires heating (see Section 18.4). Aluminum sheets can be cut with scissors into convenient sizes for TLC plates. Glass sheets are cut with a special diamond-tipped tool.

18.2   Sample Application

For TLC analysis, dissolve 10–20 mg of the solid in 1 mL of solvent.

The sample must be dissolved in a volatile organic solvent; a very dilute (1–2%) solution works best. Because the atmosphere in the developing chamber must be saturated with solvent vapor, the solvent needs a high volatility so that it will evaporate easily at room temperature. Anhydrous reagent-grade acetone or ethyl acetate is commonly used. If you are analyzing a solid, dissolve 10–20 mg of it in 1 mL of the solvent. If you are analyzing a nonvolatile liquid, dissolve about 10 μL of it in 1 mL of the solvent.

Micropipets for Spotting TLC Plates

Commercial micropipets are available in 5- and 10-μL sizes and work well for applying samples onto plastic-backed plates. Glass- and aluminum-backed plates require micropipets of a smaller internal diameter. Narrow capillary tubes of 0.7-mm internal diameter are commercially available.

A micropipet can also be made easily from an open-ended, thin-walled, melting-point capillary tube. The capillary tube is heated at its midpoint. A microburner is ideal because only a small flame is required, but a Bunsen burner may be used. (If you do not know how to use a microburner or Bunsen burner, consult your instructor.) The softened glass tube is stretched and drawn into a narrower capillary.

SAFETY PRECAUTION

Be sure there are no flammable solvents nearby when you are using a microburner or Bunsen burner.

While heating the tube, rotate it until it is soft on all sides over a length of 1–2 cm. When the tubing is soft, remove it from the heat and quickly draw out the heated part until a constricted portion 4–5 cm long is formed (Figure 18.3). After cooling the tube for a minute or so, score it gently at the center with a file and break it into two capillary micropipets. The diameter at the end of a micropipet needs to be tiny—just a little larger than the diameter of a human hair, about 0.2–0.3 mm. The break must be a clean one, at right angles to the length of the tubing, so that when the tip of the micropipet is touched to the plate, liquid is pulled out by the adsorbent.

FIGURE 18.3
Constricted capillary tube.

4–5 cm

Spotting a TLC Plate

Carefully apply tiny spots of the dilute sample solution with a micropipet near one end of the plate. Keeping the spots small ensures the cleanest separation. It is important not to overload the plate with too much sample, which leads to large tailing spots and poor separation.

Do not use a pen to mark TLC plates because components of the ink separate during development and may obscure the samples.

Preparing the plate. Before spotting a TLC plate, measure 1.0 cm from the bottom edge of the plate and *lightly* mark both edges with a 0.3-cm or shorter pencil mark (Figure 18.4). The imaginary line between these marks indicates the compound's starting point for your analysis after the TLC has been completed.

Number of lanes per plate. If you are using 2.5 × 6.7 cm TLC plates, you can apply two spots to one plate (Figure 18.4); the spot in each lane should be one-third of the distance from the side of the plate. Three lanes require a 2.9-cm-wide plate. The spots become larger by diffusion during development, and if they are too close to each other or to the edge of the plate, the chromatograms will likely be difficult to interpret.

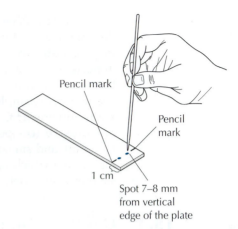

Pencil mark

Pencil mark

1 cm

Spot 7–8 mm
from vertical
edge of the plate

FIGURE 18.4 Spotting a
thin-layer plate.

Applying the samples. Fill the micropipet by dipping one end of the
capillary tube into the solution to be analyzed. Only 1–5 μL of the
sample solution are needed for most TLC analyses. Hold the micropipet vertically and apply the sample by **touching the micropipet
gently and briefly to the plate** on the imaginary line between the
two pencil marks (Figure 18.4). It is important to touch the micropipet to the plate very lightly so that no hole is gouged in the adsorbent, and to remove it quickly so that only a very small drop is left
on the adsorbent. The spot should be no more than 2 mm in diameter
to avoid excessive broadening during the development. If you apply
very small spots, you will probably need to apply more sample by
touching the micropipet to the plate a second time *at exactly the same
place.* Allow one spot to dry before applying the next. The spotting
procedure may be repeated numerous times, if necessary.

Testing the amount of sample to spot. You can quickly test for the
proper amount of solution to spot on the plate by spotting two different amounts on the same plate. If you have used plates with a
fluorescent indicator, use a UV lamp to visualize the spots (see Section 18.4) before developing the plate to determine the best sample
amount. Otherwise, develop the plate as directed in Section 18.3 and
decide which spot gives better results.

Using known standards. If available, include an authentic standard
on the TLC plate for comparison. **If two compounds travel up
the plate the same distance, they may be the same compound;** if
the distances differ significantly, they most definitely are not the
same compound. A useful method for determining whether two
compounds are identical is to co-spot both compounds in the same
place. Create three lanes on the plate and apply one compound on
the left and middle lanes and the second compound on the middle
and right lanes. The middle lane is the co-spot, where you apply
both samples directly on top of one another. After the plate is developed, the three lanes will look identical if the compounds are identical. If the compounds are not identical, the middle lane should show
some separation or elongation of the two co-spotted compounds.

Accurate record keeping. Accurate record keeping is essential while doing a TLC analysis. **Before spotting the plate, draw a sketch in your notebook** of the TLC plate with a line drawn across it to indicate the initial position of the sample. Set up a key underneath the sketch with the position and name of each sample that will be spotted. For example, if you are analyzing a reaction mixture using a 2.9-cm-wide TLC plate, you could use the notation **rx** (reaction mixture), **co** (co-spotting of the reaction mixture and the starting material), and **sm** (starting material). Most samples are colorless, so identifying which sample is spotted in a specific position is impossible without a detailed record.

18.3   Development of a TLC Plate

Development of a TLC plate is carried out in a closed developing chamber containing a developing solvent. If a developing solvent is not specified for the system you are analyzing, read Section 18.7 on how to choose a suitable one before starting your TLC analysis.

Preparing the Developing Chamber

To ensure good chromatographic resolution, **the developing chamber must be saturated with solvent vapors** to prevent the evaporation of solvent from the TLC plate as the solvent rises up the plate. **If the solvent mobile phase evaporates, the compounds in the sample can end up unseparated near the top of the TLC plate** because sample spots will continue to rise up the plate even though the solvent front might not rise as the solvent evaporates. Insert a piece of filter paper three-quarters of the way around the inside of the developing chamber to help saturate its atmosphere with solvent vapor by wicking solvent into the upper region of the chamber (Figure 18.5). The paper wick should be a little shorter than a TLC plate so the plate does not touch the paper. After adding the correct amount of developing solvent, shake the capped TLC chamber briefly to ensure that the paper wick is saturated with solvent.

The solvent depth in the developing chamber must be less than the height of the spots on the TLC plate.

    Use enough developing solvent to allow a shallow layer (3–4 mm) to remain on the bottom after the closed chamber has been

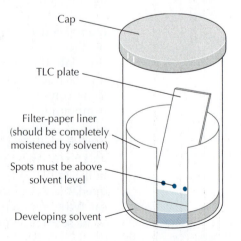

Cap

TLC plate

Filter-paper liner
(should be completely
moistened by solvent)

Spots must be above
solvent level

Developing solvent

FIGURE 18.5 Developing a TLC plate.

shaken in order to saturate the filter paper with the solvent. If the solvent level in the jar is too high, the spots on the plate may be below the solvent level. Under these conditions, the spots leach into the solvent, thereby ruining the chromatogram.

Carrying Out the TLC Development

Do not touch the adsorbent side of the TLC plate with your fingers. Hold the plate by the top edge with a pair of tweezers.

Uncap the developing chamber and carefully place the TLC plate inside with a pair of tweezers, taking care that it is level and not touching the paper wick. Recap the chamber and allow the solvent to move up the plate. The adsorbent will become visibly moist. **Do not lift or otherwise disturb the chamber while the TLC plate is being developed.**

The development of a chromatogram usually takes 5–10 min if the chamber is saturated with solvent vapor. When the solvent front is 1–1.5 cm from the top of the plate, remove it from the developing chamber with a pair of tweezers and immediately mark the adsorbent at the solvent front with a pencil. The final position of the solvent front must be marked before any evaporation occurs. To analyze the chromatogram, you need an accurate measurement of the distance the compounds have traveled up the TLC plate relative to the distance the solvent has traveled. Allow the developing solvent to evaporate from the plate before visualizing the results.

S A F E T Y   P R E C A U T I O N

Evaporate the solvent from a developed chromatogram in a fume hood.

18.4 Visualization Techniques

Chromatographic separations of colored compounds usually can be seen directly on the TLC plate, but colorless compounds require indirect methods of visualization. Fluorescence and visualization reagents are commonly used to visualize TLC plates.

Fluorescence

The simplest visualization technique involves the use of adsorbents that contain a fluorescent indicator. The insoluble inorganic indicator rarely interferes with the chromatographic results, and it makes visualization straightforward. When the output from a short-wavelength ultraviolet lamp (254 nm) is used to illuminate the adsorbent side of the plate in a darkened room or dark box, the plate fluoresces visible light.

S A F E T Y   P R E C A U T I O N

Never look directly at an ultraviolet radiation source. Like the sun, UV radiation can cause eye damage.

The separated compounds usually appear as dark spots on the fluorescent field because the substances forming the spots usually quench the fluorescence of the adsorbent, as shown in Figure 18.6a. Sometimes substances being analyzed are visible by their own fluorescence and produce a brightly glowing spot. Outline each spot with a pencil while the plate is under the UV source

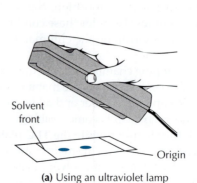

(a) Using an ultraviolet lamp

(b) Using an ultraviolet lamp with dark box

FIGURE 18.6 UV visualization.

to give a permanent record, which will allow you to analyze your chromatogram.

Visualization Reagents

Not all substances are visible on fluorescent silica gel, so visualization by one of the following methods may be necessary.

Dipping reagents for glass or aluminum plates. Glass or aluminum TLC plates can be dipped briefly in visualizing reagent solutions containing reagents that react to form colored compounds upon heating. Alternatively, the TLC plates can be sprayed with the visualizing reagent. Visualization occurs when you heat the dipped or sprayed TLC plates with a heat gun or place them on a hot plate for a few minutes. Three common visualizing solutions are *p*-anisaldehyde, vanillin, and phosphomolybdic acid.* The colors fade with time, so outline the spots with a pencil soon after the visualization process.

SAFETY PRECAUTION

Sulfuric acid and phosphomolybdic acid are highly corrosive substances and iodine vapor is toxic and corrosive. Wear gloves and work in a hood while using visualization reagents. Use tweezers to hold the TLC plate.

If you use *p*-anisaldehyde, vanillin, or phosphomolybdic dipping reagents, consult your instructor for directions on how to dispose of the TLC plates and any excess reagents. **Never put TLC slides that have come into contact with dipping reagents in your lab notebook.**

**p-Anisaldehyde visualizing solution: 2 mL of *p*-anisaldehyde in 36 mL of 95% ethanol, 2 mL of concentrated sulfuric acid, and 5 drops of acetic acid.*

Vanillin visualizing solution: 6.0 g of vanillin in 100 mL of 95% ethanol and 1.0 mL of concentrated sulfuric acid. Store the vanillin reagent in an amber-colored bottle covered with aluminum foil; discard the solution when it acquires a blue color.

Phosphomolybdic acid visualizing solution: 20% phosphomolybdic acid by weight in ethanol.

Iodine visualization. Another way to visualize colorless organic compounds uses their adsorption of iodine (I_2) vapor. A plastic wash bottle containing a thin layer of iodine crystals is used for this method.

Lay the TLC plate on a clean piece of paper or paper towel. Hold the tip of the wash bottle containing the iodine about 1 cm above the plate and gently squeeze the sides of the bottle as you move it from the bottom to the top of the plate; repeat the motion two or three times. The spots on the plate should appear within 30–60 sec. The reaction of the substances with iodine vapor produces yellow-brown colored spots. If no spots appear, repeat the application of iodine vapor several more times. The colored spots disappear within a few minutes, so outline them with a pencil immediately after they appear. The spots will reappear if the plate is again treated with iodine vapor.

Further information on visualization reagents. Consult the references at the end of Chapter 18 for detailed discussions of visualization reagents.

18.5 Analysis of a Thin-Layer Chromatogram

Once the spots on the chromatogram are visualized, you are ready to analyze the chromatographic results. To do this, you must determine the **ratio of the distance each compound has traveled on the plate relative to the distance the solvent has traveled.**

Determination of the R_f

Under a constant set of experimental conditions, a given compound always travels a fixed distance relative to the distance traveled by the solvent front (Figure 18.7). This ratio of distances is called the R_f *(ratio to the front)* and is expressed as a decimal fraction:

$$R_f = \frac{\text{distance traveled by compound}}{\text{distance traveled by developing solvent front}}$$

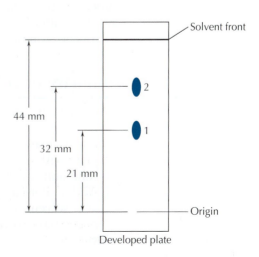

FIGURE 18.7
Measurements for the R_f value.

The R_f value for a compound depends on its structure as well as the adsorbent and mobile phase used. It is a physical characteristic of the compound, just as its melting point is a physical characteristic. When a chromatogram is done, calculate the R_f value for each substance and record the experimental conditions. The important data that need to be recorded include the following:

- Brand, type of backing, and adsorbent on the TLC plate
- Developing solvent
- Method used to visualize the compounds
- R_f value for each substance

Calculation of an R_f Value

To calculate the R_f value for a given compound, measure the distance the compound has traveled from where it was originally spotted, and the distance the solvent front has traveled from where the compound was spotted (see Figure 18.7). Measure from the center of the spot. The best data are obtained from chromatograms in which the spots are less than 5 mm in diameter. If a spot shows "tailing," measure from the densest point of the spot. The R_f values for the two substances shown on the developed TLC plate in Figure 18.7 are calculated as follows:

$$\text{Compound 1:} \quad R_f = \frac{21 \text{ mm}}{44 \text{ mm}} = 0.48$$

$$\text{Compound 2:} \quad R_f = \frac{32 \text{ mm}}{44 \text{ mm}} = 0.73$$

Identical R_f Values

When two samples have identical R_f values, you should not necessarily conclude that they are the same compound without doing further analysis. There are perhaps 100 possible R_f values that can be distinguished from one another, whereas there are more than 10^8 known organic compounds. Further analysis by infrared (IR) or nuclear magnetic resonance (NMR) spectroscopy would be needed to prove whether the compounds are identical. You can also conclude that the samples are different compounds if subsequent TLC analyses with different developing solvents reveal different R_f values for each sample.

18.6   Summary of TLC Procedure

1. Obtain a precoated TLC plate of the proper size for the developing chamber.
2. Lightly mark the edges of the origin line with a pencil. Spot the plate with a small amount of a 1–2% solution containing the compounds to be separated.
3. Add a filter-paper wick to the developing jar. Then add a suitable solvent, cap the jar, and shake it briefly to saturate the paper with solvent and the air in the chamber with solvent vapors.

4. Using tweezers, place the spotted TLC plate into the developing jar, taking care that it doesn't touch the wick, and quickly recap the jar.
5. Develop the chromatogram until the solvent front is 1–1.5 cm from the top of the plate.
6. Mark the solvent front immediately after removing the plate from the developing chamber. Use tweezers to remove and hold the plate.
7. Visualize the chromatogram and outline the separated spots.
8. Calculate the R_f value for each compound.

18.7 How to Choose a Developing Solvent When None Is Specified

Finding a Suitable Developing Solvent

Chromatographic behavior is the result of competition between the stationary phase (adsorbent) and the mobile phase (developing solvent) for the compounds being separated.

Solvent considerations. In general, you should use a nonpolar developing solvent for nonpolar compounds and a polar developing solvent for polar compounds. Selecting a suitable solvent is often, however, a trial-and-error process, particularly if a mixture of solvents is required to give good separation. A solvent that does not cause any compounds to move from the original spot is not polar enough, whereas a solvent that causes all the spotted material to move with the solvent front is too polar (Figure 18.8a and b). An appropriate solvent for a TLC analysis gives R_f values of 0.20–0.70, with ideal values in the range 0.30–0.60, as shown in Figure 18.8c.

With a silica gel plate, nonpolar hydrocarbons should be developed with hydrocarbon solvents, but a mixture containing an alcohol and an ester might be developed with a pentane/ethyl acetate mixture. Highly polar solvents are seldom used with silica gel plates, except in the case of reverse-phase TLC.

Testing developing solvents. If you know the compounds in the mixture you want to separate, use Table 18.1 to select the solvents to test. This table shows the relative polarity of common TLC developing

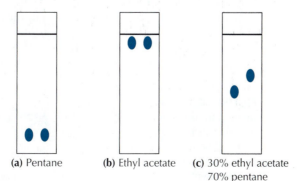

FIGURE 18.8 TLC results with different developing solvents.

(a) Pentane   (b) Ethyl acetate   (c) 30% ethyl acetate
                                                      70% pentane

TABLE 18.1	Relative polarities of common TLC solvents and organic compounds	
Common developing solvents	Increasing polarity	Organic compounds by functional group class
Alkanes, cycloalkanes		Alkanes
Toluene		Alkenes
Dichloromethane		Aromatic hydrocarbons
Diethyl ether		Ethers, halocarbons
Ethyl acetate		Aldehydes, ketones, esters
Acetone (anhydrous)		Amines
Ethanol (anhydrous)		Alcohols
Methanol		Carboxylic acids
Acetonitrile		
Water		

solvents and organic compounds by functional group class. If the composition of the mixture is unknown, begin by testing with a nonpolar solvent such as pentane and then with a medium-polarity solvent such as ethyl acetate. When testing mixed solvents, you might start by testing a 50:50 mixture (by volume, or v/v) to see how much separation occurs and how far up the plate the two compounds travel. If they travel more than halfway up the plate, test a solvent mixture with a higher percentage of pentane; conversely, if they travel less than halfway up the plate, test a solvent mixture with a higher percentage of ethyl acetate.

Very polar compounds. Some compounds, such as carboxylic acids and amines, hydrogen bond strongly to the TLC plate and migrate very little or else develop streaks on the TLC plate. In these cases, it can help to add small amounts of concentrated acetic acid (for carboxylic acids) or ammonium hydroxide (for amines) to the developing solvent. For example, a mixture of 98:2 (v/v) ethyl acetate/acetic acid is routinely used for developing plates spotted with carboxylic acids. Before you use acetic acid or ammonium hydroxide, you need to ensure that these additives are miscible with your developing solvent.

Rapid Method for Testing Developing Solvents

As a rapid way to determine the best TLC developing solvent among several possibilities, spot three or four samples along the length of the same plate (Figure 18.9). Fill a micropipet with the solvent to be tested and gently touch one of the spots. The solvent will diffuse outward in a circle, and the sample will move out with it. Mixtures

FIGURE 18.9 Rapid method for determining an effective TLC solvent: (a) good development; (b) and (c) poor development.

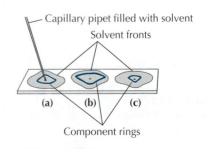

of compounds will be partially separated and approximate R_f values can be estimated. Ideal R_f values should be in the range 0.30–0.60.

Consider the separation of an alcohol and an ester. Start with a relatively nonpolar solution of 90:10 (v/v) pentane/ethyl acetate. If the R_f values are below 0.2, test a second spot with 70:30 (v/v) pentane/ethyl acetate, then test other spots with 50:50 (v/v) pentane/ethyl acetate and with pure ethyl acetate. If an ethyl acetate system does not produce R_f values in the satisfactory range, select a more polar solvent system such as a mixture of diethyl ether and acetone and repeat the test with various proportions.

18.8 Using TLC Analysis in Synthetic Organic Chemistry

When synthesizing an organic compound, you may have multiple compounds in a reaction mixture, and the starting reagents may be the only known compounds available. TLC analysis is extremely useful both in determining when the limiting reagent is consumed—thus, the reaction is complete—and in ascertaining how many compounds are formed during the reaction.

Following the Course of a Reaction

Spot a TLC plate with the limiting reagent in one lane and the reaction mixture in another lane. Run an initial TLC on the reaction mixture as soon as all reagents are combined. Withdraw samples of the reaction mixture from the reaction flask with a long micropipet at periodic intervals and analyze them by TLC. The reaction is complete when the lane of the reaction mixture no longer shows a spot with the same R_f as the limiting reagent in the other lane.

How Many Products Are Formed in the Reaction?

TLC analysis can be used to determine how many products are present in a reaction mixture when multiple products are formed. Again, spot one lane with the limiting reagent for reference. You may need to test developing solvents of different polarities to ascertain how many compounds are in the mixture because all the compounds present may not separate completely in every solvent.

18.9 Sources of Confusion and Common Pitfalls

I Can't Get My Solid Compound to Stick to the Plate

Samples must be dissolved in a volatile organic solvent to make a dilute solution. When you apply the solution to the plate, the solvent evaporates and a very small amount of the sample adheres to the stationary phase.

My Capillary Doesn't Work

Did you put your finger on the top of the capillary tube when you tried to withdraw a solution of your sample solution? If so, the air inside the capillary could not be displaced by the solution, and the solution could not be drawn into the capillary. Alternatively, by pressing too hard on the plate you may have clogged the end of the capillary with silica gel.

**I Don't See Any
Spots on the
Developed Plate**

There are a number of reasons why you might not see any spots on your TLC plate.

Not enough sample was on the plate. Check the boiling point of your compound: If it is volatile, it may have evaporated off the plate before you developed it. Liquids with boiling points below 160°C or solids that sublime might vaporize before the plate is visualized. In addition, you may have applied too little material because your spotting solution was too dilute. You may need to spot multiple times with a dilute solution to get enough material on the plate, and you can often check if there's enough by visualizing under UV light *before* you develop the plate. You can also use multiple lanes on the same plate, spotting a different number of times in each lane. After developing the plate, you can choose which spotting level gives the best outcome.

You dissolved the sample off the plate in the developing jar. Check the solvent level in the developing jar. Was the solvent high enough to submerge the origin line containing the spots? If so, the spots probably leached into the developing solvent instead of moving up the plate as the solvent ascended.

You are not visualizing properly. Most UV lamps have two switches: one for short-wavelength light and one for long-wavelength light. Short-wavelength light is necessary for visualizing TLC plates. Check that you selected the correct switch. Also, the spots will be visible only if you irradiate the side of the plate containing the TLC adsorbent.

A dipped plate was not heated long enough. A few minutes of heating are necessary to visualize the spots when *p*-anisaldehyde, vanillin, or phosphomolybdic acid visualizing solutions are used.

**All My Spots Are at
the Solvent Front**

Two factors can produce spots near the solvent front. One is that the atmosphere within the TLC developing chamber was not fully saturated with the developing solvent during the development of the chromatogram because no saturated paper wick of the correct size was used. The second factor is that the developing solvent was too polar.

**The R_f Values of My
Spots Are Not the
Same as Reported**

The purity of the developing solvent is an important factor in the success of a TLC analysis and in obtaining reproducible R_f values. The presence of a soluble impurity can dramatically affect the developing power compared with that of the pure solvent. For example, the presence of water in acetone changes its developing power appreciably, and therefore the R_f values will differ from values obtained with pure acetone.

**The R_f Values Are
Very Similar**

If the R_f values for two compounds are very similar—within ± 0.05—then another solvent or mixture of solvents can be tested in order to distinguish between them, or co-spotting can be used.

My Spots Look Like Big Blobs and They Run Together

The developed TLC plate may show very large spots, two spots that overlap at the center of the plate, or a spot that shows a long oval tail instead of being circular. Tailing spots, in particular, lead to poor reproducibility of R_f values. These problems are likely to arise because too large a sample of the spotting solution was applied to the TLC plate. Prepare another plate using smaller spots and less over spotting. If the spots are still too large or if they tail, prepare a more dilute spotting solution.

Can I Get Quantitative Information from a TLC Analysis?

The size and intensity of the spots can be used as a rough measure of the relative amounts of the substances, but these parameters can be misleading, especially with fluorescent visualization. Some organic compounds interact much more intensely with ultraviolet radiation than others, making one spot appear to be more concentrated than another but not reflecting their relative quantities. Quantitative information is not one of the strengths of thin-layer chromatography.

Further Reading

Fried, B.; Sherma, J. *Thin-Layer Chromatography: Techniques and Applications;* 4th ed.; Chromatographic Science Series, Vol. 81, Marcel Dekker: New York, 1999.

Hahn-Deinstrop, E. *Applied Thin-Layer Chromatography: Best Practices and Avoidance of Mistakes;* 2nd ed.: Wiley, New York, 2007.

Sherma, J.; Fried, B. (Eds.) *Handbook of Thin-Layer Chromatography;* 3rd ed.; Chromatographic Science Series, Vol. 89, Marcel Dekker: New York, 2003.

Touchstone, J. C. *Practice of Thin Layer Chromatography;* 3rd ed.; Wiley: New York, 1992.

Questions

1. When 2-propanol was used as the developing solvent, two substances moved with the solvent front ($R_f = 1$) during TLC analysis on a silica gel plate. Can you conclude that they are identical? If not, what additional experiment(s) would you perform?

2. The R_f value of compound A is 0.34 when a TLC plate is developed in pentane and 0.44 when the plate is developed in diethyl ether. Compound B has an R_f value of 0.42 in pentane and 0.60 in diethyl ether. Which solvent would be better for separating a mixture of A and B by TLC? Explain.

3. A student needs to analyze a mixture containing an alcohol and a ketone by silica gel TLC. After consulting Table 18.1, suggest a likely developing solvent.

4. Consider the compounds 4-*tert*-butylcyclohexanol and 4-*tert*-butylcyclohexanone. Using silica gel plates, which of the two compounds will have the higher R_f value when ethyl acetate is the developing solvent? Which will have the higher R_f value when dichloromethane is the developing solvent? Give your reasoning in each case.

LIQUID CHROMATOGRAPHY

If Chapter 19 is your introduction to chromatographic analysis, read the Essay "Modern Chromatographic Separations" on pages 253–254 before you read Chapter 19.

Overview of Liquid Chromatography (LC)

Liquid chromatography (LC) and high-performance liquid chromatography (HPLC) are important chromatographic methods that are used for separations of organic compounds. Liquid chromatography, also called column chromatography, is generally used to separate compounds of low volatility, but unlike thin-layer chromatography (TLC), liquid chromatography can be carried out with a wide range of sample quantities—ranging from a few micrograms for HPLC up to 10 g or more for column chromatography. Most liquid chromatography is carried out under partition conditions.

In liquid chromatography the stationary phase is a *solid adsorbent with a liquid coating* packed into a column. An *elution solvent* serves as the mobile phase. As the elution solvent travels down the column, separation occurs by the many equilibria of the compounds between the stationary and mobile phases. The relative polarities of these two phases determine the order in which compounds in the sample elute from the column. Figure 19.1 illustrates how a mixture of two compounds separates on a chromatographic column. With a polar adsorbent such as silica gel, the compound represented by A, which elutes first, is less polar than compound B.

Liquid chromatography at atmospheric or somewhat higher pressure is used for the purification of samples that require only modest resolution. Gravity LC at atmospheric pressure can be time consuming if you are separating gram quantities of a mixture; however, samples less than 200 mg can often be chromatographed efficiently on small columns, usually using a modest pressure to increase the rate of flow. *Flash chromatography* is LC that can be done in a flash. It either uses pressure to help drive the eluent down the column or involves the chromatography of small samples on small columns.

19.1 Adsorbents

Most chromatographic separations use silica gel ($SiO_2 \cdot xH_2O$) because it allows the separation of compounds with a wide range of polarities. Aluminum oxide (alumina, $Al_2O_3 \cdot xH_2O$) can also be used for separations of compounds of low to medium polarity. Both adsorbents produce a polar stationary phase and both are generally used with nonpolar to moderately polar elution solvents as the mobile phase.

Gravity and flash chromatography both use relatively large—greater than 37 μm—adsorbent particles, which allow a reasonably fast flow of the mobile phase. In HPLC, much smaller adsorbent particles are used, which requires high pressure to force the elution solvent through the column.

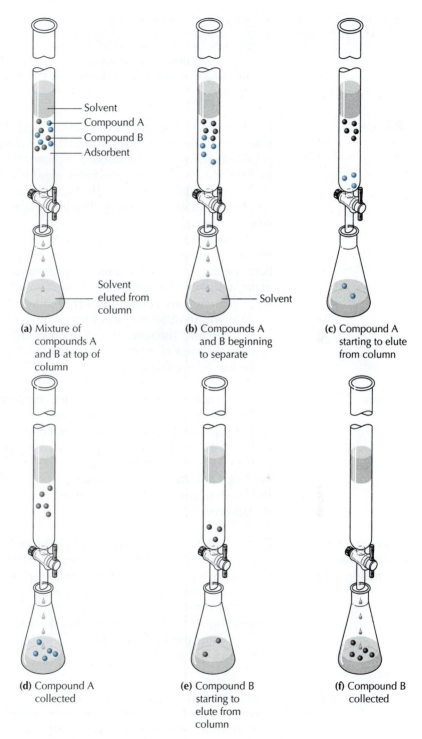

Solvent
Compound A
Compound B
Adsorbent

Solvent
eluted from
column

(a) Mixture of
compounds A
and B at top of
column

Solvent

(b) Compounds A
and B beginning
to separate

(c) Compound A
starting to elute
from column

(d) Compound A
collected

(e) Compound B
starting to
elute from
column

(f) Compound B
collected

FIGURE 19.1 Stages in liquid chromatographic separation of a mixture containing two compounds. Compound A moves faster than compound B, which is more strongly adsorbed on the stationary phase.

Silica Gel

For flash chromatography 38–63-μm silica gel (230–400 mesh) is used, whereas in gravity LC 63–210-μm (70–230 mesh) particle size silica gel is usually used as the stationary phase. The smaller particle size in flash chromatography produces more effective separations but slower eluent flow rates. Thus flash chromatography is used either to separate small samples on small columns or with greater than atmospheric pressure.

Particles of silica gel have 10–20% adsorbed water by weight on their surface. Compounds separate by partitioning themselves between the elution solvent mobile phase and the strongly adsorbed water, which is the stationary phase. The adsorptive properties of silica gel may vary considerably from one manufacturer to another or even within different lots of the same grade from one manufacturer. Therefore, the solvent system previously used for a particular analysis may not work exactly the same way for another separation of the same sample mixture.

Alumina

Activated alumina, made explicitly for chromatography, is available commercially as a finely ground powder in neutral (pH 7), basic (pH 10), and acidic (pH 4) grades. Different brands and grades vary enormously in adsorptive properties, mainly because of the amount of water adsorbed on the surface. The strength of the adsorption holding a substance on aluminum oxide depends on the strength of the bonding forces between the substance and the polar surface of the adsorbent.

Adsorbents for Reverse-Phase Chromatography

Reverse-phase chromatography is used most often for HPLC separations (see Section 19.8). In reverse-phase HPLC, the liquid stationary phase is less polar than the mobile phase, and elution of the more polar compounds occurs first, with the less polar compounds adsorbed more tightly to the stationary phase. For reverse-phase chromatography, the surface of silica particles is rendered less polar by replacing the hydroxyl groups with alkoxy groups and long-chain alkyl groups (C_{12}–C_{18}).

19.2   Elution Solvents

The *elution solvents* used to dislodge compounds adsorbed on an LC column are usually made increasingly polar as the separation progresses. Nonpolar compounds bind less tightly than polar compounds on a polar adsorbent and dislodge more easily with nonpolar solvents. Therefore, the nonpolar compounds in a mixture exit from the column first. The more polar compounds must be eluted, or washed out of the column, with more polar solvents.

Purity of Elution Solvents

Elution solvents for column chromatography must be rigorously purified and dried for best results. Small quantities of polar impurities can radically alter the eluting properties of a solvent. For example, the presence of water in a solvent can significantly increase its eluting power. Wet acetone may have an eluting power greater than anhydrous ethanol.

Selecting an Elution Solvent

Silica gel usually works well as the adsorbent for separating most organic compounds, and therefore TLC on silica gel plates (see Section 18.7) can be used to determine a good solvent system for separating a mixture. Although not a perfect match, the separation on a silica gel TLC plate with a particular solvent or combination of solvents generally reflects the separation that the mixture will undergo with a silica gel column. A solvent that moves the desired compound with an R_f of approximately 0.3 should be a good elution solvent.

Before running a flash or gravity column, determine the TLC characteristics of the sample's components. Although it is desirable to have an R_f difference of ≥ 0.2 for the compounds being separated, it is possible to separate compounds with an R_f difference of ~ 0.15. Useful elution solvent mixtures include petroleum ether (30°–60°C) or pentane mixed with one of the following: diethyl ether, ethyl acetate, or anhydrous acetone. The composition of the elution solvent can be changed during the course of eluting the column.

Table 19.1 lists common elution solvents and organic compounds by functional group class in order of increasing polarity. There is no universal series of eluting strengths because this property depends not only on the activity of the adsorbent but also on the compounds being separated.

19.3 Determining the Column Size

The size of the column used for an LC separation depends on how much material you want to separate. After deciding which adsorbent you want to use, you must decide how much to use. In general, for a moderately challenging separation, you should use about 10–20 times as much silica gel by weight as the material to

TABLE 19.1 Relative polarities of common LC solvents and organic compounds on silica gel

Common elution solvents	Increasing polarity	Organic compounds by functional group class
Alkanes, cycloalkanes		Alkanes
Dichloromethane		Halocarbons
Diethyl ether		Ethers
Acetone (anhydrous)		Ketones
Ethyl acetate		Esters
Ethanol (anhydrous)		Alcohols
Methanol	↓	Carboxylic acids

be separated. Use more adsorbent for a difficult separation, less for an easy one. If too little adsorbent is used, the column will be overloaded and the separation will be poor. If too much adsorbent is used, the chromatography will take longer, require more elution solvent, and be no more efficient.

A height of 10–20 cm of silica gel often works well, and an 8:1 or 10:1 ratio of the adsorbent height to the inside column diameter is normal. Thus, a 1.0–2.5-cm column diameter is common for liquid chromatography on silica gel. A short, fatter column often produces worse separation, while a tall, thinner column can retain the compounds so tenaciously that the polar solvents required for their elution do not discriminate well between the various compounds on the column.

Table 19.2 provides column and solvent dimensions for preparation of a flash silica gel column that is 12–15 cm in height and can accommodate a range of sample sizes from 40 mg to 2.5 g. This table shows that a smaller column diameter requires the collection of smaller fraction sizes. In addition, the smaller the difference in R_f values, the smaller the size of the sample that can be chromatographed. Elution fractions are usually analyzed by TLC or GC.

Amount of Adsorbent and Calculation of Column Diameter

If you are carrying out a chromatographic separation on a 1.0-g sample, 15 g of silica gel would be appropriate. Silica gel has a bulk density of about 0.3 g/cm³, so 15 g would occupy a volume of 45–50 cm³ (45–50 mL); this quantity is called the *column volume.* Aiming for a column height of 15 cm of silica gel, you can calculate the inside diameter of the necessary chromatography column. The column of silica gel is a cylinder with a volume of $\pi r^2 h$; if $h = 15$ cm and $V = 50$ cm³, then $r = 1.0$ cm. Thus, a chromatography column with a 2–2.5-cm inside diameter would be appropriate. Common inside diameters for commercially available glass columns used in flash chromatography are 1.0, 1.5, 1.9, and 2.5 cm.

Approximately two column-volumes of elution solvent above the adsorbent are usually used to push the liquid through the silica gel column in gravity LC. Therefore, the chromatographic separation of 1.0 g of material on silica gel would require a glass column 2 cm in diameter and 40 cm long. Either a commercial

TABLE 19.2	Column dimensions and solvent volumes for flash chromatography			
Column diameter, mm	Volume of eluent, mL	Typical sample size, mg		Recommended fraction size, mL
		$\Delta R_f \geq 0.2$	$\Delta R_f \geq 0.1$–0.2	
10	100	100	40	5
20	200	400	160	10
30	400	900	360	20
40	600	1600	600	30
50	1000	2500	1000	50

Source: Still, W. C.; Kahn, M.; Mitra, A. *J. Org. Chem.* **1978,** *43,* 2923–2925.

chromatography column of 2.5-cm diameter and 30-cm length or one of 1.9-cm diameter and 40-cm length would be appropriate for the separation of a 1.0-g sample.

19.4 Flash Chromatography

Miniscale Liquid Chromatography

Unless small columns are used for the separation of small samples, gravity liquid chromatography can be time consuming because the solvent flow through the column can be very slow. In many chemistry labs, it has been largely replaced by flash chromatography. The separation or purification of less than 200-mg samples on 1.0- or 1.5-cm diameter columns, however, can often be done at atmospheric pressure. With larger samples more adsorbent is required and pressure may be necessary to push the elution solvent through the stationary phase. The total time to prepare and elute a column can be less than 30 min. It is important for you to read and understand Sections 19.1–19.3 on fundamental aspects of LC before you use flash chromatography.

The smaller particle size of the silica gel in flash chromatography may require pressures of up to 20 pounds per square inch (psi) to increase the flow rate of the elution solvent through the column. For smaller columns of less than 2.5-cm diameter, a one-way atomizer bulb fitted with plastic or rubber tubing and a one-hole stopper fit snuggly into the top of the column provides safe and sufficient pressure to drive the eluent through the column (Figure 19.2). If a solvent reservoir is attached to the top of the column, a few pumps on the atomizer bulb can provide pressure for the collection of several fractions. **Care should be taken to slowly twist off the rubber stopper when fresh solvent needs to be added.** A quick release of the pressure can cause the adsorbent column to crack.

Another type of chromatography column uses a source of nitrogen gas or compressed air to increase the column flow rate, and consists of a glass column topped by a variable bleed device that is used as a flow controller for the gas pressure. The bleed device has a Teflon needle valve that controls the pressure applied to the top of the solvent (Figure 19.3).

Preparation of a Flash Column

If silica gel is your adsorbent, use 230–400-mesh silica gel that is specifically designed for flash chromatography.* You can measure the amount you plan to use either by volume using a dry graduated cylinder (density ~ 0.3 g/mL) or by weight. After selecting a chromatography column and measuring the requisite amount of adsorbent, you are ready to prepare the column. The packing of a column is just as crucial to the success of the chromatographic separation as is the choice of adsorbent and elution solvents. If the adsorbent has cracks or channels or if the top surface is not flat, separation can be poor.

*Aldrich and other suppliers indicate whether the silica gel is suitable for flash chromatography.

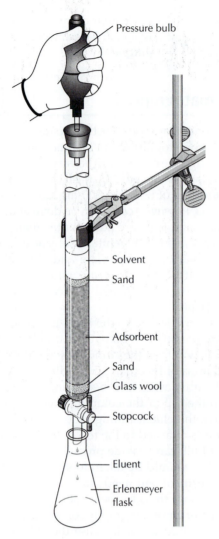

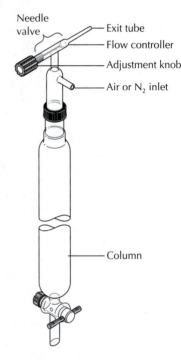

FIGURE 19.2 Flash chromatography column equipped with an atomizer bulb.

FIGURE 19.3 Apparatus for flash chromatography.

Figure 19.4 shows additional solvent above a completed chromatographic column. It is essential that the column never be allowed to dry out once it has been prepared; the solvent level should never be allowed to fall below the top of the sand above the adsorbent. If the adsorbent becomes dry, it may pull away from the walls of the column and form channels. **Once you begin a chromatographic separation, it is best to finish it without interruption.**

Clamp the chromatography column in an upright position on a ring stand or vertical support rod, and with the stopcock closed, fill it approximately halfway either with the first developing solvent you plan to use or with a less polar solvent. Add a small piece of glass wool as a plug and push it to the bottom of the column with a long glass rod, making sure all the air bubbles are out of the glass wool. Cover the glass wool plug with 2–3 mm of 50–100-mesh clean

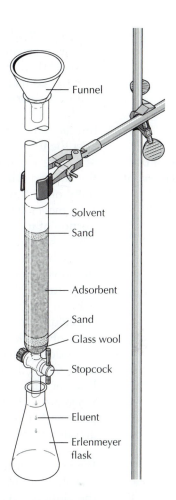

Funnel

Solvent

Sand

Adsorbent

Sand

Glass wool

Stopcock

Eluent

Erlenmeyer
flask

FIGURE 19.4 A completed chromatographic column.

white sand. The glass wool plug and sand serve as a support base to keep the adsorbent in the column and prevent it from clogging the stopcock. Columns equipped with a sintered glass frit at the bottom are available and very convenient, obviating the need for glass wool and sand. The adsorbent can be added to the column by either the dry adsorbent method or the slurry method.

Dry adsorbent method. Place a powder funnel in the top of the column, and with the stopcock closed, pour the adsorbent slowly into the solvent-filled column. Take care that the adsorbent falls uniformly to the bottom. Do not add the adsorbent too quickly or clumping may occur. The adsorbent column should be firm, but if it is packed too tightly, the flow of elution solvents may become too slow.

The top of the adsorbent must always be horizontal. Gently tap on the side of the column as the adsorbent falls through the solvent to prevent the formation of bubbles. If large bubbles or channels develop in the column, discard the adsorbent and repack the column. Irregularities in the adsorbent column may cause poor separation because part of the advancing sample may move faster than the rest. The time spent repacking will be much less than the time wasted trying to make a poor column function efficiently.

Slurry method. If you are using a solvent more polar than an alkane in packing the column, you may need to prepare a slurry of the adsorbent and solvent in an Erlenmeyer flask by **slowly** adding the requisite amount of adsorbent to an excess of solvent. The use of a slurry prevents the formation of clumps or gas bubbles in the column, which can form from the heat produced by the interaction between more polar solvents and the surface of the adsorbent.

Even packing of the adsorbent is essential to ensure that no cracks, air bubbles, or channels form while preparing the column.

Place a powder funnel in the top of the column and half fill the column with the same solvent used to prepare the slurry. Partially open the stopcock so that the solvent drains slowly into an Erlenmeyer flask. Swirl the flask containing the slurry and pour a portion of it into the column. Tap the side of the column constantly while the slurry is settling. Swirl the slurry thoroughly before adding each portion to the column. Add more solvent as needed so that the solvent level never falls below the level of the adsorbent at any time during the packing procedure. The solvent drained from the column can be reused for this purpose. Once all the adsorbent is in the column, return the collected solvent to the column once or twice to firmly pack the adsorbent and then close the stopcock.

Adding a second layer of sand (2–3 mm) at the top of the silica gel and leveling it with gentle tapping provides the finishing touches. The layer of sand protects the adsorbent from mechanical disturbances when new solvents are poured into the column during the separation process. Allow solvent to drip through the stopcock until only a small amount of solvent is above the sand and then close the stopcock.

Preparation and Application of the Sample

Prepare a concentrated solution of a solid or viscous oil sample dissolved in the elution solvent. If the sample is not very soluble in the elution solvent, use a small amount of a more polar solvent or preadsorb the sample on 1–2 g of silica gel. A neat (without solvent) liquid sample can be added to the column directly or preadsorbed on a small amount of silica gel.

Sample solution. The solvent used in packing the column or another solvent of similar polarity is preferred and the aim is to have a 25% or greater concentration of the sample. If the sample's components do not dissolve in the first elution solvent, a small amount of a more polar solvent may be used to prepare the solution. In this case especially, the sample solution should be as concentrated as possible. Poor separation will occur if the sample volume is too large—the compounds will begin to move down the column while some of the sample is still entering at the top.

Draw the sample solution or the neat liquid sample into a 9-in Pasteur pipet, hold the pipet with the tip just above the level of the sand, and add the sample **one drop at a time to the center of the sand.**

Reopen the stopcock and allow the upper level of the sample solution to just reach the top of the sand; then close the stopcock again.

Sample adsorbed on silica gel. Instead of preparing a sample solution, you can preadsorb the sample onto a small amount of silica gel, remove the solvent, and using a diagonally folded weighing paper, carefully pour the dry mixture onto the top of the column. For a miniscale sample, add 1–2 g of silica gel to a solution of the sample, remove the solvent using a rotary evaporator (see Section 11.3), and carefully add the dry powder to the top of the column.

Elution of the Column

After the sample is on the column, fill the column with your first elution solvent, using a chromatography funnel. This funnel has a closed bottom and small holes in the stem wall, which provide a gentle flow of solvent down the wall of the column that does not disturb the packing of the sand and adsorbent (Figure 19.5). If you are using a column such as the one in Figure 19.2, insert the one-hole stopper with the atomizer attached and pressurize the column so that the stationary phase is packed tightly. When the solvent has just reached the level of the sand (or the adsorbent if you applied the sample preadsorbed on silica gel), close the stopcock, and gently twist off the rubber stopper to relieve the pressure.

If you are using the flash column shown in Figure 19.3, insert the flow controller, and with the needle valve open, gently turn on the flow of pressurized gas. Control the pressure by placing a finger (wear gloves) over the end of the exit tube, and manipulate the pressure so that the stationary phase is tightly packed. When the solvent has just reached the level of the sand (or the adsorbent if you applied the sample preadsorbed on silica gel), close the stopcock and remove the flow controller.

Fresh solvent needs to be added to the top of the column continuously during the elution process, but never allow the top surface of the adsorbent to be disturbed by the addition of solvents. Also **never let the column run dry**—the solvent must never go below the level of the top column layer.

Fill the column with your elution solvent and adjust the pressure so that the solvent level drops at a rate of approximately 5 cm/min. If you have made a smaller chromatography column that doesn't need additional pressure, the optimum flow rate is 2–3 mL/min. A greater solvent height above the adsorbent layer provides a faster flow rate through a gravity LC column. If the flow is too slow, poor separation may result from diffusion of the compound bands as they travel down the column. A reservoir at the top of a column can be used to maintain a proper height of elution solvent above the adsorbent so that an adequate flow rate is maintained. A separatory funnel makes a good reservoir if you are not using additional pressure.

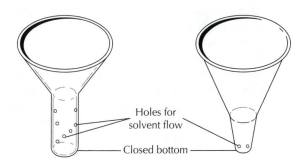

FIGURE 19.5
Chromatography funnels.

Holes for solvent flow

Closed bottom

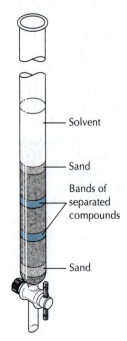

Solvent

Sand

Bands of
separated
compounds

Sand

FIGURE 19.6
Chromatography
column during elution.

The funnel can be filled with the necessary amount of solvent and clamped directly above the column. The stopcock of the separatory funnel can be adjusted so that elution solvent flows into the column as fast as it flows out at the bottom.

Elute the compounds in the sample by using a series of increasingly polar elution solvents. The less polar compounds elute first with the less polar solvents. Polar compounds usually come out of a column only after a switch to a more polar solvent. As the elution proceeds, the compounds in the mixture separate into a series of bands in the column (Figure 19.6). With colorless compounds, the bands are usually invisible.

Changing elution solvents during a separation. A mixture of two solvents is commonly used for elution. If the change of solvent is made too abruptly, enough heat may be generated from adsorbent-solvent bonding to cause cracking or channeling of the adsorbent column. In some cases, a low-boiling elution solvent may actually boil on the column. The bubbles that form will degrade the efficiency of the column. Adding small amounts of a polar solvent to a less polar one increases the eluting power in a gentle fashion. For example, the development of the column can begin with pentane, and if nothing elutes from the column with this solvent, a 2–5% solution of diethyl ether in pentane can be used next, followed by a 10% solution of diethyl ether, then a 25% solution of diethyl ether, and then pure diethyl ether for the most polar compounds.

Size of eluent fractions. The size of the eluent fractions collected at the bottom of the column depends on the particular experiment. If the separated compounds are colored, it is a simple matter to tell when the different fractions should be collected. When you are collecting colorless compounds, use the recommended fraction sizes in Table 19.2. Collect eluent fractions until all the solvent you planned to use has passed through the column or until fraction monitoring indicates that the desired components have been eluted.

With an efficient column, each compound in the mixture elutes separately. After one compound has come through the column, there is a time lag before the next one appears. Because most organic compounds are colorless and not easily visualized on the column, generally no attempt is made to time fractions to the elution of the components in the mixture. Rather, a rack of test tubes or small Erlenmeyer flasks is used to collect fractions of equal volume, and TLC or GC is used to determine which fractions to pool after complete elution of the column. On the other hand, when compounds in the mixture are colored, you can often watch the bands elute through the column, and therefore switch fractions based on this visual clue. Occasionally with highly crystalline compounds, crystals can form on the tip at the bottom of the column as the compound elutes and solvent evaporates. It is best to wash these crystals as they form into the collection fraction using a few milliliters of clean elution solvent so they do not contaminate later eluting compounds.

Removing the Adsorbent from the Column

When you are finished eluting the sample from the column, allow any remaining solvent to drain out. Then empty the chromatography tube by opening the stopcock, inverting the column over a beaker, and using gentle air pressure at the tip to push out the adsorbent.

Recovery of Separated Compounds

Determine the purity of each fraction by GC analysis (Chapter 20) or TLC (Chapter 18) and combine the fractions containing each pure compound. Recover the compounds by evaporation of the solvent. Evaporation methods include using a rotary evaporator (see Section 11.3) or blowing off the solvent with a stream of nitrogen or air in a fume hood.

19.5   Microscale Liquid Chromatography

When you are separating samples of 100 mg or less using columns having internal diameters of 1.0 cm or less, microscale flash LC methods, which can be much simpler to use, can save you time and effort.

19.5a Preparation and Elution of a Microscale Column

A column suitable for separating 50–100 mg of a mixture can be prepared in a large-volume Pasteur pipet (5¾ in).* Regular-size 5¾-in Pasteur pipets can be used to separate a 10–30-mg sample. **Prepare the sample solution and assemble all equipment and reagents for the entire chromatographic procedure before you begin to prepare the column.** The entire procedure of preparing the column and collecting the fractions must be done without interruption.

Alternatively, you can use a smaller version of a miniscale LC column (see Section 19.4). A 1-cm-diameter column that is 20 cm long works well to separate 100–150 mg of a mixture.

Preparation of the Sample

Dissolve the mixture being separated in a small test tube using 0.5–1 mL of the elution solvent or another solvent that is less polar than the elution solvent. Stopper the tube until you are ready to apply the sample to the column.

Alternatively, add 300 mg of silica gel to the sample solution, and in a fume hood, evaporate the solvent by warming the sample container in a hot-water bath while stirring the mixture with a microspatula to prevent bumping. The resulting dried solid is ready for addition to the column.

Test Tubes for Sample Collection

Label a series of 10 test tubes (13 × 100 mm) for fraction collection. Pour 5 mL of elution solvent into one test tube and mark the liquid level on the outside of the tube. Place a corresponding mark on the outside of the other nine test tubes. Label each test tube with its fraction number.

* Available from Fisher Scientific, catalog item 22-378-893; the pipets have a capacity of 4 mL.

Packing the Column

Pour about 50 mL of pentane (or other nonpolar solvent) into an Erlenmeyer flask and stopper the flask. Pack a small plug of glass wool into the stem of the Pasteur pipet microscale column using a wooden applicator stick or a thin stirring rod (Figure 19.7, Step 1). Clamp the pipet in a vertical position. Add a 2–3-mm layer of sand. Place a test tube or small Erlenmeyer flask under the pipet column to collect the solvent that will drip out.

Place 1.7–1.8 g of silica gel adsorbent in a 50-mL Erlenmeyer flask and then add approximately 15 mL of pentane to make a thin slurry. Transfer the adsorbent slurry to the Pasteur pipet column using a 9-in Pasteur pipet (Figure 19.7, Step 2). Continue adding slurry until the column is two-thirds full of adsorbent. Fill the column four to five times with pentane to pack the adsorbent well. The eluted pentane can be reused for this purpose. **Do not let the solvent level fall below the top of the adsorbent.** After the adsorbent is packed, add a 1–2-mm layer of sand above the adsorbent by letting it settle through the solvent. The sand helps to prevent disturbance of the surface of the column when the elution solvent is added. If you preadsorbed your sample on silica gel, pour it slowly into the column from a diagonally folded weighing paper.

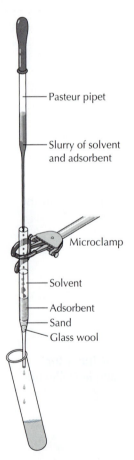

Pasteur pipet

Slurry of solvent and adsorbent

Microclamp

Solvent

Adsorbent

Sand

Glass wool

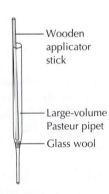

Wooden applicator stick

Large-volume Pasteur pipet

Glass wool

FIGURE 19.7 Setting up a microscale column.

1. Pack glass wool plug in large-volume Pasteur pipet.

2. Add slurry of solvent and adsorbent.

Addition of the Sample and Elution of the Column

Allow the solvent level to almost reach the top of the adsorbent and place the test tube labeled "Fraction 1" under the column. If you are adding a solution of your sample, draw it into a 9-in Pasteur pipet; hold the pipet tip just above the surface of the sand, and add the sample one drop at a time to the center of the column.

When the entire sample is on the column, use a 9-in Pasteur pipet to add the elution solvent by **gently running it down the interior wall of the pipet**. Maintain a column of solvent above the silica gel while you collect fractions of approximately 2–4 mL in the 10 labeled test tubes.

Recovery of Separated Compounds

Determine the purity of each fraction by GC analysis (Chapter 20) or TLC (Chapter 18) and combine the fractions containing each pure component. Recover the compounds by evaporation of the solvent either by using a rotary evaporator (see Section 11.3) or by blowing off the solvent with a stream of nitrogen or air in a fume hood.

19.5b Preparation and Elution of a Williamson Microscale Column

The Williamson microscale chromatography apparatus consists of several pieces fitted together. Before you prepare the column, collect all the reagents and equipment you will need for the entire procedure. Prepare 10 test tubes for sample collection as directed on page 281.

Preparation of the Sample

Dissolve the mixture being separated in a small test tube using 1 mL of the elution solvent or another solvent that is less polar than the elution solvent. Stopper the test tube until you are ready to apply the sample to the column.

Alternatively, add 300 mg of silica gel to the sample solution. In a fume hood, evaporate the solvent by warming the mixture in a hot-water bath while stirring with a microspatula to prevent bumping. The resulting dried solid is ready for addition to the column.

Packing the Column

Assemble the plastic funnel, glass column, Buchner microfunnel with a polyethylene frit, and plastic stopcock as shown in Figure 19.8. With the stopcock closed, fill the column with pentane (or other nonpolar solvent) nearly to the top. Weigh approximately 3.0–3.5 g of silica gel adsorbent in a tared 50-mL beaker. Add enough pentane to make a thin slurry and swirl the beaker gently to thoroughly wet the adsorbent. Gently swirl the beaker to suspend the adsorbent and pour the mixture into the funnel. Place an Erlenmeyer flask under the column and open the stopcock to collect the solvent as it drains. Use a few milliliters of solvent to rinse the remaining adsorbent from the flask and add the slurry to the funnel. Tap the side of the column gently to help pack the adsorbent. Close the stopcock when the solvent level is just slightly above the top of the adsorbent.

Addition of the Sample

If you prepared a solution of your sample, draw it into a 9-in Pasteur pipet, hold the pipet tip just above the surface of the adsorbent, and

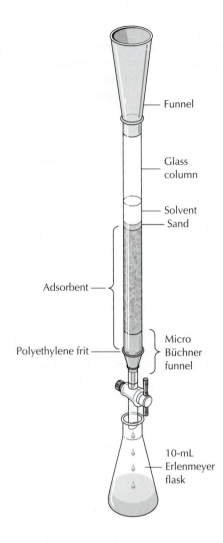

Funnel

Glass
column

Solvent
Sand

Adsorbent —

Polyethylene frit —

Micro
Büchner
funnel

10-mL
Erlenmeyer
flask

FIGURE 19.8
Williamson microscale
column.

add the sample one drop at a time to the center of the column. Open
the stopcock slightly to drain the solvent to *just above* the top of the
adsorbent.

For a sample adsorbed on silica gel, drain the solvent to *exactly*
the top of the adsorbent and then close the stopcock. Place the
sample mixture on diagonally folded weighing paper and transfer
it slowly into the funnel at the top of the column. After the sample is
applied, add a 1–2-mm layer of white sand to the column.

**Elution of the
Column**

Fill the column with elution solvent by allowing the liquid to run
slowly down the side of the funnel, open the stopcock, and begin
collecting 2–4-mL fractions in your labeled test tubes. Do not allow
the solvent level to fall below the top of the column at any time during the elution. Continue to add solvent while collecting fractions.

**Recovery of
Separated
Compounds**

Determine the purity of each fraction by GC analysis (Chapter 20)
or TLC (Chapter 18) and combine the fractions containing the pure
compounds. Recover the compounds by evaporation of the solvent.

Evaporation methods include using a rotary evaporator (see Section 11.3) or blowing off the solvent with a stream of nitrogen or air in a fume hood.

19.6   Summary of Liquid Chromatography Procedures

1. Prepare a properly packed column of adsorbent.
2. Carefully add the sample mixture to the column as a small volume of solution or liquid, or as a solid adsorbed on silica gel.
3. Elute the column with progressively more polar solvents.
4. Collect the eluted compounds in fractions from the column.
5. Evaporate the solvents to recover the separated compounds.

19.7   Sources of Confusion and Common Pitfalls

I Added Solvent to the Solid Adsorbent to Prepare My Slurry and the Solvent Started to Boil! What's Going on?

Exothermic interactions between polar solvents and the surface of the adsorbent can produce substantial heat and, especially with lower-boiling solvents, can cause the solvent to boil. The effect can be quite dramatic when, for instance, diethyl ether (bp 35°C) is added to silica gel. Always mix the solvent and adsorbent slowly, whether packing your column by the slurry method or by the dry adsorbent method.

Why Didn't My Column Separate the Components in My Crude Product? No Purification Occurred!

A number of factors could contribute to poor separation of the components in a mixture.

Polarity of elution solvent. If the elution solvent is too polar, the sample mixture will elute too quickly and poor separation will result. If the solvent is not polar enough, the sample will elute too slowly and the bands of compounds will broaden by diffusion, resulting in poor separation, along with a waste of time and solvent. An elution solvent that produces an R_f of about 0.30 for the desired compound, with good separation from other components in the mixture ($\Delta R_f \geq 0.20$), is best.

Preparing the sample improperly. Achieving a good separation with a chromatography column depends on how the sample mixture is prepared and applied to the column.

Overloading the column. If the amount of sample is too large for the amount of adsorbent used in packing the column, the column will be overloaded, and incomplete separation of the mixture's components will occur. Determine the correct amount of adsorbent by using the information in Table 19.2 for flash chromatography.

Solvent used to dissolve the sample is too polar and/or too much solvent was used to prepare the sample solution. Ideally, the sample should be dissolved in a minimal amount of the elution solvent, or in a *less* polar solvent. If too much solvent is used to dissolve the sample, the excess will behave as an elution solvent and start to carry the mixture's components down the column. Separation will be incomplete because the entire sample was not on the column before its components started to move through the stationary phase. Likewise, if the sample is dissolved in a solvent that is *more* polar than the elution solvent, the sample components can co-elute, carried along by this polar solvent. The result is a very quick elution and poor separation because of insufficient interaction with the stationary phase. Occasionally, a sample is only partially soluble in the elution solvent at the high concentration desired for sample application. In this case, a more polar solvent should be added dropwise until the sample *just* dissolves, keeping the quantity of a polar solvent to a bare minimum. Better yet, this would be a good situation to apply the sample preadsorbed onto a small amount of silica gel, as described on pages 278–279.

Imperfections in the column. For a chromatography column to work successfully in separating a mixture, the adsorbent must be packed uniformly without air bubbles, gaps, or surface irregularities. If the packing is not satisfactory, different compounds in the sample will co-elute and the separation will be incomplete.

Nonhorizontal bands. Nonhorizontal bands result if the adsorbent surface at the top of the column is not flat and horizontal, if the column is not clamped in a perfectly vertical position, or if the sample is not evenly applied to the column (Figure 19.9a). If nonhorizontal bands are present, poor separation can result because the lower part of one band may co-elute with the upper part of the next band.

Channeling. If a depression or other irregularity is present at the top of the adsorbent surface, if cracks occur in the adsorbent, or if an air bubble is trapped in the column, part of the advancing front of a band will move ahead of the rest of the band—a process called channeling (Figures 19.9b and 19.9c). Channeling allows the bottom of one band to co-elute with the top of a lower band. Channeling is commonly

FIGURE 19.9 Problems that occur as a result of a poorly packed column.

(a) Nonhorizontal bands

(b) Channeling caused by irregular surface

Air bubble

(c) Channeling caused by air bubble

caused by applying the sample to the column too vigorously. A liquid sample should be applied with a 9-in Pasteur pipet one drop at a time to the center of the column, with the tip of the pipet just above the adsorbent surface. Even with a layer of stabilizing sand at the top of the absorbent, the column surface is very easily disturbed. If the sample solution is quickly squirted from the application pipet, it can easily bore a deep depression in the absorbent (Figure 19.9b), providing an excellent opportunity for channeling.

I've Carefully Packed my Column, Prepared and Applied my Sample Solution, and I'm Ready to Run my Column. What Can Go Wrong During the Elution Process?

The column can become dry. If the solvent level falls below the top of the adsorbent, it can become dry and pull away from the column wall. The channels that form compromise the effectiveness of the column. Be sure that the adsorbent is covered with solvent throughout the chromatographic procedure. Have all elution solvents at hand before starting the elution so the separation can be completed without interruption.

The solvent polarity can be changed too quickly. The polarity of the elution solvent often needs to be increased as the elution proceeds; however, the increase in polarity must be made gradually. If the polarity change is made too rapidly, enough heat may be generated from adsorbent-solvent bonding to cause gas bubbles that lead to channeling or even open cracks in the adsorbent column. The first change in polarity should add only 2–5% of the more polar solvent to the original elution solvent.

Diffuse bands or tailing can occur. If the elution solvent flows through the column too slowly or is not polar enough to displace the desired compounds at a reasonable rate, poor separation may result from diffusion of the bands. The optimum flow rate is about 2–3 mL/min for a gravity column and 5 cm/min for a flash column.

Running a successful purification by column chromatography requires care and attention and it can seem that there are a thousand points where things can go wrong. However, the technique can be surprisingly forgiving and separations are sometimes successful even when a column goes dry or a crack forms in the adsorbent.

19.8   High-Performance Liquid Chromatography

High-performance liquid chromatography (HPLC) is one of the most widely used analytical separation techniques in organic chemistry. It allows analyses to be completed quickly with superior separation and sensitivity compared with other liquid chromatography methods. In this regard HPLC is comparable to gas chromatography (see Chapter 20). Like GC, HPLC utilizes small samples and is often used to analyze mixtures. Unlike GC, however, HPLC can be used equally well with volatile and nonvolatile compounds. Because of its high cost and demanding instrumental requirements, HPLC is not nearly as common as GC in organic laboratory courses, but virtually all organic chemistry research labs have access to HPLC instruments.

HPLC Columns and Injection Systems

HPLC is carried out with packed columns rather than the open-tubular columns used in GC capillary columns (see Section 20.2). Diffusion in liquids is much slower than diffusion in gases, so as molecules pass through an HPLC column in the mobile liquid phase, they cannot diffuse quickly enough for effective adsorption equilibria to occur with a liquid stationary phase coating only the column wall. The liquid stationary phase in packed HPLC columns has a particle size of only 3–10 μm. This small particle size produces tremendously efficient partition of compounds between the mobile phase and the liquid stationary phase on the very large surface area of the particles. Particles of this small size pack very tightly, however, which severely restricts the flow of solvent through the column. Consequently, pressures of 50–200 atm are required to force solvent through an HPLC column at a reasonable rate.

The instrumentation for HPLC consists of a column, a sample injection system, a solvent reservoir, a pump, a detector, and a recorder or computer readout. Figure 19.10 is a diagram of a typical HPLC setup, and a picture of an HPLC setup is shown in Figure 19.11.

At the beginning of an HPLC run, you would typically use an automated injection system (autosampler) to inject a tiny amount of sample solution into the column. There is generally a short *guard column* in position before the more expensive main column. The guard column retains fine particles and strongly adsorbed compounds that would degrade the main column; it must be replaced periodically.

The length of the main column can range from 5 to 30 cm, with an inner diameter of 1–5 mm for analytical HPLC of 0.01–1.0-mg samples. HPLC columns usually have a liquid stationary phase that is covalently bonded to microporous spherical silica (SiO_2) particles. These particles are permeable to solvent and have very large surface areas.

Reverse-Phase HPLC Columns

Most HPLC is done using *reverse-phase* chromatography, in which the mobile phase is polar and the stationary phase that covers the surface of the silica particles is a thin layer of a nonpolar organic compound. **In reverse-phase chromatography, more polar compounds**

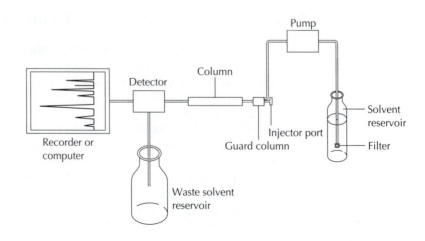

FIGURE 19.10
Schematic representation of a typical high-performance liquid chromatograph.

FIGURE 19.11
A high-performance
liquid chromatograph.

Courtesy of Shimadzu Corporation, Kyoto, Japan

elute first because the mobile-phase solvent is more polar than the nonpolar stationary phase. If inorganic salts and buffers are present in the sample, they are eluted very quickly. A generalized diagram of how a hydrophobic organic stationary phase is covalently bonded to the silica is shown in Figure 19.12.

The most popular bonded stationary phase in reverse-phase HPLC columns is the nonpolar C_{18} octadecyl group, which adsorbs organic compounds by van der Waals interactions. Other R groups, such as $(CH_2)_7CH_3$ and $(CH_2)_3C_6H_5$, can also be used. Reverse-phase columns are especially useful in separating moderately polar to polar compounds, but they can be used to separate most organic compounds.

Detectors

Simple HPLC systems use a fixed-wavelength, low-pressure 254-nm mercury vapor ultraviolet lamp as the detector; however, the most common type of detector is the sensitive photodiode-array UV/visible detector (see Chapter 25 for the principles and practice of UV spectroscopy). Diode-array detectors use 500–2000 individual detectors, each covering a discrete spectral region of 1–2 nm, to accumulate an entire UV spectrum almost simultaneously as each compound emerges from the column. Analog signals from the detector are then digitized for computer manipulation. The only limitation of the photodiode-array detector is that compounds must have measurable UV absorbance above 210 nm to be detected. However, a majority of organic compounds fulfill this criterion.

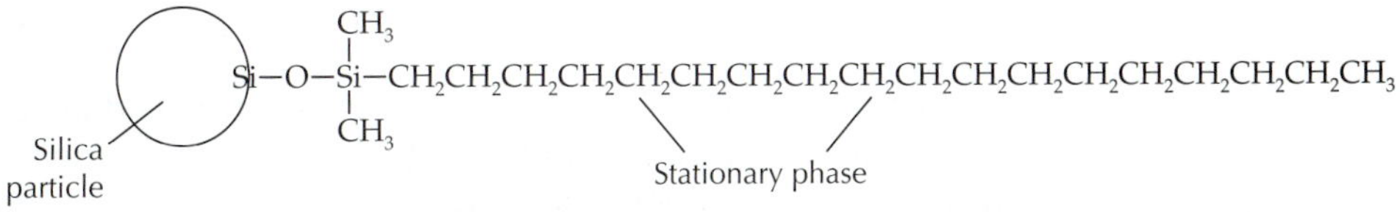

$$\underset{\substack{\text{Silica}\\\text{particle}}}{\bigcirc}\!\!\text{Si}-\text{O}-\underset{\underset{CH_3}{|}}{\overset{\overset{CH_3}{|}}{\text{Si}}}-\underset{\text{Stationary phase}}{CH_2CH_2CH_2CH_2CH_2CH_2CH_2CH_2CH_2CH_2CH_2CH_2CH_2CH_2CH_2CH_2CH_2CH_3}$$

FIGURE 19.12 One mode of covalent attachment of a common liquid stationary phase to a microporous silica particle in reverse-phase HPLC.

HPLC is very useful for quantitative analysis if standards are available for constructing a calibration curve for the dependence of the detector signal upon concentration. The measurements should be carried out under conditions where the measured absorbance is less than 1.0 and definitely no greater than 2.0. The photodiode-array detector generally has a good linear range over five orders of magnitude in which the Beer-Lambert law is followed (see Section 25.1).

Sometimes refractometry detectors are used for HPLC. These detectors measure changes in the refractive index of the eluent as a sample's components move off the column and through the detector. Refractometry detectors are not as sensitive as diode-array UV detectors and cannot easily be used with gradient elution; however, they bypass the requirement that the compounds being analyzed must absorb UV light.

HPLC Solvents

The two most useful elution solvents for reverse-phase HPLC are methanol and acetonitrile ($CH_3C\equiv N$), which are usually mixed with water. Neither of these polar organic solvents absorbs UV radiation above 210 nm, so either one can be used with a photodiode-array UV detector. Combinations of $CH_3C\equiv N$ or CH_3OH with water are sufficient to separate most organic compounds. HPLC columns are easily degraded by dust and particles in the sample or the solvent. Consequently, the pressure necessary to push the solvent through the column can double during the life of a column because of progressive clogging. To minimize this problem, the solvent, which is stored in the solvent reservoir, is passed through a 0.5-μm pore filter before being pumped through the injector port.

Solvents used for HPLC must be of high purity because impurities can degrade the column by irreversible adsorption onto the stationary phase. Before use, solvents must also be purged with helium or by a vacuum to remove dissolved air. Dissolved O_2 absorbs ultraviolet radiation in the 200–250-nm wavelength range, which interferes with UV detectors.

Many HPLC instruments can accommodate a gradient elution system, allowing the composition of the solvent to be changed during the course of a separation. During gradient elution, the mobile phase is changed from a more polar solvent, which is less able to move compounds through the column, to a less polar solvent; this change gives improved sensitivity and shorter analysis times.

Sample Preparation

The ideal solvent for sample preparation is the same solvent as that used for the mobile liquid phase. Approximately 10–150 μL of a very dilute solution (0.0001–0.001 M) are normally used for the injection sample. Prepare a solution of 1 mg or less of the sample in approximately 5 mL of solvent. Filter the sample solution through a micropore filter of about 0.5-μm pore size to remove any solid impurities that could clog the HPLC column. The filtration is done by taking up about 1 mL of the sample solution into a syringe and injecting it through a micropore filter into a small vial. After filtration, cap the vial with a rubber septum. Place the vial in the correct position in the HPLC instrument, and then the automatic injection system takes over when the chromatography run is initiated. Consult your instructor about specific operating procedures for the HPLC instrument in your laboratory.

Further Reading

Harris, D. C. *Quantitative Chemical Analysis*; 8th ed.; W. H. Freeman and Company: New York, 2011.

Kromidas, S. *Practical Problem Solving in HPLC*; Wiley-VCH: New York, 2000.

Meyer, V. R. *Practical High-Performance Liquid Chromatography*; 4th ed.; Wiley: New York, 2004.

Miller, J. M. *Chromatography: Concepts and Contrasts*; 2nd ed.; Wiley: New York, 2005.

Skoog, D. A.; Holler, F. J.; Crouch, S. R. *Principles of Instrumental Analysis*; 6th ed.; Thomson Brooks/Cole: Pacific Grove, CA, 2007.

Snyder, L. R.; Kirkland, J. J.; Dolan, J. W. *Introduction to Modern Liquid Chromatography*; 3rd ed.; Wiley: New York, 2010.

Snyder, L. R.; Kirkland, J. J.; Glajch, J. L. *Practical HPLC Method Development*; 2nd ed.; Wiley: New York, 1997.

Still, W. C.; Kahn, M.; Mitra, A. "Rapid Chromatographic Technique for Preparative Separations with Moderate Resolution"; *J. Org. Chem.* **1978**, *43*, 2923–2925.

Questions

1. Once the adsorbent is packed in a liquid chromatography column, it is important that the level of the elution solvent not drop below the top of the adsorbent. Why?

2. What precautions must be taken when you introduce a mixture of compounds to be separated onto a liquid chromatography adsorbent column?

3. What effect will the following factors have on a liquid chromatographic separation?

   (a) too strong an adsorbent; (b) collection of large elution fractions; (c) very slow flow rate of the mobile phase.

4. Arrange the following compounds in order of decreasing ease of elution from a column of silica gel: (a) 2-octanol; (b) 1,3-dichlorobenzene; (c) *tert*-butylcyclohexane; (d) benzoic acid.

5. Why do silica gel columns with smaller particle sizes produce more effective chromatographic separations?

CHAPTER

20

GAS CHROMATOGRAPHY

Few techniques have altered the analysis of volatile organic chemicals as much as gas chromatography (GC), also called gas-liquid chromatography (GLC). Before GC became widely available just over 50 years ago, organic chemists usually looked for ways to convert liquid compounds into solids in order to analyze them. Gas chromatography changed all that by providing a quick, easy way to analyze volatile organic mixtures both qualitatively and quantitatively. In addition, GC has a truly fantastic ability to separate complex mixtures.

If Chapter 20 is your introduction to chromatographic analysis, read the Essay "Modern Chromatographic Separations" on pages 253–254 before you read Chapter 20.

Gas chromatography does, however, have limitations. It is useful only for the analysis of small amounts of compounds that have vapor pressures high enough to allow them to pass through a GC column, and, like thin-layer chromatography (TLC) (see Chapter 18), gas chromatography does not identify compounds unless known samples are available. Coupling a gas chromatograph with a mass spectrometer (GC-MS) combines the superb separation capabilities of GC with the superior identification methods of mass spectrometry (see Section 24.1).

Overview of Gas Chromatography

GC is an example of partition chromatography, where the compounds being analyzed adsorb on the stationary phase, which consists of a nonvolatile liquid, usually a polymer with a high boiling point. The mobile phase is an inert gas, generally helium or nitrogen. Unlike LC and TLC, where the mobile phase actively competes with the stationary phase for the compounds being analyzed, in GC the mobile phase does not interact with the compounds. The inert gas simply carries them down the column when they are in the vapor state.

In capillary columns, the stationary phase is a thin, uniform liquid film applied either to the interior wall of a long, narrow capillary tube or to a thin layer of solid support lining the capillary tube. In either case, a clear channel through the center is left for passage of a carrier gas and molecules of the sample (Figure 20.1a). For older, packed-column chromatographs, the liquid is coated on a porous, inert solid support that is then packed into a tube (Figure 20.1b).

When the mixture being separated is injected into the heated injection port, the components vaporize and are carried by the carrier gas into the column, where separation occurs. The compounds in the mixture partition themselves between the gas phase and the liquid phase in the column, in an equilibrium that depends on the temperature, the rate of gas flow, and the solubility of the components in the liquid phase (Figure 20.2).

Mixtures separate during gas chromatography because their compounds interact in different ways with the liquid stationary phase. A GC column has thousands of theoretical plates as a result

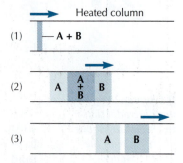

FIGURE 20.2 Stages in the separation of a two-component (**A, B**) mixture as it moves through a packed column.

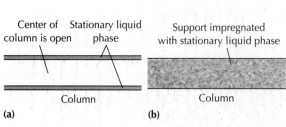

(a)

Center of column is open   Stationary liquid phase

Column

(b)

Support impregnated with stationary liquid phase

Column

FIGURE 20.1 View of (a) a wall-coated open tubular capillary column and (b) a packed column.

FIGURE 20.3 A gas chromatograph.

Courtesy of Shimadzu Corporation, Kyoto, Japan

of the huge surface area on which the gas and liquid phases can interact (see Section 12.4, page 189, for a discussion of theoretical plates). The partitioning of a substance between the liquid and gas phases depends on both its relative attraction to the liquid phase and its vapor pressure. The greater a compound's vapor pressure, the greater its tendency to go from the liquid stationary phase into the mobile gas phase. So, in the thousands of liquid-gas equilibria that take place as substances travel through a GC column, a more volatile compound spends more time in the gas phase than does a less volatile compound. In general, lower-boiling compounds with higher vapor pressures travel through a GC column faster than higher-boiling compounds. A picture of a gas chromatograph is shown in Figure 20.3.

20.1   Instrumentation for GC

The basic parts of a gas-liquid chromatograph are as follows:

- Source of high-pressure pure *carrier gas*
- *Flow controller*
- Heated *injection port*
- *Column* and *column oven*
- *Detector*
- *Recording device* or *data station*

These components are shown schematically in Figure 20.4.

A small syringe is used to inject the sample through a sealed rubber septum or gasket into the stream of carrier gas in the heated injection port (Figure 20.5). The sample vaporizes immediately and the carrier gas sweeps it into the column—a metal, glass, or fused-silica tube that contains the liquid stationary phase (Figure 20.6). The column is enclosed in an oven whose temperature can be regulated from just above room temperature to greater than 200°C. After

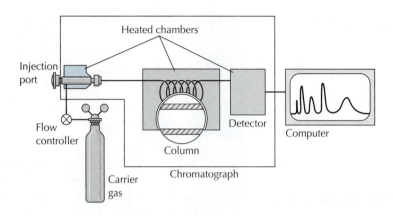

FIGURE 20.4
Schematic diagram of
a gas-liquid
chromatograph.

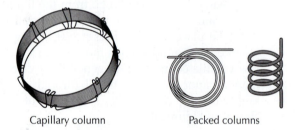

FIGURE 20.5 Injection
port during sample
injection.

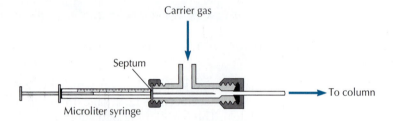

FIGURE 20.6
GC columns.

the sample's components are separated by the column, they pass into a detector, where they produce electronic signals that can be amplified and recorded.

20.2 Types of Columns and Liquid Stationary Phases

A gas chromatograph can have either capillary or packed columns. Capillary columns, also called open tubular columns, have an interior diameter of only 0.2–0.5 mm and a length of 10–100 m. A packed column typically has an interior diameter of 2–4 mm and a length of 2–3 m. Capillary columns usually give much better separation than packed columns. The greater length of capillary columns and the better diffusion of sample molecules between the liquid and gas phases provide more theoretical plates. Capillary columns not only give better separations, they also do it in much less time.

Types of Columns

Capillary columns. Several types of capillary columns are available. In a *wall-coated open tubular column (WCOT)*, the liquid phase coats the interior surface of the tube, leaving the center open. In a *support-coated open tubular column (SCOT)*, the liquid phase coats a thin layer of solid support that is bonded to the capillary wall, again leaving the center of the column open.

Packed columns. The solid support in packed columns (and SCOT capillary columns) consists of a porous, inert material that has a very large surface area. The most commonly used substance is calcined diatomaceous earth, which contains the crushed skeletons of algae. Its major component is silica (SiO_2). The efficiency of separation increases with decreasing particle size, which provides an expanded surface area. With packed columns, however, there is a practical lower limit to the particle size because increased gas pressure is necessary to push the mobile phase through a column packed with smaller particles.

Nature of the Liquid Stationary Phase

The liquid stationary phase interacts with the substances being separated by a number of intermolecular forces: dipolar interactions, van der Waals forces, and hydrogen bonding (see the Essay "Intermolecular Forces in Organic Chemistry" on pages 127–131). These intermolecular forces contribute to the relative volatility of the adsorbed compounds and play important roles in the separation process.

As a general rule, a liquid phase provides the best separation if it is chemically similar to the compounds being separated. Nonpolar liquid coatings are used to separate nonpolar compounds, and polar liquid phases are best for separating polar compounds. In part, this rule is simply a manifestation of the adage "like dissolves like." Unless the sample dissolves well in the liquid phase, little separation may occur as the sample passes through the column. Table 20.1 lists some commonly used liquid stationary phases for both packed and capillary columns and gives their chemical composition.

Silicones, or polysiloxanes, are polymers with a silicon/oxygen backbone, which can have variation in the R groups attached to the silicon atoms. If all the R groups are methyl, the liquid phase is nonpolar. Substituting benzene rings (phenyl groups) for 5–10% of the methyl groups increases the polarity somewhat. Substitution of other functional groups for the methyl groups of polydimethylsiloxane provides a wide variety of stationary phases suited to almost any application.

Polyethylene glycol, commonly called Carbowax, and diethylene glycol succinate are polymers frequently used as liquid phases for separating polar compounds, which they dissolve in part by being good hydrogen bond acceptors.

Useful Temperature Range of a Liquid Phase

An important characteristic of a liquid phase is its useful temperature range. A stationary phase cannot be used under conditions where it decomposes or in which its vapor pressure is high enough so that it vaporizes from the column. All liquid stationary phases evaporate, or "bleed," if they are heated to a high enough temperature; this vaporized material then fouls the detector. Therefore, GC columns have specified temperature maxima.

TABLE 20.1	Common GC liquid stationary phases

Polarity of column	Maximum temperature (°C)	Chemical composition
Nonpolar	225	$$\left[\!\!-O-\underset{\underset{R}{|}}{\overset{\overset{R}{|}}{Si}}-O-\underset{\underset{R}{|}}{\overset{\overset{R}{|}}{Si}}-O-\!\!\right]$$ R = CH$_3$ Polydimethylsiloxane (methyl silicone)
Medium polarity	300	$$\left[\!\!-O-\underset{\underset{R}{|}}{\overset{\overset{R}{|}}{Si}}-O-\underset{\underset{R}{|}}{\overset{\overset{R}{|}}{Si}}-O-\!\!\right]$$ R = CH$_3$ or C$_6$H$_5$ Polymethylphenylsiloxane (methylphenyl silicone) Typically, 5–50% of the R groups are phenyl
Polar	250	$-O-CH_2-CH_2-O-CH_2-CH_2-O-$ Polyethylene glycol (Carbowax)
	200	$-O-CH_2-CH_2-O-CH_2-CH_2-O-\overset{\overset{O}{\|\|}}{C}-CH_2-CH_2-\overset{\overset{O}{\|\|}}{C}-O-$ Diethylene glycol succinate (DEGS polyester)

Selecting a Liquid Phase

The proper choice of a liquid stationary phase is often a trial-and-error process. Published experimental procedures often specify the type of column used for a GC analysis, but when you are designing your own experiment, you might have to make your own choices. Tables of appropriate liquid phases for specific classes of compounds can be found in the references at the end of the chapter.

20.3   Detectors

Two kinds of detectors are most often used in gas chromatographs: *flame ionization detectors* and *thermal conductivity detectors*. As you will see, the function of a detector is to "sense" a material and convert the sensing into an electrical signal.

Flame Ionization Detectors (FIDs)

Flame ionization is a highly sensitive detector system that is commonly used with capillary columns, where the amount of sample reaching the detector is substantially less than the amount emerging from a packed column. In a flame ionization detector, the organic compounds leaving the column are burned in a hydrogen/air flame

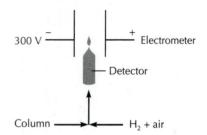

FIGURE 20.7 Flame
ionization detector.

(Figure 20.7). The combustion process produces ions that alter the
current output of the detector.

$$H_2 + O_2 + \text{organic} \xrightarrow{\Delta} CO_2 + H_2O + 2(\text{ions})^+ + (\text{ions})^- + e^-$$

$$\Sigma(\text{ions})^- + \Sigma e^- \rightarrow \text{electric current}$$

In the chromatograph, the electrical output of the flame is fed to an
electrometer, where the response can be recorded.

**Thermal
Conductivity
Detectors (TCDs)**

The older thermal conductivity detectors operate on the principle
that heat is conducted away from a hot body at a rate that depends
on the composition of the gas surrounding it. In other words, heat
loss is related to gas composition. The electrical component of a thermal conductivity detector is a hot wire or filament. Most of the heat
loss from the hot wire occurs by conduction through the gas and
depends on the rate at which gas molecules can diffuse to and from
the metal surface. Helium, the carrier gas often used with thermal
conductivity detectors, has an extremely high thermal conductivity.
Larger organic molecules are less efficient heat conductors because
they diffuse more slowly. With only carrier gas flowing, a constant
heat loss is maintained and there is a constant electrical output.
When an organic compound reaches the detector, the gas composition changes and causes the hot filament to heat up and its electrical
resistance to increase. The change in electrical resistance creates an
imbalance in the electrical circuit that can be recorded.

In practice, the filament of a thermal conductivity detector,
a tungsten/rhenium or platinum wire, operates at temperatures
from 200°C to over 400°C. An enlarged view of a common thermal
conductivity detector is shown in Figure 20.8. Thermal conductivity
detectors have the advantages of stability, simplicity, and the option
of recovery of the separated materials, but they have the disadvantage of low sensitivity. Because of their low sensitivity, they are
unsuitable for use with capillary columns.

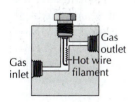

FIGURE 20.8 Thermal
conductivity detector.

20.4 Recorders and Data Stations

The recorded response of the detector's electrical signal as the
sample passes through it over time is the *chromatogram.* A typical
chromatogram for a mixture of alcohols, which plots the intensity

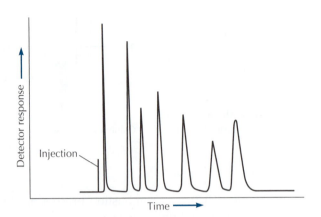

FIGURE 20.9 GC of a complex mixture of alcohols.

of the detector response against time, is shown in Figure 20.9. It shows the changes in the electrical signal as each component of the mixture passes through the detector. Notice that the later peaks are somewhat broader. This pattern is typical; the longer a compound remains on the column, the broader its peak will be when it passes through the detector.

Most modern gas chromatographs are equipped with a computer-based data station that allows manipulation of the results and their display on the recorder. Not only can the computer print out the chromatogram, but it automatically prints out a table containing the following data:

- Retention time in minutes for each peak
- Area under each peak
- Percentage of the total area

Retention Time

Under a definite set of experimental conditions, a compound always travels through a GC column in a fixed amount of time, called the *retention time.* The retention time for a compound, like the R_f value in thin-layer chromatography, is an important number, and it is reproducible if the same set of instrumental parameters is maintained from one analysis to another.

Figure 20.10 shows how retention times are determined. The distance from the time of injection to the time at which the peak maximum occurs is the retention time for that compound. Most computer-based data stations label the top of each peak on the chromatogram. If you are not using a data station, you can determine the retention time by measuring the distance from the injection to the peak on the chromatogram and dividing it by the recorder chart speed.

The retention time depends on many factors, starting with the compound's structure. After that, the kind and amount of stationary liquid phase used in the column, the length of the column, the carrier gas flow rate, the column temperature regimen, the solid support, and the column diameter are most important. To some extent, the sample size can also affect the retention time. **Always record these experimental parameters when you record a retention time in your lab notebook.**

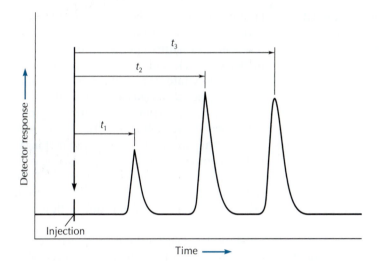

FIGURE 20.10
Measuring retention
times.

Peak Areas

If you are not using a modern, computer-based data station, see Section 20.8 for the determination of peak areas. If you are using a GC instrument with a data station, tick marks on the chromatogram show the limits of what is included in each calculated area printed out in the data table and a two- or three-letter code on the table of results tells which statistical method was used to calculate each peak area.

There may be small peaks that are not included in the data table because their areas are smaller than the area rejection setting of the data station. This feature makes it possible to ignore the noise that is present in any gas chromatogram. If it is important to know the area of a small peak, the area rejection setting can easily be changed.

Most computer-based data stations present data to many signifi-cant figures past the decimal point. In fact, the data are not nearly as precise as the number of significant figures implies and they cannot be duplicated to such a precise extent. **You should report the areas on data station printouts to only three or at most four significant figures**.

If a solvent is included in the sample being analyzed, its area may be a large part of the total integration area. If you are interested in only the relative percentages of two peaks on the chromatogram, you can calculate their relative amounts by using only their two areas, as well as their sum.

20.5 GC Operating Procedures

Modern GCs have great analytical power, but they are also complex. In order to operate a GC, you need to learn the functions of many buttons, switches, and dials, and you need to learn the sequential steps in the procedure for readying the gas chromatograph for an analysis. Your instructor or lab technician will probably already have set a number of the instrumental parameters, but you should always check that they have been set correctly. Your instructor will show you how to do these operations; the procedures vary for dif-ferent instruments.

Turning on the GC and Adjusting the Carrier Gas

First make sure that the chromatograph and the detector are heated and that the carrier gas is on and its pressure is properly set. The necessary pressure depends on the instrument and columns you are using, so check with your instructor before changing the pressure setting. Flow rates for capillary columns are often about 20 cm/sec if helium is the carrier gas and about 10 cm/sec for nitrogen. A proper flow rate of the carrier gas is important for effective GC separations.

When using helium as the carrier gas with a packed column that is 2 m long and 3 mm in diameter, a flow rate of 20–30 mL/min is common; for a 6-mm diameter column of the same length, 60–70 mL/min is typical. A convenient measure of the carrier gas flow rate in a packed-column chromatograph is made at the exit port by using a soap-film (bubble) flowmeter.

Choosing the Correct GC Column and Temperatures

Most modern gas chromatographs have two different columns, only one of which is operational at any time. You can activate the column of your choice with the flick of a switch. Decide whether a polar or nonpolar column is needed to separate the sample being analyzed and send the signal for that column to the detector. You also need to make sure the GC column oven temperature is set properly for your sample and that the detector and the injector port are at the correct temperatures. Temperature equilibration of the column can require 20–30 min on some chromatographs and only 3–4 min on others.

The column temperature can be programmed to increase during an analysis on modern capillary-column GCs. This feature gives the instrument far greater flexibility compared with the older isothermal gas chromatographs where a constant column temperature is used. Having the option of temperature programming allows you to begin a GC run at 50°C or so and then increase the column temperature at a selected rate per minute until it reaches a selected maximum temperature. Using temperature programming allows efficient and quick separation and analysis of organic mixtures whose components have widely different volatilities.

Turning on the Detector

If you are using a flame ionization detector, the hydrogen and air tanks must be regulated with the correct flow rates, and the flame must be lit. It's likely that your instructor will carry out this operation.

Before the sample is injected, the detector circuit must be balanced and the proper sensitivity (attenuation) chosen for the analysis. If you are using a thermal conductivity detector, you must turn on the inert carrier gas flow 2–3 min before the detector current is turned on. The thin metal filament of the detector can oxidize and burn out in the presence of oxygen, much like a tungsten lightbulb.

Sample Preparation and Amount Needed

Only volatile organic compounds can be injected into a gas chromatograph. Injection of organic or inorganic salts, acids, or bases can destroy the effectiveness of GC columns. When the instrument is ready and the sample is prepared, you can inject the sample. Gas chromatographs take very small samples; if too much is injected, poor separation will occur from overloading the column. **Injecting**

the proper amount of sample is the most important operation in obtaining a useful gas chromatogram. Consult with your instructor about sample preparation and size for the chromatographs in your laboratory.

Capillary-column GC. For a capillary-column GC, the sample must be in a dilute solution. A 2–5% solution in a volatile solvent, such as diethyl ether, works best. Usually one drop of a liquid or 20–50 mg of a solid sample diluted with 1 mL of the solvent is sufficient. Then only 0.5–1.0 μL of this dilute sample solution is injected into the GC with a microliter syringe. Even this amount of sample can overload a capillary column, so the injected mixture is split into two highly unequal flows and the smaller one is actually introduced into the column. A split ratio of 1:50 is not uncommon.

For some capillary chromatographs, it may be necessary to pull the plunger back until the entire sample is inside the syringe barrel before inserting the needle. Ask your instructor if this step is necessary for the chromatographs in your laboratory.

Packed-column GC. For a packed-column GC, 1–3 μL of a volatile mixture are directly injected through the rubber septum with a microliter syringe.

Injection Technique

Proper injection technique is important if you want to get well-formed peaks on the chromatogram. Using both hands, insert the needle all the way into the injection port and immediately push the plunger with a smooth, rapid motion (Figure 20.11). **Withdraw the syringe needle immediately after completing the injection.** This procedure ensures that the entire sample reaches the column at the same time and that there is minimal disturbance of the gas flow. If your GC is equipped with a computer-driven, automatic

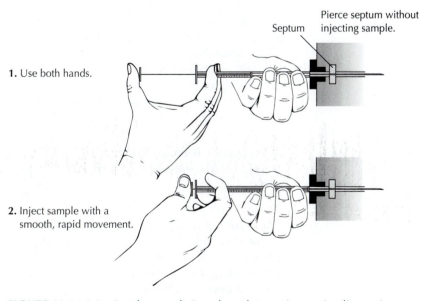

1. Use both hands.

2. Inject sample with a smooth, rapid movement.

Septum

Pierce septum without injecting sample.

FIGURE 20.11 Injecting the sample into the column using a microliter syringe.

digital integrator, simply press the start button after withdrawing the syringe needle.

If you are using a noncomputerized packed-column GC, the time of injection can be recorded in several ways. A mark can be made on the recorder base line just after the sample has been injected, but this action may be difficult to reproduce. If the GC has a thermal conductivity detector, a better method is to include several microliters of air in your syringe. The air is injected at the same time as the sample, and it comes through the column very quickly as the first tiny peak. Retention times can then be calculated using this air peak as the injection time. It is worth noting that an air peak cannot be used to mark the injection time with a flame ionization detector because air does not burn and thus gives no peak.

Completion of a Chromatographic Separation

After injection, wait for the peaks to appear on the moving chromatogram. If you are analyzing mixtures with a known number of components, you need to wait only until the last compound has come through the column. If the analysis involves an unknown mixture, it is sometimes difficult to know exactly how long to wait before injecting another sample because compounds with unexpectedly long retention times may still be present in the column. Determining the total analysis time for unknown mixtures is a matter of trial and error. Refer to Section 20.4 for interpretation of retention times and integration data on computer-driven data stations.

Keeping Microliter Syringes Clean

A microliter syringe has a tiny bore that can easily become clogged if it is not rinsed after use. If viscous organic liquids or solutions containing acidic residues are allowed to remain in the syringe, you may find that it is almost impossible to move the plunger. For this reason, a small bottle of acetone is often kept beside each GC instrument. Filling the syringe once or twice with acetone will normally suffice to clean it, if done directly after an injection.

During a series of analyses, it is unnecessary to rinse the syringe with acetone after each injection. This practice may even cause confusion if traces of acetone show up on the chromatogram. For multiple analyses, it is best to rinse the syringe several times with the next sample to be analyzed before filling the syringe with the injection sample.

When you have finished your injections, thoroughly rinse the microliter syringe with acetone.

Record Keeping

Attach your GC printouts firmly in your lab notebook, along with a notation of the experimental conditions under which the chromatograms were run. Record the following experimental parameters:

- Injection port temperature
- Column temperature and programmed temperature ramp (if applicable)
- Detector temperature
- Carrier gas flow rate
- Injection sample size
- Length of column and identity of its liquid stationary phase

20.6   Sources of Confusion and Common Pitfalls

Because of the complexity of modern GCs, many factors require careful attention in order to get good results. Using a GC requires thinking and problem-solving skills. Mastering the operation of a gas chromatograph—with the various adjustments of the column, injector port, and detector temperatures, the carrier gas flow rate, the hydrogen/air fuel mixture, and the sensitivity controls—can seem formidable. Yet it is worth the challenge because there are few other ways to get quantitative data on the composition of organic mixtures quickly.

A number of the instrumental parameters are likely to have been set by your instructor or lab manager, but you should always check to ensure that they have been set correctly. It pays to be careful and systematic in setting up the chromatograph, because if a key factor is overlooked, you must decide how long to wait before you abort a questionable experimental run that is already underway. **Remember that compounds from an earlier aborted run may still be in the GC column.** They may then come through the detector at unexpected times in the next chromatographic run.

I Injected My Sample but No Peaks Have Appeared on the Chromatogram. What Could Have Gone Wrong?

A lack of peaks is most often due to a failure to inject sufficient sample. The following two situations could also be responsible.

Injecting the sample into the wrong injection port. Most modern GCs have two columns, with a separate injection port for each. A lack of peaks in your chromatogram may be the result of the GC being activated to respond to the output from the wrong column, or perhaps you injected your sample into the wrong column.

Using a clogged syringe. It is easy for the very narrow bore microliter syringe to become clogged, and determining whether it is drawing properly can sometimes be difficult. Try drawing in cleaning solvent (e.g., acetone) to the full capacity of the syringe, and then eject the contents onto a piece of paper towel or tissue (keep the needle tip close to the paper). The tissue or towel should show a small wet spot if the syringe is functioning properly. The use of packed columns makes it easier to know if the syringe is working properly because a larger sample volume is injected. To prevent clogging, remember to rinse the syringe several times with cleaning solvent after you have finished your GC analyses.

My Peaks Run Together and I Don't Have Good Separation of the Components. What Went Wrong?

Poor resolution of the components of your mixture can be the result of a number of factors, including problems with sample injection and improper instrumental parameters. Developing good injection technique with a microliter syringe is probably the biggest challenge for the GC beginner.

Overloading the column. Too much sample can cause very broad peaks and poor separation of the components. This can be the result of either a sample that is too concentrated or the injection volume being too large. The correct size of the injections and concentration of the sample are crucial to success in GC.

Pushing the syringe plunger too slowly. If the leading edge of the sample reaches the column before the entire sample has vaporized in the injector port, the chromatogram will show multiple overlapping peaks and the run must be repeated.

Problems with instrument parameters. Poor separation can be the result of the column temperature being set too high, causing the sample to run through the column too quickly with insufficient interaction with the stationary phase. Also, the wrong type of liquid stationary phase may have been used, requiring a column change. Adjust only one parameter at a time until you have achieved a good separation of the mixture.

My Chromatogram Shows More Peaks Than I Expect. What Should I Make of This Result?

There are a number of things that could have caused you to see additional peaks on your chromatogram.

Making a second injection before the first chromatogram is complete. If you make an injection but then decide that something was not quite right, it is tempting to stop the run and immediately inject again, which will result in extra peaks. **Remember that once material is on the GC column, you cannot stop the run—the compounds must be allowed to fully elute before the next injection is made.** Injecting a second sample too soon can cause the components from the two injections to intermingle, resulting in a doubling of the total number of peaks.

Observing overlapping and repeating peaks. If the sample is not injected immediately after the needle has been inserted into the injection port, there can be vaporization of the sample from the needle into the hot port before the rest of the sample is injected. This results in the sample entering the column in portions instead of in a single burst. The effect can be mitigated by drawing some air into the needle after filling the syringe with sample in order to provide a buffer of air between the sample and the injection port.

Observing trace impurities. Capillary-column GC is very sensitive and it is common to see many small peaks on your chromatogram from trace impurities, even if you are analyzing a "pure" compound. There are virtually always tiny amounts of impurities in pure compounds. A GC chromatogram can be a vivid reminder of the immense size of Avogadro's number. Many trillions of molecules pass through the detector of a GC in every chromatographic run. If the detector is sensitive enough, the trace impurities will show up. Often, you can safely ignore them.

20.7　Identification of Compounds Shown on a Chromatogram

GC analysis can quickly assess the purity of a compound, but as with thin-layer chromatography, a compound cannot be identified by GC unless a known sample is available to use as a standard. Comparison of retention times, peak enhancement, and spectroscopy are among the methods used to identify the components of a mixture.

Comparison of Retention Times

One method of identification compares the retention time of a known compound with the peaks on the chromatogram of the sample mixture. If the operating conditions of the instrument are unchanged, a match of the reference compound's retention time to one of the sample peaks may serve to identify it. This method will not work for a mixture in which the identity of the components is totally unknown, because several compounds could have identical retention times.

Peak Enhancement

When mixtures containing known compounds are being analyzed, *peak enhancement* serves as a method for identifying a peak in the chromatogram. The sample being analyzed is "spiked" with a drop of the known compound and the mixture injected into the chromatograph. If the known compound that is added is identical to one of the compounds in the original mixture, its peak area is enhanced relative to the other peaks on the chromatogram (Figure 20.12).

Mass Spectrometry

Positive identification of the compounds in a completely unknown mixture requires the pairing of GC methods with a spectrometric method such as mass spectrometry (MS), where a mass spectrometer serves as the GC detector. In a GC-MS the two instruments are interfaced so that the separated components pass directly from the chromatograph into the spectrometer (see page 444).

20.8   Quantitative Analysis

Gas-liquid chromatography is particularly useful for quantitative analysis of the components in volatile mixtures. A direct comparison of relative peak areas on the chromatogram often gives a good approximation of relative amounts of the compounds.

Determination of Peak Areas

One great advantage of GC over other chromatographic methods is that approximate quantitative data are almost as easy to obtain as information on the number of components in a mixture. If we assume equal response by the detector to each compound, then the relative amounts of compounds in a mixture are proportional to their peak

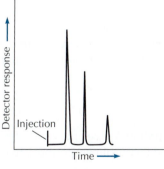

(a) Original chromatogram

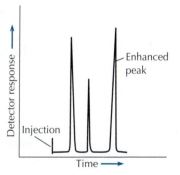

(b) Chromatogram after addition of a known compound identical to a compound in the sample

FIGURE 20.12
Identification by the peak enhancement method.

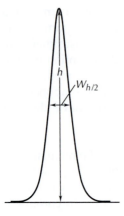

FIGURE 20.13
Determining peak area: h = height; $W_{h/2}$ = width at half-height.

areas. Most peaks are approximately the shape of either an isosceles or a right triangle, whose areas are simply $A = \frac{1}{2}$ base × height. Measuring the base of most GC peaks is difficult because abnormalities in their shapes usually occur there. A more accurate estimate of peak area is A = height × width at half-height (Figure 20.13).

Digital integrators, common on most modern chromatographs, determine peak areas electronically. Chromatograms produced by these recorders include a table of data that lists both retention times and relative peak areas.

Internal normalization is the easiest method for calculating the percentage composition of a mixture. **The percentage of a compound in a mixture is its peak area divided by the sum of all peak areas.** If you have a two-component mixture, use the following equations:

$$\% \text{ compound 1} = \frac{\text{area}_1}{\text{area}_1 + \text{area}_2} \times 100$$

$$\% \text{ compound 2} = \frac{\text{area}_2}{\text{area}_1 + \text{area}_2} \times 100$$

Relative Response Factors

For accurate quantification of a GC analysis, the response of each component to the detector must be determined from known samples. Analysis of standard mixtures of known concentration must be carried out and a correction factor, called a *response factor (f)*, must be determined for each compound. The area under a chromatographic peak, A, is proportional to the concentration, C, of the sample producing it; the response factor is the proportionality constant.

$$A = fC \qquad (1)$$

Response factors can be determined as either weight factors or mole factors, depending on the units of concentration used for the standard samples.

For a two-component system, the response-factor equation for each component is

$$A_1 = f_1 C_1 \qquad (2)$$

$$A_2 = f_2 C_2 \qquad (3)$$

The relative response factor of compound 1 to compound 2 can be determined by dividing equation 2 by equation 3:

$$\frac{A_1}{A_2} = \frac{f_1}{f_2} \times \frac{C_1}{C_2} \qquad (4)$$

Rearranging equation 4 gives the ratio of response factors, f_1/f_2, the relative response factor of compound 1 to compound 2:

$$\frac{f_1}{f_2} = \frac{A_1}{A_2} \times \frac{C_2}{C_1} \qquad (5)$$

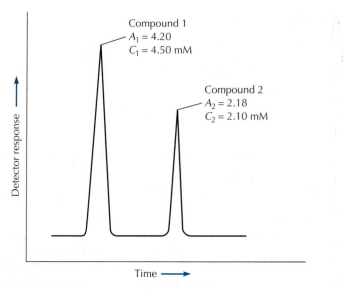

FIGURE 20.14 Chromatogram of a standard mixture containing known concentrations of two compounds.

Using data from the chromatogram shown in Figure 20.14 as an example, equation 5 can be used to calculate the molar response factor of compound 1 relative to compound 2; compound 2 is arbitrarily assigned a response factor of 1.00.

$$\frac{f_1}{f_2} = \frac{4.20}{2.18} \times \frac{2.10}{4.50} = \frac{0.899}{1.00}$$

Therefore, the molar response factor for compound 1 is 0.899 relative to 1.00 for compound 2.

Once relative molar response factors have been determined, the composition of a mixture can be calculated from the areas of the peaks on a chromatogram. Table 20.2 shows how molar response factors (designated M_f) can be used to determine the corrected mole percentages of a sample containing compound 1 and compound 2; Table 20.2 also compares these results to the uncorrected composition. The differences between the uncorrected and corrected compositions illustrate the necessity of using response-factor corrections for accurate quantitative analysis.

| TABLE 20.2 | Molar percentage data for a two-compound mixture uncorrected and corrected for molar response factors, M_f |||||

Compound	Area (A) (arbitrary units)	Uncorrected % (A/118.4) × 100	M_f	A/M_f	Corrected mol % (A/M_f) × (100/124.0)
Compound 1	50.2	**42.4**	0.899	55.8	**45.0**
Compound 2	68.2	**57.6**	1.00	68.2	**55.0**
Total	118.4	100	—	124.0	100

Further Reading

Grob, R. L.; Barry, E. F. *Modern Practice of Gas Chromatography*, 4th ed.; Wiley: New York, 2004.

Miller, J. M. *Chromatography: Concepts and Contrasts*, 2nd ed.; Wiley: Hoboken, NJ, 2005.

Skoog, D. A.; Holler, F. J.; Crouch, S. R. *Principles of Instrumental Analysis*, 6th ed.; Thomson Brooks/Cole: New York, 2007.

Questions

1. Why is a GC separation more efficient than a fractional distillation?
2. What characteristics must the liquid stationary phase have?
3. How do (a) the flow rate of the carrier gas and (b) the column temperature affect the retention time of a compound on a GC column?
4. Describe a method for identifying a compound using GC analysis.
5. Describe a method for identifying a compound purified by and collected from a gas chromatograph.
6. If the resolution of two components in a GC analysis is mediocre but shows some peak separation, what are two adjustments that can be made in the operating parameters to improve the resolution (without changing columns or instruments)?
7. Suggest a suitable liquid stationary phase for the separation of (a) ethanol and water; (b) cyclopentanone (bp 130°C) and 2-hexanone (bp 128°C); (c) phenol (bp 182°C) and pentanoic acid (bp 186°C). Explain your reasoning.
8. A GC analysis of a mixture containing two compounds dissolved in pentane produced the following peak areas: compound 1 = 23.2, compound 2 = 56.3, and pentane = 203.5. Relative M_f values are 0.85 for compound 1 and 1.00 for compound 2. Calculate the corrected mol % for compounds 1 and 2 in the mixture.

5

Spectrometric Methods

Essay—The Spectrometric Revolution

Throughout the study of organic chemistry, you are asked to think in terms of molecular structure because structure determines the properties of molecules. The connection between structure and reactivity is a central principle of organic chemistry. An experienced chemist can anticipate many of the physical and chemical properties of various compounds by simply looking at their structures.

A few decades ago, the structure of an organic compound was discovered largely by time-consuming and sometimes ambiguous chemical methods. It took decades to figure out the structures of important compounds such as cholesterol and morphine. Modern organic spectrometric methods have revolutionized how the structures of complex organic molecules are determined. These techniques are based in large part on spectroscopy, the absorption of radiation from various portions of the electromagnetic spectrum.

Different spectroscopic techniques provide different snapshots of molecules because the energy of radiation affects the type of molecular excitation that takes place. For example, infrared (IR) radiation excites the vibrational modes of molecules, whereas higher energy ultraviolet and visible (UV-VIS) radiation excites valence electrons into higher energy states. If you were to irradiate your finger briefly with IR radiation, you would feel the heat from this light source but it would not produce long-term damage to your tissue. In contrast, UV radiation causes electronic excitations that can damage your DNA.

Combining different information from several spectroscopic techniques gives a more complete picture of the molecule, enabling us to deduce the structure. What used to take years or months can now often be done in a few days. For organic molecules with molecular weights of 300 or less, the job can often be done within an hour or two. The spectroscopic revolution has had a pronounced effect on how organic and biological chemistry are done. It has allowed chemists to spend more time on the fascinating and important issues of molecular recognition and transformation.

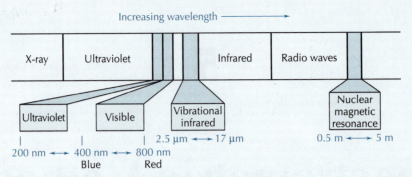

Portions of the electromagnetic spectrum used in organic chemistry.

Infrared Spectroscopy (IR)

The infrared region of the electromagnetic spectrum causes molecules to vibrate and heats up the irradiated matter. IR absorption is responsible for the greenhouse effect. Carbon dioxide, methane, and water molecules in the atmosphere absorb IR radiation, resulting in more heat at the surface of the earth. In chemical analysis, IR spectroscopy provides valuable information concerning the functional groups present in molecules. In addition, IR spectra can be used as fingerprints to identify particular compounds.

Nuclear Magnetic Resonance Spectroscopy (NMR)

NMR spectroscopy is arguably the most useful technique for organic structure analysis. It uses radio waves in the presence of a strong magnetic field. NMR is also at the heart of magnetic resonance imaging (MRI), a powerful and noninvasive medical diagnostic probe of soft tissue. NMR provides snapshots of the protons (^{1}H nuclei) or carbons (^{13}C nuclei) in organic structures. By understanding how the energy of NMR signals and their coupling patterns relate to chemical structure, we can determine the structures of organic compounds and complex biopolymers, such as nucleic acids, proteins, and carbohydrates. As you use this powerful technique to determine structure, you will see why NMR is the major focus of this spectrometric methods section.

Ultraviolet and Visible Spectroscopy (UV-VIS)

Ultraviolet and visible spectroscopy continue to be important methodologies in organic chemistry, but less for structure determination than for the analyses of organic and biochemical mixtures, especially in high-performance liquid chromatography (HPLC) detectors.

Mass Spectrometry (MS)

MS differs from the spectroscopic methods discussed in Part 5 because it does not involve the interaction of electromagnetic radiation with molecules. Instead, molecules are ionized by a variety of techniques and the resultant ions are manipulated by electric and/or magnetic fields. Because the mass of the ions affects their interaction with these fields, MS allows chemists to determine the molecular weight of a compound, and high-resolution MS can determine a compound's molecular formula as well. The fragmentation pattern of an ionized molecule also provides data that can assist in the identification of the compound. Like IR spectroscopy, MS can be used to provide a fingerprint that can pin down the structure of a molecule. MS is particularly useful when complex samples are separated in a gas chromatograph and a mass spectrometer is used as the detector (GC-MS).

21

INFRARED SPECTROSCOPY

Infrared (IR) spectroscopy is one of the two important spectroscopic techniques for determining the structures of organic molecules. It provides a rapid and effective method for identifying the presence or absence of simple functional groups. When infrared energy is passed through a sample of an organic compound, absorption bands are observed. The *mid-infrared*, extending from 4000 to 600 cm^{-1}, is the region of most interest to organic chemists because it is where absorptions from typical organic compounds appear. The positions of these IR absorption bands have been correlated with types of chemical bonds, which can provide key information about the nature of functional groups in the sample.

If Chapter 21 is your introduction to spectroscopic analysis, read the Essay "The Spectrometric Revolution" on pages 309–310 before you read Chapter 21.

21.1   IR Spectra

In an IR spectrum, energy measured as frequency or wavelength is plotted along the horizontal axis, and the intensity of the absorption is plotted along the vertical axis. There are several different formats for plotting the data, depending on the scales used for the axes. Figure 21.1 shows examples of the IR spectra of cyclopentanone recorded on two different IR spectrometers.

The horizontal scale in Figure 21.1a is linear in wavelength of the infrared radiation, which is the default axis used by older IR spectrometers. Many of the original libraries of infrared spectra were plotted using this format. The horizontal scale in Figure 21.1b is linear in wavenumbers, the standard frequency scale for infrared radiation used by most modern IR spectrometers. Microcomputers incorporated into modern IR spectrometers can quickly interchange data between the two formats. The shapes of the absorption bands appear quite different in Figures 21.1a and 21.1b, but their actual positions in the spectrum are the same. In the two IR spectra of cyclopentanone, the major absorption band appears at 5.72 μm in Figure 21.1a and at 1747 cm^{-1} in Figure 21.1b. This IR band is characteristic of the carbonyl group (C=O), one of the major functional groups in organic chemistry.

21.2   Molecular Vibrations

The atoms making up a molecule are in constant motion, much like balls at the ends of springs. Covalent bonds act as the springs that connect the atomic nuclei. The movements of the atoms relative to each other can be described as vibrations, and infrared spectroscopy has been called *vibrational spectroscopy*. The photons of IR radiation absorbed by an organic molecule have just the right amount of energy to stretch or bend their covalent bonds. The energy of infrared radiation is on the order of 8–40 kJ/mol (2–10 kcal/mol). This is not enough energy to break a covalent bond, but it is enough to increase the amplitude of bond vibrations. When infrared radiation

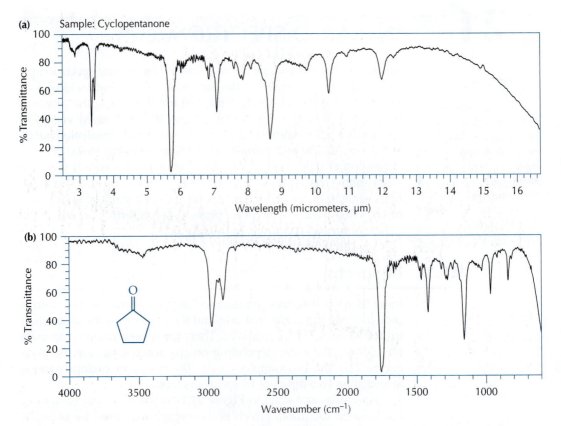

FIGURE 21.1 Infrared spectra of cyclopentanone recorded with (a) the horizontal (energy) scale linear in wavelength (micrometers) and (b) the horizontal scale linear in wavenumbers (frequency).

is absorbed, the sample becomes warm as its molecules increase their kinetic energy. This is how infrared heat lamps work.

An absorption band appears in an infrared spectrum at a frequency where a molecular vibration occurs in the molecule. Energy levels of molecular vibrations are quantized, which means that only infrared energy with the same frequencies as the molecular vibrations can be absorbed. The energy levels available to a molecular vibration are expressed as

$$E = h\nu_0 (v + \frac{1}{2}) \quad \text{for} \quad v = 0,1,2,3 \dots$$

where h = Planck's constant and ν_0 = the zero-point vibrational level of the bond. The energy (ΔE) of the absorbed radiation that will promote a vibration of frequency (ν) from one energy level to the next energy level is

$$\Delta E = h\nu$$

The frequency (ν) and wavelength (λ) of light are related by

$$\nu = c/\lambda$$

where c = the speed of light in a vacuum. Substituting this relationship into the equation for the absorbed radiation yields

$$\Delta E = hc(1/\lambda)$$

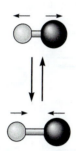

FIGURE 21.2
Fundamental stretching vibrational mode of a diatomic molecule.

The quantity $(1/\lambda)$ is called the *wavenumber* $(\bar{\nu})$ and is usually expressed in units of reciprocal centimeters (cm^{-1}). A wavenumber defines the number of wave crests per unit length. It is proportional to the frequency as well as the energy of an IR absorption.

$$\Delta E = hc\,(\bar{\nu})$$

An IR absorption band is often called a *peak,* and its maximum is defined as the position of maximum absorption in wavenumber units. Frequency in units of wavenumbers, cm^{-1}, and wavelengths in units of micrometers, μm (10^{-6} meters, called microns in the older literature), can be interconverted by the following relationship:

$$cm^{-1} = \frac{10,000}{\mu m}$$

Fundamental Molecular Vibrations

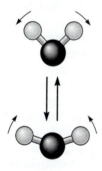

FIGURE 21.3
Fundamental bending vibrational mode of a triatomic molecule.

There are two kinds of fundamental molecular vibrations: stretching and bending. In a *stretching vibration,* the distance between two atoms increases and decreases in a rhythmic manner, but the atoms remain aligned along the bond axis. Figure 21.2 shows a *symmetric stretching vibration* in which the atoms stretch in and out simultaneously. In a *bending vibration,* the positions of atoms change relative to the bond axis, as shown in Figure 21.3. A nonlinear molecule made up of n atoms has $3n - 6$ possible fundamental stretching and bending vibrations.

EXERCISE

Water (H_2O) is a nonlinear molecule consisting of three atoms. (a) How many fundamental vibrations does it have? (b) Describe them.

Answer: (a) Water has three fundamental vibrations. Two are stretching vibrations and one is a bending vibration. (b) The vibrations are shown in Figure 21.4. The first is a symmetric stretching vibration. The second stretching vibration is an **asymmetric stretching vibration** in which one hydrogen atom moves out as the other hydrogen atom moves in. The bending vibration involves a kind of scissoring motion in which the H–O–H bond angle changes back and forth.

For molecules containing many atoms, there are numerous fundamental vibrations. The stretching and bending vibrations of a methylene (CH_2) group are shown in Figure 21.5.

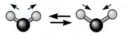  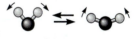

Symmetric stretching   Asymmetric stretching   Scissoring

FIGURE 21.4 The three fundamental vibrational modes of water.

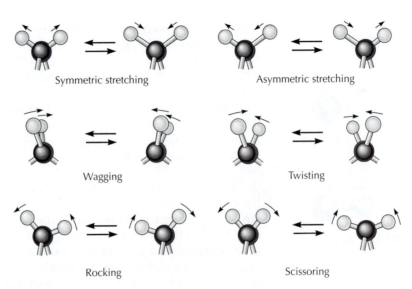

FIGURE 21.5 Vibrational modes of the methylene group (CH_2).

Complexity of IR Spectra

Organic compounds, which contain 10 or 20 atoms or more, can manifest substantial numbers of IR peaks, and their spectra can be complex. The total number of observed absorption bands is generally different from the total number of possible fundamental vibrations. Some fundamental vibrations are not IR active and do not absorb energy. However, additional absorption bands, which occur as a result of overtone vibrations, combination vibrations, and the coupling of vibrations, more than make up for the decrease.

Overtone bands may be observed when fundamental vibrations produce intense absorption peaks. An overtone peak results from the change of more than one vibrational energy level when infrared energy is absorbed. They are usually much less intense than fundamental vibrations, and they generally appear at frequencies a little less than twice the fundamental frequencies. In Figure 21.6 the weak peak at 3395 cm^{-1} is an overtone of the strong C=O stretching vibration at 1705 cm^{-1} in the IR spectrum of benzaldehyde.

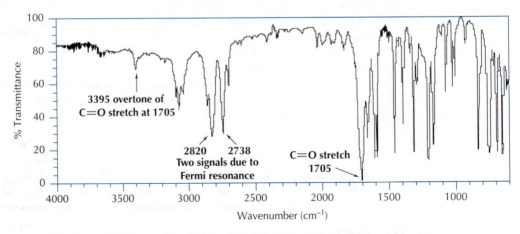

FIGURE 21.6 Overtones and Fermi resonance in the IR spectrum of benzaldehyde.

A coupling interaction called *Fermi resonance* most often occurs between fundamental and overtone vibrations that are at nearly the same frequency. Fermi resonance is particularly apparent in the C–H stretching vibration of aldehydes. The interaction of the overtone and the fundamental vibration causes the intensity of the fundamental to decrease and the intensity of the overtone band to increase. This results in two peaks of roughly equal intensity in the IR spectrum, as shown by the two peaks at 2820 and 2738 cm^{-1} in Figure 21.6.

Combination bands appear at frequencies that correspond to sums and differences of two or more fundamental vibrational frequencies. The intensities of combination bands are usually less than the intensities of fundamental vibrations.

Fortunately, many of the peaks in an IR spectrum can usually be ignored. The large number of fundamental vibrations, their overtones, and combinations of vibrations make it far too difficult to understand quantitatively entire IR spectra of most organic compounds. But as you will see, **IR spectra can easily yield a great deal of qualitative information about functional groups.** Moreover, the complexity of an IR spectrum imparts a unique pattern for each compound, allowing the spectrum to be used as a fingerprint for identification.

Correlation of Peaks with Specific Bond Vibrations

The absorptions corresponding to specific molecular vibrations appear in definite regions of the IR spectrum, regardless of the particular compound. For example, the stretching region of O–H bonds in all alcohols appears at nearly the same frequency. In the same way, the C=O vibrations of all carbonyl compounds appear within a narrow frequency range.

There are four important factors that determine the frequency and intensity of IR peaks:

- Type of vibration—stretching or bending
- Strength of the bond connecting the atoms, particularly the bond order
- Masses of the atoms attached by the covalent bonds
- Electronegativity difference between the atoms or groups of atoms in a bond

Type of vibration. In general, the stretching of covalent bonds takes more energy than bending vibrations. Thus, stretching vibrations appear at higher frequencies.

Type of vibration	Frequency (cm^{-1})
C–H stretching	3000–2850
–CH_2–bending	1480–1430

Bond order. Bond order is simply the amount of bonding between two atoms. For example, the bond order between carbon atoms increases from one to two to three for ethane (CH_3–CH_3), ethene (ethylene, CH_2=CH_2), and ethyne (acetylene, HC≡CH), respectively. In general, the higher the bond order, the greater the energy required to stretch the bond. Higher bond order produces a higher-frequency IR absorption.

Bond order	Type of bond	Stretching frequency (cm^{-1})
1	C–C, C–O, C–N	1300–800
2	C=C, C=O, C=N	1900–1500
3	C≡C, C≡N	2300–2100

Atomic mass. The frequency of the IR absorption also relates to the atomic masses of the vibrating atoms. Covalent bonds to hydrogen occur at high frequencies compared to bonds between heavier atoms—just as a light weight on a spring tends to oscillate faster than a heavy weight.

Type of bond	Stretching frequency (cm^{-1})
O–H	3650–2500
N–H	3500–3150
C–H	3300–2850

Electronegativity differences and peak intensities. Bond polarity does not significantly affect the position of an IR absorption, but it greatly influences the intensity of IR peaks. If a stretching or bending vibration induces a significant change in the dipole moment, an intense IR peak will result. Thus, when bonds are between atoms with different electronegativities, such as C–O, C=O, and O–H, the IR stretching vibrations are very intense. A symmetric molecule such as ethylene, on the other hand, does not show any absorption band for the C=C stretching vibration.

The intensity (peak size) of an IR absorption can be reported in terms of either transmittance (T) or absorbance (A). *Transmittance* is the ratio of the amount of infrared radiation transmitted by the sample to the intensity of the incident beam. Percent transmittance is $T \times 100$.

A properly prepared sample produces an IR spectrum in which the most intense peak nearly fills the vertical height of the chart. However, it is important that the strongest peak in an IR spectrum is above 0% transmittance (5–10% is good) so that the frequency of the peak maximum can be measured accurately. In practice, peak intensities are reported in a qualitative fashion depending on their magnitude. The most intense peaks are termed *strong (s)*; smaller peaks are called either *medium (m)* or *weak (w)*. Peaks can also be described as *broad (br)* or sharp.

21.3   IR Instrumentation

There are two major classes of instruments used to measure IR absorption: Fourier transform (FT) spectrometers and dispersive spectrometers. *Dispersive spectrometers* were developed first and for a long time were the standard infrared instruments. The advent of powerful and inexpensive microcomputers has allowed the development of *Fourier transform infrared (FTIR) spectrometers*.

Dispersive Spectrometers

In a dispersive IR spectrometer, the source of radiation, often a heated filament, provides a beam of IR radiation that is split into two beams. The beams are directed by mirrors through both sample and reference cells. The sample and reference beams are alternately selected for measurement by means of a special rotating sector mirror, which allows the selected beam components to be recombined into a single beam. This beam is then focused onto a diffraction grating, which separates the beam into a continuous band of infrared frequencies. A slit allows only a narrow range of these frequencies to reach the detector. By continuously changing the angle of the diffraction grating, the entire infrared spectrum can be scanned.

Fourier Transform Spectrometers

FTIR spectrometers gather data for all IR wavelengths at the same time. A picture of a modern FTIR spectrometer, as well as a simplified diagram, is shown in Figure 21.7. The diagram in Figure 21.7b shows that infrared radiation from a heated source is directed to a *beam splitter,* a thin film of the element germanium sandwiched between two highly polished plates of potassium bromide. The beam splitter separates the radiation into two beams. One beam is reflected off the beam splitter and directed to a fixed mirror. The other beam is transmitted through the beam splitter and directed to a moving mirror, which is controlled by a laser. The mirrors reflect

(a)

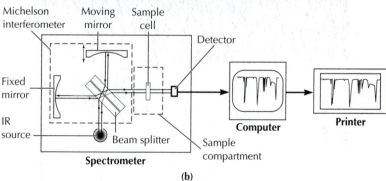

(b)

FIGURE 21.7 (a) An infrared spectrometer. (b) Diagram of a single-beam FTIR spectrometer. The interior of the instrument is often isolated from the ambient environment by purging with dry nitrogen or dry, carbon dioxide-free air.

their respective beams of infrared energy back to the beam split-ter, where the beams recombine. The two beams travel different distances to the mirrors, so their frequencies become out of phase. The constructive and destructive combination of the out-of-phase frequencies produces an *interferogram.* The beam splitter and mir-ror assembly is known as a ***Michelson interferometer.***

The interferogram, which contains information about every infrared frequency, is an array of signal intensities that reveals the difference in the two optical paths. The beam of infrared energy, encoded as an interferogram, is directed through a sample to the detector. On interacting with the sample, specific frequencies of infrared energy are absorbed through excitation of molecular vibrations. Fourier transform mathematics is then used to sort out the frequencies of infrared energies encoded in the modified interferogram. The result is an infrared spectrum plotted as an array of intensities versus frequencies measured in cm^{-1}.

In actual practice, two scans are required—a scan of the empty sample compartment referred to as the ***background scan*** and a scan with the sample in the beam of infrared energy. The background scan contains signals due to water vapor and gaseous carbon diox-ide in the atmosphere, as well as the emission profile of the source and the film coatings of the optics. The background spectrum is sub-tracted from the sample spectrum to produce a spectrum displaying only absorptions due to the sample. The steps involved in creating a spectrum from the data are outlined in Figure 21.8.

Although it is more complicated than dispersive IR spectros-copy, there are numerous advantages to the FTIR method. Results of multiple scans can be combined to average out random noise, and excellent spectra can be obtained rapidly from very small samples.

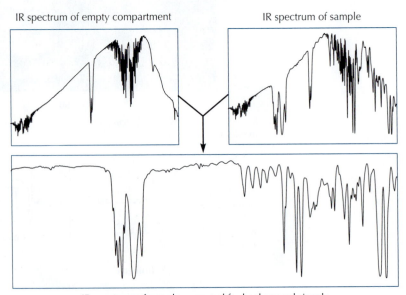

IR spectrum of empty compartment   IR spectrum of sample

IR spectrum of sample corrected for background signals

FIGURE 21.8 The collection and processing of data required for the creation of an infrared spectrum with a single-beam FTIR spectrometer.

FTIR spectrometers have few mirror surfaces, and because more energy gets to the detector, they are much more sensitive. Also, the resolution of the spectrum from an FTIR spectrometer is much higher. FTIR data are digitized; the quality of a spectrum can often be improved by baseline correction or the subtraction of peaks resulting from impurities.

21.4   Operating an FTIR Spectrometer

An FTIR spectrometer is a robust, modern instrument with many capabilities, but it must be used with care and respect. The most difficult step in taking the IR spectrum is often the preparation of the sample.

If you are using the attenuated total reflectance (ATR) accessory, see Section 21.6; otherwise use the following operating procedure.

1. Prepare the sample. A number of different methods for preparing samples for transmittance IR spectra are described in Section 21.5.
2. Briefly open the sample compartment and confirm that there is nothing in the sample beam. Close the compartment.
3. Run a background scan. The data are collected, processed, and stored in the instrument's computer memory. The instrument indicates when this operation is completed.
4. Briefly open the sample compartment and place the sample in the sample beam. Close the compartment.
5. Run a sample scan. The data are collected and processed. The background scan is automatically subtracted from the sample scan. The result, an infrared spectrum of the sample, is displayed on the monitor.
6. Use the instrument's software to mark the frequency of each major peak in the region of 4000–1500 cm^{-1}. Having the exact frequencies (wavenumbers) of these peaks on the printed spectrum can be helpful in analyzing it.
7. Format the spectrum and print out a copy for analysis and inclusion in your laboratory notebook.

21.5   Sample Preparation for Transmission IR Spectra

IR spectra can be obtained from liquid, solid, or gas samples. Almost all the IR spectra shown in this book are *transmission spectra.* Solid and liquid compounds are often prepared as thin films that allow infrared radiation to pass through them. Various additional methods for preparing samples of solids and liquids for transmission IR spectra are also described in this section. A newer method for obtaining IR spectra—attenuated total reflectance (ATR)—works in quite a different manner and makes the preparation of IR samples, particularly solids, much easier (see Section 21.6). Gas samples, which require a special cell for sampling, are encountered infrequently in organic chemistry and are not included in our discussion.

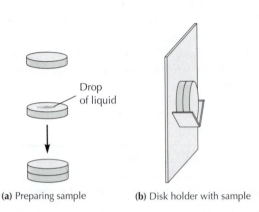

(a) Preparing sample   (b) Disk holder with sample

FIGURE 21.9 Preparation of thin-film sample for IR spectroscopy.

FIGURE 21.10 IR sample disk holder for low-viscosity liquids. The holder slips into a bracket on the IR spectrometer.

Sample Cells for Transmission IR Spectra

The windows of the sample cells used for transmission spectra must be transparent to IR radiation in the mid-infrared region. Because glass absorbs IR radiation, it cannot be used. Most cells are made from alkali halides, in particular polished sodium chloride disks that, for the most part, are transparent in the mid-infrared region.

It is important to be aware that **alkali halide sample cells are very susceptible to water damage,** and care must be taken to ensure that all samples are completely dry. Water etches and clouds the surface of cells and disks, rendering them useless. Also, touching the polished surfaces of salt disks leaves indelible fingerprints from skin moisture and oils. **NaCl disks should be handled only by the edges.** The disks are much softer than glass and they break easily if dropped even a short distance. When preparing an IR sample, avoid touching the polished surface of a sodium chloride disk with a glass pipet because the pipet will nick and scratch the surface. The only way to remove nicks, scratches, and fingerprints is to repolish the disk.

Thin Films for Liquid Compounds

A thin film pressed between NaCl disks is the most convenient method for preparing a liquid for IR analysis (Figure 21.9). A drop of *neat* sample (liquid with no added solvent) is placed on one disk; the other disk is placed on top of the drop. The disks are gently rotated and then gently squeezed together to form a film approximately 0.01 mm thick. The sample in the NaCl disk sandwich is placed in a holder that is subsequently positioned in the sample compartment of the IR spectrometer. When the sample has a low viscosity, the holder shown in Figure 21.10 is a better choice because it keeps the sample film tightly in contact with the salt disks.

SAFETY PRECAUTION

Wear gloves and handle all solvents in a fume hood.

Steps in Preparing and Using a Thin Film

1. **Clean the disks with a dry solvent**—acetone or dichloromethane.
2. Place a folded tissue on the hood bench and place one disk on top of the tissue pad.

3. Using a Pasteur pipet, place one drop of the liquid sample on the center of the disk. Be careful not to touch the surface of the disk with the pipet.

4. Place the second disk on top of the first and gently rotate it; then gently press the disks together.

5. Obtain the IR spectrum.

6. Clean the disks with a **dry** solvent—acetone or dichloromethane. Store the disks in a desiccator to protect them from moisture.

Cast Films for Solid Compounds

A thin film of solid can be prepared by placing a drop of a concentrated solution of the compound in the center of a clean sodium chloride disk and allowing the solvent to evaporate. The best solvent to use for this solution is one that has a high vapor pressure at room temperature and that does not dissolve NaCl. Diethyl ether, dichloromethane, and ethyl acetate work well; methanol, ethanol, and water must be avoided. For best results the salt disk must have a smooth, polished surface because scratched and pitted disks lead to uneven distribution of the sample.

SAFETY PRECAUTION

Wear gloves and handle all solvents in a fume hood.

Steps in Preparing and Using a Cast Film IR Sample

1. In a small test tube, prepare 0.3–0.5 mL of a 10–20% sample solution in a volatile organic solvent. Stopper the test tube.

2. Clean a NaCl disk with a **dry** solvent—acetone or dichloromethane, but **not** methanol or ethanol.

3. Place a folded tissue on the hood bench. Place the clean disk on top of the tissue pad. Make sure the disk is level.

4. Using a Pasteur pipet, place one drop of the sample solution at the center of the disk. Be careful not to touch the surface of the disk with the pipet.

5. Allow the solvent to completely evaporate. It may be necessary to repeat steps 4 and 5 up to four or five times to build up a film of the compound thick enough to produce an acceptable IR spectrum.

6. When the solvent has evaporated completely, place a clean second disk on top of the first and gently rotate it; then gently press the disks together.

7. Place the NaCl disk assembly in a sample holder like that shown in Figure 21.9 or 21.10.

8. Obtain the IR spectrum.

9. Clean the disks with a **dry** solvent—acetone or dichloromethane. Store the disks in a desiccator to protect them from moisture.

10. If your sample compound is especially valuable, you can wash the sample from the NaCl disk into the sample test tube and then evaporate the solvent from the remaining solution to recover the compound.

Mulls for Solid Compounds

A *mull* used for IR samples is not a true solution but is a fine dispersion of a solid organic compound in a viscous liquid. The most common liquids used for IR mulls are Nujol (a brand of mineral oil

TABLE 21.1	Absorption regions of common mulling compounds
Compound	Absorption region (cm^{-1})
Fluorolube	1300–1080
	1000–920
	910–870
	<670
Nujol	3000–2800
	1490–1450
	1420–1360
	750–720

that is a mixture of long-chain alkanes) and Fluorolube (a mixture of completely fluorinated alkanes). The fluorinated mulling substances are often used for more polar compounds. Unfortunately, neither Nujol nor Fluorolube is transparent over the entire IR region. Both display IR peaks that may obscure peaks due to the dispersed compound (Table 21.1). The spectrum of Nujol exhibits only C–H stretching and bending absorptions. Thus, Nujol does not obscure most IR peaks due to the functional groups found in organic compounds. However, the preparation of a good Nujol mull requires care and practice to prevent the problems discussed in Section 21.11, page 342.

SAFETY PRECAUTION

Wear gloves and handle all solvents in a fume hood.

Steps in Preparing and Using a Mull

1. Using a small agate or nonporous ceramic mortar and pestle, grind 10–15 mg of the solid until the sample is exceedingly fine and has a caked, glassy appearance. Use a small flat spatula to scrape the ground solid from the surface of the mortar.
2. Add one drop of mulling liquid to the ground solid in the mortar. **Be careful!** Err on the side of adding too little mulling liquid because it is impossible to remove it if you add too much. Grind the mixture to make a uniform paste with the consistency of toothpaste; it should not be grainy but must not be runny.
3. Transfer the paste to the center section of a NaCl disk with a small flat spatula, as in Figure 21.9. Press the disks together gently, rotate the top disk, and place them in a sample holder.
4. Obtain the IR spectrum.
5. Clean the disks with a **dry** solvent—acetone or dichloromethane. Store the disks in a desiccator to protect them from moisture.
6. Clean the equipment you used for grinding the sample.

Two Additional Methods for Solid Compounds—KBr Disks and Sample Cards

Potassium bromide (KBr) does not absorb mid-region IR radiation. Thus, a solid compound can be prepared for IR spectroscopy by grinding the sample with anhydrous KBr powder and pressing the mixture into a thin, transparent disk. Potassium bromide disks are excellent for IR analysis, but their preparation is challenging and requires great care. It may take several attempts to prepare KBr disks that are suitable for IR analysis, especially if you have not made them before. Consult your instructor if you need to prepare a KBr disk.

A relatively new innovation in IR spectroscopy is the use of a disposable sampling card. First the sample card is scanned in an FTIR spectrometer and its IR spectrum is saved in the instrument's memory, and then the sample is applied to the inert, microporous matrix in the middle of the card. Liquids are applied neat (without solvent). Solids are applied in solution, and the solvent is allowed to evaporate. The card is placed in the sample beam and scanned. The spectrum of the blank sample card is then subtracted from that of the card containing the applied compound by software provided with the FTIR instrument.

Polyethylene or polytetrafluoroethylene is usually used for the solid support matrix on sample cards. Polyethylene has strong absorptions in the regions 2918–2849 cm^{-1}, 1480–1430 cm^{-1}, and 740–700 cm^{-1}. Polytetrafluoroethylene has strong absorptions in the regions 1270–1100 cm^{-1} and 660–460 cm^{-1}. The infrared peaks of the sample card matrix may obscure peaks due to your sample.

21.6   Sample Preparation for Attenuated Total Reflectance (ATR) Spectra

FTIR instruments are extremely sensitive, and FTIR techniques using *attenuated total reflectance (ATR)* sampling accessories make obtaining IR spectra of solids much easier. Liquid samples can also be used for ATR spectra. When an ATR accessory is used, it is unnecessary to prepare Nujol mulls or cast films, or even use NaCl disks.

With ATR, the infrared radiation is passed through an infrared transmitting crystal with a high refractive index, which allows the radiation to reflect within the crystal. Zinc selenide (ZnSe) is the most common ATR crystal material, although it scratches quite easily and must be handled with care. Diamond is the best crystal material for ATR because of its durability, but the original expense is greater.

A single-reflection ATR accessory, such as the one shown in Figure 21.11, often works best for the IR spectra of solids. The solid is finely powdered and then pressed into intimate contact with the top surface of the crystal by screwing down a pressure tip onto the sample. The effects of poor contact, which produces a weak spectrum, are greater at higher frequency. After entering the ATR crystal, the beam of infrared energy interacts with the surface of

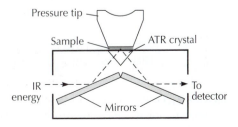

FIGURE 21.11 Cross section of a single-reflection attenuated total reflectance (ATR) accessory.

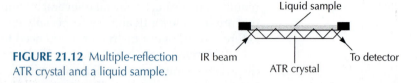

FIGURE 21.12 Multiple-reflection ATR crystal and a liquid sample.

the solid sample, effectively penetrating a small distance (0.5–5 μm) into the sample before being reflected. The IR beam becomes attenuated (becomes less intense) in regions of the IR spectrum where the sample absorbs. The beam then exits from the opposite end of the crystal and passes to the IR detector where a spectrum is generated.

For high-quality IR spectra of liquids, multiple-reflection ATR—with a long crystal and a trough that can be filled with the liquid sample—is often used. With each reflection, the IR beam becomes attenuated in regions of the IR spectrum where the sample absorbs. The IR beam reflects off the liquid five to 10 times, as shown in Figure 21.12. As with all FTIR spectra, a background scan must be made before the sample is scanned.

An IR spectrum from an ATR accessory is similar to a transmission IR spectrum, but there are some differences. The frequencies of the absorptions are the same, but the relative intensities of the peaks may differ. A comparison of the transmission spectrum and the ATR spectrum of solid polystyrene is shown in Figure 21.13. The differences occur because lower-frequency IR energy penetrates farther into the sample than higher-frequency IR energy. Because the lower frequencies interact with more sample, their absorbance bands are more intense.

Software is available that can correct for the different intensities at different wavelengths. This software produces IR spectra that more closely resemble transmission spectra, which makes it easier to compare ATR and transmission spectra.

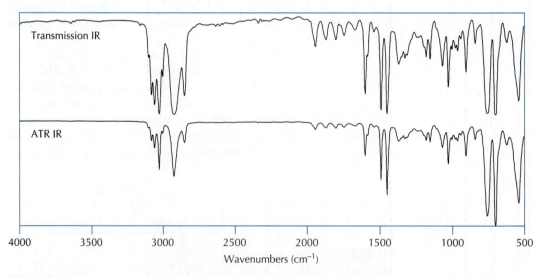

FIGURE 21.13 Comparison of (top) a transmission spectrum of polystyrene with (bottom) an ATR spectrum of polystyrene.

Steps in Obtaining an ATR-IR Spectrum of a Solid

An ATR-IR spectrum of a liquid can be obtained by carefully placing a small amount of the liquid on the ATR crystal. Omit steps 3 and 4.

1. Carefully clean the surface of the ATR crystal with a lint-free tissue or a gun oil wipe, followed by a dry lint-free tissue. Then run a background scan.
2. Place a small amount of finely powdered solid sample on the ATR crystal. Use just enough sample to cover the crystal area. The sample height should not be more than a few millimeters. If you are using a ZnSe ATR crystal, use a wooden stick or other nonabrasive tool for this operation because a metal spatula can easily scratch the crystal surface.
3. Lower the pressure tip so that it is in contact with the solid. (**Note:** To avoid contamination of the tip, place a small piece of paper between the tip and the sample.)
4. Apply a pressure of approximately 10 psi to the sample. The mechanism and appropriate pressure vary for different ATR accessories, so find out from your instructor the procedure for your accessory.
5. Obtain the IR spectrum.
6. Raise the pressure tip from the sample. Gently wipe the sample from the crystal and from the pressure tip with a tissue. Then wipe the crystal and pressure tip with a tissue wetted with methanol or 2-propanol, and gently wipe with a dry lint-free tissue to remove any remaining solvent.

21.7 Interpreting IR Spectra

Confirming the identity of a compound is one of the most important uses of IR spectroscopy. Because of the numerous vibrations of a typical organic molecule, **no two compounds are known to have identical IR spectra.** The unique pattern of each compound allows an IR spectrum to be used as a fingerprint for identification. Databases of IR spectra of known compounds can be searched for a match with the spectrum of an unknown compound—an identification method frequently used in forensic and quality control laboratories. Comparing an IR spectrum obtained in the laboratory to the spectra available in a compendium of IR spectra can be useful;* however, it can also be very time consuming and may not be worth the effort.

In addition, there is often no sample spectrum available for comparison, and it is necessary to interpret the IR spectrum.

The most basic interpretation consists of an inventory of the functional groups in the molecule. Systematic examination of the IR spectrum and identification of absorption peaks due to fundamental stretching vibrations are used to construct a functional group inventory. Combining IR data with structural information from other techniques, such as nuclear magnetic resonance (NMR) spectroscopy, usually allows unequivocal assignment of a molecular structure to an organic compound.

*The Aldrich Library of FT-IR Spectra; 2nd ed.; Aldrich Chemical Company: Milwaukee, WI, 1997; 3 volumes.

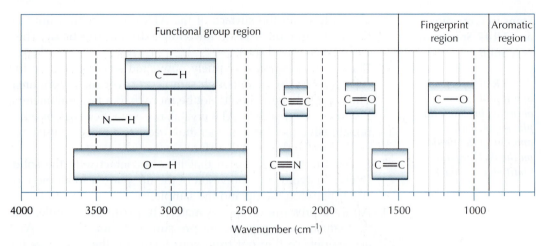

FIGURE 21.14 Approximate regions of chemical bond stretches in an IR spectrum.

Regions of the IR Spectrum

As shown in Figure 21.14, an IR spectrum can be broken down into three regions:

- Functional group region
- Fingerprint region
- Aromatic region

Functional group region. The functional group region (4000–1500 cm^{-1}) provides unambiguous, reasonably strong peaks for most major functional groups. Figure 21.14 shows the approximate regions in which peaks appear as a result of important bond-stretching vibrations. Structurally similar compounds that contain the same functional groups are virtually identical in this region. For example, the absorption peaks for the carbonyl groups of 2-butanone and 3-hexanone both appear at the same frequency, 1715 cm^{-1}. Since many functional groups show IR peaks in the 4000–1500 cm^{-1} region, **both the presence and absence of peaks in this region are significant.** The absence of an appropriate IR peak in the functional group region argues against the presence of that functional group, except in the rare cases when a stretching vibration has no associated dipole change.

Fingerprint region. The fingerprint region (1500–900 cm^{-1}) is normally complex because of the many bending vibrations and combination bands that appear here. Before the development of NMR spectroscopy, much effort went into analyzing and assigning characteristic vibrations in this region. NMR spectroscopy now provides detailed structural information more directly and reliably. Except for a few intense absorptions, such as C–O stretching vibrations and a few other corroborating IR peaks, the fingerprint region is now primarily used for fingerprint pattern matching.

Aromatic region. The aromatic region (900–600 cm^{-1}) provides information about the substitution pattern of benzenes and other aromatic compounds, and it also provides corroborating information about the structure of alkenes.

EXERCISE

All organic compounds have IR absorptions because of C–H and C–C stretching and bending vibrations. For each of the following compounds, identify the additional bond-stretching vibrations that should be observed. Using Figure 21.14 as a guide, identify regions of the IR spectrum where you would expect to see characteristic absorptions for each compound.

(a) 2-propanol	(c) phenylethyne	(e) 4-methylphenylamine
(b) propanoic acid	(d) 1-hexene	(f) benzonitrile

Answer

(a) 2-propanol (O–H, 3650–2500 cm^{-1}; C–O, 1300–1000 cm^{-1})

(b) propanoic acid (O–H, 3650–2500 cm^{-1}; C=O, 1850–1650 cm^{-1} C–O, 1300–1000 cm^{-1})

(c) phenylethyne (C≡C, 2250–2100 cm^{-1}; C=C, 1680–1440 cm^{-1})

(d) 1-hexene (C=C, 1680–1440 cm^{-1})

(e) 4-methylphenylamine (N–H, 3550–3150 cm^{-1}; C=C, 1680–1440 cm^{-1})

(f) benzonitrile (C≡N, 2280–2200 cm^{-1}; C=C, 1680–1440 cm^{-1})

FOLLOW-UP ASSIGNMENT

Using Figure 21.14 as a guide, identify regions of the IR spectrum in which you would expect to see characteristic functional group absorptions for each of the following compounds: (a) cyclopentanone, (b) methyl acetate, (c) methoxybenzene, (d) acetamide, (e) 1-aminohexane.

Where to Begin

An efficient approach to interpreting an IR spectrum usually starts with a survey of the 4000–1500 cm^{-1} functional group region and the creation of an inventory of bond types present in the molecule. This inventory allows you to get a good idea of which functional groups are in the compound and which are not.

The functional group region can be subdivided into narrower frequency regions that are characteristic of specific bond types. Table 21.2 lists the positions of characteristic IR absorption peaks of various functional groups. It is fairly accurate for strong (s) and broad (br) peaks. However, because the intensities of IR absorptions can vary a good deal, the use of Table 21.2 has limitations, particularly for peaks listed as medium (m) and weak (w) intensity. Useful IR corroborating peaks of the C–H out-of-plane bending vibrations

TABLE 21.2   Characteristic infrared absorption peaks of functional groups

Vibration	Position (cm^{-1})	Intensity[a]
Alkanes		
C–H stretch	2990–2850	m to s
C–H bend	1480–1430 (CH$_2$), 1395–1340 (CH$_3$)	m to w
Alkenes		
=C–H stretch	3100–3000	m
C=C stretch	1680–1620 (nonconj.)[b], 1650–1600 (conj.)[b]	w to m
=C–H bend	1000–665 (see Table 21.3 for detail)	s
Alkynes		
≡C–H stretch	3310–3200	s
C≡C stretch	2250–2100	s to w

(Continued)

T A B L E   2 1 . 2   (Continued)

Vibration	Position (cm^{-1})	Intensity[a]
Aromatic Compounds		
C−H stretch	3100–3000	m to w
C=C stretch	1620–1440	m to w
C−H bend	880–680 (see Table 21.3 for detail)	s
Alcohols		
O−H stretch	3650–3550 (non H-bonded), 3550–3200 (H-bonded)	s to m, br
C−O stretch	1300–1000	s
Ethers		
C−O stretch	1300–1000	s
Amines		
N−H stretch	3550–3250 (1° two peaks), (2° one peak)	br, m
C−N stretch	1250–1025 (alkyl), 1350–1250 (aromatic)	m
Nitriles		
C≡N stretch	2280–2200	s
Isocyanates		
N=C=O stretch	2275–2230	s
Aldehydes		
=C−H stretch	2900–2800 and 2800–2700, Fermi doublet	w
C=O stretch	1740–1720 (non–conj.), 1715–1680 (conj.)	s
Ketones		
C=O stretch	1725–1705 (non–conj.), 1700–1650 (conj.)	s
Esters		
C=O stretch	1765–1735 (non–conj.), 1730–1715 (conj.)	s
C(=O)−O stretch	1260–1230 (acetates), 1210–1160 (all others)	s
Carboxylic Acids		
O−H stretch	3200–2500	br, m to w
C=O stretch	1725–1700 (non–conj.), 1715–1680 (conj.)	s
C−O stretch	1300–1000	s
Amides		
N−H stretch	3500–3150 (1° two peaks), (2° one peak)	m
C=O stretch	1700–1630	s
N−H bend	1570–1515 (primary), 1640–1550 (secondary)	s
Anhydrides		
C=O stretch	1850–1800 and 1790–1740	s
C−O stretch	1300–1000	s
Acid Chlorides		
C=O stretch	1815–1770	s
Imines		
R$_2$C=N−R stretch	1690–1630	m to s
Nitro Compounds		
NO$_2$ stretch	1570–1490 and 1390–1300	s
Sulfoxides		
S=O stretch	1070–1030	s
Sulfonates		
S=O stretch	1375–1335 and 1195–1165	s
Phosphine Oxides		
P=O stretch	1200–1130	s
Alkyl Halides		
C−F stretch	1400–1000	s
C−Cl stretch	785–540	s

a. s = strong; m = medium; w = weak; br = broad. b. non-conj. = non-conjugated; conj. = conjugated.

of alkenes and aromatic compounds are listed in Table 21.3. Besides correlating stretching and bending vibrations with a frequency (wavenumber), it is important to consider the general appearance of the signal. Is it sharp? Is it broad? Is it weak? Is it strong? As in the analysis of other experimental data, you must think about the significance of your conclusions, rather than assuming that an algorithm will lead to the correct answer every time.

TABLE 21.3   **Out-of-plane C–H bending vibrations of alkenes and aromatic compounds**

Structure		Position (cm^{-1})
Monosubstituted	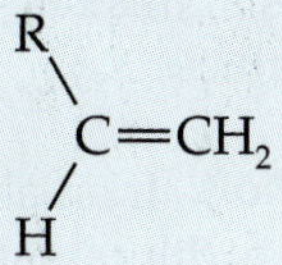 R–CH=CH$_2$ (R and H on C)	997–985, 915–905
1,2-Disubstituted (E)	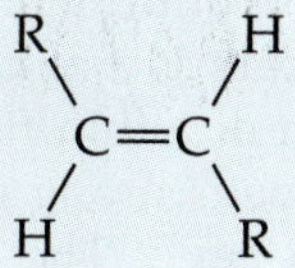 RHC=CHR (E)	980–960
1,2-Disubstituted (Z)	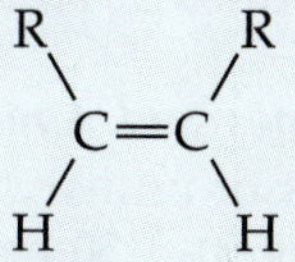 RHC=CHR (Z)	730–665
1,1-Disubstituted	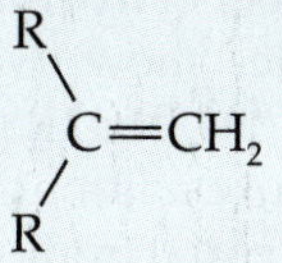 R$_2$C=CH$_2$	895–885
Trisubstituted	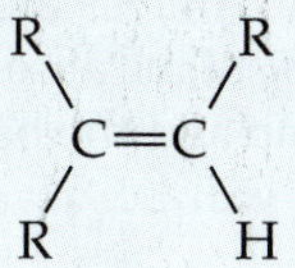 R$_2$C=CHR	840–790
Monosubstituted aromatic	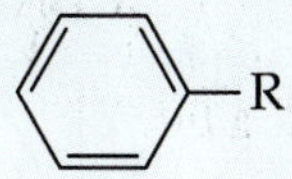 C$_6$H$_5$–R	770–730, 720–680
1,2-Disubstituted aromatic	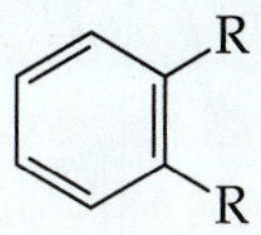 1,2-C$_6$H$_4$R$_2$	770–735
1,3-Disubstituted aromatic	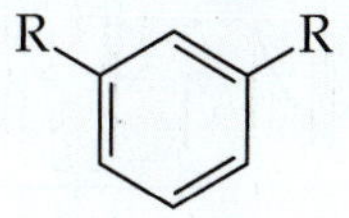 1,3-C$_6$H$_4$R$_2$	810–750, 725–680
1,4-Disubstituted aromatic	 1,4-C$_6$H$_4$R$_2$	860–780

21.8   IR Peaks of Major Functional Groups

We start our discussion of the major peaks in IR spectra and the functional group vibrations that produce them at the high-frequency (left) side of IR spectrum, where stretching vibrations appear, and move down the frequency scale to the right portion of the spectrum. In the following pages, each important IR region is described and examples of spectra illustrating the fundamental stretching and bending peaks are given.

O–H Stretch of Alcohols and N–H Stretch of Amines and Amides (3650–3150 cm^{-1})

Alcohols and phenols show strong IR peaks due to oxygen-hydrogen bond stretching, and amines and amides show medium intensity IR peaks due to nitrogen-hydrogen bond stretching. The appearance of peaks in this region is highly varied, which can actually add to their usefulness.

Alcohols. If an alcohol is prepared for IR analysis in any form other than a dilute solution, the hydroxyl group hydrogen bonds with neighboring molecules and the signal caused by the O–H stretch appears as a broad band between 3550 and 3200 cm^{-1}. The IR spectrum of a thin film of 2-propanol, shown in Figure 21.15, exhibits a broad, strong O–H stretching absorption with the maximum at 3365 cm^{-1}. The C–O stretching vibration in Figure 21.15 is also identified at 1131 cm^{-1}.

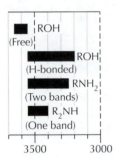

Amines and Amides. The medium intensity N–H stretching vibrations of primary and secondary amines and amides also appear in the 3550–3150 cm^{-1} region. The number of signals depends on the number of hydrogen atoms attached to nitrogen. Primary amines and amides show two peaks, and secondary amines and amides show only one. The IR spectrum of 1-aminobutane is shown in Figure 21.16. Because it is a primary amine, there are two absorptions (at 3369 and 3293 cm^{-1}) from symmetric and asymmetric H–N–H stretching vibrations. Amines are capable of hydrogen bonding, so the position and shape of the absorption may vary. The higher the concentration of the amine and the better it can hydrogen bond, the broader the absorption will be. Hydrogen-bonding shifts N–H stretching absorptions

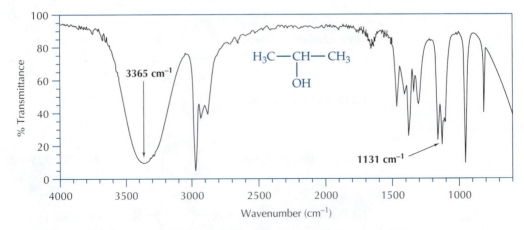

FIGURE 21.15 IR spectrum of 2-propanol (thin film).

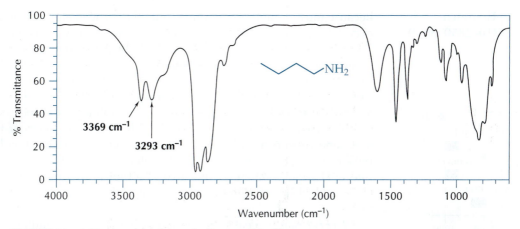

FIGURE 21.16 IR spectrum of 1-aminobutane (thin film).

to lower frequencies. Alkyl amines are stronger bases than aromatic amines and tend to form stronger hydrogen bonds.

O–H Stretch of Carboxylic Acids (3200–2500 cm⁻¹)

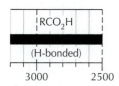

As a result of extensive intermolecular hydrogen bonding, carboxylic acids generally show an unusually broad O–H stretching absorption, with the band often tailing from about 3200 cm⁻¹ all the way down to 2500 cm⁻¹. The intensity of this band is medium to weak. The spectrum of propanoic acid shown in Figure 21.17 illustrates this behavior; the O–H stretching band is so broad that the sharper C–H stretch at approximately 3000 cm⁻¹ is superimposed on it. This superimposition is not uncommon with the O–H and C–H stretching vibrations of carboxylic acids. The structure of the intermolecular hydrogen-bonded dimer of propanoic acid is

$$CH_3CH_2C\overset{\displaystyle O\text{-----}H\text{—}O}{\underset{\displaystyle O\text{—}H\text{-----}O}{}}CCH_2CH_3$$

The C=O and C–O stretching vibrations for propanoic acid are also indicated in Figure 21.17.

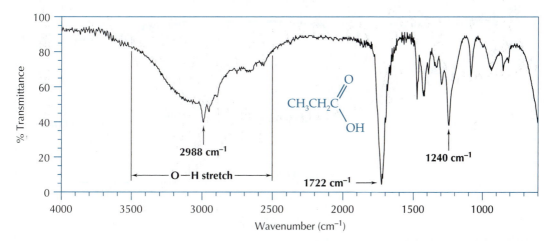

FIGURE 21.17 IR spectrum of propanoic acid (thin film).

C–H Stretch
(3310–2850 cm⁻¹)

Because most organic compounds contain hydrogen atoms, you can expect to find C–H stretching signals in most IR spectra. The position of the C–H stretch depends on the hybridization of the carbon atom to which the hydrogen is bound.

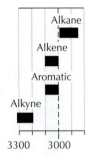

sp Hybridization. If the carbon atom is *sp* hybridized, the C–H peak appears near 3300 cm⁻¹. A good example is found in the spectrum of phenylacetylene shown in Figure 21.18. The C–H stretch of the acetylene appears at 3277 cm⁻¹. This peak could be confused with a peak resulting from an O–H or N–H stretch, were it not for its shape. The C–H peak is much sharper than the typical hydrogen-bonded O–H or N–H stretch found in this region.

sp² Hybridization. Peaks that occur when hydrogen atoms are attached to *sp²*-hybridized carbon atoms of alkenes and aromatic compounds appear in the region 3100–3000 cm⁻¹. In the spectrum of phenylacetylene (see Figure 21.18), the aromatic C–H stretching vibrations appear from 3066 to 3006 cm⁻¹. In the spectrum of 1-hexene shown in Figure 21.19 the vinyl-hydrogen stretch appears at 3084 cm⁻¹.

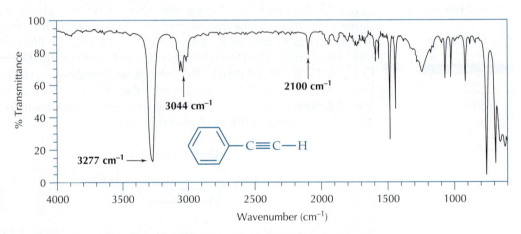

FIGURE 21.18 IR spectrum of phenylacetylene (thin film).

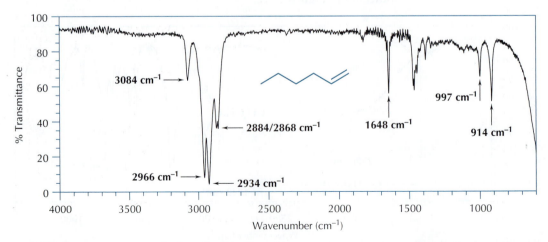

FIGURE 21.19 IR spectrum of 1-hexene (thin film).

sp³ Hybridization. Hydrogen atoms attached to sp^3-hybridized carbon atoms exhibit absorption peaks in the 2990–2850 cm^{-1} region. There are usually several alkyl C–H stretching vibration peaks in an IR spectrum. In the spectrum of 1-hexene, there are four distinct peaks from 2966 to 2868 cm^{-1} because of C–H stretching vibrations of hydrogen atoms attached to sp^3 carbon atoms.

C≡C and C≡N Stretch (2280–2100 cm⁻¹)

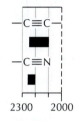

Only the triple bonds of nitriles and alkynes have absorptions in this region. If it is a strong absorption, it is likely to be the C≡N stretching vibration of a nitrile, such as the peak at 2230 cm^{-1} in the spectrum of benzonitrile (Figure 21.20). The difference in electronegativity between carbon and nitrogen leads to a highly polarized C≡N bond and thus a strong absorption. Alkynes have weak- to medium-intensity absorption bands in this region. The C≡C bond stretch in phenylacetylene is the small peak at 2100 cm^{-1} (see Figure 21.18).

C=O stretch (1850–1630 cm⁻¹)

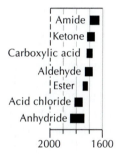

If there is a C=O present in the molecule, there will be a strong, sharp absorption band in the 1850–1630 cm^{-1} region. Good examples of C=O stretching are the strong band at 1747 cm^{-1} in the spectrum of cyclopentanone (see Figure 21.1) and at 1722 cm^{-1} in the spectrum of propanoic acid (see Figure 21.17). If there is no strong band in the 1850–1630 cm^{-1} region, there is no C=O in the molecule. The exact position of the signal within this region, however, depends on what type of functional group contains the C=O group.

Functional group	Example	C=O stretch (cm⁻¹)
Amides	Acetamide	1681
Ketones	Acetone	1715
Carboxylic acids	Propanoic acid	1722
Aldehydes	Acetaldehyde	1727
Esters	Methyl acetate	1745
Acid chlorides	Acetyl chloride	1806
Acid anhydrides	Propanoic anhydride	1827 and 1766

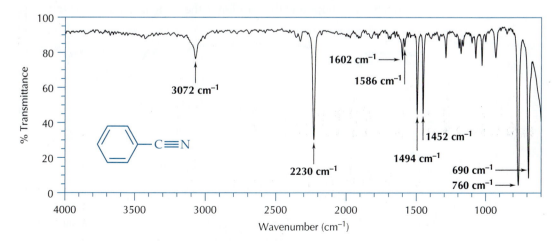

FIGURE 21.20 IR spectrum of benzonitrile (thin film).

Notice that acid anhydrides are characterized by the presence of two peaks in the C=O stretching region. These peaks arise from symmetric and asymmetric C=O stretching vibrations.

Symmetric C=O stretch   Asymmetric C=O stretch

Effect of ring strain and conjugation. Ring strain causes deviations from the position of the C=O band for cyclic carbonyl compounds. The position of the absorption band moves to a higher frequency, indicating that the strength of the bond has increased. Compare the C=O absorption bands of acetone (1715 cm^{-1}) and methyl acetate (1745 cm^{-1}) to the carbonyl absorption bands of the cyclic compounds listed here.

Ketone	C=O stretch (cm^{-1})	Ester	C=O stretch (cm^{-1})
Cyclopropanone	1818	—	—
Cyclobutanone	1783	Propanolactone	1840
Cyclopentanone	1747	4-Butanolactone	1770
Cyclohexanone	1716	5-Pentanolactone	1730
Cycloheptanone	1702	6-Hexanolactone	1732

Conjugation with a C=C double bond or with an aromatic ring decreases the bond order of the C=O slightly and causes the position of a C=O peak to move to a lower frequency by 20–30 cm^{-1}. We can see this effect by comparing the position of the C=O peak of 4-methylpentan-2-one (Figure 21.21) to that of the conjugated compound, 4-methyl-3-penten-2-one (Figure 21.22). The absorption band is shifted from 1719 cm^{-1} to 1695 cm^{-1}. A similar shift is observed when the C=O group is conjugated with a benzene ring. The C=O peak for acetone appears at 1715 cm^{-1}, whereas the C=O peak for acetophenone appears at 1690 cm^{-1}.

Corroborating IR peaks. Because the C=O bond is present in many functional groups and the position of the stretching vibration is affected

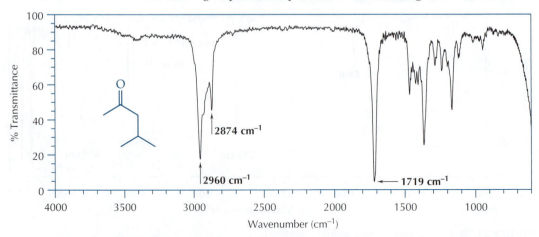

FIGURE 21.21 IR spectrum of 4-methylpentan-2-one (thin film).

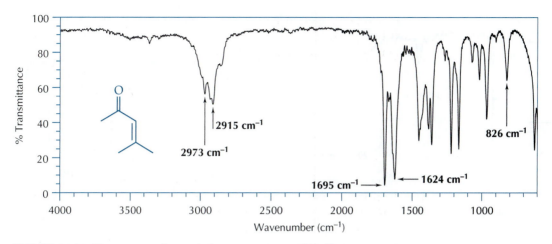

FIGURE 21.22 IR spectrum of 4-methyl-3-penten-2-one (thin film).

by many variables, it can be difficult to differentiate between carbonyl-containing functional groups by the C=O stretching frequency alone. It is usually necessary to identify other peaks in the IR spectrum to ascertain the identity of the functional group exhibiting the C=O stretch.

Primary and secondary amides exhibit N–H stretching absorption peaks in the 3500–3150 cm⁻¹ region, two peaks for primary amides and one for secondary amides. Amides also show strong peaks associated with N–H bending vibrations in the 1640–1515 cm⁻¹ region.

Carboxylic acids exhibit an extremely broad absorption band between 3200 and 2500 cm⁻¹ because of hydrogen-bonded O–H stretching vibrations (see Figure 21.17).

Aldehydes exhibit two weak but very distinct absorption peaks in the C–H stretching region (2900–2800 cm⁻¹ and 2800–2700 cm⁻¹). The two peaks are an example of a Fermi doublet as a result of the interaction of the fundamental stretching vibration of the aldehyde C–H bond with an overtone band. The characteristic aldehyde C–H bands at 2815 cm⁻¹ and 2743 cm⁻¹ are evident in the spectrum of cinnamaldehyde shown in Figure 21.23. You can also see sp^2 C–H stretching vibrations in the 3100 cm⁻¹ region and the C=C stretching vibration at 1627 cm⁻¹ in Figure 21.23.

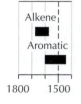

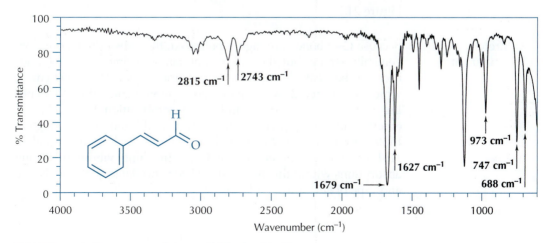

FIGURE 21.23 IR spectrum of cinnamaldehyde (thin film).

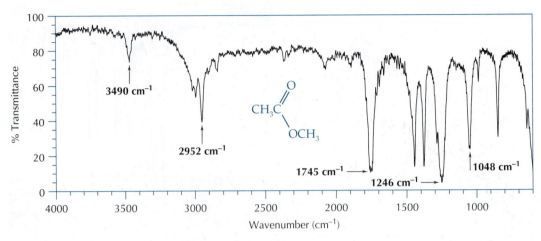

FIGURE 21.24 IR spectrum of methyl acetate (thin film).

Esters exhibit a very strong band in the C–O stretching region (1300–1000 cm^{-1}). In the spectrum of methyl acetate, shown in Figure 21.24, C–O stretching vibrations appear at 1246 and 1048 cm^{-1}.

C=C Stretch (1680–1440 cm^{-1})

Absorptions in the 1680–1440 cm^{-1} region occur because of C=C bonds in alkenes and in aromatic compounds. Their intensities vary from weak to medium. A typical absorption of this type is the band at 1648 cm^{-1} in the spectrum of 1-hexene (see Figure 21.19). The position and intensity of the peak are affected by conjugation. The C=C stretching absorption in 4-methyl-3-penten-2-one (see Figure 21.22) appears at 1624 cm^{-1}, and its intensity is significantly stronger than the intensity of the band in 1-hexene.

Aromatic compounds have four peaks in this region near 1600, 1580, 1500, and 1450 cm^{-1}. The first two peaks are generally weak and the second two are generally moderate in intensity. The peak at 1450 cm^{-1} can be obscured by CH$_2$ bending vibrations if an alkyl group is present. These peaks are evident in the spectra of phenylacetylene (see Figure 21.18) and benzonitrile (see Figure 21.20).

C–O Stretch (1300–1000 cm^{-1})

Because C–O bonds are highly polarized, their absorption bands are generally strong; but the assignment can sometimes be ambiguous because the peaks occur in the fingerprint region (1500–900 cm^{-1}), which is cluttered with many absorption bands due to bending vibrations, overtone bands, and combination bands. Esters, ethers, and alcohols, however, show useful bands in this region. In the spectrum of 2-propanol (see Figure 21.15), the signal at 1131 cm^{-1} is attributed to the C–O stretching vibration. Strong absorptions within this region have been correlated with the degree and type of substitution of alcohols.

Type of alcohol	C–O stretch, cm^{-1}
RCH$_2$—OH Primary	1080–1000
R \ HC—OH / R' Secondary	1130–1000
R'' | R—C—OH | R' Tertiary	1210–1100
(aromatic ring)—OH Aromatic	1260–1180

Esters often exhibit two C–O stretching vibrations, one for the C–O bond to the carbonyl carbon and one for the C–O bond to the carbon of the alcohol group. In the spectrum of methyl acetate (see Figure 21.24), the bands appear at 1246 cm^{-1} and 1048 cm^{-1}, respectively.

NO$_2$ Stretches (1570–1490 cm^{-1} and 1390–1300 cm^{-1})   Aromatic nitro groups have two distinctive absorptions due to symmetric and asymmetric O–N–O stretches. These are usually the most intense peaks in the spectrum. In the spectrum of 3-nitrotoluene shown in Figure 21.25 the signals appear at 1532 cm^{-1} and 1355 cm^{-1}.

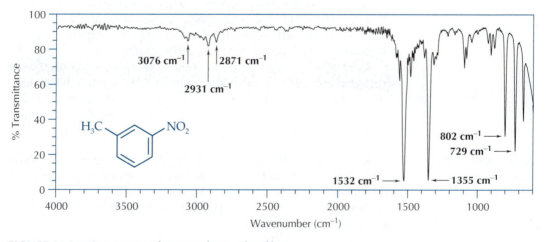

FIGURE 21.25 IR spectrum of 3-nitrotoluene (thin film).

Symmetric stretch   Asymmetric stretch
of nitro group   of nitro group

Useful Corroborating Peaks (1000–600 cm⁻¹)

An infrared spectrum can be highly cluttered with peaks, and not every one can be easily or directly correlated to a specific vibration. However, there are some peaks, in addition to the fundamental IR stretching vibrations, that can provide structural information. The number and arrangement of substituents on a C=C bond can often be determined from the presence of strong signals below 1000 cm⁻¹; these occur because of out-of-plane C–H bending vibrations. Table 21.3 on page 329 summarizes these diagnostic peaks in the region of 1000–600 cm⁻¹.

Absorptions at 997 and 914 cm⁻¹ in the spectrum of 1-hexene (see Figure 21.19) are characteristic of a monosubstituted alkene. In the spectrum of cinnamaldehyde (see Figure 21.23), the *trans*-disubstituted C=C bond is indicated by the absorption at 973 cm⁻¹. The trisubstituted alkene in 4-methyl-3-penten-2-one is indicated by the absorption appearing at 826 cm⁻¹ (see Figure 21.22). Absorptions at 760 and 690 cm⁻¹ in the spectrum of benzonitrile (see Figure 21.20) and at 747 and 688 cm⁻¹ in the spectrum of cinnamaldehyde (see Figure 21.23) are characteristic of a monosubstituted benzene ring. The two peaks at 802 and 729 cm⁻¹ in Figure 21.25 are characteristic of a 1,3-disubstituted benzene ring.

21.9   Procedure for Interpreting an IR Spectrum

IR spectroscopy is an important tool for determining which functional groups are in a molecule. For most organic compounds, this information alone is not sufficient to unequivocally determine the structure. However, the inventory of functional groups coupled with other data, particularly NMR and mass spectrometry, leads to a definitive elucidation of a compound's structure.

After you have interpreted numerous IR spectra, the need for a structured approach in compiling an inventory of functional groups may not be very great. But in the beginning, a general method that provides a structured and logical approach is helpful in learning to interpret an IR spectrum.

Step 1. Check the 1850–1630 cm⁻¹ Region

A strong peak in this region indicates the presence of a carbonyl group. If there are no strong peaks, no C=O group is present, and you should proceed to Step 2. If a C=O group is present, use peaks in other regions of the IR spectrum to identify the specific type of carbonyl functional group:

- Two strong peaks centered near 1800 cm⁻¹ indicate an acid anhydride group.
- Two weak absorption peaks in the region 2900–2700 cm⁻¹ indicate an aldehyde group.

- An extremely broad band extending from 3200 to 2500 cm^{-1} indicates a carboxylic acid group.
- Two strong absorption peaks in the region 1300–1000 cm^{-1} indicate an ester group.
- One or two medium-intensity peaks in the 3500–3150 cm^{-1} region indicate an amide group.

In the absence of any of these observations, a single strong C=O stretching absorption near 1700 cm^{-1} probably indicates a ketone. A single strong absorption near 1800 cm^{-1} probably indicates an acid chloride.

Step 2. Check the 3550–3200 cm^{-1} Region

The presence of a strong, broad signal indicates the hydroxyl group of an alcohol. There should be an accompanying strong peak due to C–O stretching in the region 1300–1000 cm^{-1}. The presence of medium-intensity band(s) in the N–H stretching region indicates an amine group. Primary amines have two peaks and secondary amines have one.

Step 3. Check the C–H Stretching Region at 3310–2850 cm^{-1}

A strong, sharp peak near 3300 cm^{-1} indicates a terminal alkyne group. There should be an accompanying weak- to medium-intensity peak due to C≡C stretching near 2200 cm^{-1}. Peaks in the region 3100–3000 cm^{-1} are a result of C–H stretching in alkenes or aromatic compounds. Corroborating peaks can narrow the choices.

- A medium-intensity peak near 1650 cm^{-1} indicates a C=C bond.
- Several weak- to medium-intensity peaks in the region 1620–1450 cm^{-1} may suggest an aromatic ring.
- If an alkene or an aromatic ring is indicated, the region 1000–600 cm^{-1} may determine the substitution pattern (see Table 21.3, p. 329).

Peaks in the region 2990–2850 cm^{-1} are caused by C–H stretching in alkyl groups.

Step 4. Check the 2280–2100 cm^{-1} Region

A strong peak near 2250 cm^{-1} indicates a C≡N group. A medium- to weak-intensity peak near 2170 cm^{-1} indicates a C≡C group.

Step 5. Check the 1300–1000 cm^{-1} Region

If there are one or two strong absorptions in this region and no signals in the O–H or C=O stretching regions, an ether group may be present.

Step 6. Prepare an Inventory of Functional Groups

Assemble a list of the functional groups indicated by the IR spectrum. If NMR data or a molecular formula are available, coordinate them with the results from IR spectroscopy. Fit the pieces together into likely chemical structures that are consistent with the data and with the rules of chemical bonding.

21.10 Case Study

In this section you will see how the information derived from an IR spectrum can help you determine the molecular structure of an organic compound. The molecular formula of the compound is

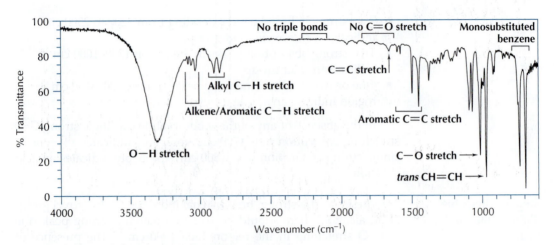

FIGURE 21.26 IR spectrum of $C_9H_{10}O$ (ATR-corrected).

$C_9H_{10}O$, and its IR spectrum, run using an ATR attachment, is shown in Figure 21.26.

Begin by surveying the 4000–1500 cm^{-1} functional group region. The general approach presented in Section 21.9 can help you create an inventory of bond types present in the molecule. This inventory allows you to get a good idea of which functional groups are and are not present in the compound.

The absence of a strong peak in the 1850–1630 cm^{-1} region indicates that no C=O groups are present. The intense, broad band at 3327 cm^{-1} is prominent in the 3650–3200 cm^{-1} region. Its intensity and position indicate the presence of a hydroxyl group. There are also strong peaks in the 1300–1000 cm^{-1} region, consistent with the C–O stretching vibration of an alcohol, although the cluttered nature of this region makes a definitive assignment of the signals difficult.

In the C–H stretching region, there are peaks from 3100 to 3000 cm^{-1} superimposed on the shoulder of the broad and intense O–H stretching peak. The peaks in this region signify C–H stretching in alkenes or aromatic compounds. The presence of a C=C bond is confirmed by a weak-intensity band at 1668 cm^{-1}. The weak to medium peaks at 1599, 1578, 1494, and 1449 cm^{-1} are consistent with the presence of an aromatic ring. There are also two peaks of medium intensity at about 2900 cm^{-1}, which signify C–H stretching vibrations of at least one alkyl group. An absence of any signals in the 2250–2100 cm^{-1} region rules out the presence of a C≡C bond.

Because the presence of a C=C bond and the presence of an aromatic ring are indicated, a check of the 1000–600 cm^{-1} region is warranted. The signal at 967 cm^{-1} indicates that the double bond is *trans*-disubstituted and the signals at 740 and 692 cm^{-1} indicate that the aromatic ring is monosubstituted (see Table 21.3).

In summary, our inventory of functional groups consists of a monosubstituted benzene ring (C_6H_5–), a *trans*-disubstituted double bond (–CH=CH–), a hydroxyl group (–OH), and at least one sp^3 carbon atom. The molecular formula of the compound is $C_9H_{10}O$. If the alkyl carbon atom is part of a methylene group, we have accounted for all the necessary atoms. There are two ways to put these pieces together:

The structure on the right can be eliminated because it is the unstable enol isomer of an aldehyde. The compound that produced the IR spectrum shown in Figure 21.26 has the structure on the left, (*E*)-3-phenyl-2-propen-1-ol, commonly called cinnamyl alcohol.

This case study was carefully chosen to show the power of infrared spectroscopy. In most cases it would be difficult, if not impossible, to reach a definitive structure for a compound given only a molecular formula and an IR spectrum unless one successfully searched a database for a match with its spectrum. However, even if a complete structure doesn't result from the assembly of an inventory of functional groups, the knowledge of which functional groups are present can be a great help in understanding the compound and its properties.

21.11 Sources of Confusion and Common Pitfalls

The three major places where mistakes can arise and cause confusion in infrared spectroscopy come from faulty sample preparation, incorrect use of an FTIR spectrometer, and the inherent complexity of IR spectra.

My Spectrum Has No Peaks! It's Just a Flat Line at 100% Transmittance. What's Wrong?

The easy explanation is that you forgot to put the sample into the IR beam. But if you did put the sample into the beam, you most likely did so before you ran a background scan and then left the sample in the beam and ran a sample scan. In that case, the background and sample scans are the same. The result of subtracting the background scan from the sample scan is equivalent to 100% transmittance over the entire wavelength range.

My Spectrum Shows Unexpected Peaks Near 2350 cm⁻¹, or I See a Peak in This Same Range That Is Going Up, Rather Than Down. What Could Cause These Peaks?

A pair of signals near 2350 cm⁻¹ in your IR spectrum, which may be either up or down in direction, are absorption peaks for carbon dioxide. In fact, it is this absorbance of IR radiation by atmospheric CO_2 that contributes to global warming.

IR absorption due to CO_2 (and water vapor) in the sample compartment atmosphere is usually handled by acquiring a background scan (see page 318) of the air in the compartment and then subtracting this background from the sample scan to generate the final spectrum. If your spectrum shows fairly strong positive peaks near 2350 cm⁻¹, it is possible that no background scan was acquired prior to the sample scan. More commonly, residual CO_2 peaks that appear in the spectrum are due to incomplete subtraction of the background. If the background scan is favored in the subtraction procedure, these peaks can even appear as "negative" CO_2 peaks. In fact, the CO_2 levels in a laboratory can vary significantly over time as the number of people (breathing!) in the room changes; it is always a good idea to run a fresh background scan if one has not been acquired within the past half-hour.

Another method for dealing with absorbances that arise from atmospheric CO_2 or water vapor is to purge the sample compartment of these gases prior to scanning the sample. Your spectrometer sample compartment may be connected to a tank of CO_2-free compressed nitrogen or purified, dry air for this purpose. If your spectrometer uses a gas purge, the compartment should be closed except when a sample is being installed or removed, and sufficient time must be allowed for the closed sample compartment to be purged with purified air or nitrogen before obtaining the IR spectrum.

My Spectrum Shows an O–H Stretch Near 3500 cm⁻¹, but My Compound Isn't an Alcohol. How Is This Possible?

The appearance of a strong, broad peak near 3500 cm^{-1} from a sample that does not contain an O–H group is almost certainly due to contamination by water. If the sample is not scrupulously dry, the suspended or dissolved water will result in bands in the O–H stretching region near 3500 cm^{-1} and in the O–H bending region near 1650 cm^{-1}. In addition to producing a spectrum with misleading absorptions, the water would also etch NaCl disks used to contain a sample. Etched disks absorb and scatter infrared radiation, and future spectra will have less-resolved IR peaks.

It is also easy for a beginner to mistake an overtone peak from a strong C=O stretch for an O–H peak. True O–H stretches give rise to very intense and broad bands near 3500 cm^{-1}, whereas C=O overtone peaks in this same region are generally much sharper and weaker. If your spectrum shows a strong carbonyl stretch, expect to see a weak C=O overtone as well. A good example is seen in the spectrum of methyl acetate (see Figure 21.24, page 336). The signal at 3490 cm^{-1} is in the region where O–H stretching absorptions appear, but the peak is clearly not an O–H stretch because of its weak intensity and the shape of the absorption. It is an overtone of the intense C=O peak at 1745 cm^{-1}.

Why Are My Peaks Broadened and Flattened Out at the Bottom?

If the sample is too thick or too concentrated, strong IR bands will "bottom out" at 0% transmittance, producing wide absorption bands from which it is impossible to determine an exact absorption frequency. Small signals will also appear to have larger significance than they deserve, often leading to erroneous assignments of the IR peaks. The remedy is to prepare a less concentrated sample or thinner film.

Why Are All the Peaks in My Spectrum Broad and Indistinct?

If you are working with a thin film, it is likely that the sample has evaporated or migrated away from the sampling region of the infrared beam. With mulls, the solid sample probably has not been ground finely enough. With an ATR spectrum, a solid sample may have had poor contact with the ATR crystal. Grind the sample into a fine powder before applying it to an ATR crystal.

My Spectrum Has a Severely Sloping Baseline. What Can I Do?

A sloping baseline, as shown in the spectrum of fluorenone in Figure 21.27, is a problem with Nujol mulls that is difficult to avoid. A severely sloping baseline can be the result of a poorly ground solid, but even with careful grinding some samples still produce spectra with sloping baselines. With the availability of digitized data on an FTIR spectrometer, the baseline can be adjusted with the instrument's software.

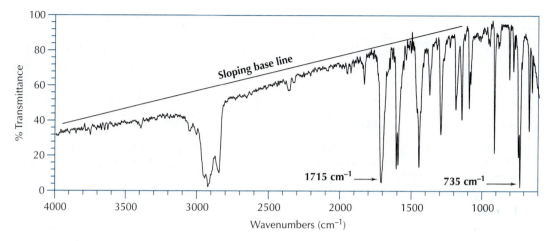

FIGURE 21.27 IR spectrum of fluorenone (Nujol mull).

My Spectrum Looks OK, but I'm Not Seeing Peaks That I Expected to See. Is My Sample Not What I Think It Is?

At times you may encounter spectra that seem internally inconsistent. For example, if you are working with the Nujol mull spectrum of a compound that you strongly suspect contains one or more functional groups, yet there are no signals in its IR spectrum indicating their presence, you have added too much mineral oil. This leads to a spectrum that is virtually indistinguishable from the spectrum of Nujol itself, shown in Figure 21.28.

On the other hand, when the IR spectrum of a symmetric or nearly symmetric compound is taken, an expected absorption peak may be missing from the spectrum because the stretching vibration is not IR-active. For example, the spectrum of diphenylethyne, shown in Figure 21.29, displays no characteristic C≡C stretch near 2200 cm^{-1}. The absence of the C≡C absorption is the result of symmetry; the C≡C bond does not have a dipole bond moment and its stretching vibration is not IR active.

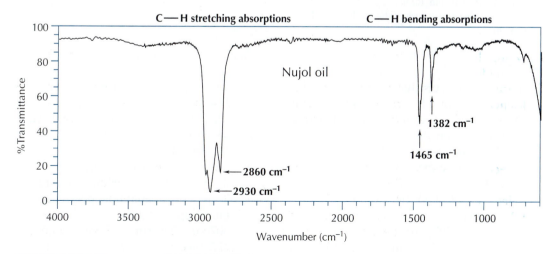

FIGURE 21.28 IR spectrum of Nujol (thin film).

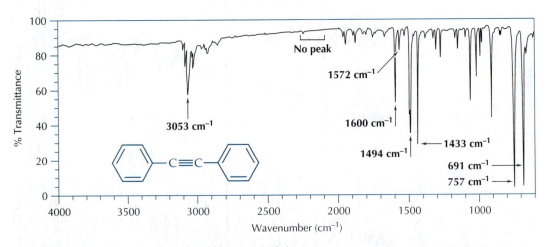

FIGURE 21.29 IR spectrum of diphenylethyne (KBr disk).

There Are Many More Peaks in My Spectrum Than I Expected. Am I Wrong About the Identity of My Sample?

Extra peaks in unexpected positions can lead to confusion, but keep in mind that the number of observed absorption bands is generally different from the total number of possible fundamental molecular vibrations. Additional peaks can result from overtone vibrations, combination vibrations, and the coupling of vibrations.

Further Reading

Crews, P.; Rodríguez, J.; Jaspars, M. *Organic Structure Analysis,* 2nd ed.; Oxford University Press: Oxford, 2009.

Silverstein, R. M.; Webster, F. X.; Kiemle, D. J. *Spectrometric Identification of Organic Compounds;* 7th ed.; Wiley: New York, 2005.

Larkin, P. *Infrared and Raman Spectroscopy: Principles and Spectral Interpretation;* Elsevier: Boston, 2011.

The Aldrich Library of FT-IR Spectra; 2nd ed.; Aldrich Chemical Company: Milwaukee, WI, 1997; 3 volumes.

Colthup, N. B.; Daly, L. H.; Wiberly, S. E. *Introduction to Infrared and Raman Spectroscopy,* 3rd ed.; Academic: Boston, 1990.

Questions

1. In each of the sets that follow, match the proper compound with the appropriate set of IR peaks and give the rationale for your assignment.

   (a) dodecane, 1-decene, 1-hexyne, 1,2-dimethylbenzene

   3311(s), 2961(s), 2119(m) cm^{-1}

   3020(s), 2940(s), 1606(s), 1495(s), 741(s) cm^{-1}

   3049(w), 2951(m), 1642(m) cm^{-1}

   2924(s), 1467(m) cm^{-1}

   (b) phenol, benzyl alcohol, methoxybenzene

   3060(m), 2835(m), 1498(s), 1247(s), 1040(s) cm^{-1}

   3370(s), 3045(m), 1595(s), 1224(s) cm^{-1}

   3330(br, s), 3030(m), 2950(m), 1454(m), 1223(s) cm^{-1}

   (c) 2-pentanone, acetophenone, 2-phenylpropanal, heptanoic acid, 2-methylpropanamide, phenyl acetate, 1-aminooctane

   3070(m), 2978(m), 2825(s), 2720(m), 1724(s) cm^{-1}

   3372(m), 3290(m), 2925(s) cm^{-1}

   3070(w), 1765(s), 1215(s), 1193(s) cm^{-1}

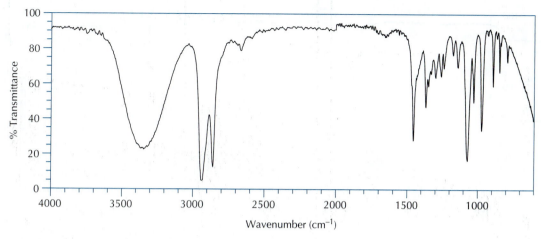

FIGURE 21.30 IR spectrum for question 2 (thin film).

3300–2500(br, s), 2950(m), 1711(s) cm^{-1}
3060(m), 2985(w), 1690(s) cm^{-1}
3352(s), 3170(s), 2960(m), 1640(s) cm^{-1}
2964(s), 1717(s) cm^{-1}

2. Treatment of cyclohexanone with sodium borohydride results in a product that can be isolated using distillation. The IR spectrum of this product is shown in Figure 21.30. Identify the product and assign the major IR bands.

3. In an attempt to prepare diphenylacetylene, 1,2-dibromo-1,2-diphenylethane is refluxed with potassium hydroxide. A hydrocarbon with the chemical formula $C_{14}H_{10}$ is isolated. The infrared spectrum exhibits signals at 3100–3000 cm^{-1} but no signals in the region 2300–2100 cm^{-1}. Is this spectrum consistent with a compound containing a carbon-carbon triple bond? Explain the absence of a signal in the 2300–2100 cm^{-1} region.

4. When benzene is treated with chloroethane in the presence of aluminum chloride, the product is expected to be ethylbenzene (bp 136°C). During the isolation of this product by distillation, some liquid of bp 80°C was obtained. Identify this product using its boiling point and the IR spectrum in Figure 21.31.

5. Consider the IR spectra shown in Figures 21.32 through 21.39 and match them to the following compounds: biphenyl, 4-isopropyl-1-methylbenzene, 1-butanol, phenol, 4-methylbenzaldehyde, ethyl propanoate, benzophenone, acetanilide. (**Note:** Some samples were prepared as thin films and others were prepared as Nujol mulls or by using an ATR accessory.)

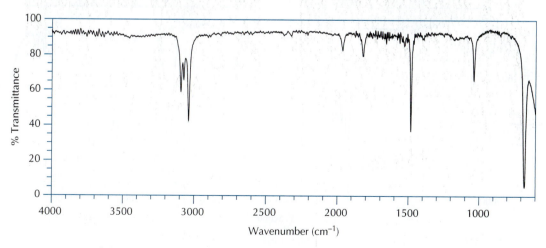

FIGURE 21.31 IR spectrum for question 4 (thin film).

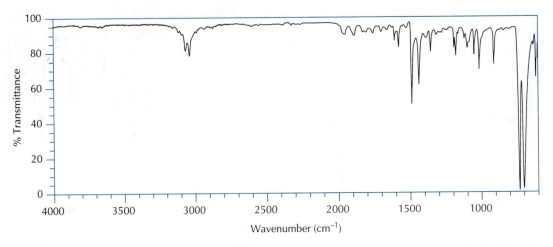

FIGURE 21.32 IR spectrum for question 5 (ATR).

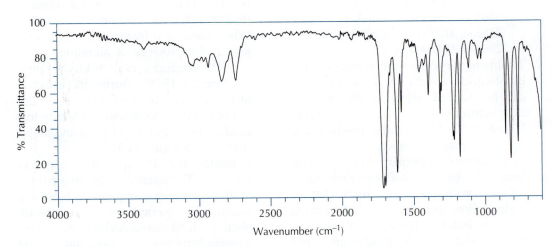

FIGURE 21.33 IR spectrum for question 5 (thin film).

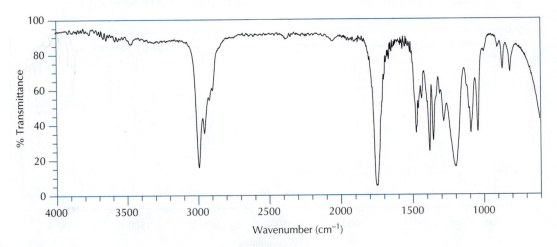

FIGURE 21.34 IR spectrum for question 5 (thin film).

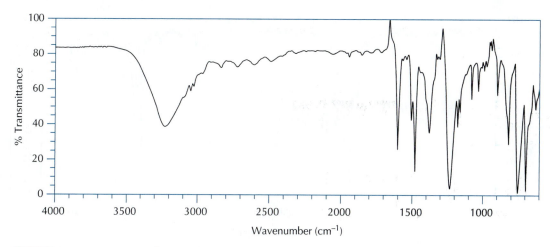

FIGURE 21.35 IR spectrum for question 5 (ATR).

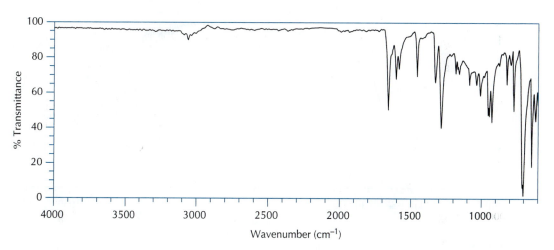

FIGURE 21.36 IR spectrum for question 5 (ATR).

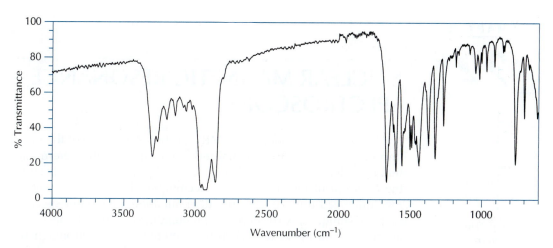

FIGURE 21.37 IR spectrum for question 5 (Nujol mull).

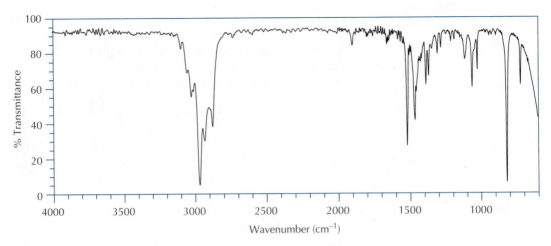

FIGURE 21.38 IR spectrum for question 5 (thin film).

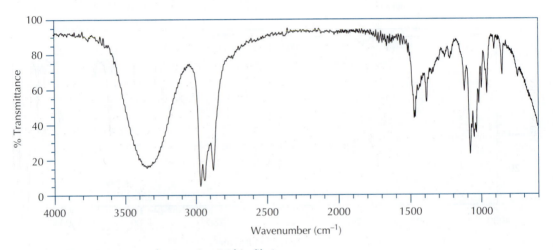

FIGURE 21.39 IR spectrum for question 5 (thin film).

CHAPTER

22

NUCLEAR MAGNETIC RESONANCE SPECTROSCOPY

If Chapter 22 is your introduction to spectroscopic analysis, read the Essay "The Spectrometric Revolution" on pages 309–310 before you read Chapter 22.

Nuclear magnetic resonance (NMR) spectroscopy is one of the most important modern instrumental techniques used in the determination of molecular structure. For the past 50 years, NMR has been in the forefront of the spectroscopic techniques that have completely revolutionized organic structure determination. Like other spectroscopic techniques, NMR depends on quantized energy changes that are induced in molecules when they interact with electromagnetic radiation. The energy needed for NMR is in the radio frequency range of the electromagnetic spectrum and is much lower energy than that needed by other spectroscopic techniques.

Nuclear Spin

The theoretical foundation for nuclear magnetic resonance arises from the *spin, I,* of an atomic nucleus. The value of I is related to the atomic number and the mass number and may be $0, \frac{1}{2}, 1, \frac{3}{2}, 2,$ and so forth. Any isotope whose nucleus has a nonzero magnetic moment ($I > 0$) is in theory detectable by NMR spectroscopy. Readily observable nuclei include $^1H, ^2H, ^{13}C, ^{15}N, ^{19}F,$ and $^{31}P.$ The most important nuclei for organic structure determination are ^1H and $^{13}C,$ both of which have spin of $\frac{1}{2}.$ ^1H NMR is the focus of this chapter and ^{13}C NMR is the focus of Chapter 23. The basic principles of NMR, which apply both to ^1H and ^{13}C NMR, are discussed in this chapter.

Any nucleus with both an even atomic number and an even mass number has a nuclear spin of 0. Because ^{12}C and ^{16}O have nuclear spins of 0, they do not produce NMR signals and do not interfere with or complicate the signals from ^1H and $^{13}C.$ In addition, ^{12}C is the major isotope of carbon and is present in almost 99% natural abundance. Therefore, the small amount of NMR-active ^{13}C does not complicate ^1H NMR spectra to any great extent.

Nuclear Energy Levels

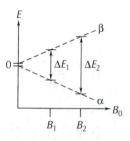

FIGURE 22.1 Influence of an external magnetic field on spin state energy levels.

There are $(2I + 1)$ energy levels allowed for a nucleus with spin of $I.$ Because ^1H and ^{13}C have spins of $\frac{1}{2},$ there are two possible energy levels for these nuclei ($2I + 1 = 2$). In the absence of an external magnetic field, the two levels are degenerate—they have the same energy. However, in the presence of an applied magnetic field, the energy levels move apart. The separation of degenerate nuclear spin energy levels by an external magnetic field is illustrated in Figure 22.1. One energy level, designated $\alpha,$ decreases in energy and the other level, designated $\beta,$ increases in energy. The difference in energy between the levels, $\Delta E,$ is directly related to the strength of the applied magnetic field, $B_0.$

In spectroscopy, the usual convention for expressing energy changes is frequency (ν), as described by Planck's law:

$$\Delta E = h\nu$$

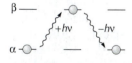

FIGURE 22.2 Excitation of a nucleus from its low-energy state to its high-energy state and emission of energy upon relaxation of the nucleus.

The change in energy of an NMR transition is extremely small by chemical standards—only about 10^{-6} kJ/mol, which corresponds to energy in the radio frequency region. With a magnetic field strength of 1.41 tesla (T), the resonance frequency for ^1H nuclei is 60 MHz; if the magnetic field strength is 7.05T, the resonance frequency is 300 MHz.

As shown in Figure 22.2, the absorption of energy can cause excitation of a nucleus from the α to the β energy level. When a nucleus in the higher-energy state drops to the lower-energy state, in a process called *relaxation,* it gives up a quantum of energy. The emitted energy, in the radio frequency region, produces an NMR signal.

Magnetic Resonance

We can think of any nucleus with a spin number greater than 0 as a spinning, charged body. The principles of physics tell us that a magnetic field is associated with this moving charge. When placed in an external magnetic field, a spinning nucleus precesses about an axis aligned in the direction of the magnetic field. The precession of a child's top about a vertical axis as it spins can be used as a mechanical model for this process. The magnetic dipole of the spinning

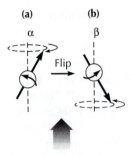

(a)   (b)

α   β

Flip

Magnetic field direction

FIGURE 22.3 Nuclear magnetic dipole (a) aligned with an external magnetic field (α) and (b) opposed to an external magnetic field (β).

nucleus shown in Figure 22.3a is aligned with the external magnetic field, whereas the magnetic dipole of the spinning nucleus shown in Figure 22.3b is opposed to the external magnetic field. Flipping the magnetic dipole from the aligned position to the opposed position requires a quantized addition of energy to the system. Absorption of energy can occur only if the system is in resonance.

For *resonance* to occur, the applied frequency (v) must be precisely tuned to the rotational frequency of the precessing nucleus. Then the nucleus can absorb a quantum of energy and flip from the lower-energy spin state (α) to the higher-energy spin state (β). The energy difference between the two spin states is very small and the number of nuclei in each spin state is nearly equal; but in the large magnetic field of a modern NMR spectrometer there are a few more nuclei, approximately 0.001%, in the lower-energy spin state than in the higher-energy spin state. Because the spin states are not equally populated, an NMR effect can be observed.

If all the ^1H nuclei in a molecule had the same resonance frequency, ^1H NMR spectroscopy would be of little use to organic chemists. In an NMR spectrometer, however, energies of the ^1H nuclei in an organic compound differ slightly because of their different structural environments, and a typical ^1H NMR spectrum is an array of many different frequencies. The same is true for ^{13}C NMR spectra.

22.1   NMR Instrumentation

The first NMR spectrometers were continuous wave (cw) instruments. The sample was irradiated with radio frequency (RF) energy as the applied magnetic field was varied. When a match between the RF energy and the energy difference between the two spin states of the nucleus—(hv) in Figure 22.2—occurred, a signal was detected. The energy required reflected the environment of the nucleus. A radio frequency receiver was used to monitor the energy changes.

Fourier Transform NMR

More recent instruments use a technique known as *pulsed Fourier transform NMR (FT NMR)*. In this technique, a broad pulse of electromagnetic radiation excites all the ^1H or ^{13}C nuclei simultaneously, resulting in a continuously decreasing oscillation caused by the decay of excited nuclei back to their stable energy distribution. The oscillating, or decaying, sine curve is called a *free-induction decay (FID)*. The FID, often referred to as a time domain signal, is converted to a set of frequencies, or a "normal" spectrum, by the mathematical treatment of a Fourier transform. The relatively simple FID of a compound with only a single frequency is shown in Figure 22.4a. You can see, however, in Figure 22.4b that the FID from a compound with two frequencies is more complex. Constructive combinations of the two frequencies produce enhanced signals, and destructive combinations give little or no signal. The Fourier transform of the FID in Figure 22.4b produces two signals.

The FIDs of most organic compounds are made up of the contributions from tens or even hundreds of frequencies and a computer program using Fourier transform mathematics is required to convert the FID to the "normal" spectrum. When the sample size is small, the acquisition of the signal from more than one pulse (or scan)

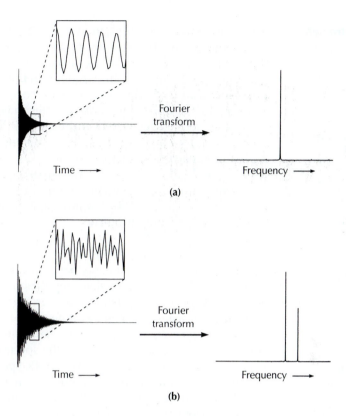

FIGURE 22.4 FID and Fourier transform of the FID of (a) one signal and (b) two signals.

is necessary to obtain NMR signals with the desired signal-to-noise ratio. Modern FT NMR spectrometers "lock" on a signal from deuterium in the NMR solvent, assuring that multiple acquisition scans are synchronized. The NMR computer programs the multiple pulses and collects the data from them. The components of an FT NMR spectrometer are illustrated in Figure 22.5.

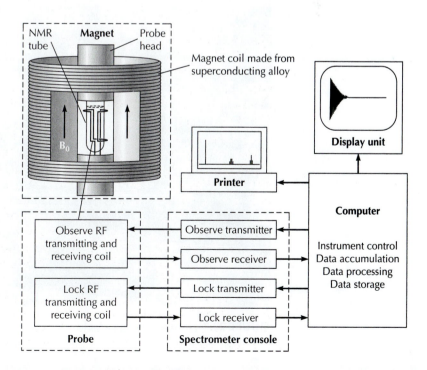

FIGURE 22.5 Block diagram of a basic FT NMR spectrometer.

NMR Spectrometers

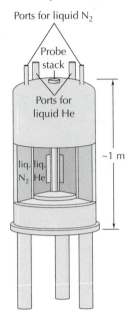

Ports for liquid N₂

Probe stack

Ports for liquid He

~1 m

liq. N₂ liq. He

FIGURE 22.6 Typical superconducting electromagnet for a 200- to 400-MHz NMR spectrometer, showing the vacuum-jacketed Dewar vessels for the liquid nitrogen and liquid helium coolants.

There are many different models of NMR spectrometers, and it is common practice to refer to them by their nominal operating frequency. Modern NMR instruments operate at substantially higher frequencies than the 60-MHz instruments that were once the standard. Research instruments routinely operate at 300–500 MHz, and many laboratories have instruments with operating frequencies of 600 and even 800 MHz. The high magnetic fields necessary for these instruments can be achieved only by using superconducting electromagnets. Because the materials used to build these magnets are superconducting only at very low temperatures, the magnets are maintained in double-jacketed Dewar vessels cooled by liquid helium and liquid nitrogen (Figure 22.6). An operating console for producing NMR spectra is shown in Figure 22.7.

There are numerous benefits to using higher frequency and field strength. High-field instruments have greater sensitivity because of a greater difference in spin state populations, which translates into stronger sample signals relative to background noise. The higher field strength also means larger energy differences between different nuclei and thus greater signal separation. The advantage of greater signal separation is evident by comparing the spectra of ethyl propanoate, $C_5H_{10}O_2$, shown in Figure 22.8. The NMR spectrum in Figure 22.8a was obtained on a 60-MHz continuous wave (cw) NMR spectrometer. The spectrum in Figure 22.8b was obtained with a 200-MHz FT NMR spectrometer. In the 60-MHz spectrum, a group of signals is centered at 1.15 ppm, which in the 200-MHz spectrum separates into two groups of signals, one centered at 1.15 ppm and one centered at 1.26 ppm.

Paul Schatz

FIGURE 22.7 NMR operating console.

(a)

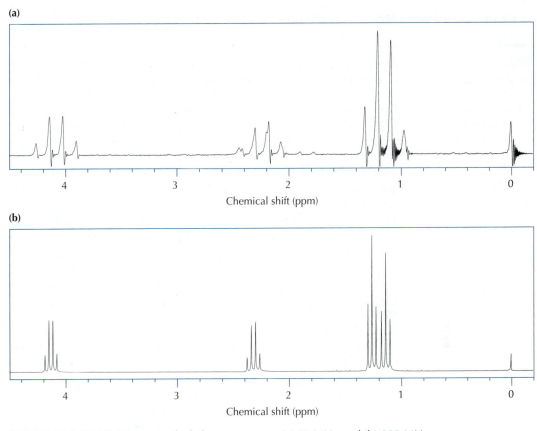

(b)

FIGURE 22.8 ¹H NMR spectra of ethyl propanoate at (a) 60 MHz and (b) 200 MHz.

22.2   Preparing Samples for NMR Analysis

Almost all NMR analysis is done using dilute solutions, whether the sample is a solid or a liquid. Concentrated solutions, undiluted liquids, and solids usually exhibit broad peaks that are not easy to interpret. Spectra with sharp, well-differentiated signals are obtained only with samples dissolved in dilute solutions. It is important to use the minimum amount of sample required for an NMR spectrum, but not much more.

NMR Solvents   Choosing an appropriate solvent for an NMR sample is important. Most of the material in an NMR tube is solvent, so ideally you want a chemically inert solvent that does not absorb energy in the magnetic field. Thus, for ¹H NMR you want a solvent with no protons. Most of the solvents used for preparation of NMR samples are deuterated forms of common solvents, such as chloroform ($CHCl_3$), acetone, and water. Although deuterium does have a magnetic moment, its signal is well removed from the region where protons absorb.

Deuterated chloroform ($CDCl_3$) is the most commonly used NMR solvent because it dissolves a wide range of organic compounds and is not prohibitively expensive. Deuterated acetone is another commonly used solvent, but it is quite a bit more expensive. Deuterated solvents are never 100% deuterated. For example, the commercial $CDCl_3$ that is commonly used for NMR samples has

TABLE 22.1	Deuterated solvents used for NMR spectroscopy		
Solvent	Structure	Residual ^1H signal (ppm)	^{13}C chemical shift (ppm)
Chloroform-d	$CDCl_3$	7.26 (singlet)	77.0 (triplet)
Acetone-d$_6$	$CD_3(C=O)CD_3$	2.04 (quintet)	29.8 (septet)
			206.5 (singlet)
Deuterium oxide	D_2O	4.6 (broad singlet)	—
Dimethyl sulfoxide-d$_6$	$CD_3(S=O)CD_3$	2.49 (quintet)	39.7 (septet)

99.8% deuterium and 0.2% protium in its molecules. The residual protons give a small peak ($CHCl_3$) at 7.26 ppm. Residual proton signals for various solvents are listed in Table 22.1. **It is important to be aware of the position of these residual signals because you do not want to confuse solvent signals with the signals of your sample compounds.**

Polar compounds. Polar chemicals, such as carboxylic acids and polyhydroxyl compounds, are usually not soluble in $CDCl_3$; however, in most cases these compounds are soluble in deuterium oxide (D_2O). If a carboxylic acid is not soluble in D_2O, it is probably soluble in D_2O containing sodium hydroxide. Adding a drop or two of concentrated sodium hydroxide solution to the sample in D_2O is usually enough to dissolve it.

Problems with the use of D_2O. The use of D_2O presents some problems. There is always a broad peak at approximately 4.6 ppm because of a small amount of HOD present in the original D_2O solvent. This solvent peak can hide important signals from the compound being analyzed. Also, D_2O may exchange with protons in the sample, producing HOD. Consider, for example, what happens when a carboxylic acid or alcohol dissolves in D_2O:

$$R-\overset{\overset{\displaystyle O}{\|}}{C}-O-H + D_2O \rightleftharpoons R-\overset{\overset{\displaystyle O}{\|}}{C}-O-D + H-OD$$

Carboxylic acid

$$R-O-H + D_2O \rightleftharpoons R-O-D + H-OD$$

Alcohol

Deuterium nuclei are "invisible" in ^1H NMR spectra, and in an NMR solution there are many more molecules of solvent D_2O than of the sample. The equilibrium positions in these reactions lie well to the right, and the hydroxyl protons of carboxylic acids, alcohols, and other compounds that can undergo H-D exchange do not appear as separate signals but instead merge into the HOD signal.

NMR Reference Calibration

Solvents used to prepare NMR samples often have a small amount of a standard reference substance dissolved in them. A reference compound is not really necessary, however, because the residual proton signal of a partially deuterated solvent can be used for reference calibration unless sample signals obscure it (Table 22.1).

Tetramethylsilane. The most common added reference compound is tetramethylsilane, $(CH_3)_4Si$. **Tetramethylsilane, usually referred to as TMS, has been so important as a reference substance in the past that the position of its signal is used to define the 0.0 ppm point on an NMR spectrum.** TMS was chosen because all its protons are equivalent, and they absorb at a magnetic field in which very few other protons in typical organic compounds absorb. TMS is also chemically inert and is soluble in most organic solvents. The amount of TMS in the solvent depends on the type of instrument being used. For a cw NMR spectrometer, the typical concentration of TMS is 1–2%. With a modern FT NMR instrument, the typical concentration of TMS is 0.1%; often TMS is not even added to the NMR solution.

NMR reference for D_2O. Tetramethylsilane is not soluble in deuterium oxide (D_2O), so it cannot be used as a standard with this solvent, and the HOD peak is too broad and variable to be a useful reference standard. The reference substance used for D_2O solutions is the ionic compound sodium 2,2-dimethyl-2-silapentane-5-sulfonate (DSS), $(CH_3)_3SiCH_2CH_2CH_2SO_3^-Na^+$. Its major signal appears at nearly the same position as the TMS absorption. A tiny amount of acetone can also be used as a reference in D_2O solutions as long as its signal does not interfere with signals from the sample. In D_2O solutions, the signal for acetone appears at 2.22 ppm.

Preparing an NMR Sample Solution

The appropriate concentration of the sample solution depends on the type of NMR instrumentation available. A sample mass of 4–20 mg of compound dissolved in approximately 0.5–0.7 mL of solvent is used to prepare a modern high-frequency FT NMR sample solution.

NMR tubes. NMR tubes are delicate, precision pieces of equipment. The most commonly used NMR tubes are made of thin glass and their rims are easily chipped if not handled carefully. **Caution:** Chipping occurs most often during pipetting of the sample into the tube and when trying to remove the plastic cap from the top of the tube.

Check solubility in deuterium-free solvent first. Before using an expensive deuterated solvent for an NMR analysis, be sure that your compound dissolves in the deuterium-free solvent. Prepare a preliminary sample using the necessary amount of solvent in a small vial or test tube. If the preliminary test is satisfactory, place the necessary amount of your sample in another vial or small test tube and add approximately 0.7 mL of deuterated solvent. Agitate the mixture to facilitate dissolution of the sample. If a clear, homogeneous solution is obtained, transfer the sample to the NMR tube with a glass Pasteur pipet.

Particulate matter in the sample solution. If a clear, homogeneous solution is not obtained, the particulate material must be removed before the sample is transferred to the NMR tube. Particulate material may contain paramagnetic metallic impurities that will produce extensive line broadening and poor signal intensity in the NMR spectrum. A convenient filter can be prepared by inserting a

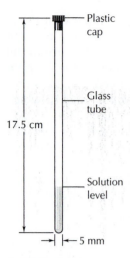

FIGURE 22.9 NMR sample tube filled to the correct height.

small wad of glass wool into the neck of a glass Pasteur pipet (see Section 9.3). The narrow end of the filter pipet is placed in the NMR tube and the sample to be filtered is transferred into the filter pipet with a second Pasteur pipet. Pressure from a pipet bulb can be used to force any solution trapped in the filter into the NMR tube.

Height of the NMR solution in the tube. Only a small part of an NMR tube is in the effective probe area of the NMR spectrometer. Typically, the height of the sample in the tube should be 25–30 mm (Figure 22.9); however, the required height can be 50–55 mm in some NMR instruments. You need to ascertain the required minimum height for the instrument you are using. Often a gauge is available in the lab for checking the solution height in the NMR tube. If the solution height is slightly short, add a few drops of solvent to bring it to the required level, but too much solution may also produce a poor-quality spectrum. Agitate the NMR tube to thoroughly mix the solution. Cap the tube and wipe off any material on the outside.

Recovery of the sample. Because none of the sample is destroyed when taking an NMR spectrum, the sample can be recovered if necessary by evaporating the solvent.

Obtaining the NMR Spectrum

Before the NMR sample tube is placed into the magnet of the spectrometer, it is often fitted with a collar that is made of a nonmagnetic plastic or ceramic material (Figure 22.10). The collar positions the sample at a precise location within the magnetic field where the RF transmitter/receiver coil is located. A depth gauge provided with the instrument is often used to set the position of the collar on the NMR sample tube. The collar can also be used to enable the sample to spin around its vertical axis once it has been placed in the magnet.

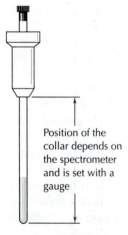

FIGURE 22.10 NMR sample tube fitted with a collar.

The magnetic field in the RF transmitter/receiver coil region must be homogeneous; that is, the strength and direction of the magnetic field must be exactly the same at every point. A homogeneous magnetic field is achieved through a complex adjustment called *shimming*, which changes the magnetic field using electrical coils with adjustable current. Even after shimming, some small magnetic field inhomogeneities may be present. Spinning the sample averages out these inhomogeneities, which allows acquisition of spectra with sharp, well-defined peaks. NMR tubes are selected for uniform wall thickness and minimum wobble. Too much sample in the tube is not only a waste of material; it also tends to make the tube top-heavy, often resulting in poor spinning performance and thus poor-quality spectra. With some modern NMR spectrometers, the magnet technology has advanced to the point that spinning the sample is not necessary.

Cleaning the NMR Sample Tube

After the spectrum has been obtained, the NMR tube should be cleaned, usually by rinsing with a solvent such as acetone and then allowing the tube to dry. Solvents cling tenaciously to the inside surface of the long, thin NMR tubes, and a long drying period or passing a stream of dry nitrogen gas through the tube is required to remove all residual solvent. If NMR tubes are not cleaned soon after use, the solvent usually evaporates and leaves a caked or gummy residue that can be difficult to dissolve.

22.3 Summary of Steps for Preparing an NMR Sample

1. Test the solubility of the sample in ordinary, nondeuterated NMR solvents. Select a solvent that dissolves the sample completely.
2. Place 4–20 mg of the sample in a clean, small vial or test tube.
3. Add 0.5–0.7 mL of the appropriate deuterated solvent.
4. Agitate the mixture in the vial to produce dissolution of the sample.
5. If there are any solids present, filter the solution through a small plug of glass wool.
6. Transfer the sample solution into a clean NMR tube using a glass Pasteur pipet.
7. Check the level of the sample in the tube. If needed, add drops of solvent to bring the solution to the recommended level for the instrument and agitate the mixture to produce a homogeneous solution.
8. Cap the NMR tube.
9. Wipe the outside of the tube to remove any material that may impede smooth spinning of the sample in the NMR instrument.

22.4 Interpreting ^{1}H NMR Spectra

The rest of this chapter will be devoted to learning how to read an NMR spectrum. It is just as important for a chemist to be able to read a spectrum as it is for a radiologist to read an MRI or CAT scan. Typically, four types of information can be extracted from a ^{1}H NMR spectrum, and all are important in determining the structure of a compound.

- *Number of different kinds of protons* in the molecules of the sample, given by the number of groups of signals (see Section 22.5)
- Relative number of protons contributing to each group of signals in the spectrum, called *integration* (see Section 22.6)
- Positions of the groups of signals along the horizontal axis, called the *chemical shift* (see Sections 22.7 and 22.8)
- Patterns within groups of signals, called *spin-spin coupling* (see Section 22.9)

22.5 How Many Types of Protons Are Present?

As the first step in analyzing an NMR spectrum, examine the entire spectrum. A common mistake is to focus on some detail, often a prominent signal, and develop an analysis from an assumption that is consistent with only that detail. Sometimes this method works, but many times it does not.

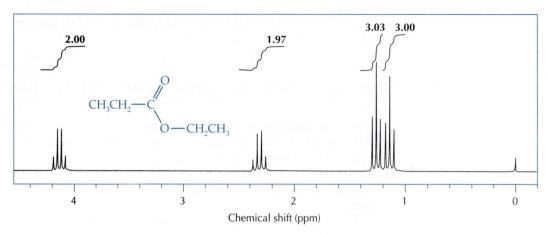

FIGURE 22.11 ¹H NMR spectrum of ethyl propanoate at 200 MHz.

Start by looking at the entire spectrum and counting the number of groups of signals. A structure consistent with the spectrum must have at least as many different kinds of protons as there are groups of signals. This number is a minimum requirement, and often, as the analysis is refined, it is possible to divide a group of signals into subsets of protons that are subtly different from each other. If you examine the 200-MHz NMR spectrum of ethyl propanoate (Figure 22.11), you will see that four groups of signals are centered at 1.15, 1.26, 2.32, and 4.13 ppm along the horizontal scale. **Note that the horizontal scale is read from right to left, with 0.0 ppm at the far right.**

22.6   Counting Protons (Integration)

Above each group of signals in the 200-MHz NMR spectrum of ethyl propanoate in Figure 22.11 is what looks like a set of steps with a number over it. **The height of each set of steps corresponds to the total relative signal intensity encompassed by the set.** The numbers are normalized to one of the signals. Software on modern digital NMR spectrometers makes normalization an easy task. Reading from right to left, the normalized integration values for the groups of signals are 3.00, 3.03, 1.97, and 2.00, respectively.

Integration values represent the relative number of each kind of proton in the molecule. If the normalization is not done correctly, the integration values will be a multiple of the true values. Also, integration values are usually not neat, whole-number ratios. Deviations from whole numbers can be as much as 10% and are usually attributed to differences in the amount of time it takes different types of excited hydrogen nuclei to relax back to their lower-energy spin states. In acquiring NMR data, it is important to allow enough time for the nuclei to relax. Otherwise, the measured integrals will not accurately reflect the relative number of protons. If the integration is done manually, you must use good judgment about where to start and stop each set of steps.

The integrals for the spectrum in Figure 22.11 are interpreted as 3:3:2:2. The two groups of signals at 1.15 ppm and 1.26 ppm, with three protons each, are produced by two groups of nearly equivalent kinds of protons. The group of signals at 2.32 ppm and the group at 4.13 ppm are each produced by a group of two equivalent protons.

- Primary hydrogens, those on a carbon atom with three hydrogens attached, are called *methyl* protons.
- Secondary hydrogens, on a carbon atom with two hydrogens attached, are called *methylene* protons.
- A tertiary hydrogen, on a carbon atom with only one hydrogen attached, is called a *methine* proton.

EXERCISE

Refer to the structure of ethyl propanoate and identify the set of protons that is responsible for each of the four groups of signals in its NMR spectrum.

$$CH_3CH_2-C\overset{\displaystyle O}{\underset{\displaystyle O-CH_2CH_3}{\big\|}}$$

Ethyl propanoate

Answer: Because there are two groups of two protons and two groups of three protons, we cannot unambiguously assign the signals without more information. But help is on the way. In the next section you will find out how to use the positions of the signals along the horizontal scale to make the necessary assignments.

22.7 Chemical Shift

An NMR spectrum is a plot of the intensity of the NMR signals versus the magnetic field or frequency. Nuclei that are chemically equivalent, such as the four protons in methane (CH_4) or the two protons in dichloromethane (CH_2Cl_2), show only one peak in the NMR spectrum; however, protons that are not chemically equivalent absorb at different frequencies. At 300 MHz, the typical range of these frequencies is about 3500 Hz. The local magnetic field experienced by the different protons in a molecule varies with different magnetic environments within the molecule.

Most importantly, the positions of the signals along the horizontal scale of an NMR spectrum, called the *chemical shifts,* can be correlated with a molecule's structure. The goal of Sections 22.7 and 22.8 is to show you how the chemical shifts can be used to determine the structures of organic compounds. **Chemical shifts are arguably the most powerful of all the information available in NMR spectroscopy.**

Chemical Shift Units (Parts per Million, ppm)

Because it is difficult to reproduce magnetic fields exactly enough for NMR spectroscopy, an internal standard is used as a reference point. The position of an NMR signal is measured relative to the absorption of the standard. Tetramethylsilane (TMS), $(CH_3)_4Si$, is the standard for ^{1}H and ^{13}C NMR. Chemical shifts are measured at a frequency (Hertz, which is abbreviated **Hz**) corresponding to a signal's position relative to TMS. It is conventional, however, to convert frequency to a value **δ** (**ppm**) by dividing the chemical shift frequency by the operating frequency of the spectrometer. This conversion produces an important result—**the chemical shift (δ) is independent of the frequency of the spectrometer.**

M = mega = million

$$\delta(\text{ppm}) = \frac{\text{frequency of the signal (in Hz, from TMS)}}{\text{applied spectrometer frequency (in MHz)}}$$

Because the frequency of an NMR spectrometer is given in megahertz (MHz), the δ values are always given in parts per million (ppm). On the chemical shift scale of an NMR spectrum, the position of the TMS absorption is at the far right and is set at 0.0 ppm. The δ values increase to the left of the TMS peak.

Consider the two NMR spectra of *tert*-butyl acetate shown in Figure 22.12. The *tert*-butyl group, which has nine equivalent protons, and the methyl group, which has three equivalent protons, give a relative integration of 3:1. In the 60-MHz spectrum (Figure 22.12a), the difference between the signals of TMS and the *tert*-butyl group is 87 Hz. In the 200-MHz spectrum (Figure 22.12b), the difference between these same signals is 290 Hz. Dividing each

(a)

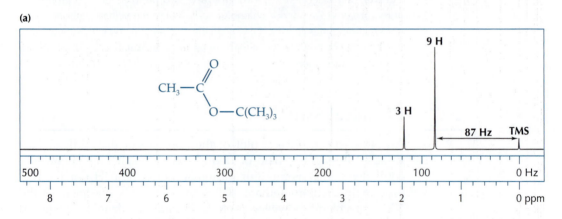

(b)

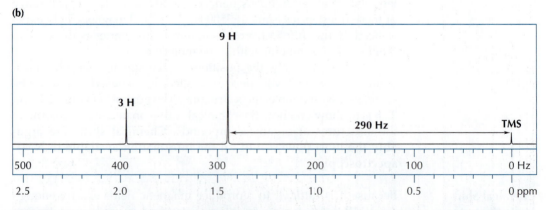

FIGURE 22.12 ^{1}H NMR spectra of *tert*-butyl acetate in the region from 0 to 500 Hz at (a) 60 MHz and (b) 200 MHz. The chemical shift of each signal is the same regardless of the spectrometer frequency.

signal's frequency by the operating frequency of the instrument, we find that the chemical shift (δ) of the *tert*-butyl protons is 1.45 ppm.

$$1.45 \text{ ppm} = 87 \text{ Hz}/60 \text{ MHz} = 290 \text{ Hz}/200 \text{ MHz}$$

The position of the signal in terms of its chemical shift (δ) is the same, regardless of the magnetic field strength. **To compare NMR spectra from different instruments, the chemical shift scales for all NMR spectra are plotted using ppm units.**

EXERCISE

On an NMR instrument operating at 60 MHz, the signal for the methyl group of *tert*-butyl acetate is shifted 118 Hz relative to the signal for TMS (see Figure 22.12a).

(a) What is the chemical shift (δ) of the methyl group signal?
(b) What is the frequency difference (in Hz) between the signal for the methyl group and the signal for TMS on an NMR instrument operating at 200 MHz (see Figure 22.12b)?

Answer: (a) δ = 118 Hz/60 MHz = 1.97 ppm
(b) ΔHz = 1.97 ppm × 200 MHz = 394 Hz

Figure 22.13 shows the approximate chemical shift regions of signals for different types of protons attached to carbon, oxygen, and nitrogen atoms. A list of chemical shifts for different types of protons is given in Table 22.2.

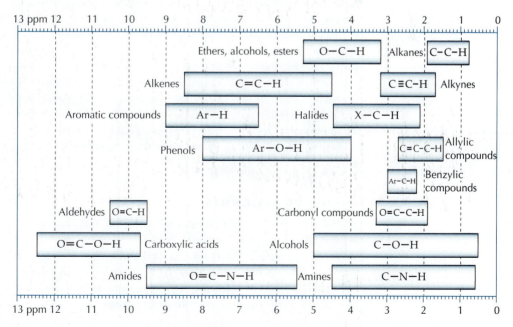

FIGURE 22.13 Approximate regions of chemical shifts for different types of protons in organic compounds.

TABLE 22.2   **Characteristic ^{1}H NMR chemical shifts in CDCl$_3$**

Compound	Chemical shift (δ, ppm)
TMS	0.0
Alkanes (C–C–**H**)	0.8–1.9
Amines (C–N–**H**)	0.6–4.5
Alcohols (C–O–**H**)	0.5–5.0
Alkenes^a (C=C–C–**H**)	1.5–2.6
Alkynes (C≡C–**H**)	1.7–3.1
Carbonyl compounds (O=C–C–**H**)	1.9–3.3
Halides (X–C–**H**)	2.1–4.5
Aromatic compounds^b (Ar–C–**H**)	2.2–3.0
Alcohols, esters, ethers (O–C–**H**)	3.2–5.3
Alkenes (C=C–**H**)	4.5–8.5
Phenols (Ar–O–**H**)	4.0–8.0
Amides (O=C–N–**H**)	5.5–9.5
Aromatic compounds (Ar–**H**)	6.5–9.0
Aldehydes (O=C–**H**)	9.5–10.5
Carboxylic acids (O=C–O–**H**)	9.7–12.5

a. Allylic protons.
b. Benzylic protons.

EXERCISE

The ^{1}H NMR spectrum of (CH$_3$)$_2$C(OH)C≡C–H shown in Figure 22.14 has three peaks. Determine the chemical shift of each peak and assign each one to the appropriate proton(s).

Answer: The respective chemical shifts are 1.41, 2.33, and 3.25 ppm. Notice that the peak integrations are 6:1:1. The peak with the relative integration of six protons must be the six protons of the two equivalent methyl groups; the chemical shift is also consistent with the range of chemical shifts for alkane protons (0.8–1.9 ppm) in Table 22.2. Integration alone cannot distinguish the protons attached to the alcohol oxygen atom and the alkyne carbon, but the chemical shift values can. The O–H proton has a wide range of chemical shifts (0.5–5.0 ppm), but the alkyne proton has a narrower chemical shift range (1.7–3.1 ppm). The peak at 2.33 ppm is within this range, but the 3.25 ppm peak lies outside of it. Thus, most likely the peak at 3.25 ppm is the O–H proton and the peak at 2.33 ppm is the alkyne proton. A method for calculating much more accurate chemical shift values is presented in Section 22.8.

FOLLOW-UP ASSIGNMENT

The label from a bottle of an ester has fallen off and two unstuck labels are found nearby, one of which likely fell off the bottle. One label shows the formula C$_6$H$_5$CH$_2$OC(=O)CH$_3$ and the other label shows the formula C$_6$H$_5$C(=O)OCH$_2$CH$_3$. The NMR spectrum of the ester is shown in Figure 22.15. Which label is the correct one for the unknown ester? Use chemical shift values to justify your answer.

Diamagnetic Shielding

The chemical shift of a hydrogen nucleus is strongly influenced by the electron density surrounding it. Under the influence of an

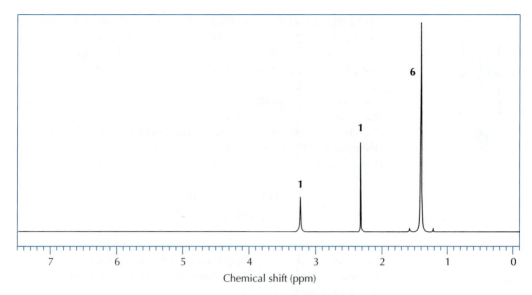

FIGURE 22.14 ^{1}H NMR spectrum of $(CH_3)_2C(OH)C\equiv CH$ at 360 MHz.

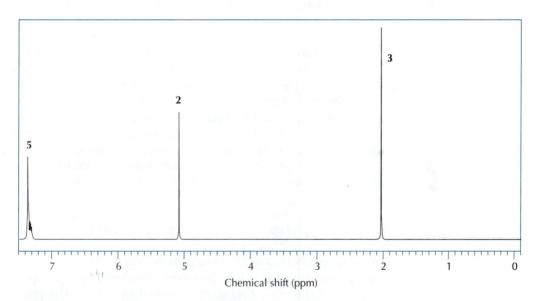

FIGURE 22.15 ^{1}H NMR spectrum of unknown ester at 360 MHz.

applied magnetic field, circulating electrons in the spherical electron cloud induce a small magnetic field opposed to the applied field, as illustrated in Figure 22.16. Thus, the effective magnetic field that a proton feels is a little less than the applied field. The electron cloud is said to shield the nucleus from the applied magnetic field, and the effect is called *local diamagnetic shielding.*

If the electron density around a proton is decreased, the opposing induced magnetic field will be smaller. Therefore, the nucleus is less shielded from the applied magnetic field, and the proton is said to be *deshielded.* With greater deshielding, the effective magnetic field felt by the proton increases, and the chemical shift of its signal increases. For example, the protons of methane resonate at 0.23 ppm.

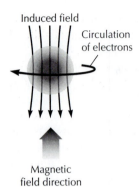

Induced field
Circulation of electrons

Magnetic field direction

FIGURE 22.16 The opposing magnetic field induced by circulation of electrons around a nucleus in an applied magnetic field. The nucleus is partially shielded from the applied magnetic field by the opposing magnetic field.

Attaching an electron-withdrawing chlorine atom to the carbon atom pulls electron density away from the electron cloud surrounding the nearby protons. Thus, the chlorine deshields the protons. The protons of chloromethane resonate at 3.1 ppm.

Magnitude of the deshielding effect. The magnitude of the deshielding effect decreases rapidly as the distance from the electron-withdrawing substituent increases. This effect is demonstrated by the decrease in the chemical shifts of methyl protons as their distance from a bromine atom increases.

CH_3Br	CH_3CH_2Br	$CH_3CH_2CH_2Br$	$CH_3CH_2CH_2CH_2Br$
2.69 ppm	1.66 ppm	1.06 ppm	0.93 ppm

Deshielding and shielding effects are additive. For instance, the chemical shift of the protons in substituted methane derivatives increases as the number of attached electron-withdrawing bromine atoms increases.

CH_4	CH_3Br	CH_2Br_2	$CHBr_3$
0.23 ppm	2.69 ppm	4.94 ppm	6.82 ppm

The additive nature of the deshielding effect is also seen as the carbon atom bearing the proton becomes more highly substituted. With the same electron-withdrawing groups nearby, tertiary hydrogen atoms have a greater chemical shift than do secondary hydrogens. Likewise, secondary hydrogen atoms have a greater chemical shift than do primary hydrogens with the same electron-withdrawing groups nearby. This trend is illustrated by the chemical shifts of the proton(s) attached to the carbon atom adjacent to a bromine atom in bromomethane, bromoethane, and 2-bromopropane.

CH_3Br	CH_3CH_2Br	$(CH_3)_2 CHBr$
2.69 ppm	3.37 ppm	4.21 ppm

Deshielding effects and electronegativity. The position of the signal for a proton attached to a carbon atom also depends on the electronegativity (χ) of the other atoms attached to carbon. The periodic trends seen in the electronegativities of elements are mirrored in the chemical shifts of methyl groups attached to these elements.

	CH_3-I	CH_3-Br	CH_3-Cl	CH_3-F
δ	2.2 ppm	2.7 ppm	3.1 ppm	4.3 ppm
χ	2.66	2.96	3.16	3.98

Similarly, as you move from left to right along a row of elements in the periodic table, the electronegativities increase and the chemical shifts of attached methyl groups also increase.

	$(CH_3)_4C$	$(CH_3)_3N$	$(CH_3)_2O$	CH_3F
δ	0.9 ppm	2.2 ppm	3.2 ppm	4.3 ppm
χ	2.50	3.04	3.44	3.98

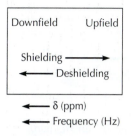

Downfield   Upfield

Shielding ⟶
⟵ Deshielding

⟵ δ (ppm)
⟵ Frequency (Hz)

FIGURE 22.17
Common NMR
terminology.

Anisotropy

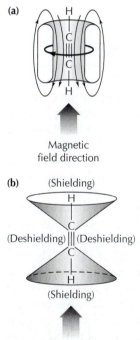

(a)

Magnetic
field direction

(b)   (Shielding)
H

C

(Deshielding) ||| (Deshielding)
C

H
(Shielding)

Magnetic
field direction

FIGURE 22.18
(a) Circulation of
π-electrons in an
applied magnetic field
induces an opposing
magnetic field that
shields the acetylenic
proton. (b) Regions
of shielding and
deshielding for
acetylene.

Summary of shielding and deshielding effects. Let's briefly summarize the effect of shielding and deshielding on the chemical shift of protons and introduce some commonly used terms (Figure 22.17). Increasing the electron density around a nucleus shields it from the applied field, making the effective field experienced by the nucleus smaller. The value of the observed chemical shift of the signal therefore decreases, and, on a typical NMR spectrum, the signal moves to the **right,** which is called an ***upfield*** shift because, at a constant frequency, a slightly higher applied magnetic field is needed for resonance to occur. Decreasing the electron density around a nucleus deshields it, causing the chemical shift to increase and moving the signal to the **left,** resulting in a ***downfield*** shift.

In molecules with filled π-orbitals, local diamagnetic shielding does not completely account for the chemical shifts observed for different protons. The shielding effect depends in part on the location of a proton relative to the induced magnetic field of the π-orbital, which is not spherically symmetrical. This effect is called ***anisotropy,*** a term that means "having a different effect along a different axis."

Consider acetylene, H−C≡C−H, for example. Although acetylene molecules are oriented more or less randomly because of rapid tumbling in solution, at any one time some of these linear molecules are lined up with the applied magnetic field. In the aligned molecules, the circulation of the electrons in the cylindrical π-orbital system of the triple bond induces a diamagnetic field, as illustrated in Figure 22.18a. This induced magnetic field opposes the applied magnetic field, shielding the acetylene proton and moving its NMR signal upfield. Regions of shielding are often represented by cones, as shown in Figure 22.18b. The chemical shift of the protons in acetylene is 1.80 ppm; it is affected by both local diamagnetic shielding effects and anisotropic shielding.

Alkenes and aldehydes exhibit strong anisotropic effects. When a π-orbital of a double bond is aligned with an applied magnetic field, the circulation of the two π-electrons induces a diamagnetic field perpendicular to the plane of the double bond. Anything in the region above and below the π-orbital is shielded; however, at the sides of the double bond the flux lines of the induced magnetic field add to the applied magnetic field, which creates a deshielding region in the plane perpendicular to the π-orbital. The shielding and deshielding regions of ethylene and formaldehyde are shown in Figures 22.19a and b. Strong anisotropic effects are demonstrated by the strongly deshielded protons of ethylene and formaldehyde.

$$H_2C=CH_2 \qquad\qquad H_2C=O$$
$$\text{5.28 ppm} \qquad\qquad \text{9.60 ppm}$$

Because of anisotropic deshielding, protons of methyl groups attached to the carbon atoms of C=C or C=O bonds appear near 2.0 ppm, whereas protons of methyl groups attached to carbon atoms of C−C or C−O bonds appear closer to 1.0 ppm.

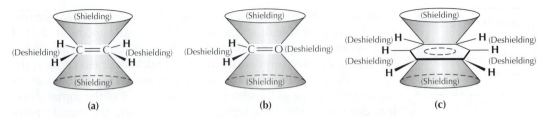

FIGURE 22.19 Regions of shielding and deshielding for (a) ethylene, (b) formaldehyde, and (c) benzene.

Protons attached to benzene rings absorb at a position even farther downfield from that of the vinyl protons in alkenes. The interactions of the six π-electrons of the aromatic ring produce a stronger anisotropic effect than that found with simple alkenes. The ring current created by the movement of these electrons induces a magnetic field, as illustrated in Figure 22.19c. The regions above and below the aromatic ring are shielded, whereas the protons at the edge of the ring are deshielded. The signal for the protons in benzene appears at 7.36 ppm, about 2 ppm downfield from the signal produced by the protons in ethylene.

22.8   Quantitative Estimation of Chemical Shifts

Much of the power of NMR spectroscopy comes from the correlation of molecular structure with positions of signals along the chemical shift scale. As you have already seen, the type of bonding and the proximity of electronegative atoms influence the chemical shift of protons.

Signals for different types of protons attached to carbon appear in well-defined regions. Tables cataloging these relationships, constructed by compiling large numbers of NMR signals from many organic compounds, contain too much data to memorize. **To master the use of NMR spectroscopy for determining molecular structures, it is important for you to be able to understand and use Tables 22.3–22.5,** which allow you to calculate estimated chemical shifts. The calculated chemical shifts can then be compared to the signals in the spectrum you are analyzing.

From the empirical correlations in Tables 22.3–22.5, it is possible to calculate the chemical shift of a hydrogen nucleus in a straightforward, additive way. The ability to add the individual effects of nearby functional groups is extremely useful because it allows an estimation of the chemical shifts for most of the protons in organic compounds.

The chemical shift values and almost all ^{1}H NMR spectra in this book are based on spectra taken on compounds in $CDCl_3$; however, if your laboratory uses deuterated acetone [$(CD_3)_2C{=}O$] as the NMR solvent of choice, the chemical shift values will be nearly the same. A sample of 30 common organic compounds showed that the chemical shifts in acetone were approximately 0.06 ppm upfield of the chemical shifts in chloroform.*

*Gottlieb, H. E.; Kotlyar, V.; Nudelman, A. *J. Org. Chem.* **1997**, *62*, 7512–7515.

TABLE 22.3 **Additive parameters for predicting NMR chemical shifts of alkyl protons in CDCl$_3$[a]**

	Base values	
Methyl	0.9 ppm	
Methylene	1.2 ppm	
Methine	1.5 ppm	

Group (Y)	Alpha (α) substituent	Beta (β) substituent	Gamma (γ) substituent
	H—C—Y	H—C—C—Y	H—C—C—C—Y
—R	0.0	0.0	0.0
—C=C	0.8	0.2	0.1
—C=C–Ar[b]	0.9	0.1	0.0
—C=C(C=O)OR	1.0	0.3	0.1
—C≡C–R	0.9	0.3	0.1
—C≡C–Ar	1.2	0.4	0.2
—Ar	1.4	0.4	0.1
—(C=O)OH	1.1	0.3	0.1
—(C=O)OR	1.1	0.3	0.1
—(C=O)H	1.1	0.4	0.1
—(C=O)R	1.2	0.3	0.0
—(C=O)Ar	1.7	0.3	0.1
—(C=O)NH$_2$	1.0	0.3	0.1
—(C=O)Cl	1.8	0.4	0.1
—C≡N	1.1	0.4	0.2
—Br	2.1	0.7	0.2
—Cl	2.2	0.5	0.2
—OH	2.3	0.3	0.1
—OR	2.1	0.3	0.1
—OAr	2.8	0.5	0.3
—O(C=O)R	2.8	0.5	0.1
—O(C=O)Ar	3.1	0.5	0.2
—NH$_2$	1.5	0.2	0.1
—NH(C=O)R	2.1	0.3	0.1
—NH(C=O)Ar	2.3	0.4	0.1

a. There may be differences of 0.1−0.5 ppm in the chemical shift values calculated from this table and those measured from individual spectra.
b. Ar = aromatic group.

Chemical Shifts of Alkyl Protons

The aggregate effect of multiple functional groups on the chemical shift of the proton(s) of an alkyl group can be determined from Table 22.3.

Base values. To use Table 22.3, begin with the base values at the top of the table. In any proposed molecular structure, primary hydrogen atoms (methyl groups) have a base value of 0.9 ppm. Secondary hydrogen atoms (methylene groups) are somewhat more deshielded, as shown by their chemical shift base value of 1.2 ppm. Tertiary (methine) hydrogen atoms have an even greater chemical shift; their base value is 1.5 ppm.

Effects of nearby substituents. The effect of each nearby substituent is added to the base value to arrive at the chemical shift of a particular proton in a molecule. If the substituent is directly attached to the carbon atom to which the proton is attached, it is called an *alpha substituent* (α). If the group is attached to a carbon atom once removed, it is a *beta substituent* (β). And if the group is attached to a carbon atom twice removed, it is a *gamma substituent* (γ).

The effect of an α substituent on the chemical shift of the proton is found by using a value from the first column in Table 22.3, and the effects of β and γ groups are found in the second and third columns, respectively. Notice that the topmost group, –R, an alkyl group, has no effect on the chemical shift other than changing the base values. When the carbon atom bearing the proton is farther away from a functional group, its effect on the chemical shift of the proton is smaller. The effect of a group more than three carbon atoms away from the carbon bearing the proton of interest is small enough to be safely ignored. There may be a difference of 0.1–0.5 ppm between the chemical shift value calculated from Table 22.3 and the measured value, but the difference is usually no greater than 0.2–0.3 ppm, close enough to figure out if a proposed structure fits the NMR spectrum.

Identifying α, β, and γ substituents. It is important in calculating estimated chemical shifts to use a systematic methodology. A good way not to forget to include all **α, β,** and **γ** substituents for each type of proton in a target molecule is to **write down all the α groups first, then all the β groups, and last the γ groups. Only then go to Table 22.3,** look up the base value and the value for each α, β, and γ substituent from the correct column, and do the necessary addition.

EXERCISE

Identify the α, β, and γ substituents for the two methylene protons and for the methyl groups attached to C-4 of 4-methoxy-4-methyl-2-pentanone.

$$H_3CO-\underset{\underset{H_3C}{|}}{\overset{\overset{H_3C}{|}}{C}}-CH_2-\underset{\underset{CH_3}{\diagdown}}{\overset{\overset{O}{\parallel}}{C}}$$

4-Methoxy-4-methyl-2-pentanone

Answer: The methylene protons have one α substituent, a –(C=O)CH$_3$ group, listed in Table 22.3 as –(C=O)R. The methylene group also has one β substituent, a methoxy group, listed in Table 22.3 as –OR. The C-4 methyl groups have no α substituents other than an alkyl group, but they do have a β substituent, the methoxy group, as well as a γ substituent, the –(C=O)CH$_3$ group.

FOLLOW-UP ASSIGNMENT

Identify the α and β substituents for the *tert*-butyl protons of the compound $(CH_3)_3C(C=O)OCH_3$.

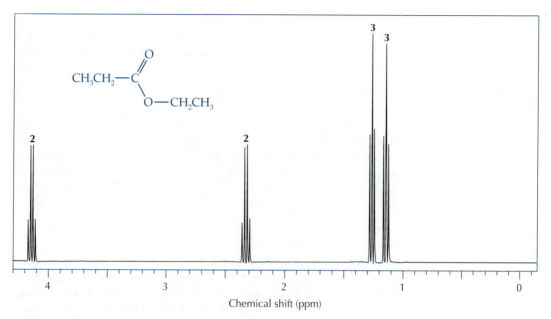

FIGURE 22.20 ^{1}H NMR spectrum of ethyl propanoate at 360 MHz.

Calculating estimated chemical shifts. Table 22.3 is laid out with carbon substituents at the top, followed by the heteroatoms—halogens, and oxygen and nitrogen substituents. To illustrate its use, let us return to the example of ethyl propanoate, whose 360-MHz NMR spectrum is shown in Figure 22.20.

WORKED EXAMPLE

The integration in Figure 22.20 shows that the signal at 4.13 ppm comes from a methylene group. Referring to Figure 22.13 or Table 22.2, the best correlation is with an α methylene group, which is directly attached to the oxygen atom of the ester group. Using Table 22.3, we can determine with more accuracy if this correlation is valid. Scan down the table to the entry fifth from the bottom to find the −O(C=O)R group. Here is the calculation:

Base value for a methylene group	1.2 ppm
Presence of the α −O(C=O)R group	2.8 ppm
Calculated chemical shift of the methylene protons	4.0 ppm

The calculated value is within 0.13 ppm of the methylene group attached to the oxygen atom, close enough to be consistent with the assignment.

The second methylene group at 2.32 ppm must be attached to the carbonyl carbon. Repeat the calculation for that methylene group, again using Table 22.3 and scanning down to the ninth entry.

Base value for a methylene group	1.2 ppm
Presence of the α −(C=O)OR group	1.1 ppm
Calculated chemical shift of the methylene protons	2.3 ppm

The estimated value of the chemical shift of the second methylene group is within 0.02 ppm of the measured value.

EXERCISE

Estimate the chemical shifts of the two different methyl groups in ethyl propanoate. Are these values consistent with the observed chemical shifts of 1.13 ppm and 1.26 ppm?

Answer: From Table 22.3, the value of the chemical shift for the methyl protons β to the $-(C=O)OR$ group is $0.9 + 0.3 = 1.2$ ppm. The chemical shift of the methyl protons β to the oxygen atom of the $-O(C=O)R$ group is $0.9 + 0.5 = 1.4$ ppm. These calculated values compare reasonably well with the measured chemical shifts of 1.13 and 1.26 ppm. Even though the estimates of the chemical shifts differ by 0.07 and 0.14 ppm from the measured values, their relative order of increasing chemical shift adds to our confidence in the assignments.

FOLLOW-UP ASSIGNMENT

The Follow-up Assignment pertaining to Figure 22.15 on page 362 asked whether the unknown ester was $C_6H_5CH_2O(C=O)CH_3$ or $C_6H_5(C=O)CH_2CH_3$. Estimate the chemical shifts of the methyl and methylene protons in each structure using Table 22.3 and see if the results confirm the prediction you made earlier.

FOLLOW-UP ASSIGNMENT

A compound, $C_7H_{14}O_2$, has the structure $CH_3(C=O)CH_2C(OCH_3)(CH_3)_2$. In its NMR spectrum are four separate signals, at 1.21 ppm, 2.08 ppm, 2.46 ppm, and 3.16 ppm, with the relative integrations of 6:3:2:3. This integration pattern is consistent with the structure of the compound having three kinds of methyl groups, one of which is duplicated, and one methylene group. Calculate the chemical shifts for each of the four kinds of protons in $C_7H_{14}O_2$, using Table 22.3, and assign them to their correct NMR signals.

Chemical Shifts of Aromatic Protons

The chemical shifts of protons on substituted benzene rings can also be calculated. To estimate these chemical shifts, the contributions of substituents shown in Table 22.4 are added to a base value of 7.36 ppm, the chemical shift for the protons of benzene dissolved in $CDCl_3$.

TABLE 22.4	Additive parameters for predicting NMR chemical shifts of aromatic protons in $CDCl_3$		
	Base value	7.36 ppm[a]	
Group	*ortho*	*meta*	*para*
—CH₃	−0.18	−0.11	−0.21
—CH(CH₃)₂	−0.14	−0.08	−0.20
—CH₂Cl	0.02	−0.01	−0.04
—CH=CH₂	0.04	−0.04	−0.12
—CH=CHAr	0.14	−0.02	−0.11
—CH=CHCO₂H	0.19	0.04	0.05
—CH=CH(C=O)Ar	0.28	0.06	0.05

(Continued)

Group	ortho	meta	para
—Ar	0.23	0.07	−0.02
—(C=O)H	0.53	0.18	0.28
—(C=O)R	0.60	0.10	0.20
—(C=O)Ar	0.45	0.12	0.23
—(C=O)CH=CHAr	0.67	0.14	0.21
—(C=O)OCH₃	0.68	0.08	0.19
—(C=O)OCH₂CH₃	0.69	0.06	0.17
—(C=O)OH	0.77	0.11	0.25
—(C=O)Cl	0.76	0.16	0.33
—(C=O)NH₂	0.46	0.09	0.17
—C≡N	0.29	0.12	0.25
—F	−0.32	−0.05	−0.25
—Cl	−0.02	−0.07	−0.13
—Br	0.13	−0.13	−0.08
—OH	−0.53	−0.14	−0.43
—OR	−0.45	−0.07	−0.41
—OAr	−0.36	−0.04	−0.28
—O(C=O)R	−0.27	0.02	−0.13
—O(C=O)Ar	−0.14	0.07	−0.09
—NH₂	−0.71	−0.22	−0.62
—N(CH₃)₂	−0.68	−0.15	−0.73
—NH(C=O)R	0.14	−0.07	−0.27
—NO₂	0.87	0.20	0.35

a. Base value is the measured chemical shift of benzene in $CDCl_3$ (1% solution).

WORKED EXAMPLE

Using Table 22.4, estimate the chemical shift of H_a in the structure of methyl 3-nitrobenzoate.

Methyl 3-nitrobenzoate

There are two functional groups that affect the chemical shift of H_a, an ortho-(C=O)OCH₃ group and an ortho-nitro group. The contribution of the ortho-(C=O)OCH₃ group to the chemical shift of H_a is the 13th item of Table 22.4; it is 0.68 ppm. The contribution of the ortho-NO₂ group, at the end of the table, is an additional 0.87 ppm. Adding these values to the base value of 7.36 ppm gives an estimated chemical shift of 8.91 ppm for H_a, compared to a measured value of 8.87 ppm.

Base value for a benzene ring	7.36 ppm
Presence of the ortho-(C=O)OCH₃ group	0.68 ppm
Presence of the ortho-NO₂ group	0.87 ppm
Calculated chemical shift for H_a	8.91 ppm
Measured chemical shift for H_a	8.87 ppm

FOLLOW-UP ASSIGNMENT

The measured chemical shifts for the remaining three aromatic protons of methyl 3-nitrobenzoate are 7.67 ppm, 8.38 ppm, and 8.42 ppm. The chemical shifts have been assigned to H_c, H_d, and H_b, respectively. Using Table 22.4, calculate the estimated chemical shifts of these three aromatic protons and justify their assignments.

Chemical Shifts of Vinyl Protons

Chemical shifts of protons attached to C=C bonds, called *vinyl* protons, can be estimated using Table 22.5. The estimated chemical shift for a vinyl proton is the sum of the base value of 5.28 ppm, the chemical shift for $H_2C=CH_2$, and the contributions for all *cis*, *trans*, and *geminal (gem)* substituents. A **geminal group** is the one that is attached to the same carbon atom as the vinyl proton whose estimated chemical shift is being calculated.

TABLE 22.5   **Additive Parameters for Predicting NMR Chemical Shifts of Vinyl Protons in CDCl$_3$**[a]

Group	Base value		5.28 ppm	
	gem	cis	trans
—R	0.45	−0.22	−0.28
—CH=CH$_2$	1.26	0.08	−0.01
—CH$_2$OH	0.64	−0.01	−0.02
—CH$_2$X (X=F, Cl, Br)	0.70	−0.11	−0.04
—(C=O)OH	0.97	1.41	0.71
—(C=O)OR	0.80	1.18	0.55
—(C=O)H	1.02	0.95	1.17
—(C=O)R	1.10	1.12	0.87
—(C=O)Ar	1.82	1.13	0.63
—Ar	1.38	0.36	−0.07
—Br	1.07	0.45	0.55
—Cl	1.08	0.18	0.13
—OR	1.22	−1.07	−1.21
—OAr	1.21	−0.60	−1.00
—O(C=O)R	2.11	−0.35	−0.64
—NH$_2$, −NHR, −NR$_2$	0.80	−1.26	1.21
—NH(C=O)R	2.08	−0.57	−0.72

a. There may be small differences in the chemical-shift values calculated from this table and those measured from individual spectra.

WORKED EXAMPLE

Styrene, the monomer from which polystyrene is made, has the formula $C_6H_5CH=CH_2$. In addition to the three signals of the protons attached directly to the benzene ring, there are separate NMR signals for the three vinyl protons at 5.25 ppm, 5.75 ppm, and 6.70 ppm. Calculate the expected chemical shifts for H_a, H_b, and H_c of styrene and assign the three measured signals to the correct protons.

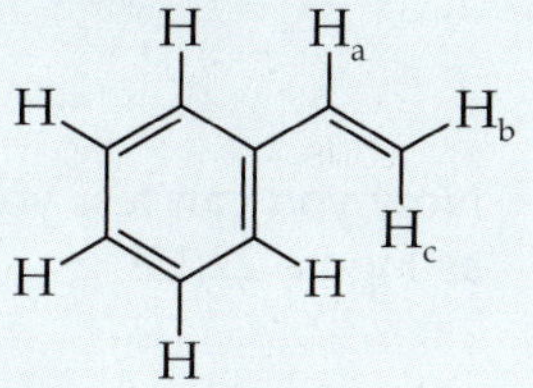

Styrene

Table 22.5 has a base value of 5.28 ppm, which will be part of the calculation for all three vinyl protons. H_a has the phenyl group (C_6H_5-) on the same carbon atom; it is a *geminal* group. The phenyl (Ar) group is about halfway down Table 22.5 and its *gem* parameter is 1.38 ppm. Here is the calculation for the chemical shift of H_a.

Base value for a vinyl proton	5.28 ppm
Presence of the *gem* C_6H_5- group	1.38 ppm
Calculated chemical shift of H_a	6.66 ppm
Measured chemical shift of H_a	6.70 ppm

In the same manner, we can calculate the chemical shifts for H_b and H_c. The phenyl group is *trans* to H_b, so –0.07 ppm must be added to the base value: 5.28 + (–0.07) = 5.21 ppm. This value fits well with the signal at 5.25 ppm in the NMR spectrum of styrene. The phenyl group is *cis* to H_c, so 0.36 ppm must be added to the base value: 5.28 + 0.36 = 5.64 ppm. It seems clear that the 5.75-ppm signal must be H_c.

EXERCISE

Consider the structure of ethyl *trans*-2-butenoate, $C_6H_{10}O_2$. Estimate the chemical shift for each of the two different vinyl protons in the molecule, using Table 22.5, and then assign H_a and H_b. The measured chemical shift values are 5.80 ppm and 6.90 ppm.

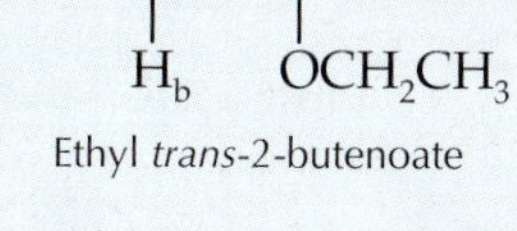

Ethyl *trans*-2-butenoate

Answer: Calculating the estimated chemical shift of H_a, we find

Base value for a vinyl proton	5.28 ppm
Presence of the *gem* —(C=O)OR group	0.80 ppm
Presence of the *cis* R group	–0.22 ppm
Calculated chemical shift of H_a	5.86 ppm
Measured chemical shift of H_a	5.80 ppm

For the estimated chemical shift of H_b, we find

Base value for a vinyl proton	5.28 ppm
Presence of the *cis* $-(C{=}O)OR$ group	1.18 ppm
Presence of the *gem* R group	0.45 ppm
Calculated chemical shift of H_b	6.91 ppm
Measured chemical shift of H_b	6.90 ppm

Using Tables 22.3–22.5 in Combination

Now you can test your skills in the use of Tables 22.3–22.5, as well as Figure 22.13.

PROBLEM ONE

Figure 22.21 is the ^{1}H NMR spectrum of a compound with the molecular formula $C_6H_{12}O_2$. It is an ester, which is one of the two isomers

$$(CH_3)_3C\,(C{=}O)\,OCH_3 \quad \text{or} \quad CH_3(C{=}O)OC(CH_3)_3$$

Calculate the chemical shifts for the two different kinds of methyl groups in each structure and then assign the NMR signals in Figure 22.21 to the appropriate methyl groups in the correct isomer.

Hint: Look first at the whole spectrum, paying attention to the integrals that are associated with the two signals. Then use the spectrum to measure each of the chemical shift values. Consider each isomer and think about the proximity of the two kinds of protons to electronegative atoms. Make a hypothesis as to which isomer seems correct. Then calculate the estimated chemical shifts for each of the two possible isomers using Table 22.3. Decide which molecular structure is correct and assign the NMR signals to the appropriate protons.

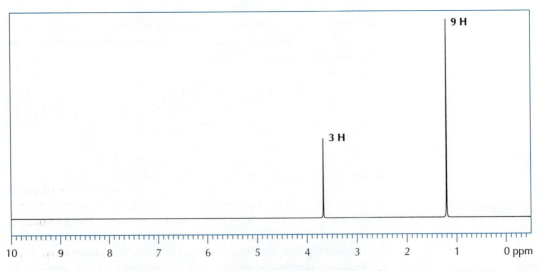

FIGURE 22.21 ^{1}H NMR spectrum of compound with the molecular formula $C_6H_{12}O_2$ at 300 MHz.

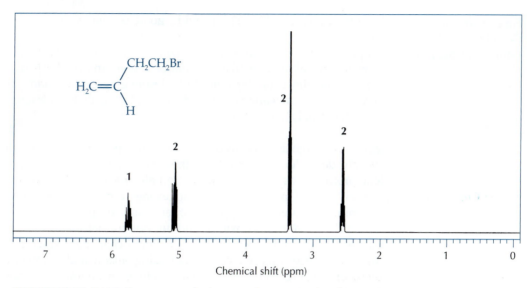

FIGURE 22.22 ^{1}H NMR spectrum of 4-bromo-1-butene simulated at 400 MHz.

PROBLEM TWO

Calculate the estimated chemical shifts of each of the five types of protons in 4-bromo-1-butene and assign each of them to their respective NMR signals. Briefly discuss any ambiguities in your assignments. The ^{1}H NMR spectrum of 4-bromo-1-butene at 400 MHz is simulated in Figure 22.22.

4-Bromo-1-butene

Hint: Decide which of the protons are alkyl protons and which are vinyl protons and then use Tables 22.3 and 22.5 to calculate the estimated chemical shifts. Assign each observed NMR signal to the appropriate proton(s).

PROBLEM THREE

trans-1-(*para*-Methoxyphenyl)propene has the following structure:

trans-1-(*para*-Methoxyphenyl)propene

Its measured NMR signals are observed at 1.83 ppm, 3.75 ppm, 6.07 ppm, 6.33 ppm, 6.80 ppm, and 7.23 ppm. Their respective integrations are 3:3:1:1:2:2. Which NMR signals are produced by alkyl protons, aromatic protons, and vinyl protons? Using Tables 22.3–22.5, calculate the estimated chemical shift for each type of proton in *trans*-1-(*para*-methoxyphenyl) propene and assign the observed chemical shifts to the correct protons.

Final Words on Calculating Estimated Chemical Shifts

In general, Tables 22.3–22.5 provide good estimates; however, a word of caution is in order. It is important to remember that the simple acyclic compounds used to generate the tables incorporate some structural features that may not be present in every situation. Anisotropic effects of rings, multiple deshielding groups, and hindered rotation can lead to estimated chemical shifts that are different from the actual chemical shifts.

Rings. The signals produced by methylene protons in rings are slightly deshielded compared to the signals produced by methylene protons in acyclic compounds. Signals produced by methylene protons in cyclopentanes and cyclohexanes typically appear near 1.5 ppm, compared to 1.2 ppm for open-chain compounds.

Cyclopropanes and oxiranes are unique in that the σ-bonding in three-membered rings has some π-orbital character, which produces an anisotropic effect and shielding above and below the plane of the ring. Chemical shifts of protons on cyclopropane and epoxide rings are approximately 1.0 ppm upfield from their acyclic counterparts.

Multiple deshielding groups. If there are multiple deshielding α substituents, especially alkoxy groups and halogen atoms, the calculated chemical shift values can sometimes differ from the measured chemical shifts by more than 1 ppm. The divergence between the actual chemical shifts and the estimates can be seen in the following series, as more methoxy groups are attached to the carbon atom of methane.

	CH_3OCH_3	$CH_2(OCH_3)_2$	$CH(OCH_3)_3$
Measured	3.24 ppm	4.58 ppm	4.97 ppm
Estimation	3.0 ppm	5.4 ppm	7.8 ppm

Hindered rotation. The estimates of chemical shifts of protons *ortho* to bulky groups on benzene rings can differ considerably from the measured values. This difference is evident in the calculation of chemical shift values for acetanilides ($Ar-NH-(C=O)CH_3$) that are substituted in one of the *ortho* positions with a substituent such as bromine, chlorine, or a nitro group. In these compounds, the chemical shift of the *ortho* proton is nearly 1 ppm downfield from the estimated chemical shift calculated from Table 22.4. Hydrogen bonding between the amide hydrogen and the *ortho* substituent impedes rotation about the C–N bond, freezing the conformation of the molecule so that the carbonyl group is located in the vicinity of the *ortho* hydrogen.

Computer Programs for Estimating ¹H NMR Chemical Shifts

Computer programs have been developed that use additivity parameters for calculating the NMR spectrum of any molecule of interest. The ChemDraw Ultra program in ChemBioOffice from PerkinElmer includes a module called ChemNMR, which estimates ¹H chemical shifts and displays the calculated NMR spectrum after the structure of a molecule is drawn. The logic of the program is a rule-based calculation of chemical shifts on structural fragments, similar to the method presented in this chapter. The ChemNMR module uses 700 base values and about 2000 increments; the calculated chemical shifts are stated to be within 0.2–0.3 ppm, roughly comparable to the use of the additivity parameters presented here.

Alternative methods for estimating chemical shifts include the Advanced Chemical Development/NMR Predictor (from Advanced Chemistry Development) and HyperChem/HyperNMR. The predicted chemical shifts are based on a large database of structures, and the database can be expanded as new compounds become available. The display can be interrogated by clicking on either the structure or the spectrum to highlight their co-relationships.

These programs are sophisticated, research-quality tools and are priced accordingly. Some institutions have negotiated site licenses making them accessible to their members.

22.9 Spin-Spin Coupling (Splitting)

The chemical shifts and integrals of NMR signals provide a great deal of information about the structure of a molecule; however, this information is often not enough to determine the structure. Closer examination of NMR signals reveals that they are generally not shapeless blobs but highly structured patterns with a multiplicity of lines. Reexamine the spectrum of ethyl propanoate shown in Figure 22.20 (see page 369). The signal at 4.1 ppm is actually a group of four peaks, as is the signal at 2.3 ppm. The signals at 1.2 and 1.1 ppm are groups of three peaks. The fine structure of these patterns is caused by interactions between the proton(s) producing the signals and neighboring nuclei, particularly other protons. The effects are small compared with those of shielding and deshielding, but analysis of the patterns provides valuable information about the local environments of protons in a molecule.

Vicinal Coupling ($^3J_{HH}$)

Vicinal protons

The interactions that cause the fine structure of NMR signals are transmitted through the bonding framework of the molecules. They are usually observable only when the interacting nuclei are near one another. The most commonly observed effects are produced by the interaction between protons attached to adjacent carbon atoms. These protons, which are separated by three bonds, are called *vicinal*, or nearby, protons.

A proton that is affected by the spin states of another nucleus is *coupled* to that nucleus and its signal is split into multiple signals. A simple example of coupling between two vicinal hydrogen atoms can be seen in the ^1H NMR spectrum of 1,1,2-tribromo-2-phenylethane shown in Figure 22.23. Both H_a and H_b have a spin of $\frac{1}{2}$ and therefore have two spin states, one aligned with the applied magnetic field and one opposed to it. In the absence of H_b, H_a would exhibit a single peak at 5.97 ppm. However, in the presence of the neighboring H_b, H_a is affected by the spin state of H_b.

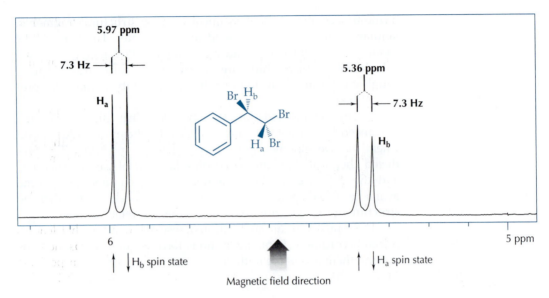

FIGURE 22.23 Section of the ¹H NMR spectrum of 1,1,2-tribromo-2-phenylethane at 360 MHz.

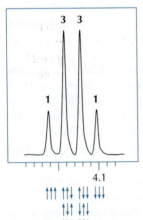

FIGURE 22.24 Signal at 4.13 ppm in the ¹H NMR spectrum of ethyl propanoate at 360 MHz.

The effective magnetic field felt by H_a increases a little when the magnetic field of H_b is aligned with the applied magnetic field. The aligned orientation leads to a slight deshielding effect, and the position of the H_a signal moves slightly downfield. The magnetic field of the spin state of H_b opposed to the applied magnetic field decreases the effective magnetic field felt by H_a, moving the H_a position slightly upfield. Thus, the signal for H_a is split into a *doublet*. Because the number of H_b nuclei in each spin state is nearly equal, two peaks of nearly equal intensity are observed.

The distance between the signals of the doublet is called the *coupling constant (J)*. When it is a vicinal (three-bond) coupling constant that involves two protons, the notation is $^3J_{HH}$. **Coupling constants are measured in Hz (cycles per second), and their values are independent of the spectrometer operating frequency**. In Figure 22.23, the value of the coupling constant is 7.3 Hz. In an analogous manner, proton H_a interacts with proton H_b, and H_b also appears as a doublet with the same coupling constant. The fact that interacting protons have coupling constants of exactly the same value is very useful for identifying which protons are coupled to one another.

Now consider a slightly more complicated pattern. An expanded section of the 360-MHz NMR spectrum of ethyl propanoate near 4.1 ppm is shown in Figure 22.24. This set of NMR signals is produced by the methylene group **b**, which has a relative integration of two protons. The protons of the methylene group are coupled to the protons of the adjacent methyl group **a**, and they split into a four-peak pattern called a *quartet*. The four peaks are produced by the three adjacent protons of the methyl group, which have the four spin states shown in Figure 22.24.

We can summarize the splitting pattern seen in Figure 22.24 as follows:

1. The three spins of the methyl protons aligned with the applied magnetic field produce the left peak.
2. The two spins of the methyl protons aligned to the applied magnetic field along with the one opposed to it produce the middle-left peak.
3. The one spin aligned with and the two spins of the methyl protons opposed to the applied magnetic field produce the middle-right peak.
4. The three spins of the methyl protons opposed to the applied magnetic field produce the right peak.

Statistically, there are three possible combinations that lead to spin states 2 and 3. Because every combination of spins has the same probability of occurring, the relative intensities of the four peaks in Figure 22.24 are 1:3:3:1. The measured coupling constant is 7.1 Hz.

EXERCISE

Analyze the **triplet** (three-peak) pattern of methyl group **a** of ethyl propanoate at 1.26 ppm, shown in Figure 22.25. The protons of this methyl group are coupled to the protons of methylene group **b**.

Answer: The key to the splitting pattern of the methyl group is the number of spin states of the methylene group to which it is coupled. As usual, the spins of the two methylene protons have an equal probability of being aligned or opposed to the applied magnetic field. Three combinations are possible. The two spins can be aligned, one can be aligned and the other opposed, or the two spins can be opposed to the applied magnetic field. There is twice the probability of one spin aligned and one opposed. This produces a triplet pattern for the nearby methyl group, with the relative intensities 1:2:1.

We can check to make sure that methyl group **a** is coupling with methylene group **b** by calculating the **coupling constant**, **J**, between them. If they are coupled, both groups of peaks must have the same **J** value. Figure 22.25 gives the positions of the peaks for the methyl group in Hz. The distance between the individual peaks must also be the same.

$$459.9 \text{ Hz} - 452.8 \text{ Hz} = 7.1 \text{ Hz}$$
$$452.8 \text{ Hz} - 445.6 \text{ Hz} = 7.2 \text{ Hz}$$

The measured coupling constant from Figure 22.26 is also 7.1 Hz. Within experimental error, the coupling constants are the same.

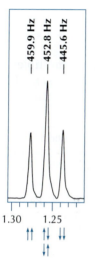

FIGURE 22.25 Signal at 1.26 ppm in the ^{1}H NMR spectrum of ethyl propanoate at 360 MHz.

Singlet	One peak
Doublet	Two peaks
Triplet	Three peaks
Quartet	Four peaks

Splitting Trees and the N + 1 Rule

A common device for predicting and analyzing the fine structure of coupling patterns is a **splitting tree**, constructed by mapping the effect of each spin-spin coupling on a signal. The splitting tree for the methylene group **b** of ethyl propanoate is shown in Figure 22.26. Notice that there are three branching sites in the tree—one set of branches for each proton in methyl group **a**, whose coupling produces the splitting tree.

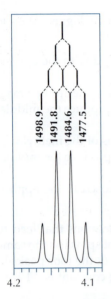

FIGURE 22.26 Splitting tree for the signal at 4.13 ppm in the ¹H NMR spectrum of ethyl propanoate at 360 MHz.

Splitting of signals. Because the three adjacent methyl protons are equivalent to one another, the methylene signal can be thought of as splitting into doublets three times. The signal is split into a doublet by the first methyl proton. The coupling with a second methyl proton splits each signal of the doublet into two signals. Because the coupling constants of these two interactions are exactly the same, the position of the high-field signal of one doublet reinforces the position of the low-field signal of the second doublet. If no more splitting occurred, the pattern would consist of three equally spaced signals, the center signal having twice the intensity of the outer two. Coupling with the third methyl proton, however, again splits each of the signals of the triplet into two signals. Because the coupling constant is the same, the signals again reinforce each other. As seen in the splitting tree in Figure 22.26, the resulting pattern is a quartet, a group of four equally spaced signals. The ratio of the intensities is 1:3:3:1.

The presence of doublets, triplets, and quartets in NMR spectra has led to the *N + 1 rule* for multiplicity: **A proton that has N equivalent protons on adjacent carbon atoms will be split into N + 1 signals.**

Pascal's triangle. The ratio of the intensities of the multiplet signals can be obtained from Pascal's triangle (Figure 22.27), a triangular arrangement of the mathematical coefficients obtained by a

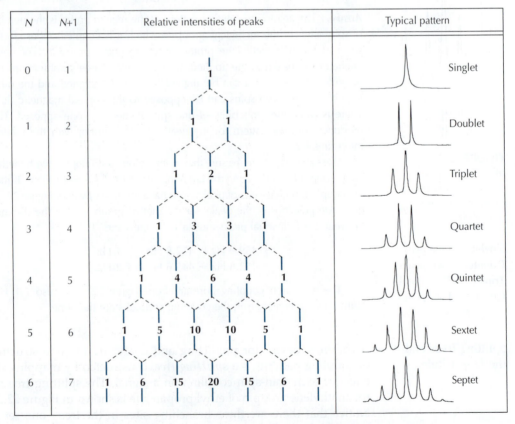

N	$N+1$	Relative intensities of peaks	Typical pattern
0	1	1	Singlet
1	2	1 1	Doublet
2	3	1 2 1	Triplet
3	4	1 3 3 1	Quartet
4	5	1 4 6 4 1	Quintet
5	6	1 5 10 10 5 1	Sextet
6	7	1 6 15 20 15 6 1	Septet

FIGURE 22.27 Pascal's triangle can be used to predict the multiplicity and relative intensities of the signal of any magnetically active nucleus coupled to N equivalent nuclei of spin $\frac{1}{2}$. It applies to both ¹H and ¹³C spectra.

binomial expansion. The $N + 1$ rule assumes that all N protons are equivalent, with equal coupling constants. If the protons are not all equivalent, the coupling constants will probably not all be equal. In that case, the total number of peaks will be greater than $N + 1$.

Returning to the 360-MHz NMR spectrum of ethyl propanoate (see Figure 22.20), you can see that the methylene signal at 4.13 ppm appears as a quartet because of the splitting by the three hydrogen nuclei of the adjacent methyl group ($N = 3$; $N + 1 = 4$). In turn, the three-proton signal at 1.26 ppm appears as a triplet because of the two hydrogen nuclei of the adjacent methylene group ($N = 2$, $N + 1 = 3$). **This triplet–quartet pattern is seen quite often and is diagnostic for an ethyl group.** A second triplet–quartet pattern occurs in the NMR spectrum of ethyl propanoate, indicative of the presence of a second ethyl group. In this case, the quartet is located at 2.32 ppm because the methylene component of the ethyl group is attached to a carbonyl group rather than to an oxygen atom.

Other Types of Coupling

Most of the observed coupling in NMR spectroscopy is a result of vicinal coupling through three bonds, $H–C–C–H$, called $^3J_{HH}$ coupling. However, coupling through one, two, four, and five bonds can also be observed. One-bond coupling occurs between ^{13}C and ^1H ($^1J_{CH}$). Because the relative abundance of ^{13}C is so small, the signals produced by this splitting are usually negligible in a ^1H NMR spectrum. With concentrated samples, it is possible to observe this splitting by turning up the amplitude, as demonstrated by the ^1H NMR spectrum of chloroform in Figure 22.28. This type of coupling is a major consideration when observing ^{13}C signals, as you will see in Chapter 23 on ^{13}C NMR spectroscopy.

Geminal coupling. Coupling through two bonds ($^2J_{HH}$), or geminal coupling, occurs between two protons attached to the same carbon atom, $H–C–H$. In many molecules, these two protons are equivalent and coupling is not observed. However, geminal coupling is frequently observed in compounds with vinyl methylene groups, $H_2C=C–$, where the two geminal protons are usually nonequivalent. The presence of a stereocenter in a molecule is also a cause of inequivalence in methylene protons, as discussed on pages 389–391.

Allylic coupling. π bonds are particularly good at transmitting coupling between protons, and coupling through four bonds ($^4J_{HH}$) is often observed in compounds containing carbon-carbon double bonds ($H–C=C–C–H$). This type of splitting is called *allylic coupling*.

Long-range coupling in aromatic rings. Not surprisingly, protons *ortho* to one another on an aromatic ring show a normal three-bond coupling. But, in analogy to allylic coupling, *meta* protons also routinely display an observable coupling ($^4J_{HH}$). Moreover, because of the efficiency of coupling through π-bonds, a five-bond coupling ($^5J_{HH}$) between aromatic protons that are *para* to one another can even be observed.

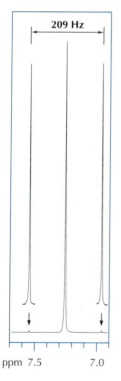

209 Hz

ppm 7.5   7.0

FIGURE 22.28

Expanded and amplified section of the ^1H NMR spectrum of chloroform at 360 MHz, showing ^{13}C splitting.

Magnitude of Coupling Constants

The magnitude of coupling constants can reveal valuable information about the structure of a molecule. The magnitude is related to the number of bonds between the interacting protons; the more bonds, the smaller the coupling constant. The size of vicinal coupling for alkyl protons ranges from about 2 to 13 Hz. The size of geminal coupling depends on bond angles and hybridization; for alkyl protons it is generally on the order of 10–16 Hz. The geminal coupling of vinyl protons is much smaller (0–3 Hz). Coupling through four or more bonds is also very small, 0–3 Hz. Typical coupling constants for various arrangements of protons are listed in Table 22.6.

If there is free rotation about a carbon-carbon single bond connecting the coupled protons, the vicinal or three-bond coupling constants are usually about 7 Hz. If rotation about the carbon-carbon bond is restricted, the coupling constant can range from 0 to 13 Hz.

TABLE 22.6   Typical proton-proton coupling constants

Arrangment of protons	J(Hz)	Arrangement of protons	J(Hz)	Arrangement of protons	J(Hz)
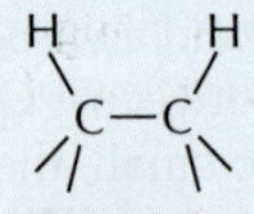 Free rotation	7	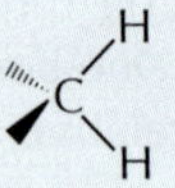	10 to 16	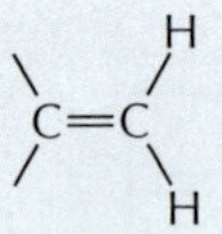	0 to 3
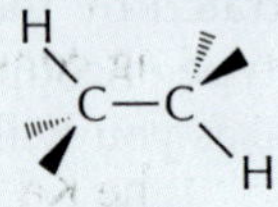 Anti	8 to 13	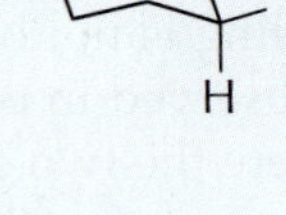	11 to 14	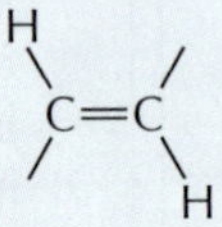	12 to 18
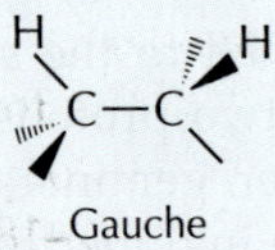 Gauche	2 to 4	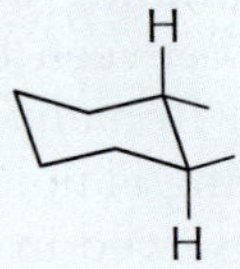	8 to 13	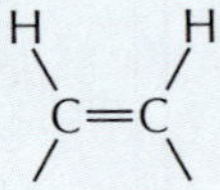	6 to 12
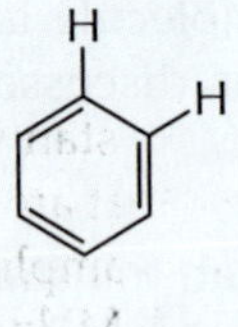	6 to 9	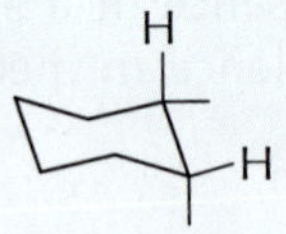	2 to 6	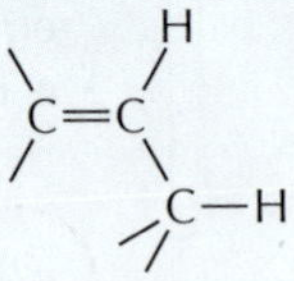	4 to 10
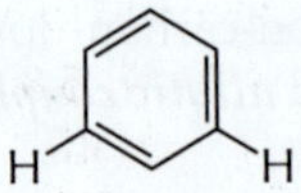	1 to 3	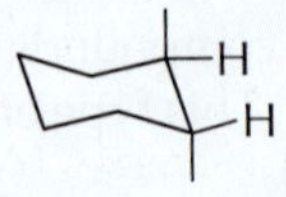	2 to 5	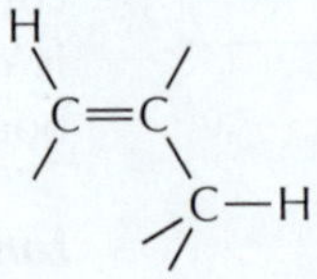	0.5 to 2
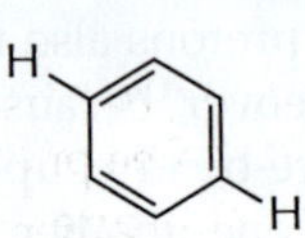	0 to 1			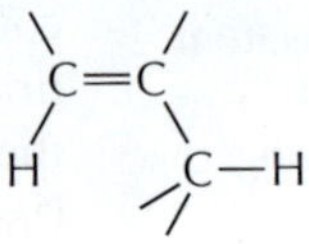	0

FIGURE 22.29
Dependence of the coupling constant on dihedral angle, ϕ, formed by two vicinal C–H bonds (Karplus relationship). Coupling constants usually fall between the two curves, which are calculated using different assumptions.

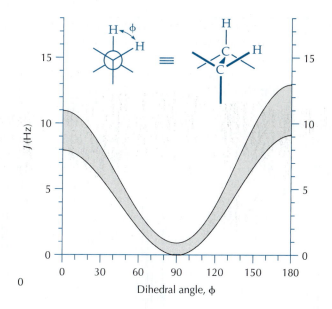

The size of a vicinal coupling constant is related to the angle ϕ on a Newman projection of the interacting protons (see Figure 22.29). This angle is called the *dihedral angle*.

In the early days of NMR spectroscopy, Martin Karplus at Harvard studied the relationship between the size of a coupling constant and the dihedral angle. His conclusions are now widely accepted and are often presented as a plot of the coupling constant versus dihedral angle. This plot is called a *Karplus curve* and is shown in Figure 22.29. The important characteristics of the Karplus relationship are the minimum value of the coupling constant at a dihedral angle of 90° and the large values of the vicinal coupling constant at dihedral angles of 0° and 180°. This dependence of the size of the coupling constant on dihedral angle often allows for the determination of double-bond geometry in alkenes. The coupling constant for *trans*-vinyl protons (dihedral angle = 180°, J = 12–18 Hz) is significantly larger than that observed for *cis*-vinyl protons (dihedral angle = 0°, J = 6–12 Hz).

Multiple Couplings

In most organic compounds there will be coupling constants that are similar in magnitude as well as coupling constants that are quite different. This situation creates patterns that are more complicated than the ones shown in Figures 22.20 and 22.23. The 360-MHz spectrum of ethyl *trans*-2-butenoate shown in Figure 22.30 is an example. The spectrum shows five groups of signals at 1.2, 1.8, 4.1, 5.8, and 6.9 ppm, with integral values of 3:3:2:1:1, respectively. Notice that Figure 22.30 has an expanded inset near every set of peaks. It is often necessary to expand regions in an NMR spectrum to more clearly reveal the detail of the splitting patterns.

The quartet pattern at 4.1 ppm, which integrates to two protons, and the triplet pattern at 1.2 ppm, which integrates to three protons, indicate that an ethyl group is incorporated into the structure of the molecule. The chemical shift of the methylene group suggests that

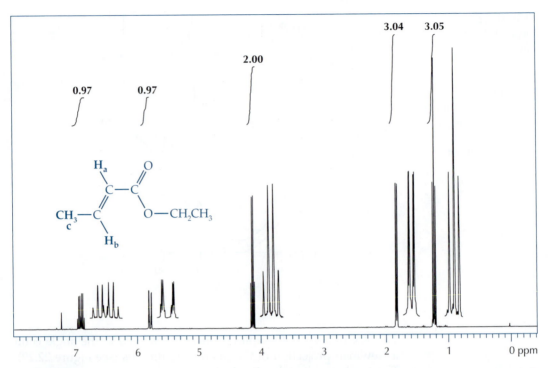

FIGURE 22.30 ¹H NMR spectrum of ethyl *trans*-2-butenoate at 360 MHz, with superimposed expanded (4×) insets adjacent to the signals.

the ethyl group is bonded to an oxygen atom. Taken together, the partial structure producing the signals at 4.1 ppm and 1.2 ppm must be $-OCH_2CH_3$.

The three-proton signal at 1.8 ppm is produced by a second methyl group. Its position is consistent with the chemical shift of a methyl group attached to a C=C bond (see Figure 22.13). This signal appears to be a doublet, which is consistent with just one proton on the adjacent carbon that is likely to be the vinyl proton on C-3. However, further expansion of this peak reveals more complex allylic coupling, which was discussed on page 381. From the chemical shift Table 22.2, we know that the two signals at 5.8 ppm and 6.9 ppm, each integrating to one proton, are produced by protons attached to the carbon atoms of a C=C bond. Taking all these observations into consideration leads us to the conclusion that the partial structure that produces the signals at 5.8, 6.9, and 1.8 ppm must be $-(C=O)-CH=CH-CH_3$.

Our analysis reveals the abundance of information available in an NMR spectrum. We have been able to assign the NMR signals without even considering the complex splitting shown for the vinyl proton signals. These patterns are more complex than the $N + 1$ rule would lead us to believe. What coupling patterns account for the complex sets of peaks centered at 6.93, 5.80, and 1.84 ppm, which are shown in expanded sections in Figure 22.31 and Figure 22.32?

The signal at 6.93 ppm is produced by the $CH_3-C\textbf{\textit{H}}=CH-$ proton (H_b in Figure 22.30). It seems to have eight peaks. If the coupling constants between the vinyl proton and the four protons on adjacent

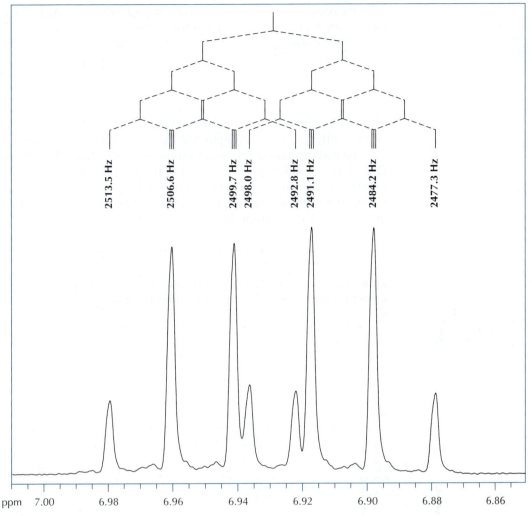

2513.5 Hz 2506.6 Hz 2499.7 Hz 2498.0 Hz 2492.8 Hz 2491.1 Hz 2484.2 Hz 2477.3 Hz

ppm   7.00   6.98   6.96   6.94   6.92   6.90   6.88   6.86

FIGURE 22.31 Splitting tree for the vinyl proton signal at 6.93 ppm in the ¹H NMR spectrum of ethyl *trans*-2-butenoate.

carbon atoms were the same, the signal should appear as a quintet ($N = 4$, $N + 1 = 5$); however, it is clearly not a five-peak pattern. If there were only coupling with the other vinyl proton signal at 5.80 ppm (H_a in Figure 22.30), the $N + 1$ rule predicts that the signal at 6.93 ppm would appear as a doublet; however, there is also coupling with the adjacent methyl group, producing what is called a **doublet of quartets**. These two overlapping quartets produce the observed eight-peak pattern in Figure 22.31. The splitting tree for the proton appearing at 6.93 ppm is shown at the top of Figure 22.31. By accurately measuring the distances between the signals, it is possible to determine the coupling constants from this doublet of quartets. First, it is necessary to convert the chemical shift (δ, ppm) to a frequency (Hz). This is done by multiplying δ by the applied magnetic field (B_0, MHz). Using the left-most peak in Figure 22.31, we have

$$J = \delta \times B_0$$
$$J = 6.982 \text{ ppm} \times 360 \text{ MHz} = 2513.5 \text{ Hz}$$

One coupling constant can be determined by measuring the distance from the outermost signal of the pattern to the adjacent signal. Using the left-most signals, we can calculate this coupling constant to be 6.9 Hz:

$$2513.5 \text{ Hz} - 2506.6 \text{ Hz} = 6.9 \text{ Hz}$$

This value is the coupling constant between the three hydrogen atoms of the methyl group at 1.84 ppm and the vinyl proton at 6.93 ppm.

The distance in Hz between the outermost signals of the pattern (2513.5 Hz – 2477.3 Hz = 36.2 Hz) is the sum of all the coupling constants. The coupling constant between the two vinyl protons can be calculated by subtracting the coupling constants of each proton in the methyl group from this sum:

$$36.2 \text{ Hz} - (3 \times 6.9 \text{ Hz}) = 15.5 \text{ Hz}$$

This large coupling constant indicates that the vinyl protons are *trans* to one another; therefore, the molecule must be an (E)-alkene.

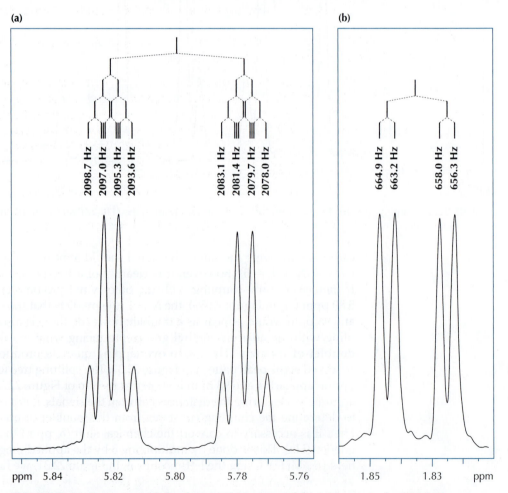

FIGURE 22.32 Expansions of signals at (a) 5.80 ppm and (b) 1.84 ppm in the ^{1}H NMR spectrum of ethyl *trans*-2-butenoate.

Expansion of the signals at 5.80 ppm and 1.84 ppm (H$_a$ and H$_c$ in Figure 22.30) reveals further fine structure (Figures 22.32a and b). At 5.80 ppm, the pattern of the NMR signal is also a doublet of quartets and the signal at 1.84 ppm is a doublet of doublets. The coupling constant for the $^4J_{HH}$ coupling is quite small, only 1.7 Hz. Until the signals are expanded, it is hardly noticeable.

Second-Order Effects

Observed splitting patterns may differ also from patterns predicted by *simple first-order coupling using the N + 1 rule* as a result of *second-order effects.* As the chemical shifts of the coupled protons become closer to one another, second-order effects become more pronounced. The usual rule of thumb is that they become apparent in a spectrum when the difference in chemical shifts (Δv, measured in Hz) is less than five times the coupling constant ($\Delta v < 5J$).

Consequences of second-order effects. Large second-order effects produce the following:

- Signal pattern intensities that are different from predicted first-order $N + 1$ values
- Additional signals beyond those predicted by simple splitting rules
- Coupling constants that cannot be directly measured from differences in signal positions

When second-order effects are small, they can be useful, such as when the differences in signal intensities produce "leaning" peaks, which indicate the relative position of a coupling partner. Look back at the quartet signal at 2.32 ppm in the 200-MHz NMR spectrum of ethyl propanoate (Figure 22.11, page 358) for an example of "leaning" peaks. The quartet pattern is not perfectly symmetrical. The right-hand peaks are slightly higher than those on the left, an indication that these protons are coupled with protons whose signals appear to the right of that pattern. In this case, the coupling partner appears at 1.15 ppm. Notice that the signal at 1.15 ppm is "leaning" to the left because its coupling partner appears downfield.

Complexities produced by second-order effects. Now examine the expanded sections of the 60-MHz and 360-MHz spectra of cinnamyl alcohol, which show the vinyl proton regions (Figure 22.33). On the 360-MHz NMR spectrum (Figure 22.33b), the individual vinyl protons appear as well-defined signals at 6.3 ppm and 6.6 ppm, separated by 100 Hz (0.28 ppm $\times$ 360 MHz); the coupling constant is 15.9 Hz. On a 60-MHz instrument these signals are separated by only 17 Hz (0.28 ppm $\times$ 60 MHz) and the coupling constant is again 15.9 Hz. When the difference in chemical shifts is nearly the same as the coupling constant between two protons, the NMR spectrum is almost useless for any analysis of NMR splitting.

Even the 360-MHz spectrum exhibits small second-order effects. The patterns at 6.3 ppm and 6.6 ppm in Figure 22.33b lean toward each other. In other words, the upfield portion of the 6.6-ppm doublet and the downfield triplet of the 6.3-ppm signal are larger than the other parts of the two patterns. On higher-field instruments, the

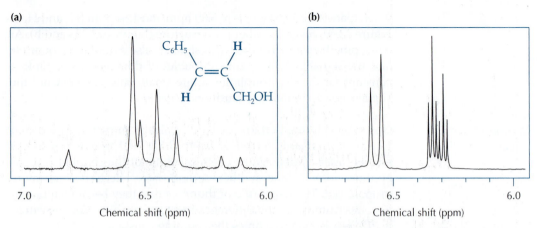

FIGURE 22.33 ¹H NMR spectra of the vinyl protons of cinnamyl alcohol at (a) 60 MHz and (b) 360 MHz.

frequency difference between signals is even larger, so that second-order effects, in many cases, become negligible.

Another dramatic example of the ability of a higher-field spectrometer to provide a much simpler spectrum compared to that acquired at a lower field is shown in the comparison of the spectra of 1-butanol acquired at 60 MHz and at 360 MHz (Figure 22.34). At

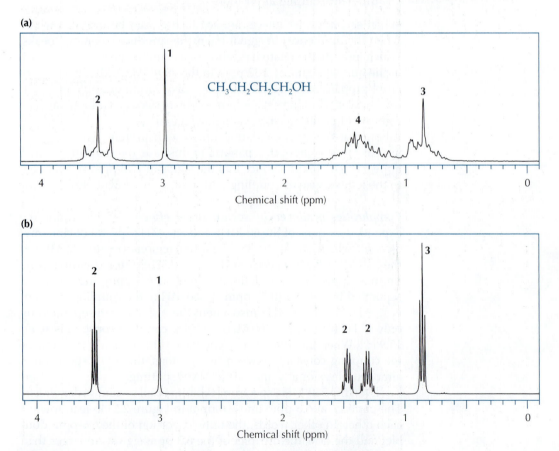

FIGURE 22.34 ¹H NMR spectra of 1-butanol at (a) 60 MHz and (b) 360 MHz.

60 MHz (Figure 22.34a), the spectrum of 1-butanol is rife with second-order effects. The 1–2-ppm region exhibits a complex multiplet integrating to four protons, which includes both of the two similar methylene groups of 1-butanol. Not only are the methylene signals overlapping, but second-order effects also result in patterns for these protons that have no relation to the $N + 1$ rule. The complex coupling pattern of these two methylene groups ($\Delta v < 5J$) cannot be analyzed by inspection. In addition, little can be learned from the methyl protons that appear at 0.7–1.0 ppm. What should be a triplet by first-order rules looks like nothing of the sort as a result of second-order effects. It is important to keep in mind, however, that the area of a signal is not affected by second-order effects, so the relative integrated areas measured for these complex peaks are still valid.

At 360 MHz, the NMR spectrum of 1-butanol is greatly simplified (Figure 22.34b). In this spectrum, the chemical shifts of the two methylene groups are different enough ($\Delta v > 5J$) so that their signals are separated into two well-defined multiplets. The signal at 1.3 ppm for the C-3 methylene group is a sextet resulting from coupling with the flanking three methyl and two methylene protons ($N + 1 = 6$), where the J values for the methyl and methylene couplings are essentially the same. The other methylene signal at 1.47 ppm is a quintet, split equally by two sets of flanking methylene protons ($N + 1 = 5$). Also, notice that the methyl group at 0.86 ppm is a well-defined triplet. Except for a little leaning observed in the coupling patterns, the spectrum is essentially first-order.

Diastereotopic Protons

Subtle structural differences between protons in a molecule may not be obvious at first glance. For example, it is easy to assume that the two protons of a methylene group are always equivalent, and in many cases they are. However, if the methylene group is next to a stereocenter, such as an asymmetric carbon atom, the two protons of the methylene group become nonequivalent. They cannot be interchanged with one another by any bond rotation or symmetry operation, and they are said to be *diastereotopic.* They have different chemical shifts, and they also couple with each other. The appearance of diastereotopic protons is common in the NMR spectra of chiral molecules, those with stereocenters.

Consider the compound 2-methyl-1-butanol:

2-Methyl-1-butanol

If there is no coupling to the hydroxyl proton, you might expect the NMR signal for the adjacent methylene protons to appear as a doublet because of coupling with the vicinal methine proton. However, the protons of the methylene group are diasterotopic, which makes the NMR spectrum of 2-methyl-1-butanol much more complex.

The spectrum shown in Figure 22.35 reveals an eight-line pattern for the C-1 methylene group of 2-methyl-1-butanol. The chemical shifts of the methylene protons are 3.4 ppm and 3.5 ppm. Because these two protons are not identical, they couple with each other, as well as with the vicinal methine proton, and each of the diastereotopic protons becomes a **doublet of doublets**. An expanded view of the C-1 methylene signals is shown in Figure 22.36. The coupling constants in one four-line set are 6.5 Hz and 10.5 Hz, and the

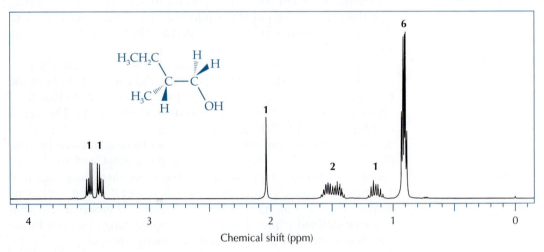

FIGURE 22.35 ¹H NMR spectrum of 2-methyl-1-butanol at 360 MHz.

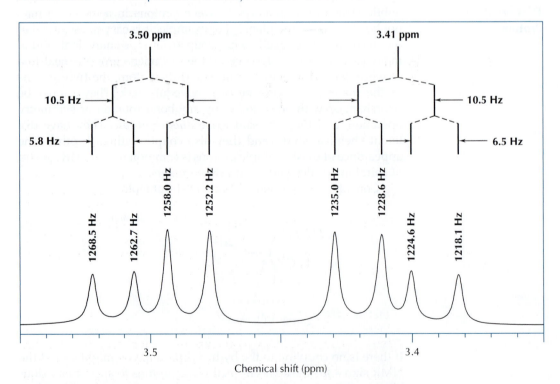

FIGURE 22.36 Splitting tree for the diastereotopic protons of 2-methyl-1-butanol.

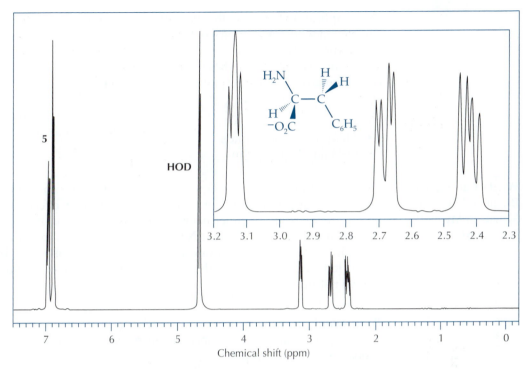

FIGURE 22.37 ^{1}H NMR spectrum of phenylalanine in D_2O at 360 MHz, showing the coupling pattern of the methylene group's diastereotopic protons.

coupling constants for the other set are 5.8 Hz and 10.5 Hz. Notice that the two halves of the eight-line pattern in Figure 22.36 lean into each other. This leaning makes the central lines more intense than the outside lines, even though the first-order splitting tree for this pattern shows equal intensities for all the lines. Using a higher-field NMR instrument for the spectrum would make all eight lines closer to equal intensity. The methylene protons of the ethyl group attached to the stereocenter are also diastereotopic, but the pattern is not distinct enough to analyze accurately.

Many important biomolecules have stereocenters and diastereotopic protons are common in molecules that have biochemical significance. For example, the amino acid phenylalanine, which is a component of virtually all proteins, has diastereotopic protons in the same way as 2-methyl-1-butanol does. Figure 22.37 shows the 360-MHz spectrum of phenylalanine taken in D_2O (pH ~ 11). The diastereotopic protons of the methylene group appear at 2.42 and 2.68 ppm. The methine proton at 3.15 ppm would be a doublet of doublets at a higher field, but in Figure 22.37 it appears as an apparent triplet with a broadened middle peak.

22.10 Sources of Confusion and Common Pitfalls

Using NMR spectroscopy to analyze the structures of organic compounds is a logical process, but sometimes interpreting a spectrum involves more than the factors of chemical shift and spin-spin splitting that we have discussed. It is important to be aware of some of the complicating factors so that you can make rational choices when confronted with unexpected, confusing, or poorly defined signals in a spectrum.

In this section we will briefly discuss four common, potentially confusing areas in the interpretation of NMR spectra:

- Broad and distorted NMR Peaks
- Overlap of NMR signals
- Mixtures of compounds
- O–H and N–H protons

What's Causing the Broad and Distorted Signals in My NMR Spectrum?

Some broad and distorted signals are inherent to the compound under analysis, as a result of second-order coupling effects, but others are the result of either poor sample preparation or improper operation or adjustment of the NMR spectrometer. It is important to avoid artifacts resulting from factors you can control and to be on the lookout for complicated spectral features that are of a fundamental nature.

Poor sample preparation. A good NMR spectrum begins with a well-prepared sample. Poor spectra can result from the following problems with the sample:

- The sample is not completely dissolved or there are insoluble impurities present.
- The sample solution is too concentrated.
- The height of the solution in the NMR tube is not correct.

Problems during data acquisition. Care must be taken when placing the sample in the magnet and when making the necessary adjustments to the spectrometer. Problems at the point of data acquisition that can lead to poor spectra include the following:

- The sample tube is not positioned properly in the spin collar.
- The sample is not spinning evenly during data acquisition.
- The NMR sample tube is spinning too rapidly during the data acquisition, leading to a vortex in the sample solution.
- The spectrometer is not tuned (*shimmed*) properly.

Magnetic field inhomogeneity. Improper spectrometer tuning deserves additional comment because of the many problems it can cause. Obtaining a good NMR spectrum requires that the magnetic field surrounding the sample be uniform. Shimming adjustments must be made on the spectrometer to achieve a homogeneous field each time a new sample is placed in the magnet. On the newest instruments, shimming is usually performed automatically under computer control, but on older spectrometers, shimming is a manual task that requires some experience to master. A spectrum acquired with a poorly shimmed magnet can show very broad peaks with strange splittings. It is often possible to determine if distortions in a spectrum are the result of poor shimming by looking closely at the TMS peak or some other peak in your spectrum that you know to be a singlet. If the TMS peak is not a sharp singlet, then poor shimming is the likely problem and all of the peaks in the spectrum will be similarly distorted. If the distortions are so bad that the spectrum is hard to interpret, the spectrum may need to be rescanned after better shimming adjustments have been made.

Spinning sidebands. Reflections of a signal, called spinning sidebands, which appear symmetrically around a peak, are also evidence of problems with magnetic field homogeneity. Sometimes the spinning sidebands are so large that a quick glance suggests that a singlet peak may be a triplet; however, it is rare that the sidebands are large enough to approach the 1:2:1 peak heights necessary for a triplet pattern produced by spin-spin coupling. You can always tell if the set of peaks is a real triplet by examining the relative heights of the three peaks. In addition, spinning sidebands will appear around all the peaks in the NMR spectrum, not just one or two of them. As the name suggests, spinning sidebands are related to the spinning of the sample in the magnet, and a defective NMR tube that is not perfectly cylindrical can cause spinning problems and be a source of spinning sidebands. Shimming adjustments required to mini-mize spinning sidebands are generally not routinely modified from sample to sample, so consult your instructor or laboratory manager if you observe troublesome spinning sidebands that are not caused by a defective NMR tube. With very modern spectrometers, sample tubes are generally not spun in the magnet, so spinning sidebands are not an issue for spectra acquired on these NMR spectrometers.

Virtual coupling. Even when a sample is prepared properly, care-fully positioned in the magnet, and all the spectrometer adjust-ments are made, unexpected spectral features that are inherent to the sample can still lead to confusion. Second-order effects, in particular, can easily bewilder someone expecting to see only $N + 1$ (first-order) coupling. A dramatic example of second-order behavior was presented in the 60-MHz spectrum of 1-butanol (Figure 22.34a). **Virtual coupling** is a particularly subtle second-order effect that results in distorted peaks with unexpectedly complex patterns. Virtual coupling arises when a proton, well separated in chemical shift from other protons, is coupled to one or a set of protons that is a member of a **strongly coupled** group of nuclei. Strong coupling occurs when protons with very similar chemical shifts are coupled to one another, so that $\Delta v < 5J$ and the regular $N + 1$ rule does not hold. Essentially, virtual coupling is observed because a proton coupled to a nucleus that is a member of a strongly coupled set behaves as if it is coupled to all the members of the strongly coupled set of protons.

The 360-MHz spectrum of 1-hexanol (Figure 22.38) provides a typical example of virtual coupling. The protons of the three methylene groups in the $CH_3-CH_2-CH_2-CH_2-$ unit of 1-hexanol constitute a strongly coupled spin system, as evidenced by their appearance as the ill-defined second-order multiplet at 1.3 ppm. The methyl proton signal at 0.9 ppm is well upfield from this mul-tiplet, and the separation between these methyl protons and C-5 methylene protons to which they are coupled is safely outside the

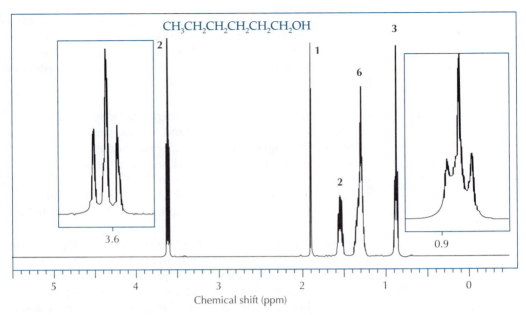

FIGURE 22.38 Distorted signals due to virtual coupling in the 360 MHz ^{1}H NMR spectrum of 1-hexanol.

$\Delta\nu < 5J$ rule of thumb for the occurrence of second-order effects. Thus, it would seem that the methyl peak should display simple $N + 1$ splitting and appear as a normal triplet. While the methyl proton signal does appear triplet-like, the expanded inset for this peak shows a badly distorted triplet due to virtual coupling. The distortion is particularly obvious when compared to the clean methylene triplet at 3.6 ppm. The effect of virtual coupling often disappears in spectra obtained from a sufficiently high-field spectrometer, where chemical shift separation will be larger, thus reducing the strong coupling.

My Spectrum Has a Splitting Pattern I Didn't Expect. How Can I Account for This?

In addition to broad and distorted NMR patterns, there are cases where two or more well-defined patterns overlap, producing what at first glance may resemble a single pattern but which, on closer examination, has coupling constants and/or signal intensities that do not correlate with a single pattern. A good example of such a deceptive pattern is the apparent quartet at 1.2 ppm in the 60-MHz NMR spectrum of ethyl propanoate. An expanded section of this 60-MHz spectrum, containing the four-line pattern, is shown in Figure 22.39a. Figure 22.39b shows the same section of the spectrum obtained on a 360-MHz NMR instrument; at the higher magnetic field, the pattern separates into two triplets, one centered at 1.13 ppm and the other centered at 1.26 ppm.

Could My Extra NMR Signals Come from a Mixture of Compounds?

Extra signals in an NMR spectrum are often a product of a mixture of the compound whose structure you want to know, along with solvents, starting materials, reaction side products, and residual proton signals from the deuterated solvent. Much frustration can be avoided by careful consideration of what might be present in the NMR tube.

(a)

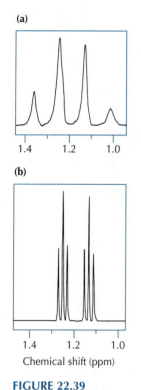

1.4   1.2   1.0

(b)

1.4   1.2   1.0

Chemical shift (ppm)

FIGURE 22.39
¹H NMR spectra of
the methyl group
signals of ethyl
propanoate at
(a) 60 MHz and
(b) 360 MHz.

TABLE 22.7	¹H NMR signals of common solvents
Solvent	**NMR signals (ppm)[a]**
Acetone	2.2 (s)
Benzene	7.4 (s)
Chloroform	7.3 (s)
Cyclohexane	1.4 (s)
Dichloromethane	5.3 (s)
Diethyl ether	1.2 (t), 3.5 (q)
Dimethyl sulfoxide	2.6 (s)
Ethanol[b]	1.2 (t), 3.7 (q)
Ethyl acetate	1.3 (t), 2.0 (s), 4.1 (q)
Hexane	0.9 (t), 1.3 (m)
Pentane	0.9 (t), 1.3 (m)
2-Propanol[b]	1.2 (d), 4.0 (m)
Methanol[b]	3.5 (s)
Tetrahydrofuran	1.9 (m), 3.8 (m)
Toluene	2.3 (s), 7.2 (m) or (s)
Water (dissolved)	1.6 (s)
Water (bulk)	4.6 (br s)

a. In CDCl₃. Multiplicity of the signals is shown in parentheses.
b. The chemical shift of the O–H proton is variable, depending on the experimental conditions.

Sources of extra signals. To determine the source of extra signals, it is important to know the history of the NMR sample. If you prepared it, you already know what solvents and reagents were present in the reaction mixture. You should also be able to answer the following questions:

• If you purified your compound by extraction, recrystallization, or chromatography, what solvents did you use?
• What solvent did you use to clean the NMR sample tube?
• What does the NMR spectrum of the starting material look like?
• What product(s) do you expect to form?

Table 22.7 lists some solvents that are common impurities in NMR samples and the chemical shift positions and multiplicity of their NMR signals.*

Example of a typical reaction mixture. The NMR spectrum shown in Figure 22.40 is an example of a typical mixture encountered in the laboratory. The material for the sample was obtained from the base-catalyzed dehydrochlorination of 2-chloro-2-methylpentane.

Cl— →(KOH, 1-propanol, heat)→ +

2-Methyl-1-pentene   2-Methyl-2-pentene

The expected products of the reaction are 2-methyl-1-pentene and 2-methyl-2-pentene, and it is quite possible that they will be

*An extensive, useful list of impurity peaks has been collected by Gottlieb, H. E.; Kotlyar, V.; Nudelman, A. *J. Org. Chem.* **1997,** *62,* 7512–7515.

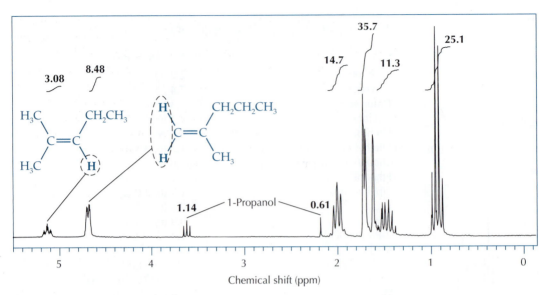

FIGURE 22.40 200-MHz ^{1}H NMR spectrum of the reaction product from dehydrochlorination of 2-chloro-2-methylpentane, a mixture of 2-methyl-1-pentene and 2-methyl-2-pentene.

contaminated with a small amount of 1-propanol, the solvent used in the reaction. The singlet at 2.2 ppm and the triplet at 3.6 ppm in Figure 22.40 are produced by the hydroxyl proton of 1-propanol and the methylene group attached to the hydroxyl group, respectively. The other signals of 1-propanol at 1.55 ppm and 0.92 ppm are obscured by signals of the two alkenes.

Even though every signal in the spectrum is not distinct, much useful information can be obtained because each component of the mixture exhibits unique features. For example, the ability of NMR to count each of the protons in every component of a mixture can be put to good use. This information allows you to tell which sets of NMR peaks relate to the same compound because each set must have integral proton ratios. It also allows you to determine the composition of the mixture.

The integrals of the vinyl-proton signals at 5.1 ppm and 4.7 ppm in Figure 22.40 can be used to determine the ratio of 2-methyl-2-pentene to 2-methyl-1-pentene in the mixture. The broad triplet at 5.1 ppm is caused by the vinyl proton attached to C-3 in 2-methyl-2-pentene. The two broad signals at 4.7 ppm are produced by the two protons attached to C-1 in 2-methyl-1-pentene. To calculate the molar ratio of the two alkenes in the product mixture, the integrals need to be normalized by dividing them by the number of protons causing the signals.

$$\frac{\text{moles of 2-methyl-1-pentene}}{\text{moles of 2-methyl-2-pentene}} = \frac{8.48/2}{3.08/1} = \frac{1.38}{1}$$

The calculation shows that the dehydrochlorination of 2-chloro-2-methylpentane produced 58% 2-methyl-1-pentene and 42% 2-methyl-2-pentene.

Why Do O–H and N–H Protons Have Such a Wide Range of Chemical Shifts? Why Don't They Follow the N + 1 Rule?

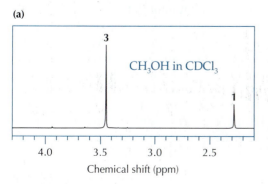

Intermolecular hydrogen bonding

Hydrogen bonding. Hydrogen bonding involving oxygen or nitrogen atoms draws electron density away from O–H and N–H protons, deshielding them and shifting their signals downfield. In dilute samples, where there is little or no intermolecular hydrogen bonding, the signals may have smaller chemical shift values. The extreme case where hydrogen bonding causes deshielding occurs with carboxylic acids, which have a chemical shift of 10–13 ppm for the O–H proton.

Because concentration and temperature affect the extent of intermolecular hydrogen bonding, the chemical shift of protons attached to oxygen and nitrogen atoms in alcohols, amines, and carboxylic acids can appear over a wide range. In fact, O–H and N–H signals often vary in two NMR spectra of the same compound in the same solvent because the concentration differs.

Proton exchange. Another potential source of confusion is the chemical exchange of O–H and N–H protons, which has two consequences for NMR spectroscopy. First, a proton may not be attached to the heteroatom long enough for coupling to occur with nearby protons. Second, the signal for the exchangeable protons can merge into a single peak with an unpredictable chemical shift.

If a proton on the oxygen atom of an alcohol exchanges rapidly, which it usually does in $CDCl_3$ that contains a small amount of water or acid, no coupling between the hydroxyl proton and protons on the adjacent carbon is observed. The hydroxyl proton signal becomes a broadened singlet. If the sample solution used to obtain the NMR spectrum is anhydrous and acid-free, however, splitting is often observed because exchange of the O–H proton is relatively slow, and the hydroxyl proton signal is split by the protons attached to the adjacent carbon atom. The choice of solvent can affect the likelihood of proton exchange. Samples of alcohols prepared in deuterated chloroform almost always show rapid proton exchange, whereas the use of dimethyl sulfoxide-d_6 suppresses it.

The NMR spectrum of methanol dissolved in $CDCl_3$ is shown in Figure 22.41a. Proton exchange is evident because both the signal produced by the methyl protons and the signal produced by the hydroxyl proton appear as singlets. In the NMR spectrum of methanol dissolved in CD_3SOCD_3, shown in Figure 22.41b, the signal

(a)

(b)

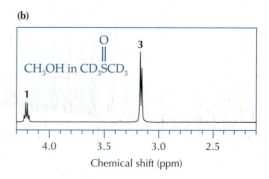

FIGURE 22.41 360-MHz ^{1}H NMR spectra of methanol in (a) deuterochloroform and (b) dimethyl sulfoxide-d_6.

produced by the methyl protons appears as a doublet and the signal produced by the hydroxyl proton appears as a quartet, as predicted by the $N + 1$ rule. Notice that the differing amounts of intermolecular hydrogen bonding in the two solvents cause very different chemical shifts for the O–H proton.

The second consequence of chemical exchange is that O–H and N–H protons can exchange so quickly that they merge into a common environment and become combined into a single "averaged" NMR peak, whose chemical shift depends on concentration, solvent, temperature, and the presence of water or acid. When NMR samples are dissolved in D_2O, all O–H and N–H protons in the compounds merge into a broadened peak at the chemical shift of HOD.

Using chemical exchange as a diagnostic probe. Chemical exchange can be used as a diagnostic probe for protons of alcohols and amines dissolved in a deuterated organic solvent. The experiment is carried out by obtaining a second NMR spectrum after adding a drop of D_2O to the NMR sample solution. Hydroxyl and amine protons in the molecule are replaced by deuterons through chemical exchange. If the compound is an alcohol or amine, the signal resulting from the exchangeable proton disappears and a new signal produced by HOD appears at approximately 4.6 ppm.

22.11 Two Case Studies

As we have discussed, NMR spectroscopy is the principal tool organic chemists use to determine the structures of organic compounds. In this section we will look at the ^1H NMR spectra of two organic compounds and show how the information derived from their spectra can help determine their molecular structures.

Four Major Pieces of Information from an NMR Spectrum

To recap, there are four major pieces of information that are used in the interpretation of a ^1H NMR spectrum of a pure compound:

- *Number of signals* tells us how many kinds of nonequivalent protons are in the molecule.
- *Integration* determines the relative number of nonequivalent protons in a ^1H NMR spectrum.
- *Chemical shift* provides important information on the environment of a proton. Downfield signals (larger ppm values) suggest nearby deshielding oxygen atoms, halogen atoms, or π-systems. Tables 22.2–22.5 and Figure 22.13 are useful aids for correlating chemical shifts with molecular structure.
- *Splitting of signals,* caused by spin-spin coupling of protons to other protons, reveals the presence of nearby protons that produce the splitting. Values of coupling constants can establish coupling connections between protons and can reveal stereochemical relationships (see Table 22.6).

Analysis of a Spectrum

First you should examine the entire spectrum without being too eager to focus on a prominent signal or splitting pattern. To ensure success, a structured and logical approach to the interpretation of an NMR spectrum is necessary. After you have interpreted numerous spectra, you can replace the structured approach below with a less formal one.

1. Make inferences and deductions based on the spectral information.
2. Build up a collection of structure fragments.
3. Put the pieces together into a molecular structure that is consistent with the data and with the rules of chemical bonding.
4. Confirm the chemical shift assignments with calculated chemical shifts based on Tables 22.3–22.5. Any inconsistencies between the values should be examined and resolved. Explanations of minor inconsistencies usually hinge on subtle structural features of the molecule.

Double-Bond Equivalents

In NMR problem sets, the molecular formula of a pure compound is often provided. If available, the molecular formula can be used to determine the *double-bond equivalents (DBE)*, which provide the number of double bonds and/or rings present in the molecule.

$$\text{double-bond equivalents (DBE)} = C + \frac{N - H - X}{2} + 1$$

C is the number of carbon atoms, H is the number of hydrogen atoms, X is the number of halogen atoms, and N is the number of nitrogen atoms. Other names for double-bond equivalents are *degree of unsaturation* and *index of unsaturation.*

Organizing the Spectral Information

We suggest the following method for organizing information from an NMR spectrum. First, prepare an informal table with the following headings:

• Chemical Shift (ppm)
• ^1H Type
• Integration
• Splitting Pattern
• Possible Structure Fragment(s)

Chemical shift. In the Chemical Shift column, list the positions (or ranges) of all the signals in the spectrum.

^1H type. Based on the chemical shifts, use Figure 22.13 and Table 22.2 as guides for entering likely structure assignments in the ^1H Type column for each NMR signal, for example, Ar–**H**, =C–**H**, –O–C–**H**, and so on. Consider any reasonable structure within the chemical shift range. A few types of protons can appear over a wide range of chemical shifts, but at this early juncture it is better to err on the side of being too inclusive. As the analysis is refined, the possibilities can be narrowed down.

Integration. Enter the value of each integral, rounded to whole numbers, in the Integration column. Remember that the integrals must add up to the total number of hydrogen atoms in the molecular formula.

Splitting pattern. In the Splitting Pattern column, enter a description of the splitting pattern. Be as precise as possible, using standard descriptive terms, such as doublet, triplet, quartet, and combinations of these terms, such as doublet of triplets. If you have processed the FID to obtain the NMR spectrum or if you have access to the actual NMR data collected on the spectrometer, rather than just being given the spectrum, you should consider expanding regions in your NMR spectrum to reveal the detail of important splitting patterns. Later in the NMR analysis, when you have a complete structure proposal to consider, you may also wish to measure some of the coupling constants. Coupling constants can be used to determine which signals are coupled to one another and in some cases to assign stereochemistry.

Possible structure fragment(s). The Possible Structure Fragment(s) column is where you pull the information together and enter all of the structure fragments that are consistent with the data for each NMR signal. Be flexible and consider all reasonable possibilities. For example, in the spectrum you may have a quartet that integrates to two protons. The quartet is probably the result of coupling to three protons with equal coupling constants. Two arrangements are consistent with this pattern, –CH$_2$–CH$_3$ and perhaps –CH–CH$_2$–CH$_2$ (boldface indicates the protons exhibiting the quartet).

Proposed Structure

Once you have constructed the table, the analysis is a matter of eliminating proposed structure fragments that are inconsistent and then putting the remaining structure fragments together into reasonable proposals for the structure of the compound. The final structure must be consistent with all the NMR data. Of particular importance is calculating the quantitative estimation of the chemical shifts (Tables 22.3–22.5). In all but the simplest cases, it is important to estimate the chemical shift for each type of proton. The estimated chemical shifts allow you to eliminate proposed structures inconsistent with the chemical shift data.

PROBLEM ONE

An organic compound has a molecular formula of C$_5$H$_{12}$O. Its 200-MHz ^{1}H NMR spectrum is shown in Figure 22.42. Determine its structure.

Double-bond equivalents. First, determine the compound's double-bond equivalents (DBE).

$$DBE = C + \frac{N - H - X}{2} + 1 = 5 + \frac{0 - 12 - 0}{2} + 1 = 0$$

Because DBE = 0, we know that the compound contains no rings or double bonds. Because the molecule has no double bonds and contains an oxygen atom, it must be either an alcohol or an ether.

Table of data from the spectrum. The data from the spectrum are summarized in Table 22.8. An assignment for the signal at 2.36 ppm is tentative because that region is normally where protons on carbons adjacent to alkenes and carbonyl groups appear, and we know from the DBE calculation that there are no double bonds in the molecule. However, the proton on the oxygen atom of an alcohol could also appear in this chemical shift region.

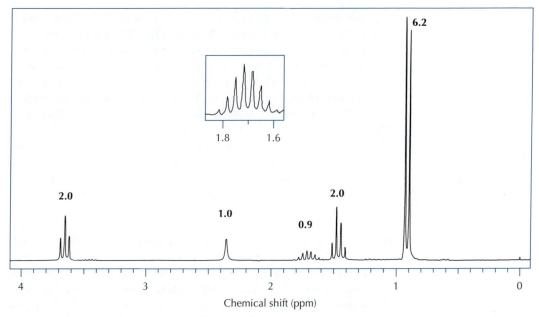

FIGURE 22.42 200-MHz ^{1}H NMR spectrum of $C_5H_{12}O$.

TABLE 22.8	Interpreted data from ^{1}H NMR spectrum (200 MHz) of $C_5H_{12}O$			
Chemical shift (ppm)	^{1}H type	Integration	Splitting pattern	Possible structure fragment(s)
0.91	C–C–H	6	Doublet	–CH(CH₃)₂
1.46	C–C–H	2	Quartet	CH₃CH₂– or –CH₂CH₂CH–
1.71	C–C–H	1	Multiplet	–CH₂CH(CH₃)₂ or –CHCH(CH₃)₂
2.36	Perhaps C–O–H	1	Broad singlet	R–O–H
3.65	O–C–H	2	Triplet	–CH₂CH₂O–

Assembling structure fragments. Two possible fragments could explain the splitting of the signal at 1.46 ppm:

CH₃–CH₂–   or   –CH₂–CH₂–CH–

The ethyl fragment has to be eliminated because there is no signal exhibiting a pattern consistent with the methyl portion of that fragment—a three-proton triplet at approximately 1.0 ppm.

   The splitting pattern at 1.71 ppm is difficult to know with certainty. Because the outermost peaks of highly split patterns are very small relative to the inner peaks, the multiplet could be either an octet or a nonet. In either case, there must be two methyl groups attached to a methine group. The downfield triplet signal at 3.65 ppm is the result of a methylene group attached to an oxygen atom.

From the analysis of Table 22.8, we can propose that the compound $C_5H_{12}O$ is an alcohol with an isopropyl group $((CH_3)_2CH-)$ or perhaps a methylene group flanked by methine and methylene groups $(-CH-CH_2-CH_2-)$. In addition, there seems to be a methylene group attached to an oxygen atom and a second methylene group $(-CH_2-CH_2-O-)$. This array of fragments consists of eight carbon atoms, 16 hydrogen atoms, and one oxygen atom. Obviously, there are some atoms common to more than one fragment.

To solve this puzzle, set out the three structure fragments side-by-side to look for possible overlap:

$$(CH_3)_2CH- \qquad -HC-CH_2-CH_2- \qquad -CH_2-CH_2-O-H$$
$$\textbf{1} \qquad\qquad\qquad \textbf{2} \qquad\qquad\qquad\qquad \textbf{3}$$

Possible structures. It looks as if fragment 2 overlaps both fragment 1 and fragment 3. Combining fragments 1 and 3 produces a five-carbon alcohol, $(CH_3)_2CH-CH_2-CH_2-OH$, 3-methyl-1-butanol. The estimated chemical shifts from Table 22.3 are shown in the following structure; as you can see, the correspondence is very good.

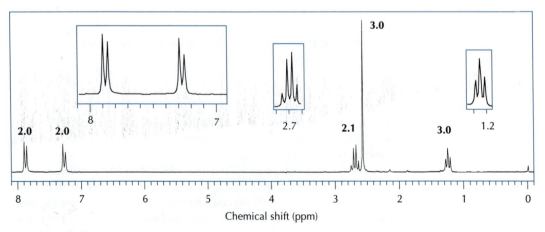

Observed chemical shifts of 3-methyl-1-butanol
(calculated estimates of chemical shifts)

PROBLEM TWO

An organic compound has a molecular formula of $C_{10}H_{12}O$. Its 200-MHz ^1H NMR spectrum is shown in Figure 22.43. Determine the structure of this compound.

FIGURE 22.43 200-MHz ^1H NMR spectrum of $C_{10}H_{12}O$.

Table of data from the spectrum. The NMR data from Figure 22.43 are summarized in Table 22.9.

Double-bond equivalents. The double-bond equivalent calculation indicates that the compound contains a combination of five double bonds and/or rings.

$$DBE = C + \frac{N - H - X}{2} + 1 = 10 + \frac{0 - 12 - 0}{2} + 1 = 5$$

Whenever there is a large DBE value, it is likely that the molecular structure of the compound incorporates one or more benzene rings because each benzene ring accounts for four DBEs (three double bonds and one ring). The NMR spectrum confirms this assumption by the presence of signals in the aromatic proton region (6.5–9.0 ppm). The total integration of the aromatic protons is four, implying that the benzene ring is disubstituted. Moreover, the symmetry of the two signals in the aromatic region, a pair of doublets, indicates two groups of equivalent protons. This pattern is possible only if the two substituents are attached to the 1- and 4-positions of the benzene ring (*para* substitution). Several **signature patterns** are observed in NMR spectra, and a pair of symmetrical doublets in the aromatic region is one of the more frequently encountered ones (see top of page 404).

TABLE 22.9 **Interpreted data from ^{1}H NMR spectrum (200 MHz) of $C_{10}H_{12}O$**

Chemical shift (ppm)	^{1}H type	Integration	Splitting pattern	Possible structure fragment(s)
1.25	C–C–H	3	Triplet	CH_3CH_2–
2.58	O=C–C–H	3	Singlet	CH_3–C=O
	or			or
	Ar–C–H			CH_3–Ar
	or			or
	C=C–C–H			CH_3–C=C
2.70	O=C–C–H	2	Quartet	CH_3CH_2–C=O
	or			or
	Ar–C–H			CH_3CH_2–Ar
	or			or
	C=C–C–H			CH_3CH_2–C=C
7.28	Ar–H	2	Doublet	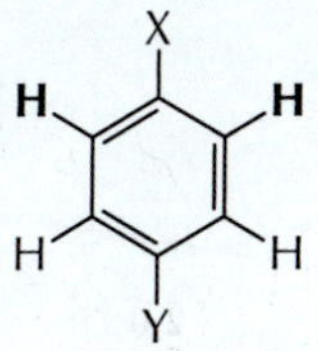
7.88	Ar–H	2	Doublet	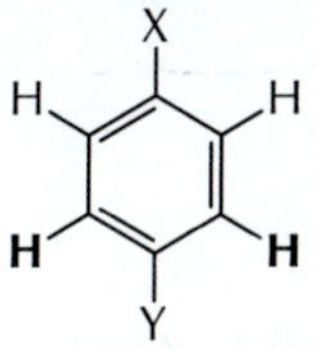

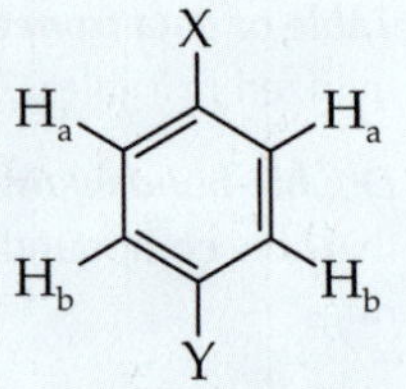

1,4 disubstitution with
two types of
aromatic protons

Possible structure fragments. The three-proton triplet at 1.25 ppm and two-proton quartet at 2.70 ppm are a signature pattern, which indicates an ethyl group. The 2.70-ppm chemical shift of the two-proton quartet indicates the environment of the methylene protons, suggesting that the ethyl group could be a part of three possible structure fragments:

$$CH_3CH_2C{=}O \qquad CH_3CH_2Ar \qquad CH_3CH_2C{=}C$$

A decision to eliminate one of these possibilities can be made reasonably easily. The compound has the molecular formula $C_{10}H_{12}O$. The benzene ring has six carbons, and the third structure fragment has four carbon atoms. It leaves no room for the methyl group appearing at 2.58 ppm, and there is no indication how the oxygen atom could be incorporated in the structure. Therefore, the only viable structure fragment options for the ethyl group are $CH_3CH_2C{=}O$ and $CH_3CH_2{-}Ar$.

The only NMR signal left to analyze is the methyl group at 2.58 ppm; based on its chemical shift, this peak must be in one of two possible structure fragments, $CH_3C{=}O$ or CH_3Ar. To summarize, $C_{10}H_{12}O$ includes a methyl group ($CH_3{-}$), an ethyl group ($CH_3CH_2{-}$), and a *para*-disubstituted benzene ring (C_6H_4). The atom count in the fragments is nine carbon atoms and 12 hydrogen atoms, leaving only one carbon atom and one oxygen atom to be accounted for. Because one more double-bond equivalent is required, the last fragment for the molecule is a carbonyl group, which is also consistent with the possible structure fragments.

Possible structures. Two possible structures are consistent with the data: 4'-methylphenyl-1-propanone and 4'-ethylphenyl-1-ethanone.

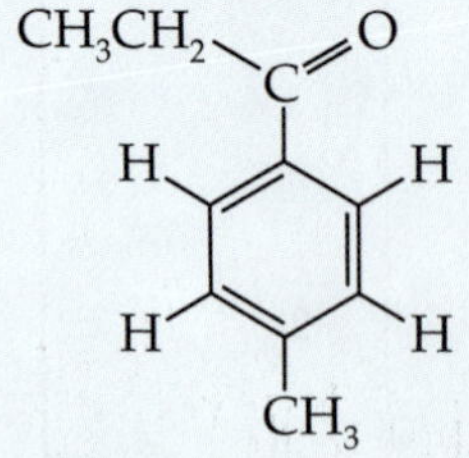

4'-Methylphenyl-1-propanone 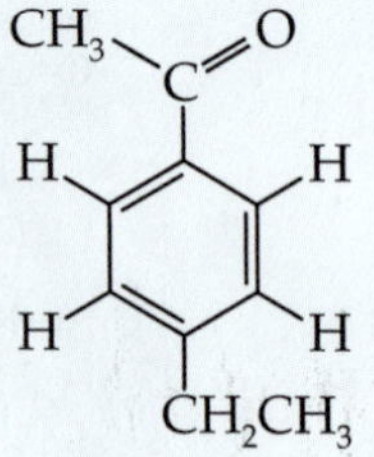   4'-Ethylphenyl-1-ethanone

FOLLOW-UP ASSIGNMENT

To discover which of the two structures is more likely, it is necessary to calculate the estimated chemical shifts for both structures, using Tables 22.3 and 22.4. Carry out these calculations and decide which structure is more likely for the compound.

Further Reading

Crews, P.; Rodríguez, J.; Jaspars, M. *Organic Structure Analysis*, 2nd ed.; Oxford University Press: Oxford, 2009.

Findelsen, M.; Berger, S. *50 and More Essential NMR Experiments—A Detailed Guide*; Wiley-VCH: Weinheim, 2013.

Friebolin, H. *Basic One- and Two-Dimensional NMR Spectroscopy*, 5th ed.; Wiley-VCH: Weinheim, 2010.

Pouchert, C. J.; Behnke, J. (Eds.) *Aldrich Library of 13C and 1H FT-NMR Spectra*; Aldrich Chemical Co.: Milwaukee, WI, 1993; 3 volumes.

Pretsch, E.; Bühlmann, P.; Badertscher, M. *Structure Determination of Organic Compounds*, 4th ed.; Springer-Verlag: New York, 2009.

Silverstein, R. M.; Webster, F. X.; Kiemle, D. J. *Spectrometric Identification of Organic Compounds*, 7th ed.; Wiley: New York, 2005.

Questions

1. Given the ^{1}H NMR spectrum and molecular formula for each of the following compounds, deduce the structure of the compound, estimate the chemical shifts of all its protons using the parameters in Tables 22.3–22.5, and assign the NMR signals to their respective protons.

   (a) $C_5H_{11}Cl$; ^{1}H NMR (CDCl$_3$): δ 3.33 (2H, s); 1.10 (9H, s)

   (b) $C_5H_{10}O_2$; ^{1}H NMR (CDCl$_3$): δ 3.88 (1H, s); 2.25 (3H, s); 1.40 (6H, s)

   (c) $C_6H_{12}O_2$; ^{1}H NMR (CDCl$_3$): δ 3.83 (1H, s); 2.63 (2H, s); 2.18 (3H, s); 1.26 (6H, s)

   (d) $C_5H_{10}O$; ^{1}H NMR (CDCl$_3$): δ 9.77 (1H, t, J = 2 Hz); 2.31 (2H, dd, J = 2 Hz, J = 7 Hz); 2.21 (1H, m); 0.98 (6H, d, J = 7 Hz)

   (e) C_4H_8O; ^{1}H NMR (CDCl$_3$): δ 5.90 (1H, ddd, J = 6, 10, 17 Hz); 5.19 (1H, d, J = 17 Hz); 5.06 (1H, d, J = 10 Hz); 4.30 (1 H, quintet); 2.50 (1H, bs); 1.27 (3H, d, J = 6 Hz)

2. The ^{1}H NMR spectrum of a compound of molecular formula $C_5H_8O_2$ in CDCl$_3$ is shown in Figure 22.44. Deduce the structure of the compound and assign its NMR signals.

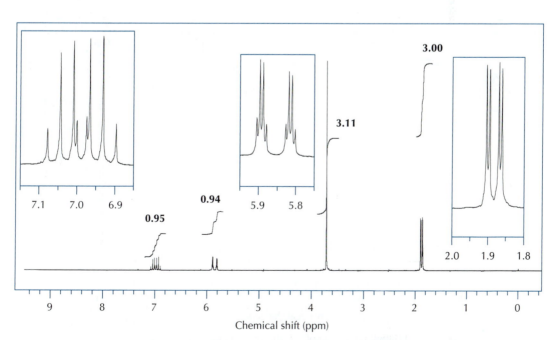

FIGURE 22.44 200-MHz ^{1}H NMR spectrum of a compound of molecular formula $C_5H_8O_2$.

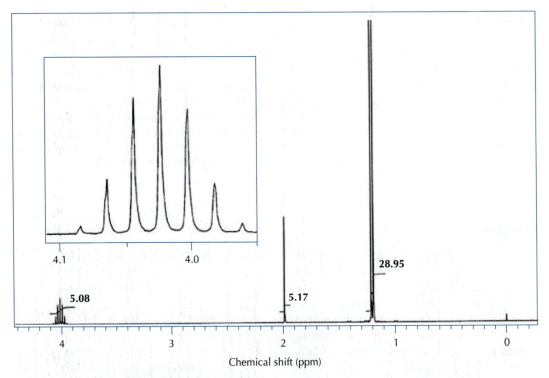

FIGURE 22.45 300-MHz ¹H NMR spectrum of a compound of molecular formula C₃H₈O.

3. A compound of molecular formula C₃H₈O in CDCl₃ produces the ¹H NMR spectrum shown in Figure 22.45. In addition, when this compound is treated with D₂O, the ¹H NMR signal at 2.0 ppm disappears and another signal at 4.6 ppm appears. Moreover, when the C₃H₈O compound is highly purified and care is taken to remove all traces of acid in the NMR solvent, the singlet at 2.0 ppm is replaced

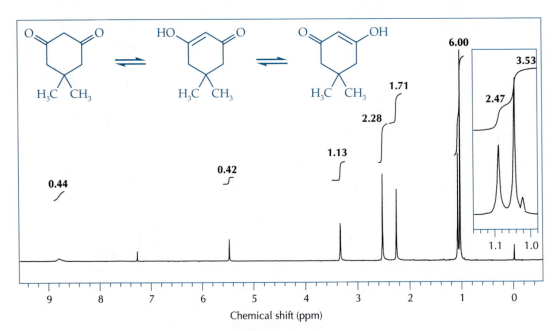

FIGURE 22.46 200-MHz ¹H NMR spectrum of 5,5-dimethylcyclohexan-1,3-dione.

by a doublet. Finally, the chemical shift of the 2.0-ppm signal is highly concentration dependent; an increase in the concentration of C_3H_8O in the NMR sample results in a downfield shift of this signal. Deduce the structure of C_3H_8O, assign its NMR signals, and explain the changes observed for the 2.0-ppm signal. Estimate the chemical shifts of its different types of protons using the parameters in Figure 22.13 and Table 22.3; compare them with those measured from the spectrum.

4. In solution, dimedone (5,5-dimethyl cyclohexan-1,3-dione) is a mixture of keto and enol isomers. The ^1H NMR spectrum of a solution of dimedone in $CDCl_3$ is

shown in Figure 22.46. In the sample, the two enol isomers are equilibrating very fast compared with the NMR time scale. Assign all the NMR signals and use NMR integrations to determine the composition of the keto/enol mixture.

5. A compound of molecular formula $C_8H_{14}O$ produces the ^1H NMR spectrum shown in Figure 22.47. Its infrared spectrum shows a strong carbonyl stretching peak, which indicates that $C_8H_{14}O$ is either an aldehyde or a ketone. Deduce the structure of the compound, estimate the chemical shifts of the different types of protons, using the parameters in Tables 22.3–22.5, and assign all the NMR signals.

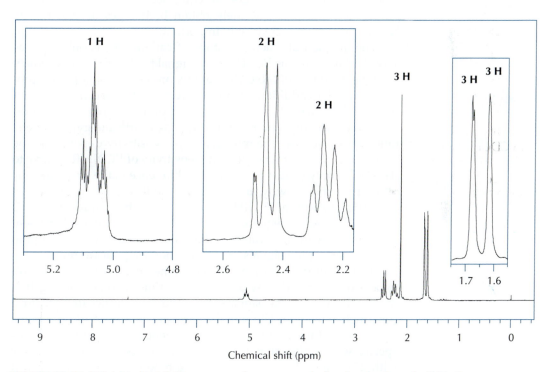

FIGURE 22.47 200-MHz ^1H NMR spectrum of a compound of molecular formula $C_8H_{14}O$.

^{13}C AND TWO-DIMENSIONAL NMR SPECTROSCOPY

If Chapter 23 is your introduction to spectroscopic analysis, read the Essay "The Spectrometric Revolution" on pages 309–310 before you read Chapter 23.

NMR topics have already been discussed in some detail in Chapter 22, which focused on ^{1}H NMR. A brief summary of NMR information that can be gleaned from a ^{1}H NMR spectrum can be found in Section 22.4. ^{13}C NMR has many similarities to ^{1}H NMR, but there are a few important differences.

^{13}C NMR gives direct evidence about the carbon skeleton of an organic molecule, so you might think that it would be the favored NMR technique for determination of molecular structures. In part, the reason ^{1}H NMR receives more attention in organic chemistry textbooks is because it was developed first. Carbon NMR became a useful and routine tool only when pulsed Fourier transform (FT) instrumentation was developed in the 1970s and 1980s. Most modern NMR spectrometers are equipped with both ^{1}H and ^{13}C probes, and chemists routinely obtain both types of spectra. ^{13}C NMR is becoming increasingly important.

The magnetically active isotope of carbon, ^{13}C, is only 1.1% as abundant in nature as ^{12}C. Thus, the signal from carbon is extremely weak because few of the carbon atoms present in a compound provide a signal. In addition, ^{13}C nuclei are much less sensitive in NMR spectroscopy than ^{1}H nuclei. An inherent property of the ^{13}C nucleus, called the *gyromagnetic ratio,* is only one-fourth the gyromagnetic ratio for ^{1}H. Because the inherent NMR sensitivity depends on the cube of the gyromagnetic ratio, the sensitivity of ^{13}C NMR relative to ^{1}H NMR is only $(0.25)^3$, or 0.016. This difference in the gyromagnetic ratio also explains why an instrument built to analyze protons at 300 MHz operates at 75 MHz, or one-fourth the frequency, for ^{13}C nuclei. The lower frequency in ^{13}C NMR usually doesn't present problems, however, because carbon shifts occur over a 200-ppm range. Overlapping signals are usually not a significant problem.

The use of pulsed FT NMR spectrometers allows NMR spectra to be acquired rapidly. By pulsing many times and adding together the NMR signals, the signals from the sample accumulate in a constructive fashion. The signal-to-noise ratio is proportional to the square root of the number of pulse sequences. Very good ^{13}C NMR spectra with a high signal-to-noise ratio can be obtained routinely with 25–50 mg of compound, using a modern high-field FT NMR instrument and a suitable number of pulses.

23.1 ^{13}C NMR Spectra

Samples are prepared for ^{13}C NMR much as they are for ^{1}H NMR (see Section 22.2). Table 22.1 (see page 354) lists the suitable solvents and their ^{13}C chemical shifts. As is the case for ^{1}H NMR, the primary standard for the ^{13}C NMR chemical shift scale is tetramethylsilane ($(CH_3)_4Si$, TMS), which is set at 0.0 ppm. TMS absorbs radio frequency radiation at a magnetic field in which few other carbon atoms in typical organic compounds absorb.

Solvent Peaks and Multiplicities

Deuterated chloroform ($CDCl_3$) has been the solvent of choice because it dissolves most organic compounds and is relatively inexpensive. Of course, each molecule of $CDCl_3$ has a carbon atom. **The solvent signal for $CDCl_3$ always appears at 77 ppm in a ^{13}C NMR spectrum.** The triplet signal of $CDCl_3$ can be used as an internal reference point for the chemical shifts of the sample signals, and it is often unnecessary to add a reference material, such as TMS, to the sample.

The spin multiplicity of ^{13}C is the same as that of hydrogen (^{1}H); thus the same splitting rules apply. For example, the ^{13}C signal of a methine group (C–H) will be split into a doublet by the attached proton. Unlike ^{1}H, however, a deuterium nucleus (^{2}H) has a spin of 1. Therefore, the ^{13}C signal of $CDCl_3$ is always a characteristic triplet in which the three lines have equal height.

Recently, in an effort to reduce the use of halogenated solvents in the organic chemistry lab, deuterated acetone has become more common as an NMR solvent. Each molecule of acetone has two different carbons, a carbonyl carbon that appears at 206 ppm and a methyl carbon at 29.8 ppm. The carbonyl carbon is well removed from most signals, except other ketone carbonyl carbon atoms, and appears as a singlet. Because each methyl carbon of deuterated acetone has three deuterium atoms attached, it appears as a seven-line pattern.

Spin-Spin Splitting

As with ^{1}H NMR, spin interactions in ^{13}C NMR are transmitted through the bonding framework of molecules. They are usually observable only when the interacting nuclei are near each other. When protons are directly attached to a ^{13}C atom, the ^{13}C–H coupling constants are very large—on the order of 150 Hz. Coupling between adjacent carbon nuclei is not observed because the probability that two attached carbons will both be ^{13}C is extremely small, about 1 in 10,000. The splitting of signals in a ^{13}C NMR spectrum can be used to identify the number of protons attached to a carbon atom. A methyl carbon signal appears as a quartet, a methylene signal appears as a triplet, a methine signal appears as a doublet, and a quaternary carbon signal appears as a singlet.

The ^{13}C spectrum of ethyl *trans*-2-butenoate shown in Figure 23.1 demonstrates these splitting patterns clearly. The carbonyl carbon at 166 ppm is a singlet because it has no attached protons. The alkene carbons at 144 ppm and 123 ppm are doublets because each one has only one attached proton. The signal at 60 ppm, produced by the methylene group attached to oxygen, appears as a triplet because it has two attached protons. The complex pattern of signals between 12 and 20 ppm is two overlapping quartets due to the two nonequivalent methyl groups.

Broadband Decoupling

In molecules containing many carbon atoms, the coupled ^{13}C spectrum can become extremely complex because of multiple overlapping splitting patterns, which are often almost impossible to interpret. The usual practice is to avoid this complexity by a technique called **broadband decoupling.** Irradiation of the sample with a broad band of energy at ^{1}H frequencies during data acquisition decouples ^{13}C from ^{1}H nuclei and collapses the ^{13}C multiplets into singlets.

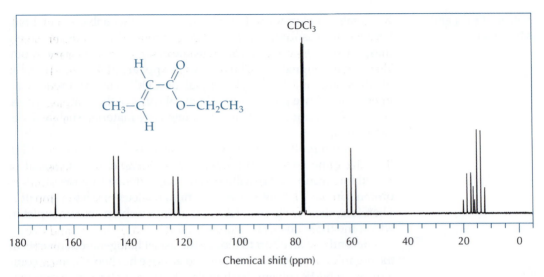

FIGURE 23.1 90-MHz ^{13}C NMR spectrum of ethyl *trans*-2-butenoate in CDCl$_3$.

The decoupled spectrum of ethyl *trans*-2-butenoate shown in Figure 23.2 consists of six sharp signals at 14, 18, 60, 123, 144, and 166 ppm, corresponding to the six different types of carbon nuclei in the molecule.

In addition to simplifying the spectrum, broadband decoupling enhances the signal-to-noise ratio and reduces the acquisition time dramatically. The coupled spectrum of ethyl *trans*-2-butenoate (Figure 23.1) was acquired in 15,000 scans (overnight); the decoupled spectrum of the same sample (Figure 23.2) was acquired in 450 scans (approximately 20 min).

In both Figure 23.1 and Figure 23.2, the signal at 166 ppm is much smaller than the other five signals. Differences in peak sizes for different ^{13}C signals are common. Peak areas for different carbon

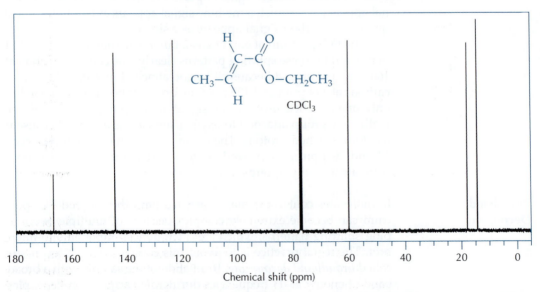

FIGURE 23.2 Broadband-decoupled 90-MHz ^{13}C NMR spectrum of ethyl *trans*-2-butenoate in CDCl$_3$.

signals vary greatly, and quantitative ¹³C integration is not easily accomplished. Excited ¹³C NMR nuclei relax back to lower-energy spin states at a slow rate. Because many scans are necessary to produce adequate signal-to-noise ratios in ¹³C spectra, the time between scans is generally set at too short a time for complete relaxation to occur. In addition, the sizes of ¹³C signals vary due to nuclear Overhauser enhancement. Integrals of routine ¹³C spectra are generally unreliable.

Nuclear Overhauser Enhancement

The size of a ¹³C signal is influenced significantly by its close proximity to protons, a phenomenon termed the *nuclear Overhauser enhancement (NOE)*. The NOE effect can produce up to a fourfold increase in the NMR signal intensity of a ¹³C nucleus when the resonance of nearby coupled protons is perturbed by broadband decoupling. This effect helps to explain why the signals of different types of carbon atoms, such as methyl, methylene, and methine carbons, can have much greater signal amplitudes than the relatively low intensity observed for quaternary carbons.

Symmetry and the Number of ¹³C Signals

The number of signals in a ¹³C NMR spectrum of a pure compound indicates the number of different types of carbon atoms in the molecule. If the number of signals is less than the number of carbon atoms, there is probably some element of symmetry, which makes some of the carbon atoms equivalent to one another. Symmetry elements, which include mirror planes and axes of rotation, can be used advantageously to select a structure from several possibilities.

Consider an aromatic compound with the molecular formula C_8H_{10}. There are four possible structures that are consistent with this formula: ethylbenzene, 1,2-dimethylbenzene, 1,3-dimethylbenzene, and 1,4-dimethylbenzene. Each structure possesses at least one plane of symmetry, and 1,4-dimethylbenzene has two planes of symmetry.

Ethylbenzene   1,2-Dimethylbenzene   1,3-Dimethylbenzene   1,4-Dimethylbenzene

Each of these compounds contains a different number of non-equivalent carbon atoms: ethylbenzene has six, 1,2-dimethylbenzene has four, 1,3-dimethylbenzene has five, and 1,4-dimethylbenzene has three. Ethylbenzene is unique because it has two ¹³C signals in the alkyl region, whereas each of the dimethylbenzenes has only one signal upfield. Obtaining the ¹³C NMR spectrum and counting the number of signals can determine the structure of the compound.

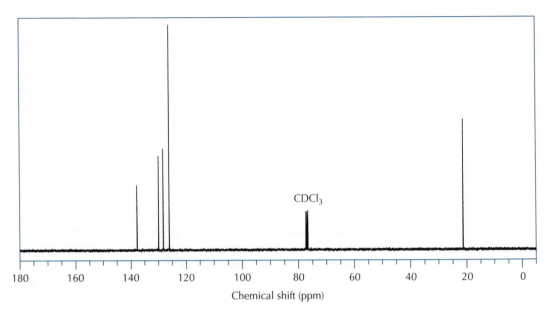

FIGURE 23.3 Broadband-decoupled 90-MHz ^{13}C NMR spectrum of an aromatic compound with molecular formula C_8H_{10} in $CDCl_3$.

EXERCISE

The ^{13}C NMR spectrum of an aromatic compound with the molecular formula C_8H_{10} is shown in Figure 23.3. What is the structure of the compound?

Answer: The ^{13}C NMR spectrum shows one upfield signal due to the carbon atoms of two identical methyl groups and four signals due to the carbon atoms in the aromatic ring. The compound that produces these five signals in the ^{13}C spectrum is 1,3-dimethylbenzene.

Summary and a Look Ahead

A typical interpretation of a simple ^{13}C NMR spectrum relies on only two kinds of information—the number of different carbon signals in the spectrum and the positions of these signals along the horizontal axis (chemical shifts). Integration of ^{13}C signals and spin-spin coupling have little importance in the interpretation. However, complex pulse sequences for ^{13}C NMR are used to determine the number of protons on carbon atoms (see Section 23.4), and spin-spin coupling is used to establish the connectivity of carbon atoms within a molecule (see Section 23.6). Sections 23.2 and 23.3 explore the topic of ^{13}C chemical shifts in some detail.

23.2 ^{13}C Chemical Shifts

The chemical shifts for carbon atoms are affected by electronegativity and anisotropy in ways similar to ^{1}H NMR (see Chapter 22, Introduction, and Section 22.7). Figure 23.4 and Table 23.1 reveal the same kind of chemical shift trends that occur in ^{1}H NMR spectroscopy. In addition, sp^2 and sp hybridization of a ^{13}C carbon nucleus have a strong deshielding effect.

Alkyl Substitution Effects

Carbon atoms of alkanes appear in the upfield region of the NMR spectrum, from approximately 5 to 40 ppm. This upfield

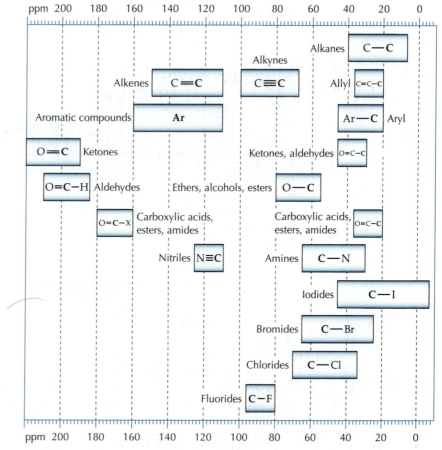

FIGURE 23.4 Approximate regions of ¹³C chemical shifts for different types of carbon atoms in organic compounds.

region can be further refined into regions for methyl, methylene, methine, and quaternary carbons, but considerable overlap occurs. In similarly substituted molecules, increasing the substitution decreases shielding and causes the chemical shift of a carbon atom to increase.

CH_4	CH_3CH_3	$(CH_3)_2CH_2$	$(CH_3)_3CH$	$(CH_3)_4C$
−2.3 ppm	7.3 ppm	15.9 ppm	25.0 ppm	27.7 ppm

Electronegativity. Signals of ¹³C atoms that are in close proximity to electronegative atoms are moved downfield by diamagnetic deshielding (see Section 22.7). Carbon atoms attached to electronegative atoms usually appear in the 30–90-ppm region. If the carbon is attached to an oxygen atom of an alcohol, ether, or ester, the typical range of the chemical shift is 55–80 ppm.

The periodic trends seen in the electronegativities (χ) of elements are mirrored in the chemical shifts of the carbons attached to these elements. Strongly electronegative halogens deshield carbon, but as the halogen atoms increase in atomic number, the deshielding of nearby carbon atoms is attenuated considerably.

TABLE 23.1	Characteristic ^{13}C NMR chemical shifts in $CDCl_3$

Compound	Chemical shift (ppm)
TMS	0.0
$CDCl_3$ (t)	77
Alkane (C–**CH$_3$**)	7–30
Alkane (C–**CH$_2$**)	15–40
Alkane (C–**CH**) and (**C–C**)	15–40
Carboxylic acids, esters, and amides (**C**–C=O)	20–35
Allyl (**C**–C=C)	20–35
Arene (**C**–Ar)	20–45
Ketones, aldehydes (**C**–C=O)	30–45
Amines (**C**–N)	30–65
Iodides (**C**–I)	–5–45
Bromides (**C**–Br)	25–65
Chlorides (**C**–Cl)	35–70
Fluorides (**C**–F)	80–95
Alcohols (**C**–OH), ethers (**C**–OR), esters (**C**–O[C=O]R)	55–80
Alkyne (**C≡C**)	70–100
Alkene (**C=C**)	110–150
Aromatic	110–160
Nitriles (**C≡N**)	110–125
Carboxylic acids, esters, and amides (**C=O**)	160–180
Aldehydes (**C=O**)	185–210
Ketones (**C–O**)	190–220

	CH_3—I	CH_3—Br	CH_3—Cl	CH_3—F
δ	−24.0 ppm	9.6 ppm	25.6 ppm	71.6 ppm
χ	2.66	2.96	3.16	3.98

Iodomethane has a ^{13}C chemical shift of −24 ppm, which is more than 20 ppm *upfield* of methane. This shielding effect has been attributed to "steric compression"; steric factors apparently cause the electrons in the orbitals of the carbon atom to become compacted into a smaller volume closer to the nucleus, thus making the nucleus more highly shielded.

Hybridization. The hybridization of a carbon atom has a dramatic effect on its chemical shift. Whereas sp^3 carbons of alkanes appear in the 5–40-ppm region, the sp carbon atoms of alkynes appear in the 70–100-ppm range, and the sp^2 carbon atoms of alkenes appear in the 110–150-ppm range. The chemical shift region for aromatic carbon atoms overlaps the alkene region, extending from 110 to 160 ppm.

Additivity of hybridization and electronegativity effects. The sp^2 carbon atom of a carbonyl (C=O) group is strongly deshielded because of its hybridization and because the carbon atom is directly attached to a strongly electronegative oxygen atom. Signals from carbonyl carbon atoms appear in the 160–220-ppm range. There are distinct differences in the shifts of carboxylic acids and their derivatives

---

TABLE 23.2 **The effects of carbonyl type and conjugation on ¹³C chemical shifts of carbonyl carbon atoms**

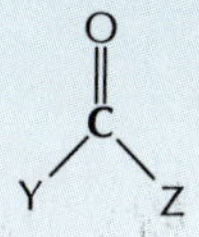

$$\underset{Y}{\overset{\displaystyle \underset{\|}{\overset{O}{}}}{C}}\diagdown_{Z}$$

Group (Y)	Group (Z)	ppm
$CH_3CH_2CH_2$	H	203
$CH_3CH=CH$	H	193
$CH_3CH_2CH_2$	CH_3	207
$CH_3CH=CH$	CH_3	197
$CH_3CH_2CH_2$	OCH_3	174
$CH_3CH=CH$	OCH_3	167

(amides and esters), which are in the 160–180-ppm range, compared with those of aldehydes and ketones, whose signals appear at 185–220 ppm. The difference is ascribed to the electron-releasing effect of an additional heteroatom (O or N) attached to the carbonyl group. The chemical shift of a ketone is shifted approximately 5 ppm downfield relative to the chemical shift of a similar aldehyde. The chemical shifts of the carbonyl carbon atoms in butanal, 2-pentanone, and methyl butanoate are found in Table 23.2.

Conjugation. In conjugated systems, the π-electron density is distributed unevenly over an extended framework. The *sp²* carbons in the middle of a conjugated system are generally shielded more than the *sp²* carbons at its extremities.

Table 23.2 shows that with α,β-unsaturated carbonyl compounds, carbonyl carbon atoms are shielded more than their saturated analogs. Carbonyl carbon atoms attached to aromatic rings are shielded to approximately the same extent. The magnitude of the shielding is approximately 10 ppm.

The β-carbon atoms of α,β-unsaturated carbonyl compounds are more deshielded than the α-carbon atoms, as shown by the following resonance structures:

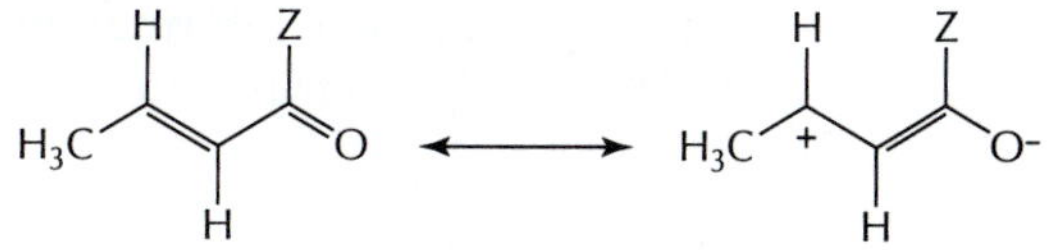

Anisotropy. As with ¹H NMR, anisotropic effects influence the chemical shifts of alkyl ¹³C nuclei adjacent to multiple bonds (see Section 22.7). Consider a comparison of the chemical shifts of the C-3 of pentane with those of 1-pentene and 1-pentyne.

$$CH_3CH_2CH_2CH_2CH_3 \qquad CH_2=CHCH_2CH_2CH_3 \qquad HC\equiv CCH_2CH_2CH_3$$

34.6 ppm   36.2 ppm   20.1 ppm

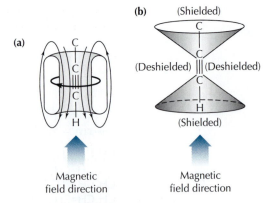

FIGURE 23.5 Two representations of the regions of shielding and deshielding for an alkyne.

C-3 is in the shielding cone of the C≡C of 1-pentyne and its chemical shift is decreased by approximately 15 ppm (Figure 23.5).

By contrast, the C-3 of an alkene is in the plane of the double bond (Figure 23.6) and the chemical shift is increased slightly relative to the alkane. The six π-electrons of an aromatic ring produce a stronger anisotropic effect than that found with simple alkenes.

The anisotropic effect can also be seen with alkyl carbon atoms adjacent to carbonyl groups, which are deshielded somewhat compared to carbons attached to the carbon of a C–O bond. The carbon atom α to the carbonyl group in 4-heptanone appears at 45.0 ppm, 5 ppm downfield from the carbon α to the C–OH group in 4-heptanol.

$$CH_3CH_2CH_2CH(OH)CH_2CH_2CH_3 \quad CH_3CH_2CH_2C(\!\!=\!\!O)CH_2CH_2CH_3$$
$$\text{40.0 ppm} \qquad\qquad\qquad \text{45.0 ppm}$$

Rings. The signals due to methylene carbons in saturated carbon rings are slightly shielded compared to the signals due to methylene carbons in acyclic compounds. In cyclopentanes and cyclohexanes, methylene carbons typically appear near 26–27 ppm, compared to 29–30 ppm for acyclic compounds.

Cyclopropanes are unique in that the σ-bonding in three-membered rings has some π-orbital character, which produces an anisotropic effect and shielding above and below the plane of the ring. Chemical shifts of carbons in cyclopropane rings are approximately 32–33 ppm upfield from their acyclic counterparts.

Recognition of trends and characteristic regions enables one to glean a great deal of information about a compound's structure from its ¹³C NMR spectrum. In general, however, only chemical shifts

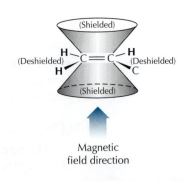

FIGURE 23.6 Regions of shielding and deshielding in an alkene.

greater than 40 ppm can be correlated unambiguously to a specific functional group.

WORKED EXAMPLE

Using Table 23.1, the signals in the broadband-decoupled ¹³C spectrum of ethyl *trans*-2-butenoate (Figure 23.2) can be assigned. The upfield signals at 14 and 18 ppm are due to the sp^3 carbon atoms of the two methyl groups. The methylene group attached to the electronegative oxygen atom—whose signal is at 60 ppm—is more deshielded than the corresponding alkyl carbon atoms attached only to other carbon atoms. The signals due to the two alkene carbon atoms appear at 123 ppm and 144 ppm. C-2 is responsible for the 123-ppm signal and C-3 for the 144-ppm signal. Finally, the ¹³C signal due to the carbonyl carbon of the ester functional group appears at 166 ppm.

Compilations of ¹³C NMR signals from a vast number of compounds have been used to develop systematic methods for estimating chemical shifts. These methods take into account functional groups, types of bonding, and steric constraints, as well as more subtle factors. Section 23.3 provides a condensed version of some of these methods.

23.3   Quantitative Estimation of ¹³C Chemical Shifts

Signals for different types of ¹³C atoms in a molecule appear in well-defined chemical shift regions, depending on their type of bonding and the proximity of nearby electronegative atoms (Table 23.1). However, you may have noticed in Section 23.2 that the chemical shifts can cover a wide spectral range for apparently similar types of carbon atoms. This uncertainty produces an undesirable degree of ambiguity in assigning peaks in ¹³C NMR spectra.

The estimated chemical shift of a ¹³C nucleus can be calculated in a straightforward, additive way from the empirical correlations in Tables 23.3–23.7. Being able to add the individual effects of nearby functional groups is extremely useful because it allows a reasonably accurate estimation of the chemical shifts for many of the carbon atoms in organic compounds. In general, calculations using these tables are accurate to within ±3%.

Chemical Shifts of Alkyl Carbons

As shown in Table 23.3, the chemical shift of methane (−2.3 ppm), with reference to TMS at 0.0 ppm, is used as the base value for alkyl ¹³C atoms. Additive parameters are then added to this value to account for chemical shift effects of nearby substituents in a molecule.

Effects of nearby substituents. The effect of each nearby substituent is added to the base value to arrive at the chemical shift of a particular ¹³C atom in a molecule. If the group is directly attached to the carbon atom, it is an *α (alpha) substituent*. If the group is attached to a carbon atom once removed, it is a *β (beta) substituent.* And if the group is attached to a carbon atom twice removed, it is a *γ (gamma) substituent.*

The effect of an α substituent on the chemical shift of the ¹³C atom is found by using a value from the first numerical column in Table 23.3, and the effect of β and γ groups are found in the second and third columns, respectively. When the substituent (Y) is farther

T A B L E   2 3 . 3　　**Additive parameters for predicting NMR chemical shifts of alkyl carbon atoms in CDCl$_3$**

Base value: CH$_4$ = −2.3 ppm

Group (Y)	alpha (α) C—Y	beta (β) C—C—Y	gamma (γ) C—C—C—Y
—C—(sp³)	9.1	9.4	−2.5
—CH=CH$_2$	20.3	6.8	−2.6
—CH=CHR (cis)	13.4	6.9	−0.3
—CH=CHR (trans)	19.3	7.0	−0.3
—CH=CR$_2$	14.4	6.9	−0.3
—C$_6$H$_5$	22.5	8.9	−2.6
—C≡C–H	5.7	7.4	−2.0
—C≡C–R	4.7	7.7	−0.2
—NH$_2$	28.7	11.3	−5.2
—NHR	40.9	5.2	−4.3
—NR$_2$	36.7	7.6	−4.6
—OH	49.4	10.1	−6.3
—O(C=O)CH$_3$	50.9	5.9	−6.2
—OR	57.4	7.2	−5.9
—I	−6.8	10.9	−1.6
—Br	20.1	10.2	−3.9
—Cl	30.0	10.0	−4.6
—F	70.5	7.8	−6.9
—(C=O)OR	20.5	2.3	−2.9
—(C=O)OH	20.5	2.0	−2.9
—(C=O)H	30.3	4.8	−2.9
—(C=O)R	30.0	1.3	−2.7
—(C=O)C$_6$H$_5$	−4.9	−0.4	−2.7

away from the ^{13}C atom, its influence becomes smaller. The effect of a group more than three carbon atoms away from the ^{13}C nucleus is small enough to be safely ignored.

Identifying α, β, and γ substituents. It is important to be systematic when calculating estimated chemical shifts. A good way to remember to include all α, β, and γ substituents for each type of carbon in a target molecule is to write down all the α groups first, then all the β groups, and finally the γ groups. Only then should you go to Table 23.3, look up the base value and the value for each α, β, and γ substituent from the correct column, and do the necessary addition.

EXERCISE

Identify the α, β, and γ substituents for carbon-4 of heptane.

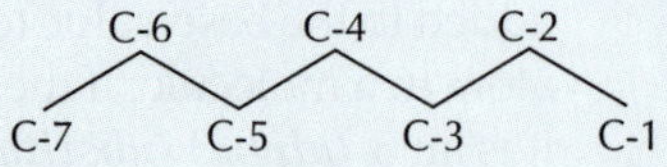

Heptane

Answer: Carbon-4 has two α substituents, the methylene groups at C-3 and C-5. Carbon-4 also has two β substituents, the methylene groups at C-2 and C-6. In addition, carbon C-4 has two γ substituents, the two methyl groups C-1 and C-7.

Calculating estimated chemical shifts. Table 23.3 is laid out with carbon substituents at the top, followed by heteroatoms—nitrogen, the halogens and oxygen—and then by carbonyl substituents. To illustrate its use, let us assign the four peaks in the ¹³C NMR spectrum of heptane, shown in Figure 23.7.

WORKED EXAMPLE

In addition to the $CDCl_3$ triplet signal at 77 ppm, Figure 23.7 shows four nonequivalent kinds of carbon atoms, which fits with the symmetry of a heptane molecule. The four signals appear at 14.1, 22.8, 29.1, and 32.0 ppm. The two methyl carbons (C-1 and C-7) are affected by one α carbon, one β carbon, and one γ carbon, and an sp^3 methyl group is the first entry in Table 23.3. Therefore, the calculated chemical shift for each methyl carbon is −2.3 + 9.1 + 9.4 − 2.5 = 13.7 ppm. This estimate is within 0.4 ppm of the experimental chemical shift of 14.1 ppm, and we can be quite confident that the signal at 14.1 ppm is due to the methyl groups of heptane.

C-4 of heptane has two α carbon substituents, two β carbon substituents, and two γ carbon substituents. The calculated chemical shift for this carbon is −2.3 + (2 × 9.1) + (2 × 9.4) + (2 × −2.5) = 29.7 ppm. This estimate compares favorably with the signal at 29.1 ppm in Figure 23.7.

FOLLOW-UP ASSIGNMENT

Two signals in Figure 23.7 remain to be assigned. Calculate the estimated chemical shifts for C-2 and C-3 of heptane and assign the 22.8-ppm and 32.0-ppm signals.

Effect of Alkyl Branching on ¹³C Chemical Shifts

To calculate the estimated ¹³C chemical shifts of branched chain alkanes, a further correction has to be taken into account. Table 23.4 shows that although these corrections are not necessary for linear alkanes, "steric" corrections become increasingly important with increased carbon branching. The steric effect can be attributed to

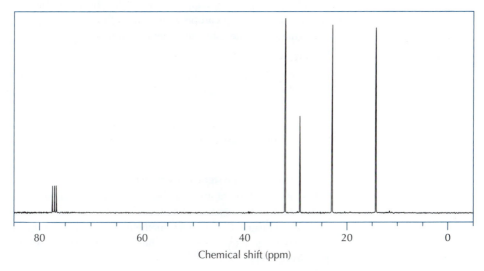

FIGURE 23.7 Broadband-decoupled 90-MHz ¹³C NMR spectrum of heptane in $CDCl_3$.

TABLE 23.4	Steric additivity parameters in ppm for predicting NMR chemical shifts of alkyl carbon atoms in CDCl$_3$			
	Type of α carbon atom			
Type of ^{13}C nucleus	Methyl	Methylene	Methine	Quaternary
Primary		0.0	−1.1	−3.4
Secondary	0.0	0.0	−2.5	−6.0
Tertiary	0.0	−3.7	−8.5	−10.0
Quaternary	−1.5	−8.0	−10.0	−12.5

intramolecular van der Waals repulsion between hydrogen atoms that are close together, which causes the electrons of C–H bonds to move away from the proton toward carbon. Nearby branching in a carbon chain always shields a ^{13}C nucleus.

These steric additivity parameters always have negative values. Note that a separate parameter must be used for each α carbon atom, and their sum is the total steric correction for the ^{13}C carbon atom whose estimated chemical shift is being calculated.

WORKED EXAMPLE

The application of the steric correction for branched alkanes can be demonstrated with the calculation for the chemical shift of carbon-2 of 3-methylhexane, whose ^{13}C NMR spectrum is shown in Figure 23.8.

3-Methylhexane

The branch point in the carbon chain of 3-methylhexane is C-3, which is α to the carbon whose chemical shift is being calculated. The other α substituent of C-2 is a methyl group, and Table 23.4 shows this has a steric parameter of 0.0 (secondary carbon with an α methyl group).

Calculation of the estimated chemical shift for C-2:

Base value	= −2.3
Two α carbon substituents (2 × 9.1)	= 18.2
Two β carbon substituents (2 × 9.4)	= 18.8
One γ carbon substituent	= −2.5
Secondary carbon atom with an α methine group (Table 23.4)	= −2.5
Total	= 29.7 ppm

The application of the steric correction to the chemical shift of a carbon atom that is the branch point itself can be shown by calculating the chemical shift for C-3 of 3-methylhexane. Here there are three α substituents to consider, one for methylene C-2, one for methylene C-4, and one for the α methyl group. For the tertiary carbon atom the steric parameters are −3.7, −3.7, and 0.0, respectively.

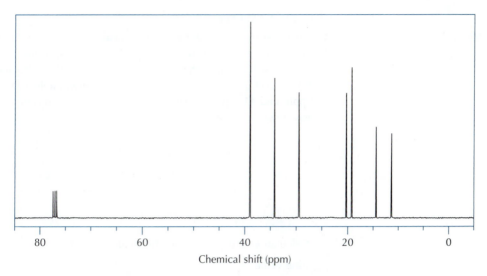

FIGURE 23.8 Broadband-decoupled 90-MHz ¹³C NMR spectrum of 3-methylhexane in CDCl₃.

Calculation of the estimated chemical shift for C-3 of 3-methylhexane:

Base value	= −2.3
Three α carbon substituents (3 × 9.1)	= 27.3
Two β carbon substituents (2 × 9.4)	= 18.8
One γ carbon substituent	= −2.5
Tertiary carbon atom with two α methylene	
groups (2 × −3.7)	= −7.4
Total	= 33.9 ppm

The measured chemical shifts of the carbon signals of 3-methylhexane shown in Figure 23.8 are 11.4, 14.4, 19.2, 20.3, 29.6, 34.4, and 39.1 ppm. We have already calculated the estimated chemical shifts for C-2 and C-3; they are 29.7 ppm and 33.9 ppm, respectively. We can be reasonably confident that the 29.6-ppm signal can be assigned to C-2 and that the 34.4-ppm signal can be assigned to C-3.

EXERCISE

Calculate the estimated chemical shifts of C-1, C-4, C-5, C-6, and C-7 of 3-methylhexane and assign the remaining signals in Figure 23.8. How good is the correspondence?

Answer: Calculation of the estimated chemical shifts using Tables 23.3 and 23.4 gives C-1 = 11.2 ppm, C-4 = 39.1 ppm, C-5 = 20.3 ppm, C-6 = 13.7 ppm, and C-7 = 19.5 ppm. The assignments are C-1 = 11.4 ppm, C-4 = 39.1 ppm, C-5 = 20.3 ppm, C-6 = 14.4 ppm, and C-7 = 19.2 ppm. The correspondence is excellent.

¹³C Chemical Shifts for Functional Groups Other Than *sp*³ Carbon

Additive parameters for a wide range of functional groups have been determined and are listed in Table 23.3. They are used in the same manner as the additive parameters for alkyl substituents. Again, the steric additivity parameters in Table 23.4 must also be part of the calculation of the estimated ¹³C chemical shifts, which are usually within 5 ppm of measured chemical shifts.

WORKED EXAMPLE

We have already seen how the ^{13}C chemical shifts of linear and branched alkanes can be estimated using Tables 23.3 and 23.4. In this example we will show how this approach can be extended to 3-methyl-1-butanol, whose ^{13}C NMR spectrum is shown in Figure 23.9. We will assign the signals at 60.2 ppm and 41.8 ppm in Figure 23.9. The hydroxyl substituent is about halfway down in Table 23.3.

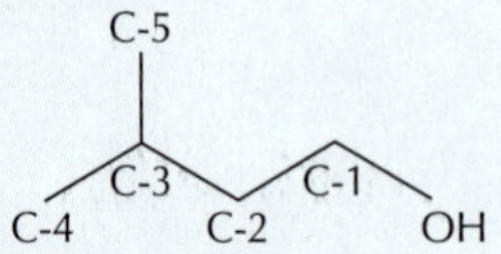

3-Methyl-1-butanol

Calculation of the estimated chemical shift for C-1:

Base value	$=-2.3$
One α hydroxyl substituent	$= 49.4$
One α carbon substituent	$=\ \ 9.1$
One β carbon substituent	$=\ \ 9.4$
Two γ carbon substituents (2 × −2.5)	$=-5.0$
Total	$= 60.6$ ppm

Calculation of the estimated chemical shift for C-2:

Base value	$=-2.3$
One β hydroxyl substituent	$= 10.1$
Two α carbon substituents (2 × 9.1)	$= 18.2$
Two β carbon substituents (2 × 9.4)	$= 18.8$
Secondary carbon atom with an α methine group	$=-2.5$
Total	$= 42.3$ ppm

We are quite safe in assigning C-1 to the 60.2-ppm signal and C-2 to the 41.8-ppm signal. Each of these calculated chemical shifts is within 3% of the measured value. In the case of 3-methyl-1-butanol, we could actually make the assignments of C-1 and C-2 using Table 23.1, knowing that the influence of an electronegative substituent is smaller the farther away it is. With a more complex molecule, however, the sole use of Table 23.1 would be problematic. Using additive substituent parameters is a much safer method.

FOLLOW-UP ASSIGNMENT

Calculate the estimated chemical shifts of C-3 and C-4 (C-5) of 3-methyl-1-butanol.

Chemical Shifts of Aromatic and Alkene Carbon Atoms

The chemical shifts of carbon atoms in substituted benzene rings can be estimated using the parameters listed in Table 23.5. Aromatic carbon atoms have sp^2 hybridization and therefore are significantly deshielded. The base value for the calculations is the ^{13}C chemical shift for benzene, which is 128.5 ppm.

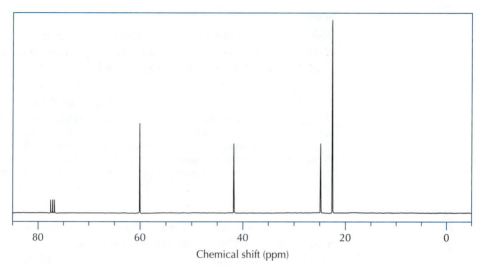

80   60   40   20   0

Chemical shift (ppm)

FIGURE 23.9 Broadband-decoupled 90-MHz ¹³C NMR spectrum of 3-methyl-1-butanol in $CDCl_3$.

TABLE 23.5	**Additive parameters for predicting NMR chemical shifts of aromatic carbons atoms in $CDCl_3$**

Base value: benzene = 128.5 ppm

Group	C-1	*ortho*	*meta*	*para*
—I	−34.1	8.9	1.6	−1.1
—Br	−5.8	3.2	1.6	−1.6
—Cl	6.3	0.4	1.4	−1.9
—F	34.8	−13.0	1.6	−1.1
—H	0.0	0.0	0.0	0.0
—(C=O)OCH₃	2.0	1.2	−0.1	4.3
—(C=O)OH	2.1	1.6	−0.1	5.2
—(C=O)H	8.2	1.2	0.5	5.8
—(C=O)CH₃	8.9	0.1	−0.1	4.4
—CH=CH₂	8.9	−2.3	−0.1	−0.8
—CH₃	9.2	0.7	−0.1	−3.0
—CH₂Cl	9.3	0.3	0.2	0.0
—C₆H₅	13.1	−1.1	0.5	−1.1
—CH₂CH₃	15.7	−0.6	−0.1	−2.8
—CH(CH₃)₂	20.2	−2.2	−0.3	−2.8
—C(CH₃)₃	22.4	−3.3	−0.4	−3.1
—NH(C=O)CH₃	9.7	−8.1	0.2	−4.4
—NH₂	18.2	−13.4	0.8	−10.0
—NO₂	19.9	−4.9	0.9	6.1
—O(C=O)CH₃	22.4	−7.1	0.4	−3.2
—OH	26.9	−12.8	1.4	−7.4
—OC₆H₅	27.6	−11.2	−0.3	−6.9
—OCH₃	31.4	−14.4	1.0	−7.7

WORKED EXAMPLE

Using methyl 3-nitrobenzoate as an example, the estimates of the chemical shifts calculated from Table 23.5 provide a useful guide for assigning the signals in the spectrum to the appropriate carbons.

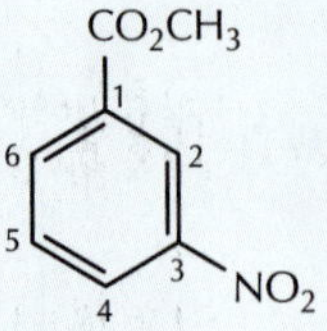

	C-1	C-2	C-3	C-4	C-5	C-6
Base value	128.5	128.5	128.5	128.5	128.5	128.5
–(C=O)OCH₃	2.0	1.2	−0.1	4.3	−0.1	1.2
–NO₂	0.9	−4.9	19.9	−4.9	0.9	6.1
Estimated (ppm)	131.4	124.8	148.3	127.9	129.3	135.8
Measured (ppm)	131.9	124.5	148.3	127.3	129.6	135.2

Additive parameters for sp^2 alkene carbon atoms are added to a base value of 123.3 ppm, the chemical shift of the carbon atoms in ethylene. There are two general types of shielding/deshielding effects of an alkene carbon atom: those that are transferred solely through the σ bond network—described as α, β, and γ—and those that are transferred through the π bond network, which can be called α′, β′, and γ′.

$$C—C—C—C{=}C—C—C—C$$
$$\gamma \quad \beta \quad \alpha \qquad\quad \alpha' \quad \beta' \quad \gamma'$$

Table 23.6 lists additive parameters for the effect of a variety of functional groups on the chemical shift of alkene carbon atoms.

WORKED EXAMPLE

Determine the substituents that must be used to calculate the estimated chemical shift of the sp^2 C-2 carbon atom in *E*-2-pentene and use Table 23.6 to calculate it.

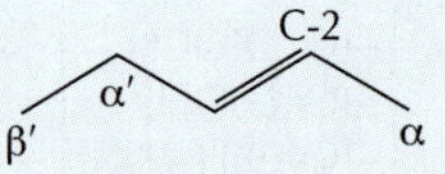

E-2-pentene

In this case, the only substituents that must be considered are sp^3 α, α′, and β′ carbon atoms, which appear at the top of Table 23.6. The chemical shift of C-2 is estimated to be $123.3 + 10.1 - 7.7 - 2.5 = 123.2$ ppm. The measured chemical shift is 123.5 ppm.

EXERCISE

Calculate the chemical shift of the C-3 carbon atom of *E*-2-pentene. The measured chemical shift is 133.2 ppm.

Answer: The estimated chemical shift of C-3 is calculated to be $123.3 + 10.1 + 7.1 - 7.7 = 132.8$ ppm.

A set of correction factors to account for steric contributions can be found in Table 23.7.

TABLE 23.6 **Additive parameters for predicting NMR chemical shifts of alkene carbon atoms in CDCl₃**

Base value: $CH_2{=}CH_2$ = 123.3 ppm

Group (Y)	Y–**C**=C–Y' α	α'	Y–C–**C**=C–C–Y' β	β'	Y–C–C–**C**=C–C–C–Y' γ	γ'
	α	β	γ	α'	β'	γ'
—C–(sp^3)	10.1	7.1	−1.5	−7.7	−2.5	1.0
—CH=CH₂	14.5	2.9	−2.4	−5.8	−0.5	1.2
—C₆H₅	13.6	4.9	−2.5	−9.6	0.7	1.1
—OH		3.9	−6.1		−0.8	3.8
—O(C=O)CH₃	17.9	−1.1	−6.7	−25.8	2.3	3.3
—OR	28.5	1.7	−4.9	−37.0	0.7	2.9
—Br	−9.4	1.0	−5.1	−1.5	2.4	4.2
—Cl	2.8	0.5		−6.1	2.7	
—(C=O)OR	3.0	−2.7	−3.4	7.3	2.2	1.9
—(C=O)OH	4.7	−2.9	−4.3	9.9	3.2	3.2
—(C=O)H	14.7		−4.1	16.2		2.0
—(C=O)R	14.1	−2.9	−3.5	5.5	3.0	1.7

Blank entries are due to lack of data.

WORKED EXAMPLE

The estimation of the ¹³C chemical shift for C-4 of 4-methyl-3-penten-1-ol provides a useful demonstration of the utility of the additive parameters in Tables 23.6 and 23.7.

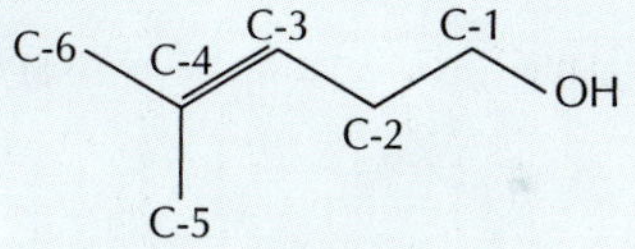

4-Methyl-3-penten-1-ol

Calculation of the estimated chemical shift for C-4:

Base value	= 123.3
Two α carbon substituents (2 × 10.1)	= 20.2
One α' carbon substituent	= −7.7
One β' carbon substituent	= −2.5
One γ' hydroxyl substituent	= 3.8
One pair of αα' cis substituents	= −1.1
One pair of αα geminal substituents	= −4.8
Total	= 131.2 ppm
Measured chemical shift	= 134.8 ppm

TABLE 23.7 **Steric additivity parameters for predicting NMR chemical shifts of alkene carbon atoms in CDCl₃**

Each pair of αα' cis substituents	−1.1 ppm
Each pair of αα geminal substituents	−4.8 ppm
Each pair of α'α' geminal substituents	2.5 ppm
Each α methine substituent	2.3 ppm
Each α quaternary substituent	4.6 ppm

EXERCISE

Using the additive parameters for alkyl carbon atoms in Table 23.3 and the parameters for alkene carbon atoms in Tables 23.6 and 23.7, calculate the estimated chemical shifts of the remaining carbon atoms of 4-methyl-3-penten-1-ol and assign them to the correct carbon atoms. How confident are you of your estimated chemical shifts? The measured chemical shifts of the four remaining carbon atoms are 17.9, 25.8, 31.6, and 120.1 ppm.

Answer: Calculation of the estimated chemical shifts using Tables 23.3, 23.6, and 23.7 gives C-2 = −2.3 + 9.1 + 14.4 + 10.1 = 31.3 ppm, C-3 = 123.3 + 10.1 + 7.1 − 6.1 − 15.4 − 1.1 + 2.5 = 120.4 ppm, C-5 = −2.3 + 13.4 + 9.4 = 20.5 ppm, and C-6 = −2.3 + 19.3 + 9.4 = 26.4 ppm. The C-2 signal appears at 31.6 ppm, the C-3 signal appears at 120.1 ppm, the C-5 signal appears at 17.9 ppm, and the C-6 signal appears at 25.8 ppm in the ^{13}C NMR spectrum. Note that the *cis*-methyl group is farther upfield than the *trans*-methyl group.

EXERCISE

The ^{13}C NMR spectrum of 2-butanone shows signals at 8, 29, 37, and 209 ppm. The ^{13}C NMR spectrum of ethyl acetate shows signals at 14, 21, 60, and 171 ppm. Assign the NMR signals to the carbon atoms in each structure.

Answer:

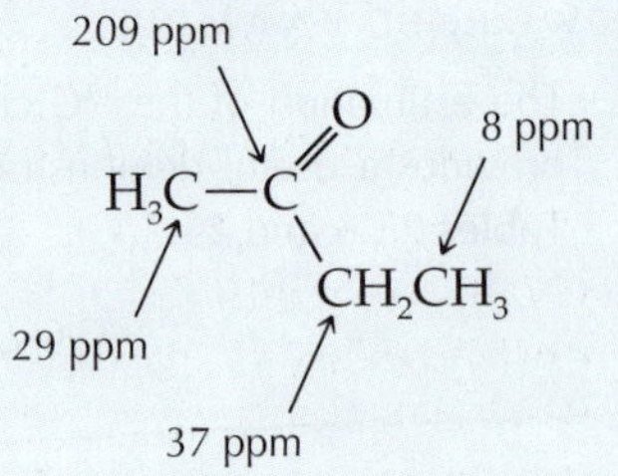

2-Butanone

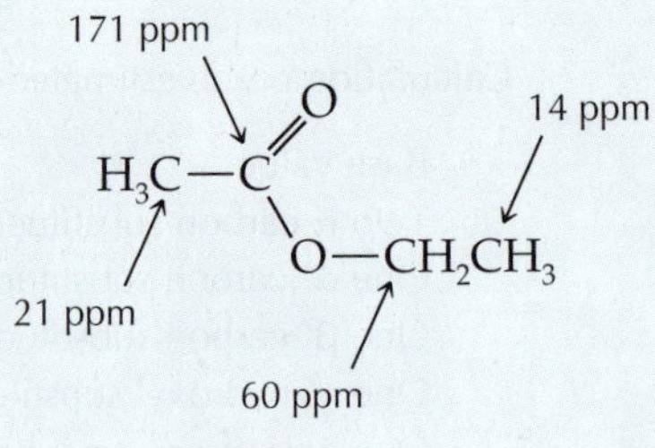

Ethyl acetate

In general, the additive parameters in Tables 23.3–23.7 provide good estimates for ^{13}C chemical shifts; however, it is important to remember that they are estimates, not precision calculations. The estimates are usually within 3% of measured values (6 ppm over a 200-ppm range), and the relative positions of the signals are usually correct.

Computer Programs for Estimating ^{13}C NMR Chemical Shifts

Computer programs have been developed that use additivity parameters for calculating the ^{13}C NMR spectrum of any molecule of interest. ChemDraw Ultra and ChemBioDraw Ultra (from PerkinElmer) include a module, called ChemNMR, which estimates ^{13}C chemical shifts and displays the calculated NMR spectrum as well as the assigned ¹³C chemical shifts after the structure of a molecule is drawn. Moving the cursor to a peak on the spectrum highlights the carbon atom in the molecule responsible for the peak, and vice versa. The logic of the program is rule-based calculation of chemical shifts on structural fragments, similar to the method presented in this chapter. To improve the accuracy of the estimates, some 4000 parameters are used.

An alternative method for estimating chemical shifts is the ACD/CNMR Predictor (from Advanced Chemistry Development). The predicted chemical shifts are based on a large database of structures (almost 200,000) with 2.5 million assigned ¹³C chemical shifts. The display can be interrogated by clicking on either the structure or the spectrum to highlight their corelationships, and the database can be expanded as new compounds become available.

These programs are sophisticated, research-quality tools and are priced accordingly. Some institutions have negotiated site licenses making the programs accessible to their members. Estimated ¹³C chemical shifts from these computer programs are usually within 3–4 ppm of the measured values, roughly comparable to the use of the additivity parameters used in this chapter.

23.4 Determining Numbers of Protons on Carbon Atoms—APT and DEPT

Typically, ¹³C NMR spectra are obtained using broadband decoupling so that the carbon signals are collapsed into singlets. The cost of this simplification is the loss of information regarding the number of protons attached to carbon atoms. Numerous techniques have been developed to supply this important information. Two commonly used experiments provided with most modern FT NMR spectrometers are APT (Attached Proton Test) and DEPT (Distortionless Enhancement by Polarization Transfer). These experiments use complex pulse sequences of radio frequency energy at observation frequencies for both ¹H and ¹³C nuclei.

APT

In a typical broadband-decoupled ¹³C NMR spectrum, each different carbon atom in the sample appears as a single positive peak. In APT spectra, CH and CH₃ carbon nuclei give positive signals, whereas quaternary and CH₂ carbon nuclei give negative signals.

In the APT spectrum of ethyl *trans*-2-butenoate, shown in Figure 23.10, positive signals at 14, 18, 123, and 144 ppm are due to the carbons of the methyl groups and the vinyl carbons. The negative signals at 60 and 166 ppm are due to the carbon of the methylene group and the carbonyl carbon. With concentrated samples, an APT spectrum can be acquired in a short time. The APT method is limited, however, because it is normally impossible to distinguish between the signals of quaternary and CH₂ carbon nuclei, or between signals of CH and CH₃ carbon nuclei.

DEPT

In a DEPT experiment, signals in the ¹³C NMR spectrum are suppressed or inverted depending on the number of protons attached to the carbon and the conditions set in the pulse program. The

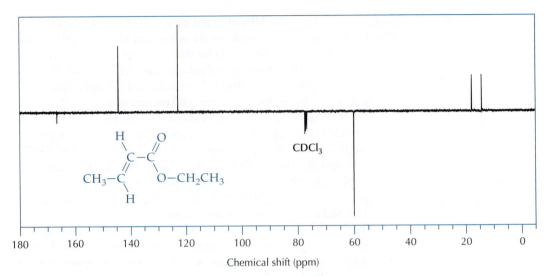

FIGURE 23.10 90-MHz ^{13}C NMR spectrum of ethyl *trans*-2-butenoate in CDCl$_3$ using an APT pulse sequence.

DEPT(45) version of the experiment provides a ^{13}C spectrum in which only carbon atoms that have protons attached to them appear. Signals due to quaternary carbons are not observed. The spectrum produced by the DEPT(90) pulse program exhibits only signals from carbon atoms that have one hydrogen attached (methine carbons). Signals due to all carbon atoms with attached protons are observed in the ^{13}C NMR spectrum from a DEPT(135) experiment; however, the signals due to carbon atoms with two protons attached (methylene carbons) are inverted. Comparing the spectra from a set of DEPT experiments allows you to determine the number of protons attached to every carbon atom in a molecule. Table 23.8 summarizes the information that can be obtained from a broadband-decoupled ^{13}C spectrum and the three DEPT experimental spectra.

The broadband-decoupled ^{13}C, DEPT(45), DEPT(90), and DEPT (135) spectra of ethyl *trans*-2-butenoate are shown in Figure 23.11. The six signals in the ^{13}C spectrum were assigned earlier in the Worked Example at the end of Section 23.2. There is no CDCl$_3$ signal at 77 ppm in DEPT spectra because the carbon nucleus in CDCl$_3$ is attached to ^{2}H, not ^{1}H.

TABLE 23.8 **Orientation of ^{13}C signals in DEPT NMR experiments**

Type of ^{13}C spectrum	CH$_3$	CH$_2$	CH	C
Broadband-decoupled ^{13}C	+	+	+	+
DEPT(45)	+	+	+	0
DEPT(90)	0	0	+	0
DEPT(135)	+	−	+	0

Type of carbon and peak direction

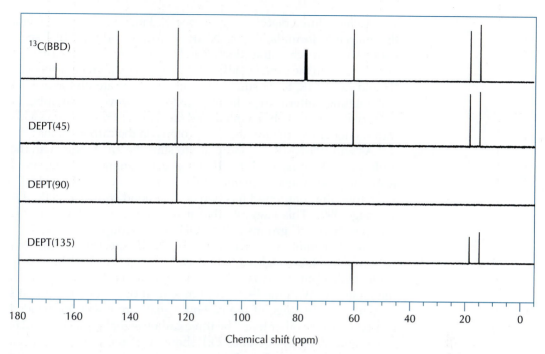

FIGURE 23.11 DEPT spectra of ethyl *trans*-2-butenoate in CDCl₃. The broadband-decoupled 90-MHz ¹³C NMR spectrum is shown at the top.

23.5 Case Study

Figure 23.12 shows the DEPT(135) and ¹³C NMR spectra for an acyclic compound whose molecular formula is $C_8H_{14}O$. The ¹³C chemical shifts are 17.6, 22.7, 25.6, 29.8, 43.7, 122.6, 132.6, and 208.4 ppm. What is the structure of $C_8H_{14}O$?

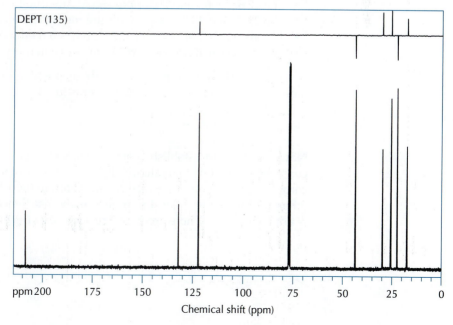

FIGURE 23.12 The 90-MHz DEPT(135) and ¹³C NMR spectra of $C_8H_{14}O$ in CDCl₃.

We have three pieces of experimental data with which to work: the molecular formula, the ^{13}C NMR spectrum, and the DEPT spectrum. An approach to this kind of problem solving was described in Section 22.11 in the context of ^{1}H NMR. There are differences between ^{1}H and ^{13}C NMR, to be sure. For example, there are no integration and splitting patterns to go by in ^{13}C NMR. However, the number of ^{13}C signals can easily be counted, and the DEPT(135) spectrum yields information about the numbers of protons on the carbon atoms.

As with ^{1}H NMR, a good place to start the analysis is with the molecular formula, $C_8H_{14}O$. If the acyclic unknown compound were fully saturated, the molecular formula would be $C_8H_{18}O$. Therefore, the compound has two double-bond equivalents (DBEs; see page 399). This suggests that it may have a C=O and a C=C group or two C=C groups and an OH or OR group.

The first thing to notice in the ^{13}C NMR spectrum is that the spectrum contains eight different carbon signals plus the $CDCl_3$ triplet at 77 ppm. There is a different signal for each carbon atom in the molecule. With the ^{13}C spectrum spread out over 200 ppm, it's easy to see that three of the carbon signals are highly deshielded compared to the other five. The three carbon signals appear at 122.6, 132.6, and 208.4 ppm. Table 23.1 shows that the 208.4-ppm peak must be due to a carbonyl group, and the 122.6- and 132.6-ppm signals are most likely due to a C=C group. It is also likely that the signal at 43.7 ppm may be due to a carbon atom α to the carbonyl group because the signals of most alkyl and allyl carbon atoms don't appear at chemical shifts greater than 40 ppm. We will delay an analysis of the four upfield ^{13}C signals until we have a better idea what the environments of these four alkyl signals might be.

The third piece of data is the DEPT(135) spectrum. Using the data in Figure 23.12, the following points can be inferred:

- The 17.6-, 25.6-, and 29.8-ppm signals are due to methyl groups.
- The 22.7- and 43.7-ppm signals are methylene groups.
- The alkene carbon atom at 122.6 ppm is a methine carbon atom.
- The signals at 132.6 and 208.4 ppm are quaternary carbon atoms.

The DEPT spectrum indicates these possible structural fragments, accounting for eight carbon atoms in all:

$$CH_3\text{—}(C\text{=}O)\text{—}CH_2\text{—} \quad \text{—}CH\text{=}CR\text{—} \quad \text{—}CH_2\text{—} \quad \text{—}CH_3 \quad \text{—}CH_3$$

Now we need to establish whether the C=O and C=C bonds are conjugated or nonconjugated. Table 23.2 shows that a conjugated carbonyl group would have its chemical shift in the 190-ppm region, whereas this one is at 208.4 ppm. It's likely that the carbonyl group is not conjugated. There are four structures that fit this data:

A   B   C   D

We can use Tables 23.3, 23.6, and 23.7 to calculate the estimated ¹³C chemical shifts to determine which of these four structures is the correct one. Begin with structures A, B, and C, which have ethyl substituents in which the methyl and methylene groups would be the most upfield signals in the ¹³C spectrum. The estimated chemical shifts for the these carbon signals are:

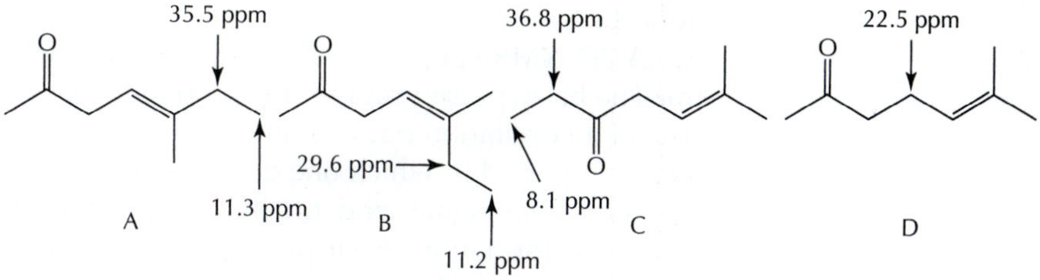

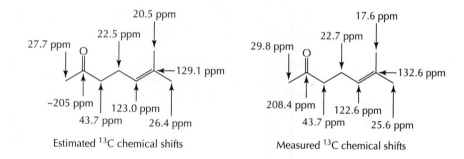

The measured chemical shifts of the methyl and methylene groups in Figure 23.12 are 17.6 ppm and 22.7 ppm. As you can see, the fit is rather poor for A, B, and C; however, the estimated chemical shift of the methylene group in D is an excellent fit.

Carrying out the calculations for every carbon atom in D gives the following result:

Estimated ¹³C chemical shifts

Measured ¹³C chemical shifts

A complete calculation for the carbon atoms of structures A, B, and C could be done to confirm this answer, but it seems unnecessary. $C_8H_{14}O$ is 6-methyl-5-hepten-2-one.

23.6   Two-Dimensional Correlated Spectroscopy (2D COSY)

The spin-spin coupling between nuclei affords a great deal of structural information. In Section 22.9, the coupling between hydrogen nuclei was discussed, and its usefulness in determining the *connectivity* within a molecule was demonstrated. Connectivity can be described as the covalent bonding network of nearby atoms within a molecule.

An alternative method of analyzing coupling within a molecule is provided by two-dimensional (2D) NMR spectroscopy. In a typical ^{1}H or ^{13}C NMR spectrum, the positions of signals along the abscissa (x-axis) of the spectrum correspond to the frequencies of the signals, which are measured as chemical shifts. The intensities of the signals are measured along the ordinate (y-axis). This typical spectrum is referred to as a one-dimensional (1D) spectrum because only one axis is a frequency axis. The most basic pulse sequence (or program) for producing a 1D NMR spectrum consists of a radio frequency excitation pulse followed by a data acquisition period.

A 2D NMR spectrum is created from a series of 1D NMR spectra. The basic pulse program for producing each 1D spectrum consists of an excitation pulse, a time delay called the evolution period, a mixing period in which one or more pulses are required to create an observable signal, and finally a data acquisition period. More detailed information about pulse programs can be found in *50 and More Essential NMR Experiments—A Detailed Guide*, by Findelsen and Berger. During the evolution period the magnetization from one nucleus is transferred to other nearby coupled nuclei. The time delay is increased by a small amount for each 1D spectrum. Since the data is time-based along both axes, it can be Fourier-transformed along both axes, so both the x-axis and the y-axis are frequency axes. The signal intensities in a 2D NMR spectrum are usually represented on the graph as a series of closely spaced contour lines, similar to a topographical map.

The most commonly utilized 2D spectroscopy experiments are (H,H) COSY (**CO**rrelated **S**pectroscop**Y**) spectra, in which both axes correspond to ^{1}H chemical shifts, and (C,H) COSY spectra, in which one axis corresponds to ^{13}C chemical shifts and the other axis corresponds to ^{1}H chemical shifts.

2D COSY spectra indicate which nearby nuclei are coupling with one another. They correlate the nuclei that are coupling partners. The correlations are shown by the presence of cross peaks, the contour line signals in 2D spectra that appear at the crossing of implicit vertical and horizontal lines connecting to the peaks on the x- and y-axes.

Two-Dimensional Homonuclear (H,H)-Correlated NMR Spectroscopy— (H,H) COSY

The 2D (H,H) COSY spectrum of ethyl *trans*-2-butenoate is shown in Figure 23.13. Here the ^{1}H NMR spectrum is displayed along both the x-axis and the y-axis. For each signal in the ^{1}H NMR spectrum of ethyl *trans*-2-butenoate, there is a corresponding peak on the diagonal that runs from the lower left corner to the upper right corner. The presence of peaks on the diagonal of an (H,H) COSY spectrum is not useful in determining coupling patterns. Peaks on the diagonal appear because the magnetization is not completely transferred between the nuclei. **It is the *off-diagonal cross peaks* that are useful in (H,H) COSY**; they appear where there is coupling between ^{1}H nuclei.

The off-diagonal cross peak that comes at the intersection of the 1.2-ppm signal in the ^{1}H spectrum on the x-axis and the 4.1-ppm signal in the ^{1}H spectrum on the y-axis indicates that these two ^{1}H nuclei are coupled. Due to the symmetry of the 2D (H,H) COSY spectrum, there is another off-diagonal cross peak at the intersection of the 1.2-ppm signal in the ^{1}H spectrum on the y-axis

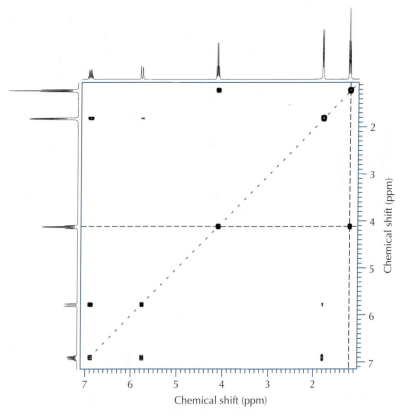

FIGURE 23.13 2D (H,H) COSY spectrum of ethyl *trans*-2-butenoate in CDCl₃. The 1D 360-MHz ¹H NMR spectra are shown at the top and left edges.

and the 4.1-ppm signal in the ¹H spectrum on the *x*-axis. There are no other off-diagonal cross peaks involving the 1.2-ppm and 4.1-ppm signals, so the protons they represent are not coupled to any additional protons.

    Figure 23.13 also shows two off-diagonal cross peaks involving the 1.8-ppm ¹H signals on the two axes. The most intense cross peak intersects with the 6.9-ppm signal. Thus, the ¹H nucleus at 1.8 ppm is coupled to the ¹H nucleus at 6.9 ppm. In addition, there is a less intense pair of off-diagonal cross peaks between the 1.8-ppm ¹H signal and the 5.8-ppm ¹H signal. Therefore, the proton at 1.8 ppm is also coupled to the proton at 5.8 ppm. The lesser intensity of the latter 2D peak suggests that the coupling constant is smaller; it is due to long-range allylic coupling. Lastly, the two off-diagonal peaks involving the 5.8-ppm signal and the 6.9-ppm signal show that these protons are coupled. In total, there are four pairs of off-diagonal cross peaks in Figure 23.13, showing four different ¹H-¹H couplings.

    Other variations of the 2D (H,H) COSY experiment give basically the same information as the experiment described. All these data confirm our original interpretation of the 1D ¹H NMR spectrum of ethyl *trans*-2-butenoate in Section 22.9 (pages 383–387), which came in part from a detailed analysis of the spin-spin splitting patterns in the spectrum.

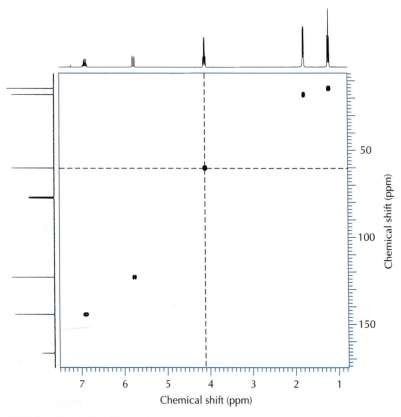

FIGURE 23.14 2D (C,H) COSY spectrum of ethyl *trans*-2-butenoate in CDCl$_3$. The 1D 360-MHz ^{1}H NMR spectrum is shown at the top edge and the 90-MHz ^{13}C NMR spectrum is shown at the left edge.

Two-Dimensional Heteronuclear (C,H)-Correlated NMR Spectroscopy— (C,H) COSY

An example of a 2D (C,H) COSY spectrum of ethyl *trans*-2-butenoate is shown in Figure 23.14. The x-axis of the 2D spectrum displays the ^{1}H NMR spectrum, and the y-axis displays the broadband-decoupled ^{13}C spectrum. **Each signal in the ^{1}H spectrum can be correlated to a signal in the ^{13}C spectrum.** These cross peaks are the result of coupling between a ^{1}H nucleus and the ^{13}C nucleus to which it is attached, where the one-bond coupling constant ($^1J_{CH}$) is very large.

The cross peak at 4.1 ppm along the chemical shift axis of the ^{1}H spectrum (x-axis) in the 2D (C,H) COSY spectrum is located at 60 ppm along the chemical shift axis of the ^{13}C spectrum (y-axis). This cross peak shows that the protons giving rise to the 4.1-ppm signal are attached to the carbon atom appearing at 60 ppm. The chemical shift of the ^{1}H spectrum thus correlates with the chemical shift of the ^{13}C spectrum. There are no cross peaks for quaternary carbon nuclei because they have no attached protons. Therefore, in Figure 23.14 there is no cross peak on the 2D spectrum for the 166-ppm peak in the ^{13}C spectrum, which is due to the carbonyl carbon atom.

There are several 2D (C,H) COSY experiments that give basically the same information. They are identified by a variety of abbreviations: HETCOR (**HET**eronuclear **COR**relation), HMQC (**H**eteronuclear **M**ultiple **Q**uantum **C**oherence), HSQC (**H**eteronuclear **S**ingle **Q**uantum **C**oherence), and others. While there are differences among the experiments, the differences are of little consequence when it comes to the interpretation of signals.

Long-Range (C,H) and (H,H) Coupling

By adjusting the evolution-period time delay in 2D NMR experiments, it is possible to identify long-range couplings over four or more bonds. Long-range couplings have also proven to be a powerful tool for determining the structures of organic compounds.

Further Reading

Breitmaier, E. *Structure Elucidation by NMR in Organic Chemistry: A Practical Guide*, 3rd ed.; Wiley: New York, 2002.

Crews, P.; Rodríguez, J.; Jaspars, M. *Organic Structure Analysis*, 2nd ed.; Oxford University Press: Oxford, 2009.

Findelsen, M.; Berger, S. *Fifty and More Essential NMR Experiments—A Detailed Guide*; Wiley-VCH: Weinheim, 2013.

Friebolin, H. *Basic One- and Two-Dimensional NMR Spectroscopy*, 5th ed.; Wiley-VCH: Weinheim, 2010.

Pouchert, C. J.; Behnke, J. (Eds.) *The Aldrich Library of ¹³C and ¹H FT-NMR Spectra*; Aldrich Chemical Co.: Milwaukee, WI, 1993; 3 volumes.

Pretsch, E.; Bühlmann, P.; Badertscher, M. *Structure Determination of Organic Compounds*, 4th ed.; Springer-Verlag: New York, 2009.

Silverstein, R. M.; Webster, F. X.; Kiemle, D. J. *Spectrometric Identification of Organic Compounds*, 7th ed.; Wiley: New York, 2005.

Questions

1. How many signals would you expect to see in the broadband-decoupled ¹³C NMR spectrum of each of the following compounds? Show your logic.
   (a) 2-pentanol, 2,2-dimethylbutane, isopropyl acetate, 2-acetoxybutane
   (b) *para*-aminobenzoic acid, methyl 2-hydroxybenzoate, 1-phenyl-2-methylpropane, 1,3-cyclopentadiene
   (c) cyclohexane, *trans*-1,4-dimethylcyclohexane, *trans*-1,2-dimethylcyclohexane
   (d)

2. Broadband-decoupled ¹³C NMR spectra for three compounds with the molecular formula C_3H_8O are shown in Figure 23.15. Deduce the structure of each compound, estimate the chemical shift of each of its carbon atoms using the additive parameters in Tables 23.3 and 23.4, and assign the NMR signals to their respective carbon atoms. The measured chemical shifts of the carbon atoms follow.
   (a) 10.2, 25.8, and 64.4 ppm
   (b) 25.3 and 64.2 ppm
   (c) 15.0, 58.2, and 67.9 ppm

3. The broadband-decoupled ¹³C NMR spectrum, DEPT(90) spectrum, and DEPT(135) spectrum of a compound with the molecular formula C_8H_{10} are shown in Figure 23.16. Determine the structure of the compound and assign its signals in the ¹³C NMR spectra. Estimate the chemical shifts of all carbon atoms using Tables 23.1, 23.3, and 23.5 and compare them with those measured from the spectrum.

4. A compound of molecular formula $C_{10}H_{14}$ produces a broadband-decoupled

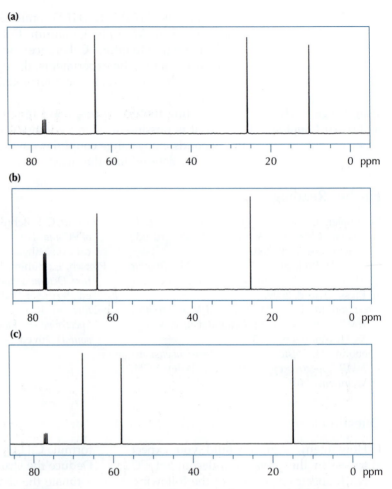

FIGURE 23.15 90-MHz ^{13}C NMR spectra of compounds with molecular formula C_3H_8O.

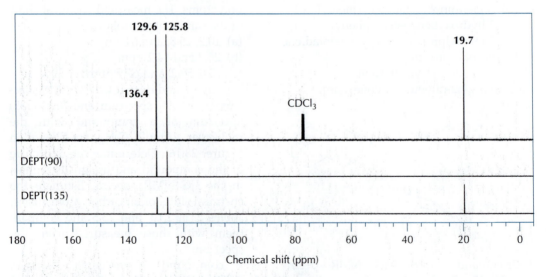

FIGURE 23.16 90-MHz ^{13}C NMR, DEPT(90), and DEPT(135) spectra of a compound with molecular formula C_8H_{10}.

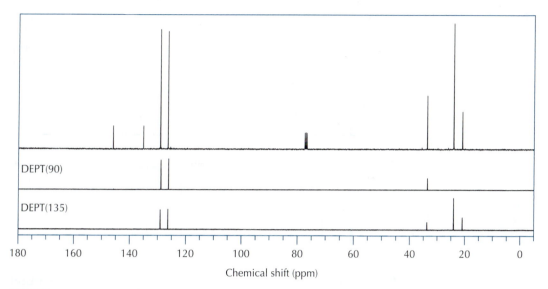

FIGURE 23.17 90-MHz ¹³C NMR, DEPT(90), and DEPT(135) spectra of a compound with molecular formula $C_{10}H_{14}$.

¹³C NMR spectrum, which has signals at 145.8, 135.1, 129.0, 126.3, 33.7, 24.1, and 20.9 ppm. The ¹³C NMR spectrum, the DEPT(90) spectrum, and the DEPT(135) spectrum are shown in Figure 23.17. Deduce the structure of $C_{10}H_{14}$, estimate the chemical shifts of all carbon atoms using the parameters in Tables 23.1–23.5, and assign all the ¹³C NMR signals.

5. Broadband-decoupled ¹³C NMR and DEPT(135) spectra for all the compounds with the molecular formula $C_4H_{10}O$ are shown in Figure 23.18. Deduce the structure of each compound, estimate the chemical shift of each of its carbon atoms using the additive parameters in Tables 23.3 and 23.4, and assign the NMR signals to their respective carbon atoms. The measured chemical shifts of the carbon atoms follow.

(a) 62.4, 34.9, 19.0, and 13.9 ppm
(b) 68.3, 32.0, 22.9, and 10.0 ppm
(c) 65.9 and 15.4 ppm
(d) 69.6, 30.8, and 18.9 ppm
(e) 74.1, 56.4, and 22.9 ppm
(f) 75.4, 59.1, 24.0, and 11.3 ppm
(g) 69.1 and 31.2 ppm

6. Ibuprofen is the active ingredient in several nonsteroid anti-inflammatory drugs (NSAIDs). The molecular formula of the methyl ester of ibuprofen is $C_{14}H_{20}O_2$. The broadband-decoupled ¹³C NMR spectrum, the DEPT(90) spectrum, and the DEPT(135) spectrum for the methyl ester of ibuprofen are shown in Figure 23.19. **Hint:** The ¹³C signal at 45 ppm is broader than the other signals in the ¹³C spectrum and resolves into two separate signals at higher resolution. Pay careful attention to the pattern of signals at 45 ppm in the DEPT(135) spectrum. The ¹H NMR spectrum of the methyl ester of ibuprofen is shown in Figure 23.20.

The 2D (H,H) COSY spectrum of the methyl ester of ibuprofen is shown in Figure 23.21, and its 2D (C,H) COSY spectrum is shown in Figure 23.22. Deduce the structure of the methyl ester of ibuprofen using the parameters in Tables 23.1 and 23.3–23.5, and estimate the chemical shifts of all carbon atoms. Assign all the ¹³C NMR signals. Show your reasoning.

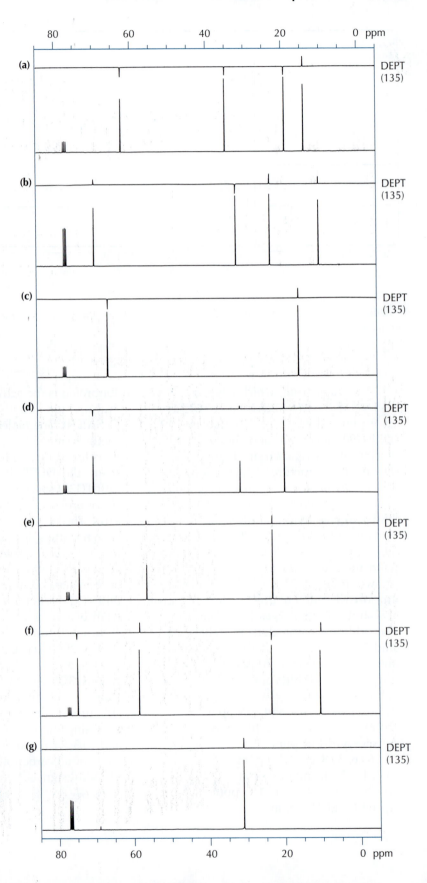

FIGURE 23.18
90-MHz ^{13}C NMR and
DEPT(135) spectra of
compounds with
molecular formula
$C_4H_{10}O$.

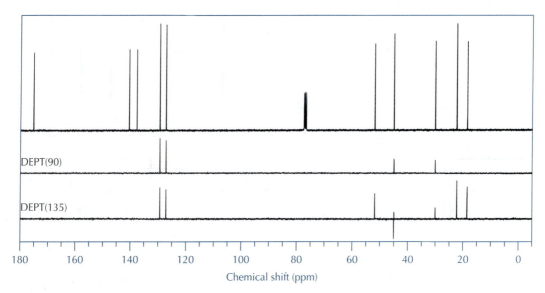

FIGURE 23.19 90-MHz ¹³C NMR, DEPT(90), and DEPT(135) spectra of the methyl ester of ibuprofen.

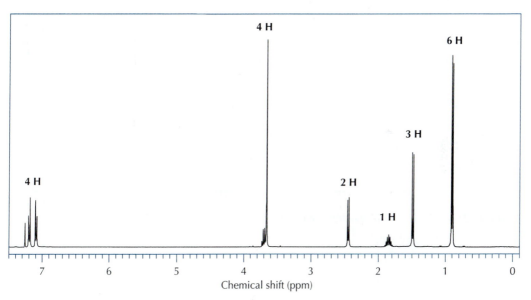

FIGURE 23.20 360-MHz ¹H NMR spectrum of the methyl ester of ibuprofen.

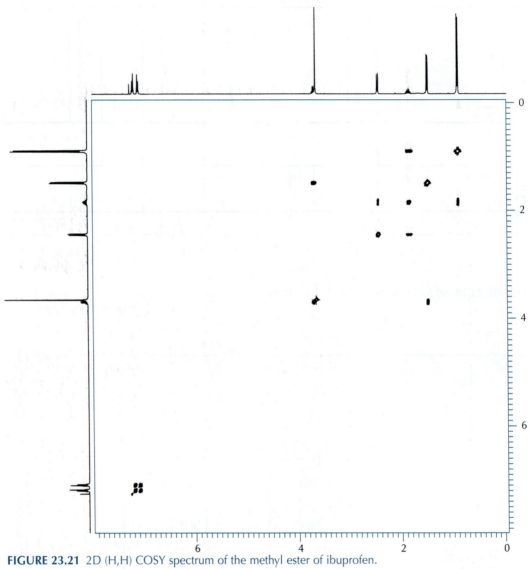

FIGURE 23.21 2D (H,H) COSY spectrum of the methyl ester of ibuprofen.

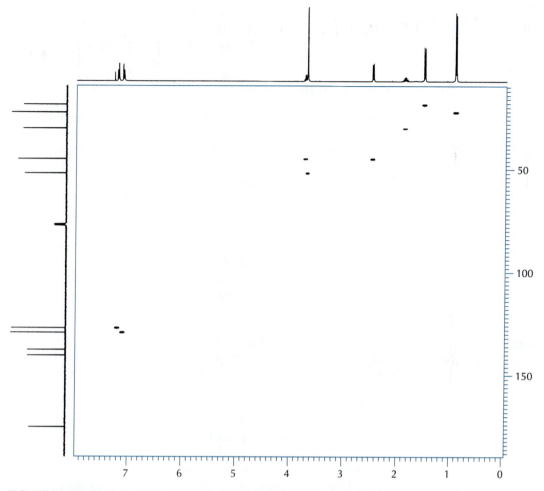

FIGURE 23.22 2D (C,H) COSY spectrum of the methyl ester of ibuprofen.

CHAPTER

24

MASS SPECTROMETRY

If Chapter 24 is your introduction to spectrometric analysis, read the Essay "The Spectrometric Revolution" on pages 309–310 before you read Chapter 24.

Most spectrometric techniques used by organic chemists involve the ability of molecules to absorb light of various energies. This yields plots of absorbance versus a spectrum of frequencies. Mass spectrometry (MS) does not involve the absorption of light. Instead, molecules are ionized using a variety of different techniques to generate either cationic or anionic species. Magnetic or electric fields are then used to manipulate the motion of these ions in a way that enables us to determine their ***mass-to-charge (m/z) ratios***. The "mass spectra" that result from these measurements are plots of the number of ions observed at each m/z value. Organic chemists primarily use MS to determine the molecular weights and molecular formulas of compounds. It is a very sensitive technique that can be carried out with microgram quantities of compounds. Fragmentation of the initially formed ions in the mass spectrometer also provides additional information that can help identify a compound or determine its structure.

24.1   Mass Spectrometers

In recent years, great strides have been made in instrumentation for mass spectrometry, and numerous types of mass spectrometers are now available. Even though they have functional differences, the basic components outlined in Figure 24.1 are common to all mass spectrometers. The ion source is where a compound is introduced into the mass spectrometer and converted into a gas-phase ion through one of a variety of ionization techniques. The mass analyzer sorts the ions by their mass-to-charge (m/z) ratios. The sorted ions generate an electric current at the detector that corresponds to the number of ions (or relative abundance of ions) of each m/z value, creating a mass spectrum. Because the charge on the ions is typically $+1$, the m/z value for the molecular ion corresponds to the molecular weight of the compound.

Electron Impact (EI) Mass Spectrometry

The classic mass spectrometer ionizes the sample by electron impact and sorts the ions with a magnetic sector mass analyzer. To better understand how all mass spectrometers work, it is worthwhile to examine this type of spectrometer in more detail.

Samples for analysis by electron impact (EI) are first vaporized. As shown in Figure 24.2, the gas-phase molecules are attracted into the ion source because it is under a vacuum. They are bombarded in the ion source by a stream of electrons with 70 electron volts (eV) of energy. A molecule struck by an external electron can become charged by either losing or gaining an electron; with electrons possessing 70 eV, the ionization produces many more positive than negative ions. The negative ions are attracted to the anode (electron trap), removing them from the ionization chamber. The positive ions are propelled toward the mass analyzer by the positively charged ($+10,000$ V) repeller plate. Additional charged plates accelerate the ions to a constant velocity and focus the ion stream into the mass analyzer.

Molecules of the vast majority of organic compounds have only paired electrons, so when a single electron is lost from a molecule, a free radical is formed. Thus, the molecular ion formed by loss of an electron is a *radical cation;* it has an unpaired electron as well as a positive charge. The cation formed from an intact molecule is called the *molecular ion (M$^{+\cdot}$).* Once formed, the highly energetic molecular ion often breaks apart, forming both charged and uncharged fragments. Uncharged molecules and fragments are removed by the

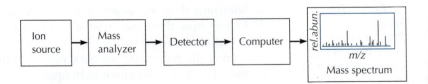

FIGURE 24.1
Basic components of a mass spectrometer.

FIGURE 24.2 Electron impact (EI) ion source. Electrons are generated by heating a metal filament and are accelerated toward the electron trap at 70 eV. The stream of electrons bombards the vaporized sample as it passes into the ion-source chamber and ionizes the sample molecules.

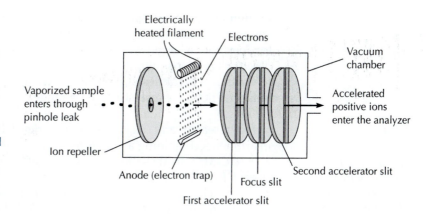

vacuum system, whereas charged ions are affected by electric and magnetic fields and are analyzed.

$$e^-$$
$$\downarrow$$
$$\text{Molecule} \longrightarrow M^{+\bullet} \xrightarrow{\text{fragmentation}} f^+ + f'^{\bullet}$$
$$\swarrow \searrow \qquad\qquad \text{Molecular ion}$$
$$e^- \quad e^-$$

A typical magnetic sector mass analyzer, shown in Figure 24.3, uses a magnetic field to sort ions with different m/z values. Applying a magnetic field perpendicular to the flight path of the ions (perpendicular to the page) causes the ions to adopt curved pathways. The amount of curvature is a function of the mass of the ion and the strength of the magnetic field. For an ion to strike the detector, it must follow a path consistent with the radius of the mass analyzer

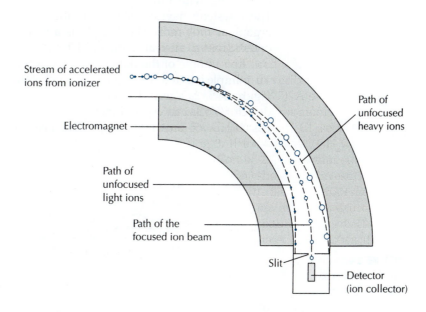

FIGURE 24.3
Magnetic sector mass analyzer.

portion of the mass spectrometer. At a particular magnetic field strength, a beam of ions with a specific m/z ratio will reach the detector. Ions with a larger m/z ratio are not deflected enough to reach the detector, and ions with a smaller m/z ratio are deflected too much to hit the detector. By systematically changing the strength of the magnetic field, you can scan over a range of m/z values.

GC-MS

If you are not familiar with gas chromatography, read "Overview of Gas Chromatography" on pages 292–293, plus Sections 20.1 and 20.2.

Many research and teaching laboratories have acquired hybrid instruments combining a *gas chromatograph and a mass spectrometer (GC-MS)*. In these instruments, small samples of the effluent stream from a gas chromatograph are directed into a mass spectrometer. The molecules in the sample are then ionized by electron impact, and the resulting ions are accelerated and passed into the mass analyzer. The result is a mass spectrum for every compound eluting from the gas chromatograph. This technique is very efficient for analyzing mixtures of compounds because it provides the number of components in the mixture, a rough measure of their relative amounts, and the possible identities of the components.

The mass analyzer in most GC-MS instruments is a *quadrupole mass filter*, diagrammed in Figure 24.4. The quadrupole filter consists of four parallel stainless steel rods. Each pair of rods has opposing direct current (DC) voltages. Superimposed on the DC potential is a high-frequency alternating current (AC) voltage. As the stream of ions passes through the central space parallel to the rods, the combined DC and AC fields affect the ion trajectories, causing them to oscillate. For given DC and AC voltages and frequencies, only ions of a specific m/z ratio achieve a stable oscillation. These ions pass through the filter and strike the detector. Ions with different m/z ratios acquire unstable oscillations, tracing paths that collide with the rods or otherwise miss the detector. Although this mass sorting method is very different from that of magnetic sector instruments, the resulting mass spectra are comparable. Quadrupole mass filters are compact and fast, making them ideally suited for interfacing with other instruments, such as gas chromatographs. These systems have excellent resolution in the mass range of typical organic compounds. The GC-MS is a major workhorse in modern organic and analytical chemistry laboratories.

An example of the data from a GC-MS is shown in Figure 24.5. The sample injected was orange oil purchased at a natural foods store. Figure 24.5a is a record of the total ion current (TIC) arriving at the detector of the mass spectrometer. This ion current corresponds to the gas chromatogram of the sample. The peak at approximately 2.5 min

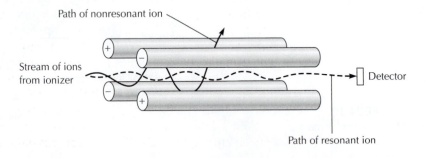

FIGURE 24.4
Quadrupole mass filter.

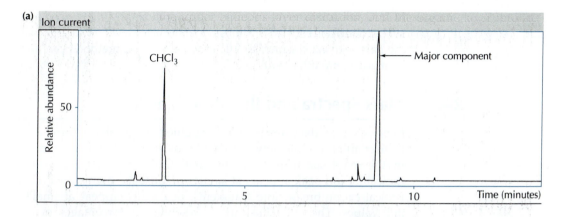

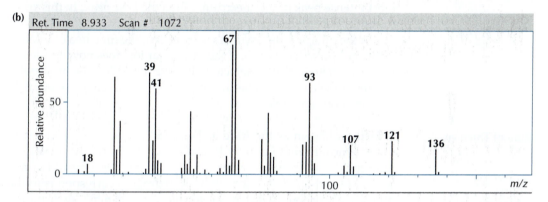

FIGURE 24.5 (a) Total ion current gas chromatogram of orange oil. (b) Mass spectrum of the major component.

was caused by some residual chloroform that was used to clean the injection syringe. The peak at 8.9 min, designated with the arrow, is the major and virtually only compound in the sample. The mass spectrum of this major component of orange oil is shown in Figure 24.5b.

Special Ionization Methods

Normally, samples are vaporized for simple mass spectrometric analysis. Thus, the electron impact technique is limited to compounds with significant vapor pressures. However, special ionization techniques, such as chemical ionization (CI), electrospray ionization (ESI), matrix-assisted laser desorption/ionization (MALDI), and atmospheric pressure chemical ionization (APCI), can be used to ionize samples directly from the solid state or from solution. These techniques make it possible to study samples that have high molecular weights and very low vapor pressures, such as proteins and peptides. Most of these ionization methods impart less energy to the molecules than conventional EI, resulting in less fragmentation of the molecular ion. The resources listed at the end of this chapter cover these ionization methods in greater detail.

ESI involves spraying a solution of sample through a small capillary to generate an aerosol of droplets, somewhat like spray-painting from a paint can. This makes it easy to couple with **liquid chromatography** (*LC-MS*). Components elute as solutions from the

LC column and flow into the ion source, where they are ionized and subsequently identified by mass spectrometry. Although ESI was the first method coupled to LC, other ionization methods can also be made compatible with LC.

24.2 Mass Spectra and the Molecular Ion

Mass spectral data are usually presented in graphical form as a histogram of ion intensity (y-axis) versus m/z (x-axis). For example, in the mass spectrum of 2-butanone shown in Figure 24.6, the molecular ion peak appears at m/z 72. Because the highly energized radical cation breaks into fragments, peaks also appear at smaller m/z values. The intensities are represented as percentages of the most intense peak of the spectrum, called the **base peak.** In this spectrum, the base peak is at m/z 43.

In the interpretation of a mass spectrum, the first area of interest is the molecular ion region. If the molecular ion does not completely fragment before being detected, its m/z provides the molecular weight of the compound (assuming $z = 1$), a valuable piece of information about any unknown.

Rule of Thirteen

A method known as the **Rule of Thirteen**[*] can be used to generate the chemical formula of a hydrocarbon, C_nH_m, using the m/z of the molecular ion. The integer obtained by dividing m/z by 13 (atomic weight of carbon + atomic weight of hydrogen) corresponds to the number of carbon atoms in the formula. The remainder from the division is added to the integer to give the number of hydrogen atoms. For example, if the molecular ion of a hydrocarbon appears at m/z 92 in its mass spectrum, we can find the number of carbon atoms in the molecule by dividing m/z by 13: $n = 92/13 = 7$ with a remainder of 1. The value of m is $7 + 1 = 8$, and the molecular formula for the hydrocarbon is C_7H_8. If the compound contains oxygen or nitrogen as well, some carbon atoms must be subtracted and the number of hydrogen atoms must be adjusted to give the appropriate

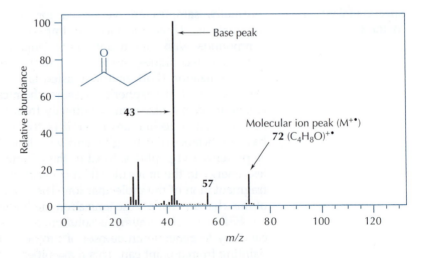

FIGURE 24.6 Mass spectrum of 2-butanone.

[*] Bright, J. W.; Chen, E. C. M. *J. Chem. Educ.* **1983**, *60*, 557–558.

m/z value. One oxygen atom is the equivalent of CH_4; that is, each oxygen and CH_4 unit accounts for 16 atomic mass units. One nitrogen atom is the equivalent of CH_2, 14 atomic mass units.

To apply the Rule of Thirteen to 2-butanone, the *m/z* of 72 is divided by 13, which gives the value of *n* in the formula C_nH_m. The remainder from the division is added to *n* to get the value of *m*. This calculation would provide the correct molecular formula if 2-butanone were a hydrocarbon. Because 2-butanone also contains an oxygen atom, an oxygen atom must be added to the formula and CH_4 subtracted. Dividing *m/z* by 13 yields $72/13 = 5$ (with a remainder of 7); C_nH_m would be C_5H_{12}. Including the presence of oxygen, the correct molecular formula of 2-butanone is $C_5H_{12}O - CH_4 = C_4H_8O$. If the number of oxygen or nitrogen atoms is unknown, a number of potential molecular formulas have to be considered. All heteroatoms in the periodic table can be set equal to a C_nH_m equivalent. If the molecular ion has an odd *m/z* value, the first non-carbon atom to consider is nitrogen. Excellent tables that list formula masses of various combinations of C, H, O, and N can be found in *Spectrometric Identification of Organic Compounds*, 7th ed., by Silverstein, Webster, and Kiemle.

Fundamental Nitrogen Rule

The fundamental nitrogen rule states that a compound whose molecular ion contains nothing other than C, H, N, or O atoms and that has an even *m/z* value must contain either no nitrogen atoms or an even number of nitrogen atoms. A compound whose molecular ion has an odd *m/z* value must contain an odd number of nitrogen atoms. The following compounds support the fundamental nitrogen rule:

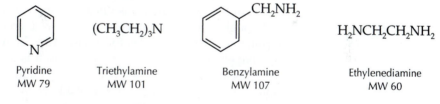

| Pyridine | Triethylamine | Benzylamine | Ethylenediamine |
| MW 79 | MW 101 | MW 107 | MW 60 |

M+1 and M+2 Peaks

Most elements occur in nature as mixtures of isotopes. Table 24.1 provides a list of isotopes for elements commonly found in organic compounds. If the molecular ion (M) signal is reasonably intense, signals for M+1 and M+2 ions can also be observed. The ratios of the intensities of the M+1 and M+2 peaks to that of the M peak depend on the isotopic abundances of the atoms in a molecule and the number of each kind of atom. Isotopes of carbon, hydrogen, oxygen, and nitrogen—the elements that make up most organic compounds—make small contributions to the M+1 and M+2 peaks, and the resulting intensities can sometimes reveal the molecular formulas of organic compounds. Using Table 24.1 for this example, the intensity of the M+1 peak relative to the intensity of the M peak in 2-butanone (molecular formula C_4H_8O) should be $(4 \times 1.08\%) + (8 \times 0.012\%) + (1 \times 0.038\%) = 4.45\%$. Useful tables listing molecular formulas and the ratios expected for these formulas can be found in *Mass and Abundance Tables for Use in Mass Spectrometry* by Beynon and Williams (Elsevier: New York, 1963).

TABLE 24.1   **Relative isotope abundances of elements common in organic compounds**

Elements	Isotope	Relative abundance	Isotope	Relative abundance	Isotope	Relative abundance
Hydrogen	^1H	100	^2H	0.012		
Carbon	^{12}C	100	^{13}C	1.08		
Nitrogen	^{14}N	100	^{15}N	0.37		
Oxygen	^{16}O	100	^{17}O	0.038	^{18}O	0.205
Fluorine	^{19}F	100				
Silicon	^{28}Si	100	^{29}Si	5.08	^{30}Si	3.35
Phosphorus	^{31}P	100				
Sulfur	^{32}S	100	^{33}S	0.79	^{34}S	4.47
Chlorine	^{35}Cl	100			^{37}Cl	32.0
Bromine	^{79}Br	100			^{81}Br	97.3
Iodine	^{127}I	100				

Adapted from J. R. de Laeter, J. K. Bohlke, P. de Bievre, H. Hidaka, H. S. Peiser, K. J. R. Rosman, and P. D. P. Taylor for the International Union of Pure and Applied Chemistry in "Atomic Weights of the Elements, Review 2000," *Pure and Applied Chemistry,* **2003,** *75*, 683–800.

Unfortunately, practical experience has shown that for many C, H, N, and O compounds, the expected ratios can be in error, which can occur for many reasons. For example, in the mass spectrometer the molecular ion may undergo ion–molecule collisions that provide additional intensity to the M+1 peak.

$$M^{+\bullet} + RH \longrightarrow MH^+ + R^\bullet$$

In addition, if the molecular ion has a low relative abundance, the precision of the M+1 data is insufficient to give reliable ratios.

Although it can be difficult to use M+1 and M+2 data to determine accurate molecular formulas, MS is highly valuable for qualitative elemental analysis. In particular, it is fairly easy to use the M+2 peak to identify the presence of bromine and chlorine in organic compounds. The appearance of a large M+2 peak in a mass spectrum is evidence for the presence of one of these elements. The relative intensities tell you which one. A good example is seen in the mass spectrum of 1-bromopropane (Figure 24.7). The two

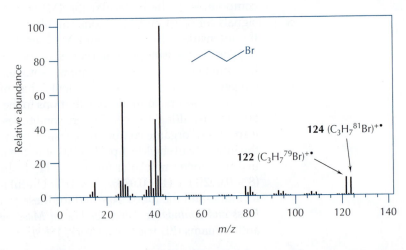

FIGURE 24.7

Mass spectrum of 1-bromopropane.

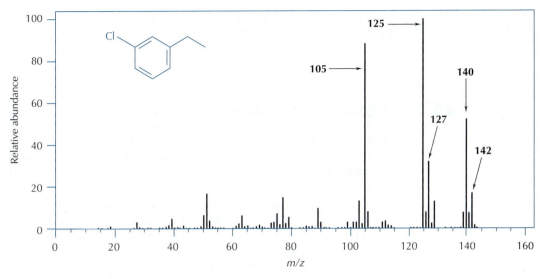

FIGURE 24.8 Mass spectrum of 3-chloroethylbenzene.

major peaks in the molecular ion region are m/z 122 and 124 with an approximately 1:1 ratio. You can see from Table 24.1 that bromine exists in nature as a mixture of ^{79}Br and ^{81}Br in a ratio very close to 1:1. The peak at m/z 122, where the bromine atom has a mass of 79, is by convention defined as the molecular ion peak (M). Although the m/z 124 peak also corresponds to an intact molecule of +1 charge, it is referred to as the M+2 peak. Isotopic contributions of carbon, hydrogen, nitrogen, and oxygen to the M+1 and M+2 peaks are comparatively small. Thus a ratio of M/(M+2) that is very close to 1:1 is a clear indication that the molecule contains a bromine atom.

A monochloro compound is expected to have an M+2 peak that is 32.0% as intense as the M peak. For example, the mass spectrum of 3-chloroethylbenzene, shown in Figure 24.8, has a peak at m/z 142 that is approximately one-third the intensity of the molecular ion peak at m/z 140. A small contribution of ^{13}C is shown in the M+1 peak at m/z 141. The Rule of Thirteen can be used to calculate the molecular formula; the carbon equivalent of ^{35}Cl is C_2H_{11}. Dividing 140 by 13 yields $140/13 = 10$ (with a remainder of 10); C_nH_m would be $C_{10}H_{20}$. Including the presence of a chlorine atom, the correct molecular formula of 3-chloroethylbenzene is shown to be $C_{10}H_{20}Cl - C_2H_{11} = C_8H_9Cl$.

Using Probability to Calculate Molecular Isotope Abundance

When multiple atoms of a particular element are present, isotope abundance and probability play a role in determining the mass and intensity of the molecular ion peaks. For example, the mass spectrum of bromine (Br_2) shows a 1:2:1 ratio of molecular ions of masses 158, 160, and 162, due to the combinations of isotopes that are possible: ^{79}Br$_2$, ^{79}Br^{81}Br, ^{81}Br^{79}Br, and ^{81}Br$_2$. In this case, the probability of having either ^{79}Br or ^{81}Br is 50% (from Table 24.1: $[^{79}Br]/([^{79}Br]+[^{81}Br]) = 100/(100 + 97.6) \sim 0.5$); therefore, each combination of two bromine atoms has a probability of $(0.5) \times (0.5) = 0.25$. Because there are two combinations of atomic isotopes that lead to the molecular isotope of mass 160, the probability of that isotope is $2 \times (0.25) = 0.5$. Therefore, the molecular isotopes 158, 160, and 162 have probabilities of 0.25, 0.5, and 0.25, respectively. Normalizing these values gives the 1:2:1 ratio that is observed. This example also illustrates that M is not necessarily the most abundant molecular ion peak; in this case, the M+2 peak is the most intense molecular ion.

24.3   High-Resolution Mass Spectrometry

In modern research laboratories, molecular formulas are usually determined by *high-resolution mass spectrometry*. The mass spectrometers used for this purpose vary in ionization method and instrument design, but all of them can measure masses to four figures beyond the decimal point. Table 24.2 provides the masses that should be used to calculate the exact mass of a molecular ion. Using carbon-12 as the standard, you can determine the exact mass of a molecule by summing the masses of the most probable atomic isotopes in the molecule. For example, the exact mass of the molecular ion of 2-butanone, using the most abundant mass isotopes, is $(4 \times 12.00000) + (8 \times 1.00783) + (1 \times 15.9949) = 72.0575$. By looking at the exact masses of molecules whose nominal molecular weight is 72, it is obvious that the correct molecular formula can be determined from the masses measured to four decimal places. An experimental m/z value within 0.003 is usually adequate for establishing

TABLE 24.2	Atomic weights and exact isotope masses for elements common in organic compounds		
Element	Atomic weight	Nuclide	Mass
Hydrogen	1.00794	^{1}H	1.00783
		D(^{2}H)	2.01410
Carbon	12.0107	^{12}C	12.00000 (std)
		^{13}C	13.00335
Nitrogen	14.0067	^{14}N	14.0031
		^{15}N	15.0001
Oxygen	15.9994	^{16}O	15.9949
		^{17}O	16.9991
		^{18}O	17.9992
Fluorine	18.9984	^{19}F	18.9984
Silicon	28.0855	^{28}Si	27.9769
		^{29}Si	28.9765
		^{30}Si	29.9738
Phosphorus	30.9738	^{31}P	30.9738
Sulfur	32.065	^{32}S	31.9721
		^{33}S	32.9715
		^{34}S	33.9679
Chlorine	35.453	^{35}Cl	34.9689
		^{37}Cl	36.9659
Bromine	79.904	^{79}Br	78.9183
		^{81}Br	80.9163
Iodine	126.9045	^{127}I	126.9045

Adapted from J. R. de Laeter, J. K. Bohlke, P. de Bievre, H. Hidaka, H. S. Peiser, K. J. R. Rosman, and P. D. P. Taylor for the International Union of Pure and Applied Chemistry in "Atomic Weights of the Elements, Review 2000," *Pure and Applied Chemistry*, **2003,** *75*, 683–800.

a molecular formula for compounds that have molecular masses below 1000.

Formula	Exact Mass
$C_2H_4N_2O$	72.0324
$C_3H_4O_2$	72.0211
$C_3H_8N_2$	72.0688
C_4H_8O	72.0575
C_5H_{12}	72.0940

There is an important caveat to this method of calculating the exact mass of the molecular ion. The exact mass reported in mass spectrometry is the mass of the most abundant molecular ion, which depends not only on the exact masses of the isotopes present but also on the probability that a heavier isotope is present in the molecule. As explained for the Br_2 example above, when two or more Br atoms are present in the same molecule, the most abundant molecular ion is the one that includes a combination of ^{79}B and ^{81}Br (not solely ^{79}Br, even though ^{79}Br is slightly more abundant). For most organic molecules that do not contain atoms with two or more high-abundance isotopes (like Br or Cl), using the exact masses of the most abundant isotopes will provide the mass of the most abundant molecular ion. However, large molecules, such as proteins that contain more than 100 carbon atoms, are more likely to contain one or more ^{13}C atoms than to be composed solely of ^{12}C atoms. Calculating the exact mass of the most abundant molecular ion for large molecules can be complicated, but there are online algorithms available to do this. For example, see the Isotope Distribution Calculator at the Scientific Instrument Services MS Tools Website: http://www.sisweb.com/mstools.htm.

24.4　Mass Spectral Libraries

When a molecular ion breaks into fragments, the resulting mass spectrum can be complex because any one of a number of covalent bonds might break during fragmentation. Examination of Figures 24.5–24.8 shows that a large number of peaks arise even with relatively low-molecular-weight organic compounds. **The array of fragmentation peaks creates a fingerprint that can be used for identification.** Modern mass spectrometers are routinely equipped with computer libraries of mass spectra (some contain hundreds of thousands of spectra) for matching purposes. Typically, a computer program compares the experimental spectrum with spectra in the library and produces a ranked *"hit list"* of compounds with similar mass spectra. The ranking is based on how close the match is in terms of the presence of peaks and their intensities. At this point, a chemist can compare the mass spectra of highly ranked compounds from the hit list with the acquired mass spectrum and determine the closest match.

The closest match does not necessarily prove the structure of a compound. Impurities can produce extra peaks in the mass spectrum and provide false hit list candidates. In addition, the compound must be in the database, which is not always the case with research samples. Two comprehensive libraries of mass spectra include a collection of electron impact mass spectra of over 212,000 compounds from the National Institute of Standards and Technology (*NIST 11, NIST/EPA/NIH Mass Spectral Library*) and the *Wiley Registry 9th Ed/NIST 2011 Mass Spectral Library,* a collection of 780,000 EI mass spectra. There are also a number of specialized mass spectral libraries available that are targeted to specific types of compounds, such as drug metabolites or steroids.

The hit list for the major component of orange oil is shown in Figure 24.9. The second column, labeled *SI* (for *similarity index*), corresponds to how well the mass spectra that are stored in the computer library match the acquired spectrum of the compound from the GC-MS. Notice that several compounds appear more than once in the list; there are several spectra for these compounds in the library because many laboratories contribute spectra to the collection. Slight differences in instrument conditions and/or configurations can lead

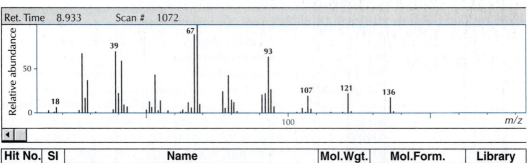

Hit No.	SI	Name	Mol.Wgt.	Mol.Form.	Library
1	94	Limonene $$ Cyclohexene, 1-methyl-4-(1-m	136	C10H16	NIST62
2	90	1,5-Cyclooctadiene, 1,5-dimethyl-	136	C10H16	NIST12
3	90	Cyclohexene, 1-methyl-4-(1-methylethenyl)-	136	C10H16	NIST12
4	87	Camphene $$ Bicyclo 2.2.1 heptane, 2,2-dim	136	C10H16	NIST62
5	86	Cyclohexanol, 1-methyl-4-(1-methylethenyl)-	196	C12H20O2	NIST62
6	86	Limonene	136	C10H16	NIST12
7	86	Cyclohexene, 1-methyl-4-(1-methylethenyl)-	136	C10H16	NIST62
8	85	D-Limonene	136	C10H16	NIST12
9	85	Bicyclo 2.2.1 hept-2-ene, 1,7,7-trimethyl- $$	136	C10H16	NIST62
10	85	D-Limonene $$ Cyclohexene, 1-methyl-4-(1-m	136	C10H16	NIST62
11	84	Limonene	136	C10H16	NIST12
12	83	D-Limonene	136	C10H16	NIST12
13	83	Cyclohexanol, 1-methyl-4-(1-methylethenyl)-	196	C12H20O2	NIST12
14	83	1,5-Cyclooctadiene, 1,5-dimethyl- $$ 1,5-Dim	136	C10H16	NIST62
15	83	Cyclohexene, 1-methyl-4-(1-methylethenyl)-	136	C10H16	NIST62
16	83	Limonene	136	C10H16	NIST12
17	82	Cyclohexene, 4-ethenyl-1,4-dimethyl- $$ 1,4-	136	C10H16	NIST62
18	82	Camphene	136	C10H16	NIST12
19	82	Cyclohexene, 1-methyl-5-(1-methylethenyl)-	136	C10H16	NIST62
20	81	2,6-Octadien-1-ol, 3,7-dimethyl-, [Z]-	154	C10H18O	NIST12
21	80	4-Tridecen-6-yne, [Z]-	178	C13H22	NIST62
22	80	.alpha.-Myrcene	136	C10H16	NIST62
23	80	Bicyclo 2.2.1 heptane, 2,2-dimethyl-3-methyl	136	C10H16	NIST62
24	80	2,6-Octadien-1-ol, 3,7-dimethyl-, [E]-	154	C10H18O	NIST12
25	80	D-Limonene	136	C10H16	NIST12

FIGURE 24.9 Hit list from a mass spectral library search for the major component of orange oil. The symbol $$ in the name denotes the start of a second name for the same compound.

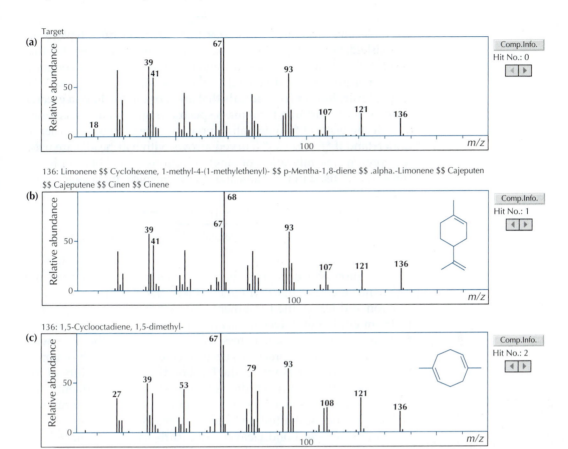

FIGURE 24.10. Computer comparison of two hit-list compounds for orange oil. (a) MS of the compound from GC-MS run. (b) MS of hit 1. (c) MS of hit 2.

to subtle differences in the acquired spectra—another reason to examine the hit list with a critical eye.

A computer screen printout for comparing spectra of the hit list candidates is shown in Figure 24.10. The spectrum of hit 1 (Figure 24.10b) is virtually identical to the mass spectrum of the sample (Figure 24.10a). The spectrum of hit 2 (Figure 24.10c) is also similar, even though the compound's structure is different. On close examination, however, some subtle differences can be discerned. A significant signal at m/z 108 is present in Figure 24.10c but not in Figure 24.10a. Also, the signal at m/z 92 in Figure 24.10a is missing in Figure 24.10c. By these observations, hit 2 can be ruled out as a match, and a tentative conclusion can be reached that the major component of orange oil is limonene. The hit list is more reliable in confirming a structural option if it is combined with other spectroscopic evidence. Infrared or NMR evidence and the history of the sample—for example, if it came from a chemical reaction—can help to ascertain the correct molecular structure.

24.5 Fragment Ions

In cases where you are working with a compound not included in a library or when the hit list does not lead to a satisfactory structure candidate, fragmentation pathways can provide important clues to the molecular structure. Numerous fragmentation rules have been established, but the topic is too broad to be adequately covered in this book. A few of the most useful fragmentation patterns for common functional groups are described in the following paragraphs. **As a general rule, ions or free radicals that are more stable have a greater probability of forming from mass spectral fragmentation reactions.** Mechanisms of fragmentation processes are sometimes easier to understand if *"fishhooks"* (curved arrows with half-heads) are used to represent the migration of single electrons. This notation is the same as that used in free-radical or photochemical processes:

$$\text{Molecule} \longrightarrow \underbrace{M^{+\bullet} = [f \frown f']^{+\bullet}}_{\substack{\text{Molecular} \\ \text{cation radical}}} \longrightarrow \underset{\text{Fragments}}{f^+ + f'^{\bullet}}$$

The equation above implies that one electron is ejected from the f–f′ bond and that the f′ fragment takes the remaining unpaired electron and leaves the f fragment with a "hole" or + charge. **The f′· fragment is not detected because it is neutral, but the charged f+ fragment is detected.** Forces that contribute to the ease with which fragmentation processes occur include the strength of bonds in the molecule (for example, the f–f′ bond) and the stability of the carbocations (f+) and free radicals (f′·) produced by fragmentation. Although these fragments are formed in the gas phase, we can still apply our "chemical intuition," based on reactions in solution.

In this type of bond cleavage, the carbocation fragment is an even-electron species, and the odd electron ends up on the free-radical fragment. Although the neutral free-radical fragment is not directly detected, its mass can be deduced by taking the difference between the mass of the molecular ion and the mass of the fragment ion. In the example below, the mass of the neutral species that is lost ($CH_3 \cdot$) can be determined in this way ($30 - 15 = 15$). When a molecular ion with an even m/z value gives a fragment ion that has an odd m/z value, a loss of a free radical by cleavage of just one covalent bond has occurred.

$$[H_3C \overset{\xi}{-} CH_3]^{+\bullet} \longrightarrow CH_3^+ + CH_3 \cdot$$
$$\underset{m/z\ 30}{} \qquad\qquad \underset{m/z\ 15}{}$$

Simple cleavage of a molecular ion that has an odd m/z value gives a fragment ion with an even m/z value:

$$(CH_3CH_2)_2 \overset{+\bullet}{N} - CH_2 \overset{\xi}{-} CH_3 \longrightarrow (CH_3CH_2)_2 \overset{+}{N} = CH_2 + CH_3 \cdot$$
$$\underset{m/z\ 101}{} \qquad\qquad\qquad \underset{m/z\ 86}{}$$

Aromatic Hydrocarbons

Aromatic hydrocarbons are prone to fragmentation at the bond β to the aromatic ring, yielding a benzylic cation that rearranges to a stable C_7H_7 aromatic carbocation called a tropylium ion.

$$m/z\ 91$$

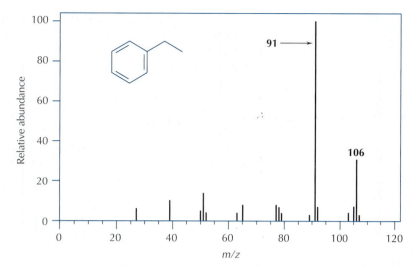

FIGURE 24.11
Mass spectrum of ethylbenzene.

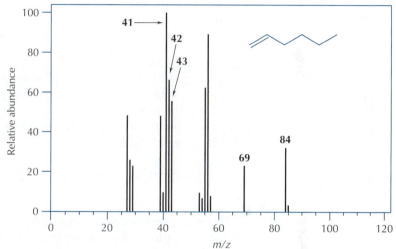

FIGURE 24.12 Mass spectrum of 1-hexene.

For mono-alkylbenzenes, the peak at m/z 91 is a very large signal, often the base peak. In the mass spectrum of ethylbenzene shown in Figure 24.11, the base peak of 91 (the tropylium ion) is the result of loss of a methyl group ($G\cdot = \cdot CH_3$).

Alkenes

Alkenes are similarly prone to fragmentation at the bond β to the double bond to give a stabilized allylic cation or allylic radical.

In the mass spectrum of 1-hexene shown in Figure 24.12, the allylic cation fragment ($CH_2=CH-CH_2^+$) is observed at m/z 41 and the propyl cation ($CH_3CH_2CH_2^+$) is observed at m/z 43.

Alcohols

Alcohols fragment easily, and as a result, the molecular ion peak is often very small. In many cases, the molecular ion is not even

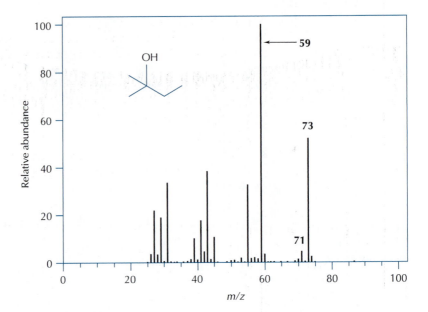

FIGURE 24.13

Mass spectrum of
2-methyl-2-butanol.

apparent in the mass spectrum. One fragmentation pathway is the loss of hydroxyl radical (·OH) to produce a carbocation. However, the most important fragmentation pathway is the loss of an alkyl group from the molecular ion to form a resonance-stabilized oxonium ion. Primary alcohols show an intense m/z 31 peak resulting from this type of fragmentation.

$$R \overset{\cdot}{\underset{\cdot}{\textstyle\diagup}} CH_2 - \overset{+\cdot}{\underset{\cdot\cdot}{O}}H \longrightarrow R\cdot + \left[CH_2 = \overset{+}{\underset{\cdot\cdot}{O}}H \longleftrightarrow \overset{+}{C}H_2 - \overset{\cdot\cdot}{\underset{\cdot\cdot}{O}}H \right]$$

$$m/z\ 31$$

The mass spectrum of 2-methyl-2-butanol shown in Figure 24.13 provides examples of the various fragmentation pathways available to alcohols. Notice that the molecular ion peak (m/z 88) is not present in the spectrum.

$$CH_3CH_2 - \underset{\underset{CH_3}{|}}{\overset{\overset{CH_3}{|}}{C}} \overset{\cdot}{\diagup} \overset{+\cdot}{\underset{\cdot\cdot}{O}} - H \longrightarrow CH_3CH_2 - \overset{\overset{CH_3}{\diagup}}{\underset{CH_3}{\diagdown}}{C}{+} \quad + \cdot\overset{\cdot\cdot}{O}H$$

$$m/z\ 71$$

$$CH_3CH_2 \overset{\cdot}{\diagup} \underset{\underset{CH_3}{|}}{\overset{\overset{CH_3}{|}}{C}} - \overset{+\cdot}{\underset{\cdot\cdot}{O}} - H \longrightarrow CH_3CH_2 \cdot + \quad \overset{\overset{CH_3}{\diagdown}}{\underset{CH_3}{\diagup}}{C} = \overset{+}{\underset{\cdot\cdot}{O}} - H$$

$$m/z\ 59$$

$$CH_3CH_2 - \underset{\underset{CH_3}{|}}{\overset{\overset{CH_3}{\wedge}}{C}} - \overset{+\cdot}{\underset{\cdot\cdot}{O}} - H \longrightarrow CH_3 \cdot + \quad \overset{\overset{CH_3CH_2}{\diagdown}}{\underset{CH_3}{\diagup}}{C} = \overset{+}{\underset{\cdot\cdot}{O}} - H$$

$$m/z\ 73$$

Other heteroatom-containing molecules undergo similar types of cleavage. Amines, ethers, and sulfur compounds can undergo fragmentations analogous to those exhibited by alcohols.

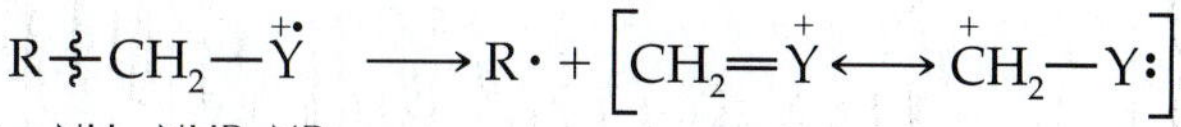

$$R \overset{\xi}{-} CH_2 - \overset{+\cdot}{\underset{\cdot\cdot}{Y}} \longrightarrow R\cdot + \left[CH_2 = \overset{+}{Y} \longleftrightarrow \overset{+}{C}H_2 - \overset{\cdot\cdot}{Y}\colon \right]$$

$Y = NH_2$, NHR, NR_2,
OR, SH, or SR

Carbonyl Compounds

Ketones and other carbonyl compounds, such as esters, fragment by cleavage of bonds α to the carbonyl group to form a resonance-stabilized acylium ion.

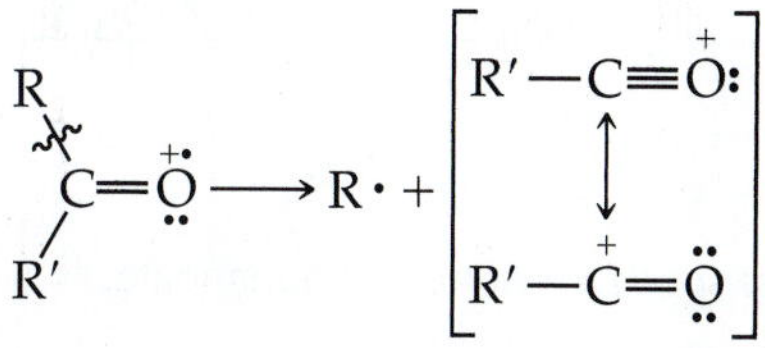

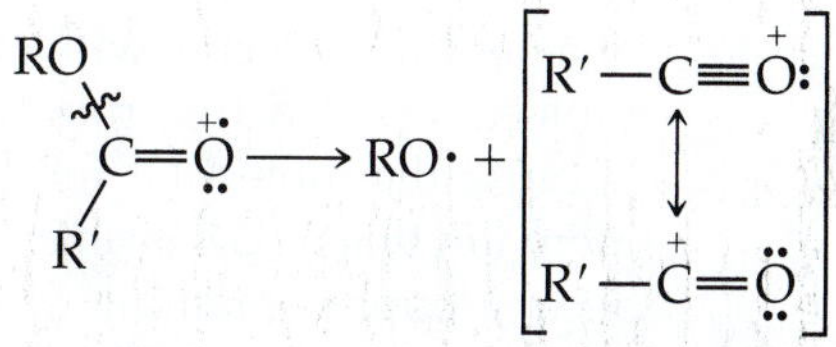

In the spectrum of 2-butanone shown earlier in Figure 24.6, we see fragmentations on both sides of the carbonyl group.

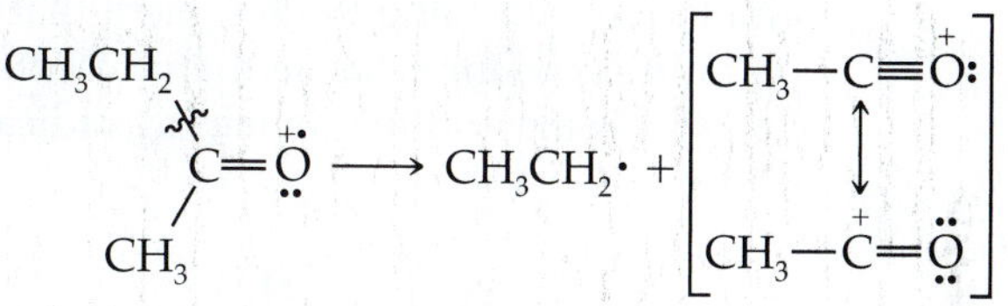

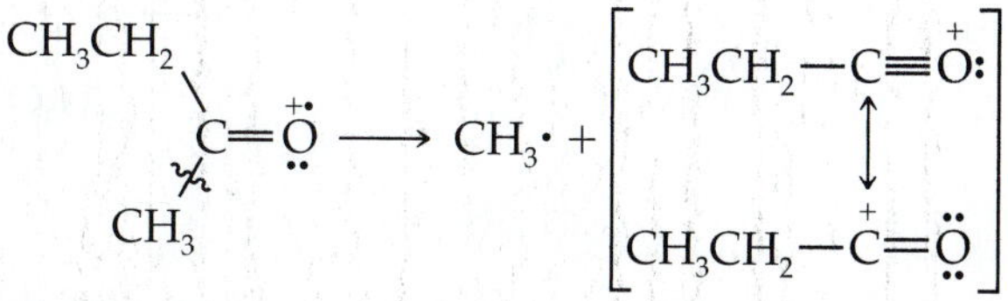

In the mass spectrum of the ester methyl nonanoate (MW 172), shown in Figure 24.14, there is a significant peak at *m/z* 141. This peak results from formation of an acylium ion by loss of a fragment with a mass of 31, corresponding to a methoxyl radical.

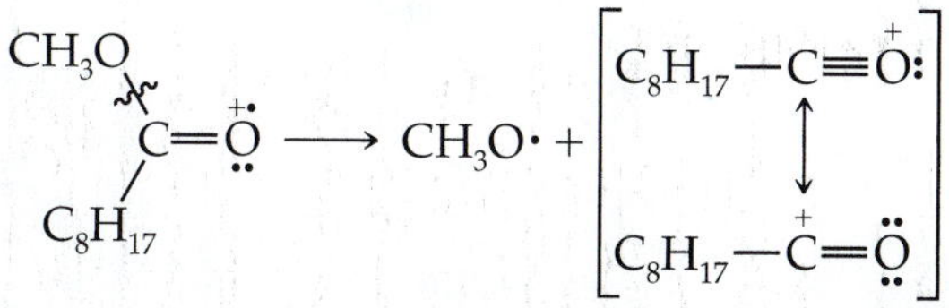

The base peak at *m/z* 74 in Figure 24.14 occurs through the loss of a fragment with a mass of 98—a mass that corresponds to the loss

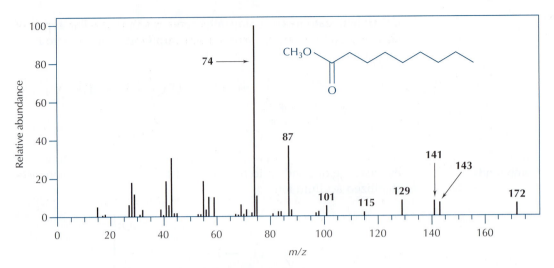

FIGURE 24.14 Mass spectrum of methyl nonanoate.

of a neutral molecule with a molecular formula C_7H_{14}. That a neutral molecule (not a free radical) is lost by fragmentation is apparent because the molecular ion has an even m/z value and gives a fragment ion that also has an even m/z value. Carbonyl compounds with alkyl groups containing a chain of three or more carbon atoms can cleave at the β bond. This pathway, called the *McLafferty rearrangement,* requires the presence of a hydrogen atom on the γ (gamma) carbon atom. Note that in this rearrangement, a neutral, closed-shell molecule is lost and the fragment that retains the positive charge is a radical cation. Because the molecule lost has an even-numbered mass, the mass of the remaining radical cation is also even.

The mass spectrum of methyl nonanoate demonstrates fragmentations also characteristic of other organic compounds with straight-chain alkyl groups. Carbon-carbon bonds can break at any point along the chain, leading to the loss of alkyl radicals.

m/z	Radical fragment lost from the molecular ion
143 (M–29)	$CH_3CH_2\cdot$
129 (M–43)	$CH_3CH_2CH_2\cdot$
115 (M–57)	$CH_3(CH_2)_2CH_2\cdot$
101 (M–71)	$CH_3(CH_2)_3CH_2\cdot$
87 (M–85)	$CH_3(CH_2)_4CH_2\cdot$

Any unsaturated functional group that contains a γ C-H group can undergo a McLafferty rearrangement. For example, the mass spectrum of 1-hexene in Figure 24.12 shows a fragment ion at $m/z = 42$, which arises from the loss of a neutral alkene.

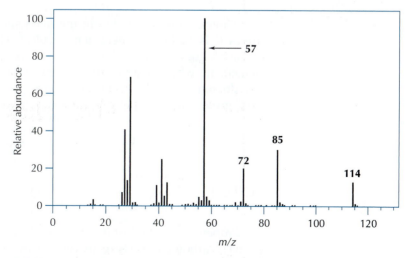

$m/z\ 84$ $\qquad$ $m/z\ 84$ $\qquad$ $m/z\ 42$

A similar fragmentation is also possible for aromatic compounds with γ hydrogen atoms.

24.6 Case Study

We have seen that if a molecular ion does not fragment completely before being detected, its m/z value provides the molecular weight of the compound, which is a significant clue to its structure. Moreover, the profile for the fragmentation of the molecular ion can establish its identity, particularly if the compound is listed in the instrument's mass spectral library. A library search usually allows you to identify the compound's structure. If the compound is not part of the library or if you do not have access to a mass spectral library, it is still possible to get useful information from the mass spectrum. Determining a structure from a mass spectrum alone, however, is challenging and in most cases requires supplementary spectroscopic information.

As an example of how to approach mass spectrometric analysis, consider the mass spectrum shown in Figure 24.15. The molecular ion peak appears at m/z 114. The even value of m/z suggests that the compound does not contain nitrogen, unless it contains more than one nitrogen atom per molecule. The absence of a significant M+2 peak rules out the presence of chlorine or bromine. The IR spectrum of the compound

FIGURE 24.15 Mass spectrum of unknown for case study.

has an intense peak at 1715 cm^{-1}, which indicates the presence of a C=O group. Knowing that we are working with a carbonyl compound suggests that the base peak at m/z 57 may be a stabilized oxonium-ion fragment, formed by α cleavage and loss of a C_4H_9 radical (M–57).

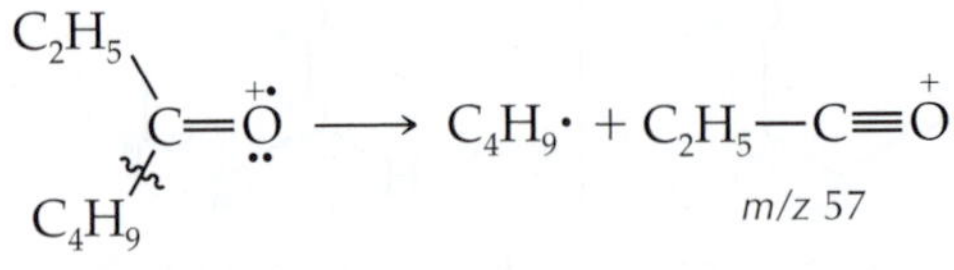

Application of the Rule of Thirteen can generate one or more potential molecular formulas for the compound. Dividing 114 by 13, we have $114/13 = 8$ (with a remainder of 10); if the compound were a hydrocarbon, its formula C_nH_m would be C_8H_{18}. Including the presence of an oxygen atom, we would have $C_8H_{18}O – CH_4 = C_7H_{14}O$. Our short list of possible molecular formulas is $C_7H_{14}O$ and perhaps $C_6H_{10}O_2$.

Next, an inventory of the significant MS peaks is put together, along with the masses lost on fragmentation.

m/z	Mass	Possible fragments
114	M	
85	M–29	$C_2H_5\cdot$
72	M–42	C_3H_6
57	M–57	$C_4H_9\cdot$

The mass spectral evidence gives no support for having more than one oxygen atom per molecule of the compound. It is likely that this carbonyl compound is a ketone because the peak at m/z 85 is consistent with an α cleavage, with loss of an ethyl radical to give a stabilized oxonium ion.

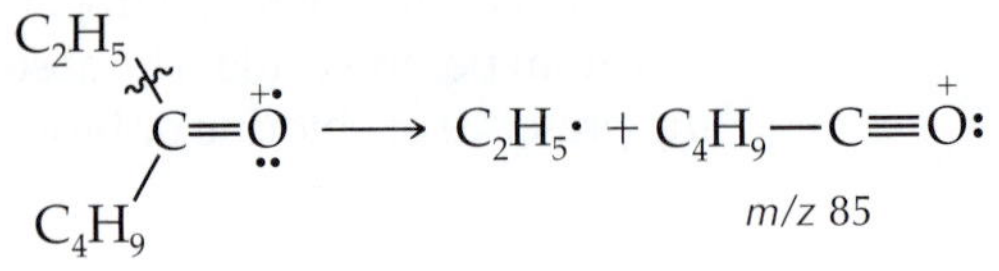

Another important clue in the mass spectrum shown in Figure 24.15 is the loss of a neutral molecule, C_3H_6, producing the peak at m/z 72. Loss of a neutral molecule results from a McLafferty rearrangement, which requires the presence of a carbonyl group and a γ hydrogen atom. This fact is also consistent with the presence of a C_4H_9 group, shown by the m/z 57 fragment ion.

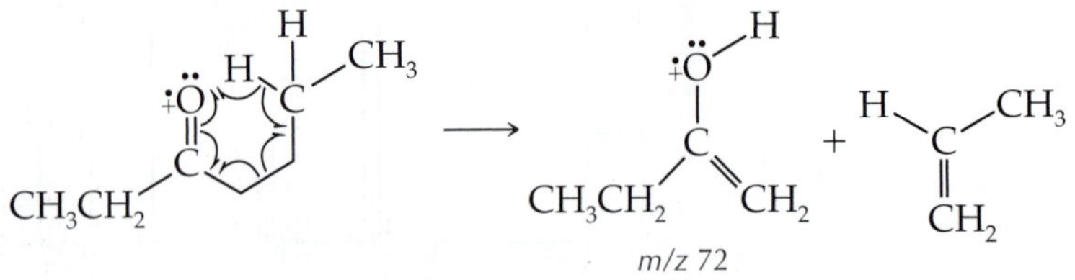

Propene could be lost if a methyl group were attached to the β or the γ carbon atom. There are two compounds consistent with all the evidence, 3-heptanone and 5-methyl-3-hexanone.

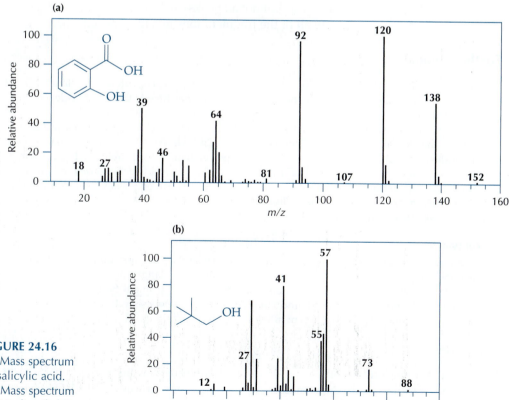

3-Heptanone

5-Methyl-3-hexanone

Reference to spectra in the *Wiley Registry of Mass Spectral Data* shows that the spectrum of the unknown is very similar to the spectra of both 3-heptanone and 5-methyl-3-hexanone. However, the spectrum of 5-methyl-3-hexanone has a signal at m/z 99 that is missing in the spectra of 3-heptanone and the unknown. The identity of the unknown is probably 3-heptanone. A ^{1}H NMR spectrum of the compound would establish the structure unambiguously.

24.7 Sources of Confusion

How Do I Know Whether the Highest-Mass Peak Is the Molecular Ion?

It is tempting to assume that the highest-mass ion identified by the computer is the molecular ion; however, some of the signals recognized by the computer are very weak and may not be significant. For example, the ions labeled $m/z = 152$ and 88 in Figures 24.16a and 24.16b, respectively, should be treated with a fair degree of skepticism. To determine if they are meaningful, look for the highest-mass peak of substantial abundance and consider whether it is likely to be the molecular ion. In general, organic compounds have

FIGURE 24.16
(a) Mass spectrum of salicylic acid.
(b) Mass spectrum of 2,2-dimethyl-1-propanol.

even-number molecular ion masses (except those containing odd numbers of nitrogen atoms) and odd-number fragment ion masses (because they usually fragment by loss of an odd-mass free radical).

The highest-mass ion of significant abundance in Figure 24.16a is $m/z = 138$; because this is an even number, a good first assumption is that it represents the molecular ion. In fact, this is a spectrum of salicylic acid, which has a molecular weight of 138. The m/z 152 peak must be due to an impurity.

In contrast, the highest-mass ion of significant abundance in Figure 24.16b is $m/z = 73$. Because this is an odd number, it is likely to be a fragment ion, and the molecular ion of this compound is very weak. This is a spectrum of 2,2-dimethyl-1-propanol, which has a molecular weight of 88. This compound loses a methyl radical very easily and therefore the molecular ion at $m/z = 88$ is very weak. Unknown compounds that give odd-numbered highest-mass ions in the mass spectrum contain either nitrogen or a fragment that results from the loss of a free radical. In the absence of a fruitful mass spectral library search, you may need to rely on additional spectroscopic techniques to positively identify the structures of unknown compounds.

I Don't Understand How Some of the Fragment Ions Are Formed

While it should be possible to explain the fragmentation leading to the base peak, you may find it hard to explain other fragment ions in the mass spectrum. Some fragments may be the result of multiple steps or they may have been formed by complex rearrangements. The references at the end of the chapter will help you understand some of these, but not all peaks can be explained. Do not feel obliged to justify all of the fragment ions in a mass spectrum.

Further Reading

Crews, P.; Rodríguez, J.; Jaspars, M. *Organic Structure Analysis*, 2nd ed.; Oxford University Press: Oxford, 2009.

Downard, K. *Mass Spectrometry: A Foundation Course*; The Royal Society of Chemistry: Cambridge, 2004.

Gross, J. H. *Mass Spectrometry: A Textbook*; Springer-Verlag: Berlin, 2004.

Lee, T. L. *A Beginner's Guide to Mass Spectral Interpretation*; Wiley: New York, 1998.

McLafferty, F. W.; Turecek, F. *Interpretation of Mass Spectra*, 4th ed.; University Science Books: Mill Valley, CA, 1993.

McMaster, M. C. GC/MS: *A Practical User's Guide*, 2nd ed.; Wiley: New York, 2008.

Silverstein, R. M.; Webster, F. X.; Kiemle, D. J. *Spectrometric Identification of Organic Compounds*, 7th ed.; Wiley: New York, 2005.

Useful Web Sites

NIST Standard Reference Database: http://webbook.nist.gov/chemistry

Spectral Database for Organic Compounds, National Institute of Advanced Industrial Science and Technology (AIST), Japan: http://riodb01.ibase.aist.go.jp/sdbs/cgi-bin/cre_index.cgi?lang=eng

Scientific Instrument Services MS Tools Website: http://www.sisweb.com/mstools.htm

Questions

1. Match the compounds azobenzene, ethanol, and pyridine with their molecular weights: 46, 79, and 182. How does the fact that in one case the molecular weight is odd and in the other two cases the molecular weight is even help in the selection process?

2. The mass spectrum of 1-bromopropane is shown in Figure 24.7. Propose a structure for the base peak at m/z 43.

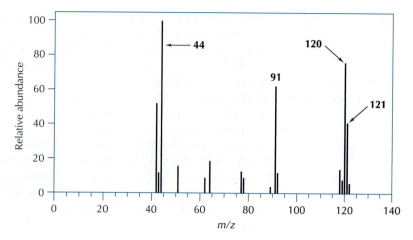

FIGURE 24.17 Mass spectrum of unknown for question 6.

3. The base peak of 1-pentanol is at m/z 31, whereas that for 2-pentanol is at m/z 45. Explain briefly.

4. Similar types of cleavages give rise to two peaks for 4-chlorobenzophenone, one at m/z 105 (the base peak) and one at m/z 139 (70% of base). A clue to their identities is the fact that the m/z 139 peak is accompanied by a peak at m/z 141 that is about one-third of the m/z 139 peak intensity. The m/z 105 peak has no such partner. What structures correspond to these peaks? Show your reasoning.

4-Chlorobenzophenone

5. What fragmentations lead to the peaks at m/z 127, 125, and 105 in Figure 24.8?

6. An unknown compound has the mass spectrum shown in Figure 24.17. The molecular ion peak is at m/z 121. The infrared spectrum of the unknown shows

Hit No.	SI	Name	Mol.Wgt.	Mol.Form.	Library
1	89	Phenol, 2-methoxy-4-(1-propenyl)- $$ Phen	164	C10H12O2	NIST62
2	88	Phenol, 2-methoxy-5-(1-propenyl)-, [E]-	164	C10H12O2	NIST12
3	87	Phenol, 2-methoxy-5-(1-propenyl)-, [E]- $$ P	164	C10H12O2	NIST62
4	86	Phenol, 2-methoxy-4-(1-propenyl)-, [E]- $$ P	164	C10H12O2	NIST62
5	86	Phenol, 2-methoxy-4-(1-propenyl)-, acetate $$	206	C12H14O3	NIST62
6	85	Eugenol	164	C10H12O2	NIST12
7	84	Eugenol	164	C10H12O2	NIST12
8	83	7-Benzofuranol, 2,3-dihydro-2,2-dimethyl- $	164	C10H12O2	NIST62
9	82	Phenol, 2-methoxy-4-(1-propenyl)-	164	C10H12O2	NIST12
10	82	Phenol, 2-methoxy-4-(1-propenyl)-	164	C10H12O2	NIST12
11	81	Eugenol $$ Phenol, 2-methoxy-4-(2-propen	164	C10H12O2	NIST62
12	81	Eugenol	164	C10H12O2	NIST12
13	80	Phenol, 2-methoxy-4-(2-propenyl)-, acetate	206	C12H14O3	NIST12
14	78	Phenol, 2-methoxy-6-(2-propenyl)- $$ o-Ally	164	C10H12O2	NIST62
15	77	3-Allyl-6-methoxyphenol $$ Phenol, 2-meth	164	C10H12O2	NIST62
16	76	Phenol, 2-methoxy-6-(1-propenyl)- $$ Phen	164	C10H12O2	NIST62
17	76	Eugenol	164	C10H12O2	NIST12
18	75	Carbofuran	221	C12H15NO3	NIST12
19	74	Phenol, 2-methoxy-4-(1-propenyl)-	164	C10H12O2	NIST12
20	73	Phenol, 2-methoxy-4-(1-propenyl)-	164	C10H12O2	NIST12
21	71	7-Benzofuranol, 2,3-dihydro-2,2-dimethyl-	164	C10H12O2	NIST12
22	71	3-Octen-5-yne, 2,2,7,7-tetramethyl-, [E]-	164	C10H12O2	NIST62
23	71	Benzene, 4-ethenyl-1,2-dimethoxy- $$ 3,4-D	164	C10H12O2	NIST62
24	71	5-Decen-3-yne, 2,2-dimethyl-, [Z]-	164	C10H12O2	NIST62
25	71	3-Octen-5-yne, 2,2,7,7-tetramethyl-	164	C10H12O2	NIST62

FIGURE 24.18 Hit list from a mass spectral library search of a major clove oil component.

a broad band of medium intensity at 3300 cm^{-1}. Determine the structure of the unknown. What fragmentations lead to the peaks at m/z 120, 91, and 44?

7. The exact m/z of a sample of aspirin was determined to be 180.0422. What is the molecular formula that corresponds to this exact mass? Show your calculation.

8. A sample of clove oil was analyzed by GC-MS. A search of the mass spectral library for a match to the mass spectrum of the major component of clove oil produced the hit list shown in Figure 24.18. A computer-screen printout comparing the mass spectra of hit 1, hit 2, and hit 6 with the mass spectrum of the major component of clove oil is shown in Figure 24.19. Which of the three hit-list candidates is the best match with the MS of the major component of clove oil? Show your reasoning.

9. For 1,2-dichlorobenzene, calculate the masses (to four significant figures) and relative abundances (to two significant figures) of the molecular ions: M, M+2, and M+4.

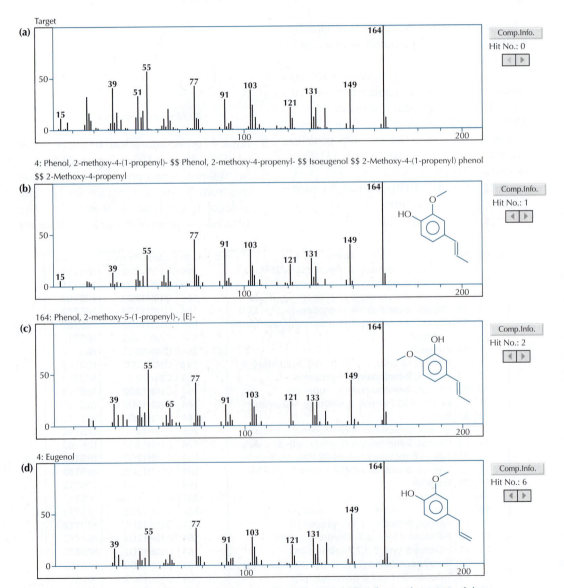

FIGURE 24.19 Computer comparison of three hit-list compounds for clove oil. (a) MS of the compound from the GC-MS run. (b) MS of hit 1. (c) MS of hit 2. (d) MS of hit 6.

25

ULTRAVIOLET AND VISIBLE SPECTROSCOPY

The absorption of ultraviolet (UV) or visible (VIS) light by organic compounds occurs by the excitation of an electron from a bonding or nonbonding molecular orbital to an antibonding molecular orbital. For example, the absorption of UV radiation can promote an electron from a π-bonding orbital into a higher energy π^*-antibonding orbital.

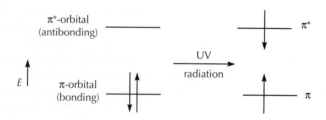

The bonding and antibonding molecular orbitals of a π-system can be depicted as follows:

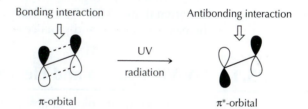

The energy required for this excitation process is comparable to the strength of a chemical bond. All organic compounds absorb UV light, but few commercial spectrometers can effectively scan the wavelengths where C–H, C–C, and nonconjugated C=C bonds absorb, due to interference from strong UV absorption by O_2 and CO_2 in the atmosphere.

The electronic transitions useful in UV spectroscopy involve the absorption of radiation of wavelengths between 200 and 400 *nanometers (nm)* or 200–400 $\times 10^{-9}$ m. In visible spectroscopy, light between 400 and 800 nm is used. In both UV and VIS spectroscopy, an electron from a π-bonding orbital or a nonbonding n-orbital is excited to an antibonding π^*-orbital. We will be concerned only with $\pi \rightarrow \pi^*$ and n $\rightarrow \pi^*$ transitions in conjugated organic compounds because these are the electronic transitions most likely to occur in the 200–800-nm region. Conjugated compounds can have either one or more pairs of alternating double bonds or else a double bond conjugated to a nonbonding pair of electrons on a heteroatom such as oxygen, nitrogen, or halogen.

Until the 1950s, the only physical method readily available to determine the structures of organic compounds was ultraviolet spectroscopy; but now, with the advent of NMR and mass spectrometry, few organic chemists rely on UV as a primary tool for structure determination. However, UV-VIS spectroscopy is an important analytical tool for the quantitative analysis and characterization of organic compounds and is of vital importance in biochemistry. In fact, the Beckman DU spectrophotometer, which became commercially available in 1940, has been cited as one of the most important instruments ever developed for the advancement of the biosciences. It has also been estimated that more than 90% of the analyses performed in clinical laboratories are based on UV-VIS spectroscopy.

The primary application of ultraviolet spectroscopy in modern organic chemistry is chromatographic analysis. High-performance liquid chromatographs (HPLCs) that are equipped with diode-array UV detectors are found in virtually all organic chemistry research labs. UV analysis is also important in the measurement of reaction rates. Students of organic chemistry encounter UV spectroscopy less often than NMR, IR, and MS, but it is important to understand the basic principles and practice of ultraviolet spectroscopy.

25.1   UV-VIS Spectra and Electronic Excitation

UV-VIS spectra are plots of absorbance (A) against the wavelength of the radiation in nanometers. The absorbance is related to concentration by the **Beer-Lambert Law**:

$$A = \log(I^\circ/I) = \varepsilon l c$$

where

> I° is the intensity of the incident light
> I is the intensity of the transmitted light
> ε is the molar extinction coefficient in $M^{-1}cm^{-1}$
> l is the length of the cell path in centimeters
> c is the sample concentration in moles/liter (M)

In UV-VIS spectroscopy, the absorbance is plotted, not the percent transmittance as in IR spectroscopy. Most importantly, the proportionality of the absorbance and the concentration is linear over a wide range of concentrations, making UV-VIS spectroscopy ideal for determining the concentration of a compound. Values of ε, *the molar extinction coefficient*, can vary from 10 to greater than 10^5. Thus, some *chromophores*, the functional groups that absorb ultraviolet or visible light, can absorb far more efficiently than others, by factors of 10^4 or greater.

WORKED EXAMPLE

The principal photoreceptor of most green plants is chlorophyll *a*, which in ether solution has a molar extinction coefficient of 1.11×10^5 cm^{-1}M^{-1} at 428 nm. If the absorbance of pure chlorophyll *a* in a 1.00-cm cell is 0.884 at 428 nm, what is the concentration of chlorophyll in the ether solution?

Use of the Beer-Lambert Law provides the answer.

$A = \varepsilon l c$ or $c = A/\varepsilon l$

$c = 0.884/(1.11 \times 10^5 \text{ cm}^{-1} \text{ M}^{-1} \times 1.00 \text{ cm})$

$c = 7.96 \times 10^{-6} \text{ M}$

π–π* and n–π* Electronic Transitions

As with IR and NMR spectroscopy, the relationship of the frequency of the absorbed radiation to the energy gap (ΔE) in UV spectroscopy is given by Planck's Law:

$$\Delta E = h\nu = hc(1/\lambda)$$

where

h = Planck's constant
c = the speed of light in a vacuum
λ = the wavelength of the radiation that is being absorbed
$\nu = c/\lambda$

The energy gap has an inverse dependence on the wavelength of absorbed light; therefore, the smaller the gap, the longer the wavelength of light that will be absorbed in the electronic transition.

Energy gaps (ΔE) for electronic transitions that occur when UV radiation is absorbed are greater for π–π* than n–π* transitions. Therefore, π–π* transitions occur at shorter wavelengths than n–π* transitions. Figure 25.1a shows the relative energy gaps for π–π* and n–π* transitions, and Figure 25.1b shows a UV spectrum that has absorbance for both types of transitions.

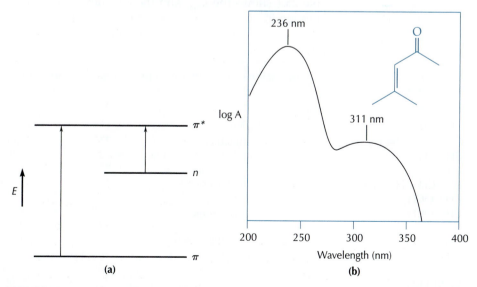

FIGURE 25.1 (a) Electronic transitions between orbital energy levels, illustrating a π–π* transition and a lower-energy n–π* transition. (b) UV spectrum of 4-methyl-3-penten-2-one in ethanol, showing a π–π* transition at λ_{max} = 236 nm and an n–π* transition at λ_{max} = 311 nm.

Electronic transitions occur much faster than the time necessary for a molecule to vibrate or rotate. Therefore, electronic excitation occurs from a range of vibrational and rotational energy levels. For this reason, when UV radiation interacts with a large population of molecules having a variety of vibrational and rotational states, it is absorbed at numerous wavelengths. In general, UV and visible radiation is absorbed in absorption bands rather than at discrete wavelengths. The absorption bands usually have a width of 10 nm or more. The wavelength of a UV absorption band is given by λ_{max}, the wavelength of maximum absorbance.

The π–π* transition at λ_{max} 236 nm in Figure 25.1b has a molar extinction coefficient (ε) of 12,800, whereas the n–π* transition at λ_{max} 311 nm has an ε value of only 59. The intensity of π–π* transitions is virtually always greater than the intensity of n–π* transitions. The n–π* absorbances are much weaker because of the unfavorable spatial orientation of orbitals containing nonbonding electrons relative to the π-orbital, which does not allow much overlap between a nonbonding orbital and the π-orbital. In a quantum mechanical sense, π–π* transitions are "allowed" and n–π* transitions are "forbidden."

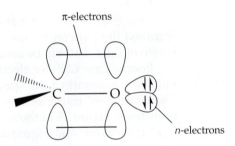

Table 25.1 shows the λ_{max} and the ε values for a variety of organic compounds.

TABLE 25.1   UV data for various functional groups in organic compounds

Compound	Name	λ_{max}[a]	ε_{max}
$CH_2=CH_2$	Ethylene	171	15,500
$CH_2=CH-CH=CH_2$	1,3-Butadiene	217	21,000
$CH_2=CH-C(CH_3)=CH_2$	2-Methyl-1,3-butadiene	222	10,800
C_5H_6	1,3-Cyclopentadiene	239	4,200
$CH_2=CH-CH=CH-CH=CH_2$	1,3,5-Hexatriene	268	36,300
CH_3COCH_3	Acetone	279	13
$CH_3COCH=CH_2$	3-Buten-2-one	217	7,100
		320	21
CH_3CONH_2	Acetamide	220	63
C_6H_6	Benzene	204	7,900
		256	200
$C_6H_5CO_2H$	Benzoic Acid	226	9,800
		272	850

a. All transitions are π–π* except for acetone, acetamide, and the longer wavelength absorption of 3-buten-2-one, which are n–π* transitions.

Conjugated Compounds

A molecular orbital diagram shows why conjugated organic compounds absorb UV radiation of longer wavelengths than nonconjugated compounds do. Ethylene has one bonding π-orbital and one antibonding π*-orbital, whereas 1,3-butadiene has two bonding π-orbitals and two antibonding π*-orbitals. The energy gap between the highest energy π-orbital of 1,3-butadiene and its lowest energy π*-orbital is much smaller than the corresponding gap for the ethylene orbitals.

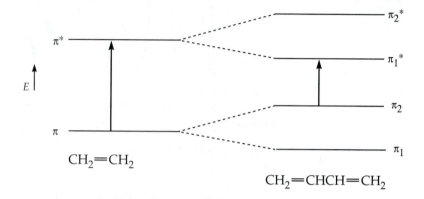

This energy difference produces a shift of the λ_{max} from 171 nm for ethylene to 217 nm for butadiene.

When a molecule contains a benzene ring, with a total of six π- and π*-orbitals, a number of electronic transitions involving similar energy changes can occur. Figure 25.2 shows the complexity of the UV spectrum of toluene ($C_6H_5CH_3$) in the 240–265-nm region.

Visible Spectroscopy

Compounds that absorb visible light are colored. Organic compounds with eight or more conjugated double bonds absorb in the visible region (400–800 nm). One example is chlorophyll *a*, the principal photoreceptor of most green plants. Figure 25.3 shows the visible spectrum of chlorophyll *a*, which has two major absorption peaks: one in the 430-nm region (violet) and the other around 660 nm (red). The exact λ_{max} values depend on the solvent in which the chlorophyll is dissolved. A complementary relationship exists between the color of a compound and the color (or wavelength)

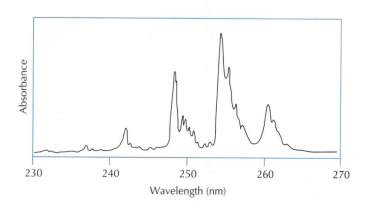

FIGURE 25.2
UV absorption spectrum of toluene.

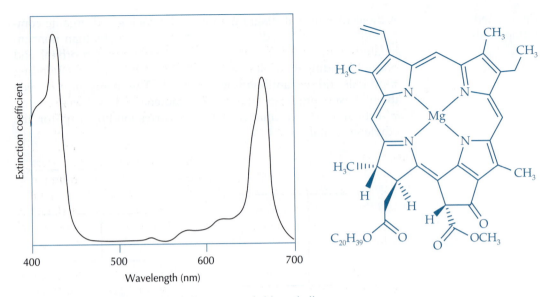

FIGURE 25.3 Visible spectrum and structure of chlorophyll *a*.

of the light it absorbs. The green light waves that are not absorbed effectively by chlorophyll are reflected back to our eyes.

The color wheel in Figure 25.4 provides another way to understand the color of compounds. For example, β-carotene, the compound that gives carrots their color, has an intense absorption in the blue-green portion of the visible spectrum (λ_{max} at 483 nm, $\varepsilon = 1.3 \times 10^5$).

β-Carotene

The color wheel reveals that β-carotene is expected to reflect (transmit) the color on the opposite side of the wheel, that is, red-orange.

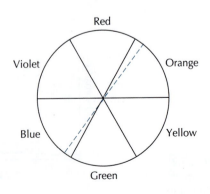

FIGURE 25.4
Color wheel. A compound that absorbs blue-green light transmits red-orange light.

25.2 UV-VIS Instrumentation

There are two major classes of UV-VIS spectrometers: *dispersive* and *multiplex diode-array* spectrometers. Dispersive spectrometers were the standard UV-VIS instruments for many years. More recently, diode-array spectrometers have become increasingly popular.

The light source in both dispersive and diode-array spectrometers is either a deuterium (D_2) discharge lamp, used for the 190–350-nm region of the spectrum, or a tungsten-halogen filament lamp, used for the 330–800-nm region of the spectrum. The tungsten-halogen lamp operates much like an incandescent tungsten light bulb. In the deuterium lamp, an electric discharge is passed through D_2 gas, which is under pressure; the gas is excited and continuous UV radiation is emitted. Often, UV-VIS spectrophotometers are equipped with both deuterium and tungsten lamps, which can be turned on or off by the flick of a switch.

Dispersive UV Spectrometers

There are a number of instrumental designs for dispersive UV spectrometers, involving mirrors, slits, and detectors. Instrumental analysis textbooks treat these designs in considerable detail. Dispersive instruments use either a single-beam or a double-beam light pathway. In both types, the light passes through a monochromator, which scans through narrow bands of separate light frequencies. In a double-beam spectrometer (Figure 25.5), after passing through the monochromator, the radiation is split into two beams and then directed by mirrors through sample and reference cells. The two beams are recombined later in the optical path. Double-beam instruments can compensate for fluctuations in the radiant output of the light source. They work well for the continuous recording of spectra.

Detectors for dispersive UV-VIS spectrometers are either photocells or photomultipliers. A photocell is the simplest kind of detector. It has a metal surface that is sensitive to light, and when radiation hits it, electrons are ejected and then converted into a signal. The Spectronic 20 is a single-beam instrument with a tungsten lamp and a photocell detector. When radiation strikes a photoreactive metal surface in a photomultiplier tube, electrons are ejected and directed to positively charged electrodes, called dynodes, which cause several more electrons to be emitted. This cascading process can be repeated

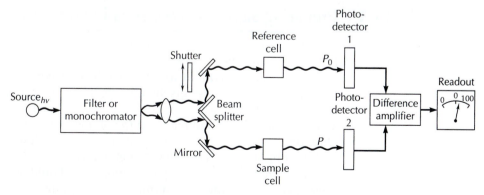

FIGURE 25.5 Schematic diagram of a double-beam spectrometer.

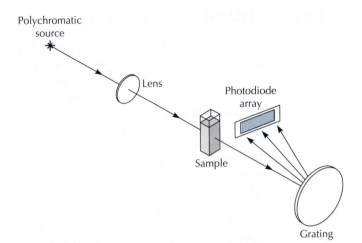

Polychromatic source

Lens

Photodiode array

Sample

Grating

FIGURE 25.6
Diagram of a
multichannel
spectrophotometer
based on a grating
and a diode-array
transducer.

several times and can lead to a great enhancement in sensitivity, up to 10^9 electrons per photon of radiation hitting the detector.

Sample concentrations obtained from UV measurements on dispersive spectrometers are often more accurate when they are obtained at the wavelength of the maximum absorbance. There is usually a small flat portion at maximum absorbance that reduces experimental error. In addition, the change in absorbance with concentration is greatest at λ_{max}.

Diode-Array UV-VIS Spectrometers

Diode-array spectrometers do not use a monochromator to scan the radiation before it passes through the sample cell. Instead, a range of light wavelengths pass through the sample. Diode-array spectrometers are single-beam instruments, which use a diffraction grating to disperse the different wavelengths of light after the light has passed through the sample. All the wavelengths are detected simultaneously on a linear array of photoreactive diodes. An electrical potential at each diode element can be converted into a digital signal. Usually, a diode-array detector has 1000–2000 elements, and each element covers a small wavelength region of the UV-VIS spectrum. Figure 25.6 is a diagram of a diode-array spectrophotometer.

25.3   Preparing Samples and Operating the Spectrometer

UV-VIS spectroscopy is often sensitive to 10^{-4}–10^{-5} M concentrations with good accuracy. Relative errors are ~1–3%; with precautions, they can be reduced to a few tenths of a percent. To obtain accurate quantitative results in UV spectroscopy, careful sample preparation is vital. The samples must be accurately weighed on an analytical balance and made up to volume in a volumetric flask. Dilutions are made by removing aliquots with volumetric pipets and diluting them in separate volumetric flasks.

After preparing your sample solution, you should obtain a complete spectrum in order to determine the wavelengths of maximum absorbance. Consult your instructor about specific operating procedures for the UV-VIS spectrometer in your laboratory.

Calibration Standards

It is important to run a set of calibration standards to ensure that the concentrations of the compounds you are working with adhere to the Beer-Lambert Law. These calibrations should be carried out under conditions where the measured absorbance is less than 1.0 and definitely no greater than $A = 2.0$. The molar absorptivity (ε) should be determined experimentally in the solvent you choose to use. The best accuracy is obtained with dilute solutions, with concentrations less than 0.01 M.

Solvents

Except when working with water, make up solutions for UV-VIS spectroscopy in a fume hood.

The solvents used in preparing solutions to be analyzed must be spectral grade. Even very small quantities of organic impurities that have high molar absorptivities can produce erroneous results.

Polar solvents stabilize antibonding π^*-orbitals more than π ground states, so the energy gap in $\pi-\pi^*$ transitions is decreased, and λ_{max} occurs at a longer wavelength in polar solvents. For $n-\pi^*$ transitions, the effect of hydrogen-bonding polar solvents is just the opposite. The energy levels of n electrons are stabilized more by hydrogen bonding than π^*-orbitals are, so the gap between n and π^* becomes greater and λ_{max} occurs at a shorter wavelength in polar hydrogen-bonding solvents. Figure 25.7 summarizes these solvent effects on the shifts of λ_{max} for $\pi-\pi^*$ and $n-\pi^*$ transitions.

Table 25.2 provides the cutoff wavelengths for standard UV solvents; below these wavelengths, the solvent absorption interferes

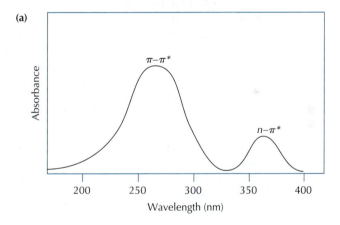

(a)

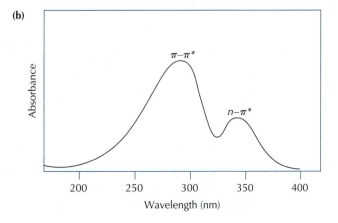

(b)

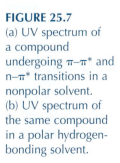

FIGURE 25.7
(a) UV spectrum of a compound undergoing $\pi-\pi^*$ and $n-\pi^*$ transitions in a nonpolar solvent.
(b) UV spectrum of the same compound in a polar hydrogen-bonding solvent.

T A B L E 2 5 . 2	Minimum wavelengths possible for use of standard UV solvents[a]
Solvent	**Low-end cutoff (nm)**
Acetonitrile	210
Cyclohexane	210
Dichloromethane	235
1,4-Dioxane	220
Ethanol	210
Hexane	220
Methanol	210
Isooctane	220
Water	205

a. These solvents can be used from the low-end cutoff up to 800 nm.

with the measurements. For example, cyclohexane can be used as a solvent from 400 nm down to 210 nm, whereas dichloromethane is not useful below 235 nm.

UV Cells

Good-quality UV transparent cells, or cuvettes, are generally made from quartz glass or fused silica and are 1.0 cm square. These cells are transparent above 200 nm and require about 3 mL of solution. Usually they come with fitted caps.

Clean cells are crucial. Before using cells with your samples, rinse them several times with pure solvent and check for absorption. Fingerprints or grease on the transparent cell surfaces must be avoided. UV cells should never be dried in an oven, as the heat may warp them. In addition, you should never use solutions of strong bases, which may etch the glass. If a double-beam spectrophotometer is being used, the reference and sample cells should be a matched pair, which allows any small solvent absorption to be erased so that it doesn't interfere with the spectral measurements. After the spectrum has been obtained, quartz cells should be cleaned immediately, usually by repeated rinsing with the solvent used for the spectrum. After rinsing, the cells should be set upside down to dry on a clean cloth or tissue.

If the wavelength region being utilized is 300–800 nm, disposable acrylic cells can be used. Below 300 nm, these disposable cells absorb too strongly to be useful.

25.4   Sources of Confusion and Common Pitfalls

Major pitfalls occur from faulty sample preparation and from incorrect use of the UV-VIS spectrometer.

My Results from One Measurement to the Next Are Not Consistent. What Am I Doing Wrong?

A number of factors can result in poor and inconsistent results including:

Dirty cells. Cells that have traces of UV-absorbing substances on their surfaces, including grease and fingerprints, can produce confusing results.

Impure solvents. If solvents are not of spectral quality or if dirty volumetric glassware is used, poor-quality spectra will be obtained.

Dust particles in the sample. If you are monitoring the continuous change in absorbance of a sample over time (for instance, in measuring the kinetics of a reaction), it is always a good idea to filter your sample solutions to remove dust particles. Dust particles in a sample will cause unpredictable changes in absorbance as they float through the light beam, making for poor and often unusable data.

Why Is My Beer-Lambert Calibration Plot Not Linear?

The nonlinearity is probably caused by using a concentration above 0.01 M for the compounds under investigation. In addition, the relationship of absorbance to concentration can become nonlinear if the measured absorbance is too high, above $A = 1.0 - 2.0$.

Why Doesn't My Molar Absorptivity ε Confirm the Published Value?

The exact value for the molar absorptivity can depend on a number of environmental factors, including the solvent used, the temperature, and other substances that may also be present in the sample solution.

Further Reading

Skoog, D. A.; Holler, F. J.; Crouch, S. R. *Principles of Instrumental Analysis*, 6th ed.; Brooks Cole: 2007.

Questions

1. Acetaldehyde shows two UV bands, one with a λ_{max} of 289 nm ($\varepsilon = 12$) and one with a λ_{max} of 182 nm ($\varepsilon = 10{,}000$). Which is the n $\rightarrow \pi^*$ transition and which is the $\pi \rightarrow \pi^*$ transition? Explain your reasoning.
2. It should not be surprising to find that cyclohexane and ethanol are reasonable UV solvents, whereas toluene is not. Why?
3. An ethanol solution of 3.50 mg/100 mL of compound X (150. g/mol) in a 1.00-cm quartz cell has an absorbance (A) of 0.972 at λ_{max} of 235 nm. Calculate its molar extinction coefficient.

4. Benzene shows more than one UV maximum. Use the orbital energy levels shown here to explain this observation.

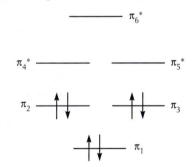

INTEGRATED SPECTROMETRY PROBLEMS

The three major spectrometric methods presented in Chapters 21–24 have revolutionized structure determinations of organic compounds. Although for the most part these methods have been considered separately, the connections were made apparent from time to time. In practice, organic chemists generally solve structural problems by using an integrated spectroscopic approach. The mass spectrum is usually a good starting point because it can provide the molecular weight, and even the molecular formula, of the compound. Next comes the IR spectrum, which provides data for the identification of the functional groups present. Finally, interpretation of the ^{1}H and ^{13}C NMR spectra usually allows the structural analysis to be completed.

Many chemists believe that NMR is the most versatile source of structural data, and we have emphasized it more than infrared spectroscopy and mass spectrometry. However, to be efficient in tackling structure determinations, organic chemists need to be proficient in all three spectroscopic methods. One method may reveal features about a compound that are not clear from another. Researchers are alert to when extra emphasis should be placed on a few pieces of data chosen from a large data set, a skill that comes from experience. The following problems highlight the use of an integrated approach to using spectrometry for organic structure determination.

1. A compound shows a molecular ion peak in its mass spectrum at m/z 72 and the base peak at m/z 43. An infrared spectrum of this compound shows, among other absorptions, four bands in the 2990–2850-cm^{-1} range and a strong band at 1715 cm^{-1}. There are no IR peaks at greater than 3000 cm^{-1}. The ^{1}H NMR spectrum contains a triplet at 1.08 ppm (3H), a singlet at 2.15 ppm (3H), and a quartet at 2.45 ppm (2H). The magnitudes of the splitting of both the quartet and triplet are identical. Deduce the structure of this compound and assign all the MS, IR, and NMR peaks.

2. The infrared spectrum of a compound is shown in Figure 26.1. Its ^{1}H NMR spectrum contains a somewhat broadened singlet at 7.3 ppm (5H), a singlet at 4.65 ppm (2H), and a broadened singlet at 2.5 ppm (1H). Deduce the structure of this compound and assign the NMR and important IR peaks.

3. A compound shows a molecular ion peak in its mass spectrum at m/z 92 and a satellite peak at m/z 94 that is 32% the intensity of the m/z 92 peak. The ^{1}H NMR spectrum contains only one signal, a singlet at 1.65 ppm. The proton-decoupled ^{13}C NMR spectrum reveals a strong peak at 35 ppm and a weaker peak at 67 ppm. Deduce the structure of this compound and assign all MS and NMR peaks.

4. The mass spectrum, infrared spectrum, and ^{1}H NMR spectrum of a compound are shown in Figures 26.2–26.4. Deduce the

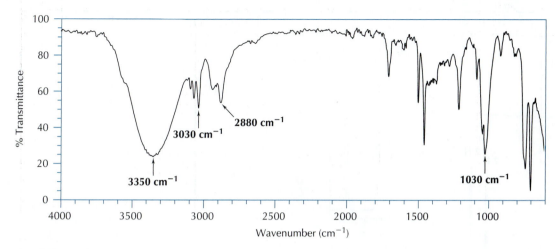

FIGURE 26.1 Infrared spectrum (thin film) of unknown compound for Problem 2.

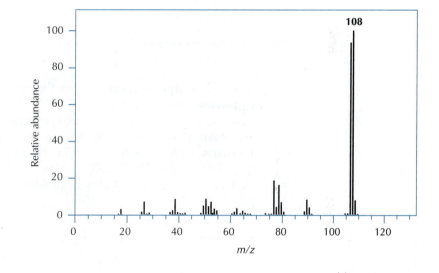

FIGURE 26.2
Mass spectrum of
unknown compound
for Problem 4.

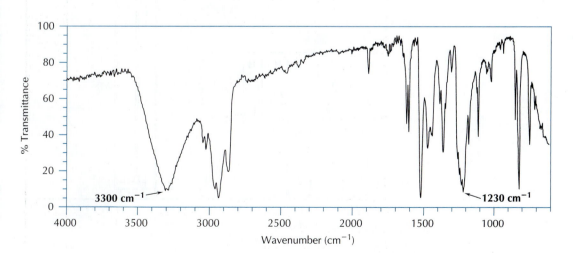

FIGURE 26.3 Infrared spectrum (KBr pellet) of unknown compound for Problem 4.

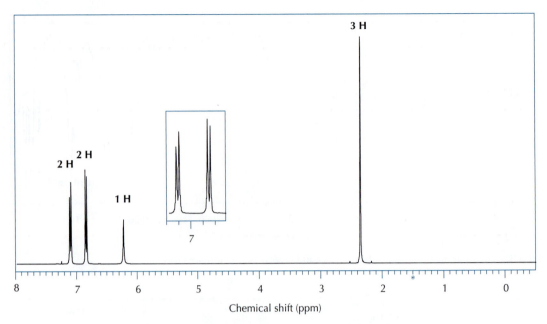

FIGURE 26.4 360-MHz ¹H NMR spectrum of unknown compound for Problem 4.

structure of this compound from the spectral data and show your reasoning.

5. Figure 26.5 shows the ¹H NMR spectrum of a compound of molecular formula $C_3H_6Cl_2$. The proton-decoupled ¹³C NMR spectrum, which has signals at 56, 49, and 23 ppm, as well as the DEPT(90) and DEPT(135) spectra, are shown in Figure 26.6. Deduce the structure of this compound and assign all NMR peaks.

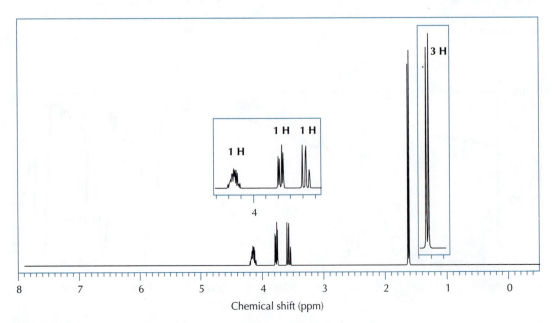

FIGURE 26.5 360-MHz ¹H NMR spectrum of unknown compound for Problem 5.

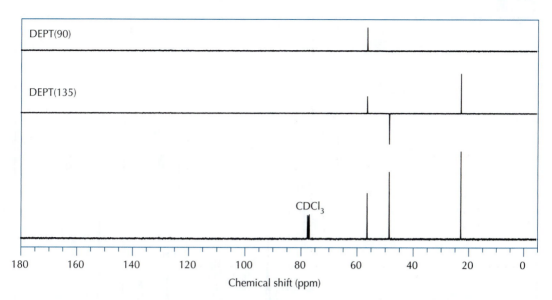

FIGURE 26.6 90-MHz ^{13}C NMR, DEPT(90), and DEPT(135) spectra of unknown compound for Problem 5.

6. The infrared spectrum of a compound of molecular formula $C_7H_{16}O$ is shown in Figure 26.7. Its 360-MHz ^{1}H NMR spectrum is shown in Figure 26.8, and its proton-decoupled ^{13}C NMR and DEPT(135) spectra are shown in Figure 26.9. The ^{13}C NMR spectrum has signals at 63, 33, 32, 29, 26, 23, and 14 ppm. Deduce the structure of this compound, assign all the NMR and important IR peaks, and explain your reasoning. Estimate the chemical shifts of all protons and carbon atoms, using Tables 22.3 and 23.3, and compare these shifts with the chemical shifts measured from the NMR spectra.

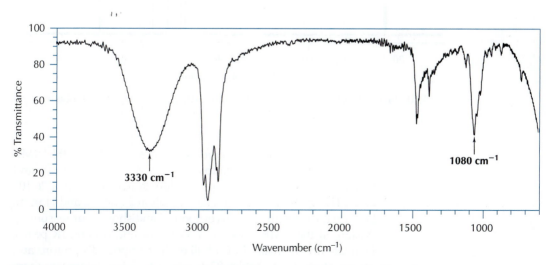

FIGURE 26.7 Infrared spectrum (thin film) of unknown compound for Problem 6.

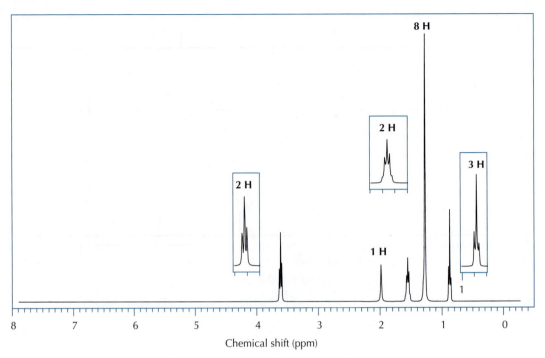

FIGURE 26.8 360-MHz ¹H NMR spectrum of $C_7H_{16}O$ for Problem 6.

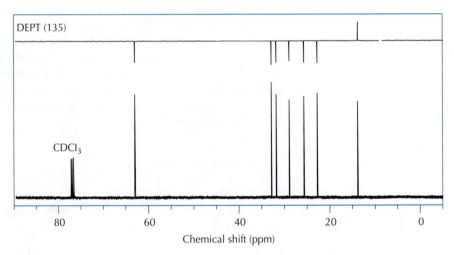

FIGURE 26.9 90-MHz ¹³C NMR and DEPT(135) spectra of $C_7H_{16}O$ for Problem 6.

7. The mass spectrum, infrared spectrum, 360-MHz ¹H NMR spectrum, and proton-decoupled ¹³C NMR and DEPT(90) and DEPT(135) spectra of a compound are shown in Figures 26.10–26.13. The ¹³C NMR spectrum has signals at 45, 44, 24, and 11 ppm. Deduce the structure of this compound and show your reasoning. Assign all the NMR peaks and all important MS and IR peaks. Estimate the chemical shifts of all of the compound's protons and carbon atoms using Tables 22.3 and 23.3–23.4 and compare them with the chemical shifts measured from the NMR spectra.

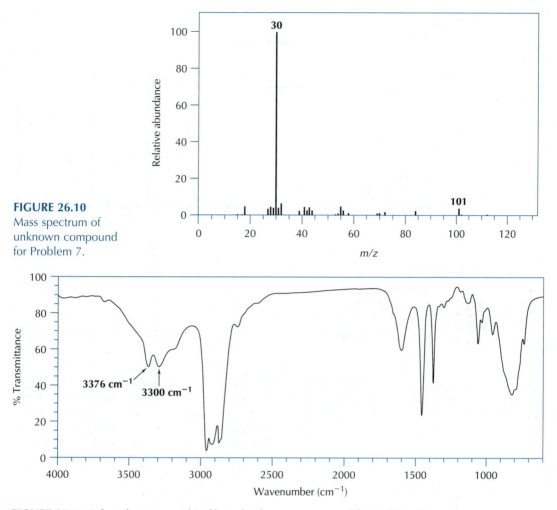

FIGURE 26.10
Mass spectrum of
unknown compound
for Problem 7.

FIGURE 26.11 Infrared spectrum (thin film) of unknown compound for Problem 7.

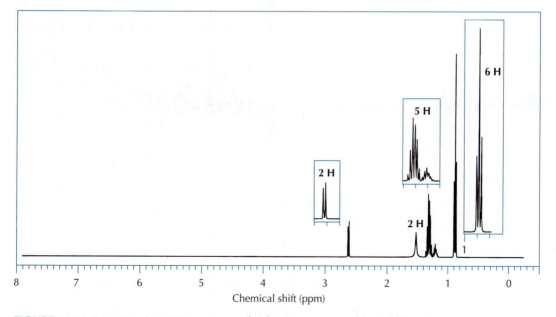

FIGURE 26.12 360-MHz ^{1}H NMR spectrum of unknown compound for Problem 7.

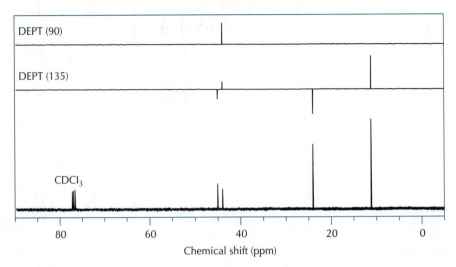

FIGURE 26.13 90-MHz ^{13}C NMR, DEPT(90), and DEPT(135) spectra of unknown compound for Problem 7.

8. The mass spectrum, infrared spectrum, 360-MHz ^{1}H NMR spectrum, and proton-decoupled ^{13}C NMR and DEPT(90) and DEPT(135) spectra of a compound are shown in Figures 26.14–26.17. The ^{13}C NMR spectrum has signals at 173, 132, 118, 65, 36, 19, and 14 ppm. The molecular ion is not discernible in the mass spectrum. Deduce the structure of this compound and show your reasoning. Assign all the NMR peaks and all important MS and IR peaks.

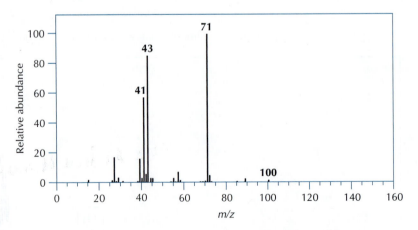

FIGURE 26.14 Mass spectrum of unknown compound for Problem 8.

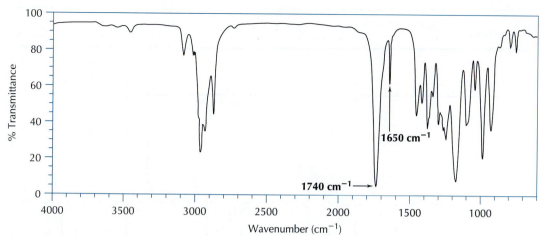

FIGURE 26.15 Infrared spectrum (thin film) of unknown compound for Problem 8.

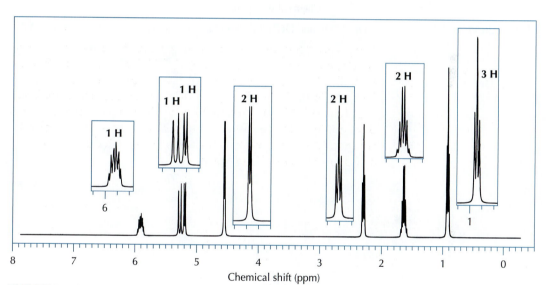

FIGURE 26.16 360-MHz ^{1}H NMR spectrum of unknown compound for Problem 8.

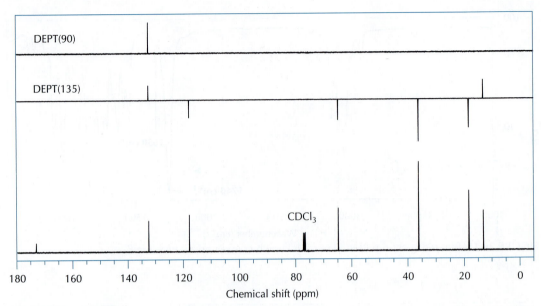

FIGURE 26.17 90-MHz ^{13}C NMR, DEPT(90), and DEPT(135) spectra of unknown compound for Problem 8.

PART 6

Designing and Carrying Out Organic Experiments

Essay—Inquiry-Driven Lab Experiments

Whereas learning in the classroom or lecture hall usually focuses on answers, science is fundamentally about questions, and the laboratory is where you learn the process of science. As the essay for Part 1 explained, the laboratory is where we learn "how we know what we know." The essay also mentioned that you may have opportunities in the lab to test your own ideas by designing new experiments. Along with making sense of experimental results, this is the basis of inquiry-driven lab work.

There are two kinds of inquiry-driven experiments: guided-inquiry and design-based experiments. In guided-inquiry, or discovery experiments, a question worthy of investigation is addressed, such as, "Among the possible outcomes, what products are formed in the reaction we carried out and what mechanism will allow us to understand how they are formed?" After carrying out the reactions, you draw conclusions that account for your experimental data in order to answer the question. You may have had opportunities already for guided-inquiry laboratory work.

Organic synthesis is usually the goal of design-based experiments. The question at the heart of these experiments is, "How can I synthesize the compound that I want to make?" In design-based experiments you will develop procedures for organic syntheses, carry out the reactions you have proposed, and determine the results of your experiments. Chapter 27 provides important practical advice on how to design your own experiments.

In both guided-inquiry and design-based experiments, you need to think about how to do experimental chemistry and how to interpret your results. These open-ended experiments involve problem solving and critical evaluation of experimental data, often in multiweek projects where you are part of a team. In fact, modern chemistry is most often done by teams of scientists. Modern instrumental methods and spectroscopic analysis will help you to evaluate your results. By carrying out inquiry-driven experiments, you will be participating in an important process, the same one carried out by chemical researchers. You will be learning chemistry the way chemists do it.

What You Will Encounter in Inquiry-Driven Lab Experiments

By taking part in a real-world lab experience, you will have to make many decisions on how to proceed. This problem solving will help you to develop your critical-thinking skills. With the help of the other members of your team, you will discuss how your experiments will be devised and have the opportunity to use the skills you have developed earlier in the organic lab. The satisfaction of applying the skills you have learned while you work with other students is like icing on the cake. Whatever results you get from your experiments, it is a win–win situation.

Your instructor will be more of a coach or a director who helps to set the stage, and your team will be encouraged to ask questions and find answers on its own. The instructor will not just be an answer-giver but will guide you to develop your own answers while probing to determine if your plans are likely to succeed and if your interpretations are consistent with the experimental evidence.

After your team has worked through the design of your proposed experiments and then the interpretation of your data, there will likely be post-lab presentations and discussions where you will present your experimental plans or your group's interpretations of the data and results. This will usually take place before your lab reports are due. It is the way science is done—experiments are first carefully planned and carried out, and then there is a research conference where the results are discussed. Only after these stages are completed does a scientist write a report, which may eventually lead to a publication that is shared with the wider scientific community.

Advantages to You and Your Classmates

The procedures used in inquiry-driven experiments are quite different from those used in verification experiments, where you are told what result you should find and your job is to show that you found it. You need to be much more fully engaged in inquiry-driven experiments, and there are several advantages for you in the process. Perhaps most important is that you are developing critical-thinking skills by using this effective learning environment, adding substantially to your education. The inquiry-driven format also allows you to tie together separate topics that you have studied. Your integration of knowledge is a powerful added benefit. Last but not least, inquiry-driven experiments are an excellent platform for success in undergraduate research, which has become a crucial part of a science graduate's CV.

In conclusion, it is worth reading statements from three students who have done guided-inquiry experiments in the past.

I learned how to analyze IR, NMR, and GC-MS simultaneously to determine the actual product formed, and that A + B doesn't always equal C; and that's OK.

I learned a lot from the post-lab discussion. I wish this was possible for each of our experiments. During lab sessions we were usually very busy trying to complete the experiments on time. It was nice to have a day where we could reflect and talk about what we learned.

I learned how chemists fit together a puzzle of data to make conclusions, and how hard it can be to do that. It was an incredible critical-thinking exercise.

27

DESIGNING CHEMICAL REACTIONS

*If Chapter 27 is
your introduction
to design-based
laboratory
experiments,
read the Essay
"Inquiry-Driven Lab
Experiments" on
pages 485–487 before
you read Chapter 27.*

As you gain experience in organic chemistry, you may have the opportunity to plan and carry out a chemical reaction based on a procedure taken from the chemical literature. You will find that literature procedures are written tersely, often with little or no explanation of the steps. In some journal articles, no experimental procedures are provided at all. The purpose of this chapter is to help you read between the lines in an experimental procedure, and to help you plan, adapt, and carry out a chemical reaction when you are not given explicit experimental directions.

Often a design-based project will focus on the synthesis of a specific target compound and you will work with a small team to design the synthetic strategy, carry out the necessary chemical reactions, purify the product(s), and demonstrate that you have synthesized what you set out to make. At the outset you will need a published synthesis, which you may be able to find by searching the chemical literature (see Chapter 28). If you cannot find one, you can look for a synthesis of a structurally similar compound to use as a guide. The material presented here is designed to provide practical advice for planning a synthesis procedure using precedents from the literature.

27.1 Reading Between the Lines: Carrying Out Reactions Based on Literature Procedures

Designing a reaction and carrying it out is often done by a small team of students and it begins with a foray into the chemical literature to find a precedent for what you hope to do. Chapter 28 provides information about using the chemical literature. Once you have identified a method for carrying out a reaction, you will need to understand the procedure thoroughly before you attempt to implement it. Discussions with other team members can be invaluable in reaching this understanding. Much is left unsaid in literature procedures, so do not be afraid to ask questions when you don't understand something. That said, you may be able to deduce much about the procedure and fill in the blanks if you approach it systematically. A reasonable approach to interpreting and implementing a literature reaction procedure is presented here.

1. Begin by analyzing the reaction. Write out the balanced equation, showing all by-products and solvents to be used, and determine which reactant is the limiting reagent.
2. Determine the potential hazards of the reagents to be used and what safety measures must be taken when setting up the reaction.
3. Consider the properties of the reagents and products of the reaction. Are any water or air sensitive (see Sections 7.3 and 7.4), or do they require any other special handling?
4. Determine how the reagents and solvents will be measured and transferred.
5. Determine how the reaction progress will be monitored.

6. Understand the steps in the workup, or devise a workup based on the balanced reaction if none is provided.
7. Determine the method to be used to purify the product.
8. Determine how you will analyze the reaction product to be sure the reaction has succeeded.

Sample Procedure from the Literature

Suppose you want to carry out the amide bond-forming reaction between the carboxylic acid group of N-acetylalanine (**1**) and the amine group of N-methylbenzylamine (**2**), to form N-acetylalanyl-N-methylbenzylamide (**3**) (equation 1). Because the names of organic compounds can be rather cumbersome, you will see that in nearly all journal articles numbers are assigned to the compounds discussed; the numbers, rather than the names, are generally used in the text to refer to the compounds. So, in this example, the goal is to synthesize amide **3**.

(– H$_2$O) (1)

1	**2**	**3**
N-Acetylalanine	N-Methylbenzylamine	N-Acetylalanyl-N-methylbenzylamide

N-Methylbenzylamine (**2**) is a secondary amine and the condensing reagent BOP-Cl is known to be a particularly good reagent for forming amide bonds between carboxylic acids and secondary amines.[1] Further literature investigation leads to a 1990 article by Colucci et al. on the synthesis of a derivative of an immunosuppressive agent, which provides a general procedure for the use of BOP-Cl for the formation of amide bonds between a range of carboxylic acids and secondary amines, as shown in equation 2.

General Procedure for Using BOP-Cl as a Condensing Reagent[2]

(BOP–Cl) (2)

Et$_3$N, CH$_2$Cl$_2$, 0 °C

Carboxylic acid	Secondary amine	Amide product

A solution of a carboxylic acid (11 mmol) and the secondary amine (10 mmol) in dichloromethane (50 mL) was cooled in an ice bath with stirring under an inert atmosphere. The cold mixture was treated with triethylamine (22 mmol), followed, in one portion,

[1] Tung, R. D.; Rich, D. H. *J. Am. Chem. Soc.* **1985**, 107, 4342–4343.
[2] This procedure is adapted from Colucci, W. J.; Tung, R. D.; Petri, J. A.; Rich, D. H. *J. Org. Chem.* **1990**, 55, 2895–2903.

by BOP-Cl (11 mmol). The mixture was stirred in the cold until judged complete by TLC analysis (SiO$_2$, 5% CH$_3$OH in CH$_2$Cl$_2$), and then it was poured into ether (3 × the reaction volume) and water (2 × the reaction volume). The organic layer was separated, and then washed with 10% aqueous KHSO$_4$, water, saturated aqueous NaHCO$_3$, water, and brine. After drying over MgSO$_4$, it was concentrated *in vacuo* and the residue was purified by flash chromatography.

Analyzing the Reaction

The first step in the analysis of the procedure is to write out the balanced equation, which may require working through the mechanism of the reaction. Most articles do not provide balanced equations or mechanisms for the reactions discussed. It is simply assumed that the reader knows the mechanisms or can figure them out. You are probably not familiar with how BOP-Cl works, but the Colluci article cites the original paper[3] describing BOP-Cl, where a partial mechanism is sketched out. Using this information, with perhaps a little help from your instructor, you can write out a likely mechanism and the balanced equation shown in Figure 27.1. Comparing the balanced reaction and the molar amounts of the reagents specified in the procedure reveals that the secondary amine (**2** in your reaction) is the limiting reagent.

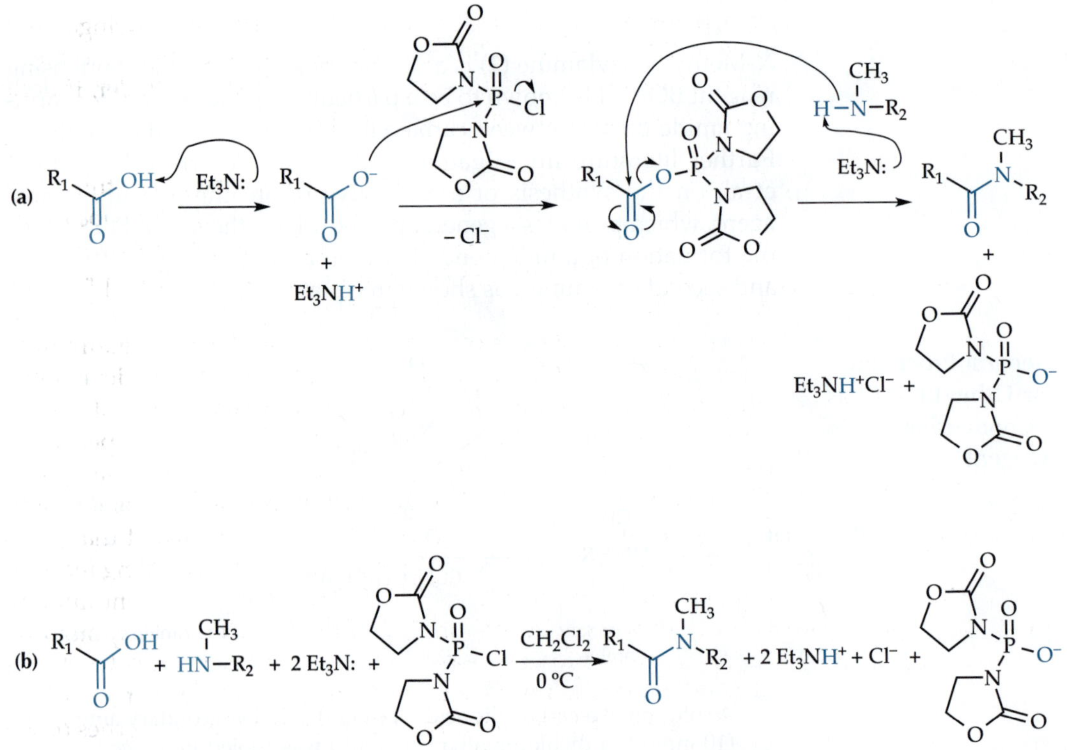

FIGURE 27.1
(a) Mechanism of the BOP-Cl mediated condensation reaction.
(b) Balanced equation for the condensation reaction.

[3] Diago-Meseguer, J.; Palomo-Coll, A. L.; Fernàdez-Lizarbe, J. R.; Zugaza-Bilbao, A. *Synthesis* **1980**, 547–551.

Identifying and Addressing Safety Concerns

Safety issues must be addressed before setting up the reaction. The MSDSs (see Section 1.7) for the reactants and solvent indicate that BOP-Cl is corrosive, *N*-methylbenzylamine is corrosive and an allergen, *N*-acetylalanine is a skin irritant, triethylamine is corrosive and toxic, and dichloromethane is a suspected carcinogen and skin irritant. Clearly, gloves should be worn while working with these chemicals and the reaction should be performed in a fume hood.

The Reaction Conditions

The procedure calls for running this reaction under an inert atmosphere. Information in the MSDS for BOP-Cl provides a clue as to why this is necessary—BOP-Cl is moisture sensitive and should be stored under an inert atmosphere. Water readily reacts with it to displace the Cl leaving group. Thus, you should run this reaction under anhydrous conditions at least (see Section 7.3), if not under a nitrogen or argon atmosphere (see Section 7.4). It would also make sense to use anhydrous dichloromethane (see Section 7.3) as the solvent.

Handling and Transferring the Reagents

What equipment will be needed to transfer the reagents to the reaction flask? *N*-Acetylalanine and BOP-Cl are solids and can be measured and transferred by the usual methods for solids (see Section 5.2). Triethylamine, *N*-methylbenzylamine, and dichloromethane are liquids and can be transferred using syringes (see Section 7.4), although the large volume of dichloromethane might be more easily transferred using a dry graduated cylinder, if done quickly to minimize exposure to moist air.

Monitoring the Reaction Progress

The procedure suggests monitoring the reaction by thin-layer chromatography (TLC) on silica gel. Section 18.8 describes how to use TLC to monitor a reaction and suggests following the consumption of the limiting reagent as the reaction proceeds. *N*-Methylbenzylamine (**2**) is the limiting reagent, so it would make sense to monitor its disappearance over time. The procedure indicates that 5% CH_3OH in CH_2Cl_2 was used as the TLC developing solvent. Keep in mind that your reactants, **1** and **2**, are not exactly the same reactants that were investigated in the Colucci paper. Thus, before beginning the reaction you will need to analyze a sample of *N*-methylbenzylamine by TLC in order to verify that this developing solvent will be effective (see Section 18.7). A solvent that gives an R_f between 0.2 and 0.7 usually works well for monitoring reaction progress, and the suggested 5% CH_3OH in CH_2Cl_2 solvent mixture is likely to work for monitoring **2**. If the CH_3OH/CH_2Cl_2 mixture, however, proves to give poor elution of **2**, other polar solvent systems (secondary amines are quite polar) should be investigated.

Generally, it is also useful to determine the TLC properties of the nonlimiting reactants so that you will know where to expect to see their spots when analyzing the reaction mixture. In this reaction, the carboxylic acid reactant **1** will be deprotonated by the basic triethylamine, present in excess in the reaction mixture. The resulting ionic salt will elute very poorly and will likely co-spot with other reaction components as a smeared-out spot at the origin of the developed

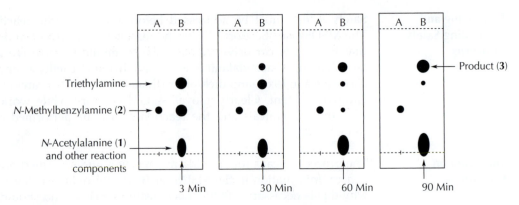

FIGURE 27.2 Progress of the reaction as monitored periodically. In each TLC plate, lane A is pure *N*-methylbenzylamine (**2**), the limiting reagent, and lane B is the reaction mixture. By 90 minutes, the limiting reagent has been consumed.

TLC plate (Figure 27.2). The BOP-Cl is also very polar and is likely to react with traces of water absorbed on the silica, so it and its phosphinate by-product will be components of the baseline spot as well.

The procedure does not indicate how the TLC spots should be visualized, so you will have to experiment with visualization techniques (see Section 18.4). The presence of the aromatic ring in **2** suggests that UV light might be an effective visualization method. Alternatively, dipping reagents, such as phosphomolybdic acid in ethanol (see Section 18.4), can be tested. A quick Google search of "TLC visualization reagents for amines" turns up pages of sources, providing the recipes for many visualization reagents. A ninhydrin-based dipping reagent appears on almost every page as a good choice for visualizing amines.

Soon after you begin the reaction, spot the reaction mixture on a TLC plate in one lane alongside the sample of *N*-methylbenzylamine (**2**) in another lane. After developing and visualizing, you should also see a spot for unreacted **2** in Lane B, lined up at the same R_f as the spot for **2** in Lane A (Figure 27.2). It is possible that the reaction product or other by-product might have the same R_f as the *N*-methylbenzylamine. This is always a danger and it may require that you investigate other developing solvents. Once you have settled on TLC conditions, you can periodically analyze the reaction mixture by TLC, looking for disappearance of the limiting reagent.

Understanding the Reaction Workup

Once the reaction is complete, the procedure involves adding the reaction mixture to ether and water and then separating the layers. Why add ether when dichloromethane is a nonpolar solvent and separates from water just fine? A seasoned synthetic organic chemist would immediately suspect that the ether is added strictly as a matter of convenience. Ether is less dense than water, while dichloromethane is more dense than water. The reaction workup calls for repeated washings (see Section 10.3) of the organic layer. If diethyl ether were not added to the reaction mixture, the dichloromethane solution would be the bottom layer in each of the extractions. This is inconvenient because after each extraction, the lower organic layer would have to be drained from the separatory funnel, the upper aqueous layer discarded, and the organic layer returned to the separatory funnel. If the reaction mixture is diluted with sufficient ether, the organic phase will be the upper layer in each extraction. Thus, the organic layer can remain in the separatory funnel throughout the successive washes, saving considerable time and minimizing losses and opportunities for spilling the reaction mixture.

The reaction workup looks rather daunting and you may wonder why there are so many aqueous washes (10% aqueous $KHSO_4$, water, saturated aqueous $NaHCO_3$, water, and finally brine). The goal of any workup is to remove as many of the reaction by-products and unreacted reagents as possible before the final purification. Most workup extractions involve aqueous solutions of differing pH and rely on the ionization of acidic or basic species in the crude reaction mixture in order to make them water soluble. To understand the workup in this procedure we need to look to the balanced reaction equation (Figure 27.1b) and consider the pK_a values of all ionizable species, as shown in Figure 27.3.

Knowing the approximate pK_a values allows us to understand each of the aqueous extraction steps in the workup:

- **10% aqueous $KHSO_4$:** The purpose of this mild acid wash (the pK_a of HSO_4^- is ~2) is to remove the basic components, triethylamine and traces of unreacted secondary amine, from the reaction mixture. The pH of a 10% $KHSO_4$ solution is about 1, well below the pK_a values of the carboxylic acid and the ammonium salts. At this pH the positively charged ammonium salt components will be extracted into the aqueous layer.
- **Water:** The purpose of this wash is to remove any lingering ionic species in the separatory funnel created by the acid wash, and to remove traces of $KHSO_4$ left behind. Removing the acidic $KHSO_4$ will minimize vigorous bubbling of CO_2 in the next step—a bicarbonate wash.
- **Saturated aqueous $NaHCO_3$:** With a pH of ~8, this basic solution will react with the excess carboxylic acid reactant and any remaining phosphinate by-product by forming water-soluble salts, which will be extracted into the aqueous layer. After this wash, the bulk of all ionizable species in Figure 27.1 will have been removed from the organic layer.
- **Water:** The second water wash is done to remove traces of the bicarbonate and any remaining traces of ionic species in the separatory funnel.

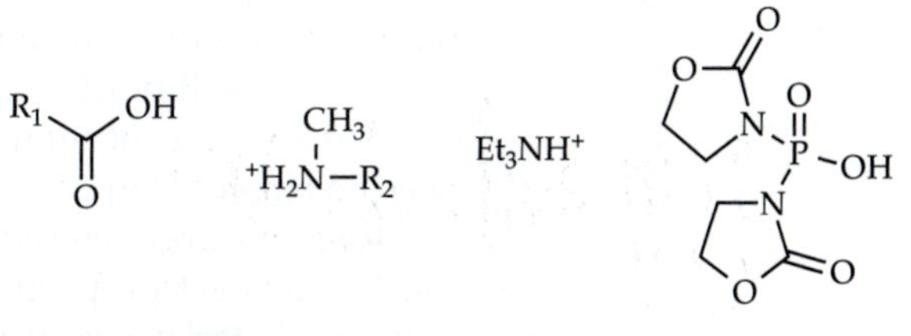

| Approximate pK_a values: | ~5 | ~10 | ~10 | ~2 |

FIGURE 27.3 Approximate pK_a values for the conjugate acid forms of the ionizable species appearing in the coupling-reaction balanced equation.

- **Brine**: The purpose of the final brine (saturated aqueous NaCl) wash is to pull water out of the organic layer. Dichloromethane, a significant part of the organic layer, is particularly prone to forming emulsions (see Section 10.3), and after all the aqueous washes, the organic layer may be quite cloudy. The brine wash will help remove the suspended water from the organic phase.

The final step of the workup calls for drying the organic layer over MgSO₄ and then concentrating *in vacuo*. The drying of the organic layer is straightforward (see Chapter 11), but what is not explicitly stated is that after drying, the drying agent is filtered from the organic layer. Concentrating the layer *in vacuo* simply means that the solvents were removed at reduced pressure by means of a rotary evaporator (see Section 11.4), leaving the crude product residue ready for final purification.

The Final Purification and Analysis of the Product

After the workup, the procedure calls for purification of the crude product by flash chromatography, but no conditions are given, presumably because the conditions varied widely for the many products prepared by this general procedure. Chapter 19 on liquid chromatography outlines how to determine the proper flash chromatography conditions for purifying a crude reaction product. Once pure, NMR (see Chapter 22) and IR (see Chapter 21) spectroscopy would be good techniques for the analysis of the final product.

Reaction processing steps will differ from reaction to reaction, but this example demonstrates the thinking process involved in working though a procedure that is new to you. The key to understanding and working with an experimental procedure is to understand the reaction and the chemical and physical properties of all the chemicals involved. If you have a good handle on the chemistry, you can fill in the missing steps in procedures that are brief and incomplete, or even devise a procedure for a reaction for which you have no procedure at all.

27.2   Modifying the Scale of a Reaction

Very often, a synthesis procedure found in the literature does not prepare the amount of compound that you wish to make. Methods from literature published prior to the 1960s and those found in *Organic Syntheses* are usually on a larger scale than most of the reactions carried out in the modern organic chemistry laboratory. These procedures will need to be scaled down. Conversely, if a synthetic method is of recent vintage, it may be on the microscale level and need to be scaled up.

At first approximation, the scale-up or scale-down is simply a matter of direct proportionality. If a procedure produces only one-half the material you want, the quantities of all the reagents and solvents should be doubled to produce enough of the product. If the amount you want is only one-tenth the amount produced in the procedure, divide the quantities of all the reagents and solvents by 10.

When scaling up or down by a large factor, however, the simple proportionality often needs to be adjusted for some of the reaction components, particularly the solvent volumes.

Also keep in mind that many published synthetic procedures report optimum product yields that were achieved only after a number of iterations. The yield on the first attempt was likely less than that reported, perhaps only 50% as much. If you propose to carry out a synthesis in three steps, lower yields may result by a factor of $50\% \times 50\% \times 50\% = 13\%$ of what has been reported. When a reaction procedure looks particularly challenging, it can be useful to try it out on a smaller scale before attempting it on the scale you need.

Before implementing any reaction procedure from the literature, you will need to analyze it, as discussed in Section 27.1. In addition, adapting a procedure for a new scale requires special attention to the following details:

- Amount of solvent to use and the size of the reaction apparatus
- How the reagents will be added
- How to determine the reaction time
- How to provide temperature control
- Adjustments to the workup and final purification

Amount of Reaction Solvent and Size of the Apparatus

In scaling down a large-scale reaction to miniscale or microscale, reducing the solvent volume by the same factor you are using to reduce the reagents may result in very small volumes of solvents. Often the scaled-down solvent volume is too small, especially for an effective reflux of a reaction mixture that needs to be heated. With reflux, almost all the solvent might vaporize, leaving little or none for dissolving the reaction components and for providing a constant reaction temperature. In such cases, extra solvent must be used for the scaled-down reaction. Conversely, when scaling up a microscale reaction by a large factor, the proportion of solvent can often be decreased, thus avoiding the use of extremely large volumes of solvent, which can be cumbersome to handle and can lead to increased waste disposal costs.

In general, the capacity of the reaction flask should be two to three times the total combined volumes of the reagents and solvents. This practice allows for the usual increase in volume if a mixture is to be heated, and it allows room for vaporization of the solvent during reflux. If the mixture is known to foam during reflux or if a gas is evolved during the reaction, a flask five or more times the volume of the reaction mixture is recommended.

Addition of Reagents

Some reactions give optimal results if one of the reagents is added gradually to the reaction mixture. With large-scale reactions, this slow addition is best accomplished using a dropping funnel for solutions and liquid reagents. For both miniscale and microscale reactions, the most convenient method is to use a syringe or Pasteur pipet to gradually drip the reagent into the reaction mixture—either directly into the flask or through the reflux condenser in a heated reaction. Care must be taken not to lose too much of the reagent on the walls of the condenser. If the reaction system is sealed to isolate it from the atmosphere, a liquid reagent or solution should be added from a syringe through a rubber septum (see Section 7.5).

Reaction Time

The time required for a scaled-up or scaled-down reaction should be approximately the same as that for the model reaction if the reactant concentrations remain the same. That being said, there can be great variation in optimal reaction times due to many variables that cannot be scaled along with the reagents, for example, heating or cooling efficiency. Because the reaction time can be unpredictable, it is a good idea to monitor the reaction progress, usually by thin-layer chromatography (see Chapter 18 and Section 27.1). Sometimes visual clues can be used to decide when to stop a reaction, for example, color change, disappearance of a solid, or appearance of a solid.

Temperature Control

Many chemical reactions require heating or cooling, and the particular equipment used for temperature control will depend on the scale of the reaction. See Chapter 6 for descriptions of the equipment available for heating and cooling miniscale and microscale reactions.

Extra care should be taken when exothermic reactions are scaled up from the microscale level. A small amount of heat evolved in a microscale reaction can turn into quite a lot of heat when the reaction is scaled up. For miniscale and microscale reactions, the surface area of the apparatus often provides efficient and rapid transfer of heat to the surrounding atmosphere and a literature procedure may not call for special cooling of a small-scale reaction mixture; however, scaling up the procedure may require cooling the reaction flask with a water or ice-water bath (see Section 6.4). Another method for controlling the temperature of exothermic reactions is by slow addition of one of the reagents to the stirred reaction mixture. If the reaction becomes too vigorous, addition can be stopped or slowed until the reaction rate subsides.

Because heat transfer is more efficient on a small scale, the reaction apparatus can often be simplified when scaling down a reaction. Scaling down may obviate the need for a cooling bath. In reactions that need to be refluxed, a microscale reaction might be carried out in a simple 20 × 150 mm test tube, where the wall of the test tube provides sufficient surface area for condensing the refluxing solvent.

Workup and Final Purification of the Products

Changing the scale of a reaction may require changes in the workup or purification of the product. Most standard aqueous extraction steps of crude reaction mixtures (see Section 27.1) can be carried out at all scales equally well, although the size of the glassware needs to be adjusted. Sometimes workup steps can be omitted, however, when working on a very small scale because the total volume of reaction by-products is so small. For instance, when a final purification is to be done by liquid chromatography (see Chapter 19), you can often omit the workup altogether and after solvent removal simply use liquid chromatography to purify the crude reaction product. The published procedures for many modern small-scale reactions do not call for any workup steps other than flash chromatography.

Because purifying by recrystallization works well at a range of reaction scales, working with small quantities of solids can be easier than working with small quantities of liquids. However, you will need to scale down the size of flasks and vacuum funnels when carrying out a recrystallization of less than 300 mg of a solid (see Section 15.6).

If you have scaled down a reaction that will produce less than 5 g of a liquid product, which is to be purified by distillation, you need to use a short-path distillation apparatus with a cow receiver and a conical distillation flask; standard taper 14/20 ground glass joints are preferable (see Section 12.3b). If you scale down a reaction even more, to produce less than 1 g of a liquid product, distillation becomes impractical. Purification by column chromatography (see Chapter 19) is often the method of choice to purify products at these small scales, whether they are liquids or solids. On the other hand, if the product is a volatile liquid, the risk of product loss due to evaporation makes column chromatography unattractive. Downsizing such a reaction to a scale where distillation becomes difficult may simply not be practical.

27.3   Case Study: Synthesis of a Solvatochromic Dye

Over 30 years ago the synthesis of a dye whose color changes dramatically when the solvent is changed was published in the *Journal of Chemical Education*.[4] This property, called solvatochromism, is not uncommon in ultraviolet and visible spectroscopy and is demonstrated in Figure 25.7. The change in solvent polarity causes a solvatochromic compound to change color. The dye—given the acronym MOED—is reported to be yellow in water solution, red in ethanol, and violet in acetone. Solvatochromism has potential applications in molecular electronics for the construction of molecular switches.

Procedure for the Synthesis of MOED

MOED
1-Methyl-4-[(oxocyclohexadienylidene)ethylidene]-1,4-dihydropyridine

1,4-Dimethylpyridinium iodide (28.4 g, 0.12 mol), freshly recrystallized (EtOH-H$_2$O, 2:1), 4-hydroxybenzaldehyde (14.5 g, 0.12 mol), and piperidine (10 mL, 0.10 mole) are dissolved in 150 mL dry ethanol and heated at reflux for 24 h. Cooling the reaction mixture yields a red precipitate, which is removed by filtration. This solid

[4] Minch, M. J.; Shah, S. S. *J. Chem. Educ.* **1977**, *54*, 709.

is suspended in 700 mL of 0.2 M KOH and heated (without boiling) for 30 min. The cool solution yields blue-red crystals, which are recrystallized three times from hot water. Yield: 22 g (86.3%), mp 220°C.

Analyzing the Procedure

1,4-Dimethylpyridinium iodide is commercially available from the Aldrich Chemical Company; however, it can also readily be synthesized from methyl iodide and 4-methylpyridine, as outlined in the published article. If you read the article by Minch and Shah in the *Journal of Chemical Education*, you might notice that there is no mention of safety considerations. This omission would not happen today, when we have learned to respect the toxicities of organic compounds. Methyl iodide, which is used in the synthesis of 1,4-dimethylpyridinium iodide, is very toxic and must be handled with caution.

Although nothing is stated in the procedure about the purification of 4-hydroxybenzaldehyde, it is well known that aldehydes undergo free-radical oxidation in the presence of molecular oxygen. Therefore, it would be best to use a new bottle of 4-hydroxybenzaldehyde that has not been opened to the atmosphere many times before. If the only available stock is an old bottle, it would be wise to take an infrared spectrum of it to make sure it has not been oxidized to 4-hydroxybenzoic acid. If oxidation has occurred, not only will the amount of available limiting reagent be reduced, which will lower the percentage yield, but 4-hydroxybenzoic acid will react with piperidine in an acid/base reaction, thereby removing some of the active catalyst.

Even though you might expect that a reaction procedure would be optimized when it is published in the *Journal of Chemical Education* and designed to be carried out by undergraduate students, it is always a good idea to check the literature cited in the article to see what conditions were used by others. For example, in the 1949 *Journal of Organic Chemistry* reference by Phillips, the heating period was only one to four hours in methanol.[5] Following the course of the reaction by thin-layer chromatography would be useful.

Scale of the Reaction

The scale of the MOED synthesis needs to be reduced to be useful in a laboratory with microscale glassware. This scale makes sense because the solutions of MOED used to study the color variation in different solvents are very dilute (5×10^{-5} M). Only a few milligrams of MOED are needed for each color experiment.

A reaction scale appropriate for microscale equipment might be one-hundredth that described in the *Journal of Chemical Education* article. The amounts of reagents would be 1,4-dimethylpyridinium iodide, 0.284 g, 1.2 mmol; 4-hydroxybenzaldehyde, 0.145 g, 1.2 mmol; and piperidine, 0.10 mL, 1.0 mmol. The amount of ethanol that would be used might be increased from the proportionate amount used for the larger scale reaction to allow for a proportionately larger vapor volume (Section 27.2); perhaps 2–4 mL of ethanol should be used. The appropriate size vessel for this microscale reaction would be a 10-mL flask.

---

[5] Phillips, A. P. *J. Org. Chem.* **1949**, *14*, 302–305.

Solvatochromism depends on the difference in dipole moments of
the MOED molecule in its ground state and excited state (see the
discussion of solvents in Section 25.3). The authors of the *Journal of
Chemical Education* article suggest that color changes are most strik-
ing when aqueous solutions of MOED in 0.01 *M* NaOH are diluted
with various portions of an organic co-solvent, producing colors
that vary across the whole visible spectrum.

An interesting path for the exploration of this synthesis might be
to use a different hydroxybenzaldehyde. One obvious molecule to
consider is 4-hydroxy-3-methoxybenzaldehyde (vanillin). Numerous 4-hydroxybenzaldehydes are available from chemical suppliers
as alternative substrates. Another path of exploration might be to
use 2-hydroxybenzaldehydes.

27.4 Case Study: Oxidation of a Secondary Alcohol to a Ketone

One experiment found in virtually all organic chemistry laboratory
programs 25 years ago was the oxidation of a secondary alcohol to a
ketone with chromium(VI), usually in the form of CrO_3 or $Na_2Cr_2O_7$.
This kind of experiment had been widely used in organic chemistry
labs since the 1940s.

In 1980 Stevens, Chapman, and Weller reported in the *Journal of
Organic Chemistry* that using "swimming pool chlorine" as the oxidizing agent is a convenient and inexpensive method of producing
ketones in good yields from secondary alcohols.[6] One of the authors
of this book was teaching a junior-level synthesis course at Carleton
College at that time and decided to use the experimental procedure
from the *Journal of Organic Chemistry* article as a way to engage
students by using synthetic reactions from the primary chemical
literature. The students were given the following procedure and no
other advice except to scale down the reaction by tenfold and use
magnetic rather than mechanical stirring.

Cyclohexanol (99.0 g, 0.988 mol) was dissolved in glacial acetic
acid (660 mL) in a 2-L three-necked flask fitted with a mechanical
stirring apparatus and thermometer. Aqueous sodium hypochlorite (660 mL of 1.80 M solution, 1.19 mol) was added one drop at
a time over 1 h. The reaction was cooled in an ice bath to maintain
the temperature in the 15°–25°C range. The mixture was stirred
for 1 h after the addition was complete. A potassium iodide–starch
test was positive. Saturated aqueous sodium bisulfite solution
(3 mL) was added until the color of the reaction mixture changed
from yellow to white and the potassium iodide–starch test was
negative. The mixture was then poured into an ice/brine mixture
(2 L) and extracted six times with ether. The organic layer was

[6] Stevens, R. V.; Chapman, K. T.; Weller, H. N. *J. Org. Chem.* **1980**, *45*, 2030–2032.

washed with aqueous sodium hydroxide (5% by weight) until the aqueous layer was basic (pH test paper). The aqueous washes were then combined and extracted five times with ether. The ether layers were combined and dried over magnesium sulfate. The ether was distilled through a 30-in Vigreux column until less than 300 mL of solution remained. The remainder was fractionally distilled through a 12-in Vigreux column. After a forerun of ether, cyclohexanone (bp 155°C) was distilled to give 92.9 g (95.8%) of a colorless liquid that had ^{1}H NMR and IR spectra and GC retention time identical with those of an authentic sample.

Experimental Results

The following week 10 students reported their results to one another. The results were not encouraging. Every student had an intense, broad peak in the O–H stretching region (~2800 cm^{-1}) of the infrared spectrum. After careful examination of their experimental results, the students realized that their product contained a significant amount of acetic acid, which had been the reaction solvent. The students had the opportunity to repeat the reaction and everyone got a high yield of pure cyclohexanone.

The problem that every student experienced in the first trial had been an incomplete extraction of acetic acid from ether into the aqueous layer. Although they had neutralized the last aqueous wash with 5% NaOH, earlier aqueous washes were still acidic. Even though the experimental procedure from the *Journal of Organic Chemistry* was more complete than many others in chemistry journals, there was still some ambiguity in the details. This situation was a classic case of the necessity of reading between the lines. To get a pure product, all the aqueous washes had to be made basic with NaOH solution before the extractions with ether were performed.

Based on their results with the NaOCl oxidation of cyclohexanol, the team recrafted the reaction to one that was less expensive and far safer and greener.[7] First, the "swimming pool chlorine," which cannot be stored from one class to the next, was replaced by household bleach (5.25% NaOCl). Then the reaction was carried out in a stirred water/cyclohexanol mixture with only enough acetic acid to provide the appropriate pH for the oxidation to proceed. The workup eliminated the need for ether extractions by using a steam distillation to separate the cyclohexanone product from the water/salt mixture. The bleach oxidation of secondary alcohols replaced the old Cr(VI) method in virtually all undergraduate organic laboratories. The development of greener reactions in academic labs has continued and the bleach oxidation of alcohols is being replaced by the use of Oxone®, which is a safer oxidant (see Section 2.3).

Further Reading

Leonard, J.; Lygo, B.; Procter, G. *Advanced Practical Organic Chemistry*, 3rd ed., CRC Press, Taylor & Francis Group: Boca Raton, Florida, 2013.

Pirrung, M. C. *The Synthetic Organic Chemist's Companion*, Wiley: Hoboken, New Jersey, 2007.

[7] Mohrig, J. R.; Mahaffy, P. G.; Nienhuis, D. M.; Linck, C. F.; Van Zoeren, C.; Fox, B. G. *J. Chem. Educ.* **1985**, *62*, 519–521.

28

USING THE LITERATURE OF ORGANIC CHEMISTRY

There is a maxim in experimental chemistry: "An hour in the library is worth a day in the laboratory." Because electronic access to chemical information has revolutionized the process of searching and retrieving the chemical literature, the words "at the computer" can largely replace "in the library"; nevertheless, the point remains valid. Experiments consume time and resources, and repeating the mistakes of an earlier researcher is counterproductive and wasteful. Before attempting any laboratory work, you should search the chemical literature for examples of the reaction you wish to carry out. If it is a chemical synthesis, you may find several different methods for preparing the desired compound or one similar to it. Compare the various methods critically and carefully in terms of scale, availability of starting materials, availability and complexity of equipment, ease of workup, and safety issues.

The body of published work accumulated over more than a century of organic chemistry is extensive and many thousands of articles are added to it each year. The vast majority of reference material is contained in primary sources, such as journal articles. These are useful references because they describe experimental details for new methods and procedures. The secondary literature consists of review articles and reference books, which compile related material from different sources. They are useful for developing proficiency in a new area of chemistry and finding well-tested synthetic procedures. Secondary sources are often helpful in the initial stages of a project, but eventually most projects will require consulting the primary literature. Identifying and accessing the information you need from this formidable array of sources is daunting, but the tools and strategies described in this chapter will make it easier. It is also helpful to talk with people at your institution, such as faculty members and librarians, who have experience with this task.

28.1 The Literature of Organic Chemistry

Traditionally, paper-based information has been accumulated by and stored in libraries as journals, textbooks, reference books, and indexes. Although most of these continue to be printed, new and updated volumes are also usually available in electronic format. Many libraries no longer collect the printed versions of some sources and subscribe to electronic versions instead. In addition, the older literature is being digitized for electronic access. Electronic access is easy and efficient, as long as your institution has subscriptions to the needed information. Your library will also have printed versions of many useful handbooks, reference books, and journals.

Journals

Journal articles that include complete experimental details, either within the article or as supplementary material, are among the best sources for experimental procedures. Review articles that appear in journals describe a collection of data with many useful references for efficient updating in an area of chemistry. Journals published by the following institutions are highly reputable and the publisher websites generally also provide a means for searching the journals by subject, title, author, and date of publication. In order to access the full articles (not just the abstracts), your institution must have a subscription or you must pay a fee for each article you access.

The American Chemical Society (ACS) (pubs.acs.org). The premier ACS journal is the *Journal of the American Chemical Society (JACS)*, which covers all areas of chemistry. The *Journal of Organic Chemistry (JOC)* and *Organic Letters* are also excellent primary sources of information in organic chemistry. *Accounts of Chemical Research* publishes short, easy-to-read review articles. *Chemical and Engineering News (C&EN)* is a weekly publication that, in addition to several in-depth articles, provides short descriptions of interesting developments in a wide range of topics in research, government, and industry. It is a great place to learn what is new in chemistry and to brainstorm research project ideas.

The Royal Society of Chemistry (RSC) (pubs.rsc.org). *Chemical Communications* contains brief articles with special merit from all fields of chemistry; *Green Chemistry* and *Organic and Biomolecular Chemistry* are self-explanatory. Historically, articles in organic chemistry were published in the *Journal of the Chemical Society, Perkin Transactions*, and the *Journal of the Chemical Society B (Physical Organic)* or *C (Organic)*. *Chemical Society Reviews* is a great source of review articles.

Wiley Online Library (onlinelibrary.wiley.com). In addition to *JACS* and *JOC*, the most important chemical journal for organic chemistry is *Angewandte Chemie International Edition,* published in English by the German Chemical Society and available online through the Wiley website. Every issue begins with a review article, which is written for, and is interesting to, a broad audience. *Chemistry—A European Journal* and the *European Journal of Organic Chemistry* are also good sources.

Thieme Chemistry (www.thieme-chemistry.com). *Synthesis, Synlett,* and *Synfacts* are devoted to chemical synthesis.

Elsevier (www.elsevier.com). *Tetrahedron* and *Tetrahedron Letters* are classic journals in organic chemistry. Procedures published in *Tetrahedron Letters* are cursory, which can make them difficult to reproduce.

Science (www.sciencemag.org) and *Nature* (www.nature.com) are world-leading science journals, and reading them will inform you of the newest developments in science and help you envision possible research directions for the future.

References for Synthesis

Many compounds that you plan to prepare will be described in reference books. Most of these sources compile information about known compounds, including their properties, methods of synthesis, and references to their original descriptions in the primary literature. They can be searched by compound name, Chemical Abstract Service (CAS) number (see Section 1.7) or Beilstein number (Section 28.2), and often by structure. Structure searches utilize Java or other plugins for translating images. Alternatively, you may need to find a procedure for performing a particular transformation, such as allylic oxidation or ring-closing metathesis. Some references are organized by reaction type, enabling precisely this type of search. The following materials have been traditionally published as books, and your library may have copies of them. Most are now available electronically, although the last several listed below are only available in printed form.

Dictionary of Organic Compounds (doc.chemnetbase.com). This easy-to-use database provides information about the properties and key literature citations for over 250,000 organic compounds.

Encyclopedia of Reagents for Organic Synthesis (onlinelibrary.wiley.com). Now available electronically as e-EROS, this contains reviews of thousands of commonly used reagents in organic chemistry.

Fiesers' Reagents for Organic Synthesis (onlinelibrary.wiley.com). This series provides information on improvements in the preparation and purification of organic compounds.

Organic Syntheses (www.orgsyn.org). This is an open-access compilation of carefully checked procedures with full experimental details. The footnotes at the end of each procedure are especially useful. The volumes can be searched using the table of contents, by structure, or by keyword.

Organic Reactions (www.organicreactions.org). Also published as books by Wiley, this series is organized by reaction type. Both a critical discussion of the reaction and a step-by-step guide for performing it are provided.

Comprehensive Organic Name Reactions and Reagents (onlinelibrary.wiley.com). This covers 701 name reactions and reagents. After describing the reaction, it provides mechanisms, applications, step-by-step experimental procedures, and references.

The following books are currently available in printed form only.

Compendium of Organic Synthetic Methods. This book by Michael B. Smith is an encyclopedic collection of tens of thousands of synthetic transformations (Vol. 12, Wiley: 2009).

Comprehensive Organic Transformations: A Guide to Functional Group Preparations. By Richard Larock, this book lists ways to carry out specific classes of reactions for the synthesis of specific functional groups and gives references to the primary journal literature (2nd ed., Wiley: 1999).

Name Reactions and Reagents in Organic Synthesis. This book by Mundy, Ellerd, and Favaloro lists reactions and reagents alphabetically and includes mechanisms, yields, and references to the primary literature (2nd ed., Wiley: 2005).

Protective Groups in Organic Synthesis. By Greene and Wuts, this book describes numerous methods for installing and removing protecting groups (4th ed., Wiley: 2007).

Advanced Textbooks

The following advanced textbooks are good resources for learning how to perform more sophisticated reactions in organic chemistry and purify reagents and solvents, for general synthetic methods, and for deducing or understanding organic reaction mechanisms. They often reference original reports from the primary literature.

1. Anslyn, E. V.; Dougherty, D. A. *Modern Physical Organic Chemistry*; University Science Books: Sausalito, CA, 2006.
2. Bruckner, R. *Advanced Organic Chemistry: Reaction Mechanisms*; Academic Press: New York, 2001.
3. Carey, F. A.; Sundberg, R. J. *Advanced Organic Chemistry. Part A. Structure and Mechanisms; Part B. Reactions and Synthesis*, 5th ed.; Kluwer Academic: New York, 2007.
4. Carruthers, W.; Coldham, I. *Some Modern Methods of Organic Synthesis*, 4th ed.; Cambridge University Press: Cambridge, UK, 2004.
5. Leonard, J.; Lygo, B.; Procter, G. *Advanced Practical Organic Chemistry*, 3rd ed.; CRC Press: Boca Raton, FL, 2013.
6. Lowry, T. H.; Richardson, K. S. *Mechanism and Theory in Organic Chemistry*, 3rd ed.; Harper & Row: New York, 1987.
7. Pirrung, M. C. *The Synthetic Organic Chemist's Companion*; Wiley: New York, 2009.
8. Smith, M. B. *Organic Synthesis*, 2nd ed.; McGraw-Hill: New York, 2002.
9. Smith, M. B. *March's Advanced Organic Chemistry*, 7th ed.; Wiley: New York, 2013.
10. Vogel, A. I.; Tatchell, A. R.; Furnis, B. S.; Hannaford, A. J.; Smith, P. W. G. *Vogel's Textbook of Practical Organic Chemistry, Including Qualitative Organic Analysis*, 5th ed.; Longman Group: London, 1989.
11. Warren, S.; Wyatt, P. *Organic Synthesis: The Disconnection Approach*, 2nd ed.; Wiley: New York, 2008.
12. Zweifel, G. S.; Nantz, M. H. *Modern Organic Synthesis*; W. H. Freeman and Company: New York, 2007.

28.2　Searching the Literature of Organic Chemistry

Because the literature of chemistry is so vast, finding specific information requires systematic searching methods. Traditionally, this has involved using printed abstracts and indexes, which compile and organize references to the published literature. These indexes, such as *Chemical Abstracts*, *Beilstein's Handbook of Organic Chemistry*, and *Science Citation Index*, are now superseded by electronic databases.

It is likely that your institution has access to one or more of the following electronic databases. In addition to what is described here, the reference books listed at the end of this section provide detailed guides for searching the chemical literature.

SciFinder

SciFinder is the electronic interface developed by Chemical Abstract Service (CAS), a division of the American Chemical Society, to provide researchers with the capability for searching Chemical Abstracts and some other index databases. In addition to *SciFinder*, CAS also provides *STN*, another interface for searching these databases. Today, most college and university libraries have subscriptions to *SciFinder* or *STN*. Consult your instructor or librarian to obtain assistance and training before undertaking an online search.

SciFinder searches the following four major databases; *STN* searches the same databases, depending on the package subscribed to by your institution:

- CAplus—the electronic successor to *Chemical Abstracts*. CAplus condenses the content of journal articles into abstracts and indexes the abstracts by research topic, author name, chemical substance or structure, molecular formula, and patent number. These abstracts cover publications from the 1800s through the present.
- CAS Registry—information about more than 73 million chemical substances. Each chemical compound is assigned a number, called a **CAS Registry Number** (often abbreviated as **CAS Number**), which can facilitate finding references to the compound.
- CASREACT—information about more than 67 million reactions and more than 13 million synthetic preparations.
- MEDLINE—a U.S. government database covering the biomedical literature since 1950.

Reaxys

This web-based tool, licensed by Elsevier, is the successor of *Cross-Fire*, which was derived from *Beilstein's Handbook of Organic Chemistry*. *Reaxys* accesses records of over 10 million organic substances. For each compound, the database contains the name(s), formula, physical properties, methods of synthesis, chemical reactions, and biological properties. Every piece of information has a reference to the primary literature so that the data can be checked. The database continues to add information on many compounds that were reported in the earlier print versions of *Beilstein's Handbook*. If *Reaxys* is available at your institution, it is worth learning how to use it effectively. In addition to CAS registry numbers, most handbooks and catalogs from commercial suppliers provide the **Beilstein Registry Numbers** for organic compounds.

Web of Science

Science Citation Index Expanded contains references to all articles published in prominent journals and also lists all the articles that were cited or referred to in current articles. It is available online through the *Web of Science* (part of the *Web of Knowledge*), which can be searched by subject, author, journal, and cited references. Very often you may identify a review article in an area that is exactly related to your project of interest but is several years old. Citation searches are invaluable for finding more recent literature related to the topic. Within *Web of Science*, you can search for the initial review article and then perform a citation search to yield a list of all the publications that have cited it.

The World Wide Web

The World Wide Web (www) is not a scientific database, but it is nonetheless useful when searching for information about chemicals and their hazards, methods of synthesis, and even the chemical literature. Bear in mind that the information published on the World Wide Web is not regulated and has not been peer reviewed. Furthermore, internet search engines are not designed to be comprehensive. Nevertheless, these searches are helpful for finding initial leads, which can then be followed up by searching for and retrieving the original publications from the primary literature.

References for Literature Searching

The following web-based and printed books contain more information about chemistry information sources, how to use them, and how to plan and carry out an online search.

1. Maizell, R. E. *How to Find Chemical Information: A Guide for Practicing Chemists, Educators, and Students*; 4th ed.; Wiley: New York, 2009.
2. Smith, M. B.; March, J. *March's Advanced Organic Chemistry*; 7th ed.; Wiley: New York, 2013, Appendix A.
3. Wiggins, G. D. *Chemical Information Sources*, http://en.wikibooks.org/wiki/Chemical_Information_Sources; 2011–2013.

28.3   Planning a Multistep Synthesis

What has been presented thus far is a brief description of the organic chemistry literature and reference materials, and chemistry search engines. To help you understand how to use these resources to plan a multistep synthesis, consider the following example. Because *SciFinder* is the most powerful tool that you will probably have available for searching the organic chemistry literature, its use is highlighted. Once you identify a synthetic route, you must evaluate it for ease of implementation and safety considerations. This analysis may require you to do further literature searching and revise the plan before you implement it. These steps are also explained in the context of the synthesis.

SciFinder is well documented and the "help" link, found at the top of every page, provides access to extensive help files, including online tutorials. Specific help files can also be accessed via the question mark icons found throughout the *SciFinder* pages. On the left side of the main page are links for selecting the type of search you want to do, with three main types of searches possible—**Research Topic, Substances**, and **Reactions**. Research Topic searches are text-based searches related to a topic, a particular author, a journal article, etc. Search types listed under Substances are useful for finding information about particular compounds. You can search by molecular formula, by chemical name, by a compound's CAS registry number, or based on a drawn chemical structure. A Reaction search allows you to search based on a drawn chemical reaction, which can be limited by specifying various criteria.

Perhaps, the most powerful feature in *SciFinder* is the ability to search by chemical structure. Clicking on the **Chemical Structure** link under Substances opens a structure editor applett with an interface for drawing chemical structures. Once the structure is drawn, you can do an **Exact Structure** search, a **Substructure** search, or a **Similarity** search. As the name implies, the Exact Structure search will find answers that relate to the exact compound drawn. The Substructure search will find answers that relate to any compound that has the substructure you drew embedded within its overall structure. The Similarity search retrieves answers for compounds that are structurally similar to the one you drew.

Identifying a Synthetic Route

As an example of the use of a Chemical Structure search, suppose you, as a member of a team of students, were charged with developing a synthetic route to molecule **1**. Your group quickly identifies the Diels-Alder reaction as a likely synthetic transformation that might be useful in the synthesis of **1**, shown retrosynthetically in equation 1.

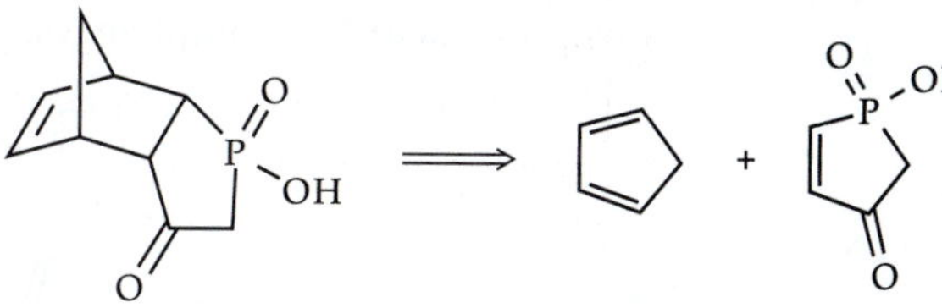

(1)

The Diels-Alder route to **1** rests on the availability of phosphinic acid **3**, or perhaps a phosphinate ester of **3**. Cyclopentadiene (**2**) is readily available, although it does require the distillation of dicyclopentadiene. Inputting the structure of **3** into the *SciFinder* structure applet and running an Exact Structure search does not turn up any answers. However, if you omit the hydroxyl hydrogen atom from the structure and repeat the search as a Substructure search, the program turns up 21 answers, organized with the best fits shown first. The first six of the 21 compounds identified in the search are shown in Figure 28.1.

The first two substances identified, **4** and **5**, look useful. To retrieve the references for these compounds, select **4** and **5** and click the **Get References** link. This results in the retrieval of four

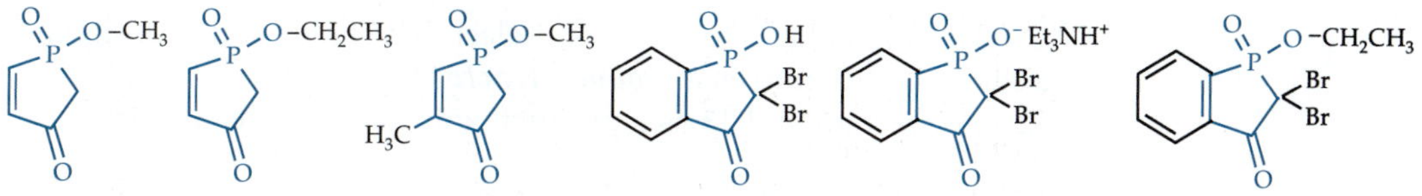

FIGURE 28.1 The first six substances identified in a *SciFinder* Substructure search. The target substructure is shown in blue in each of the compounds identified.

references. If your library has online access to the journals that appear in a reference identified by *SciFinder*, you can click on the **Full Text** link associated with the reference and the program will take you to the electronic version of the article if it is available. In this example, one of the articles[1] identified discusses the use of the methyl ester **4** as the dienophile in a Diels-Alder reaction, which is encouraging; however, this article does not describe the synthesis of methyl phosphinate, **4**. Fortunately, another article retrieved by *SciFinder* provides a synthesis of **4** from **10**, as shown in equation 2.[2] The synthesis of compound **10** is not described in this paper, but the authors cite a paper[3] for its synthesis.

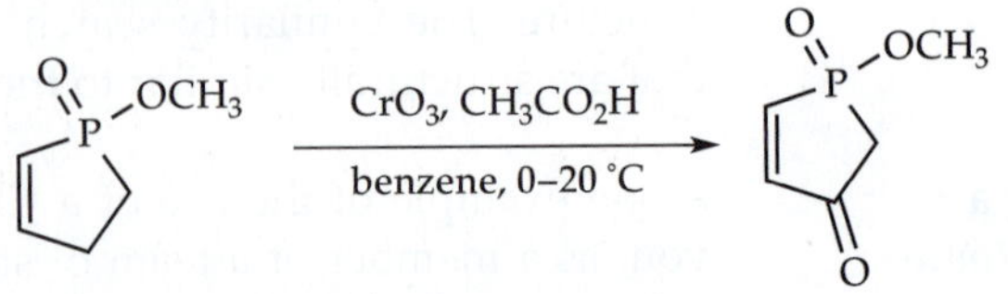

$$\text{(2)}$$

10   **4**

Based on this *SciFinder* search, your group can propose the synthesis shown in Figure 28.2 to give the Diels-Alder adduct **11**. The only step remaining would be to convert methyl ester **11** into the phosphinic acid **1** to complete the synthesis.

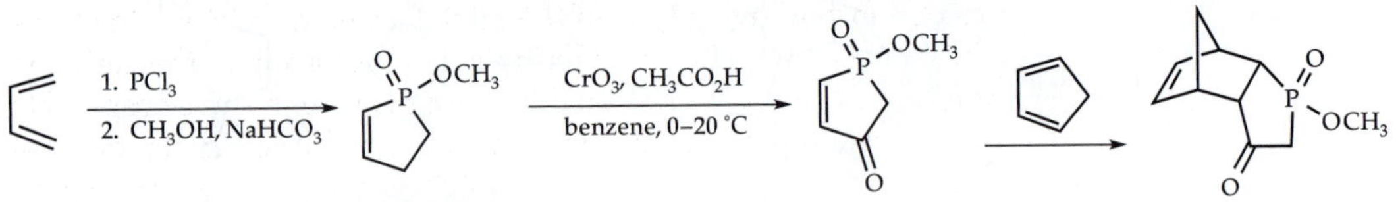

10   **4**   **11**

FIGURE 28.2 A proposed synthesis.

Performing a **Reaction Structure** search in *SciFinder* is one way to look for a method to convert the methyl ester to the acid. To do so, click on the Reaction Structure link to open the structure editor applet. In the editor, you can draw a generic form of a phosphinate methyl ester (**12**) and a corresponding generic form of a phosphinic acid (**13**), as shown in equation 3.

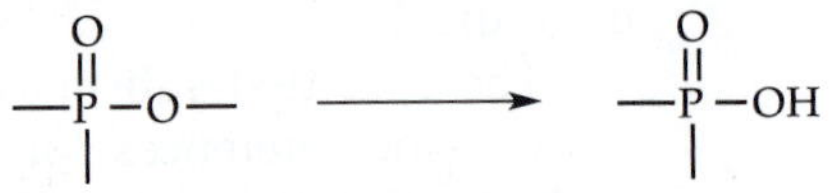

$$\text{(3)}$$

12   **13**

The applet has a tool you can select that allows you to designate a selected structure as a *product, reactant, reagent,* etc. Designating **12** as a reactant and **13** as a product, and selecting the Substructure

[1] Kashman, Y.; Awerbouch, O. *Tetrahedron Lett.* **1973**, 3217–3220.

[2] Reddy, V. K.; Rao, L. N.; Maeda, M.; Haritha, B.; Yamishita, M. *Heteroat. Chem.* **2003**, *14*, 320–325.

[3] Quin, L.D.; Gratz, J.P.; Barket, T.P. *J. Org. Chem.* **1968**, *33*, 1034–1041.

search type, turns up a huge number of possible reactions. Although there are too many answers to sort through them all, you can quickly look at a few to get some ideas. For instance, the first reaction displayed in the search is shown in equation 4.

$$H_3C-\overset{\overset{\displaystyle O}{\|}}{\underset{\underset{\displaystyle CH_2OH}{|}}{P}}-OCH_2CH_3 \quad \xrightarrow[H_2O]{KOH} \quad H_3C-\overset{\overset{\displaystyle O}{\|}}{\underset{\underset{\displaystyle CH_2OH}{|}}{P}}-OH \qquad (4)$$

The entry for this answer indicates that this base-mediated hydrolysis reaction was run at room temperature for 18 hours, and it provides the reference with experimental details.[4] This procedure looks attractive, but the answer set offers many other reagents that have also been reported to hydrolyze phosphinate esters.

On the left side of the *SciFinder* window is a tool that allows you to analyze the results of the search. For instance, you can limit it to particular publication years or a range of years, to a particular journal, etc. In the present example, a useful way to analyze it is by the reagents used to mediate this transformation. Choosing this option provides a list of all the reagents used that turned up in the search, ordered by how often they appear in the answer set. For example, HCl is used in more of the reactions identified than any other reagent; it turns up in 2843 of the 6287 answers found. So, you might also consider an acid hydrolysis of the methyl ester, rather than the base-mediated process. In fact, given that the protons α to the carbonyl group are quite acidic, activated by both the ketone and phosphinate moieties, the hydroxide hydrolysis may not be the best choice because it may produce undesirable by-products. It would be worth exploring a few of the references where acidic conditions were used for this type of reaction for the final step in your proposed synthesis of **1**.

Evaluating the Proposed Route

Before settling on a final synthesis scheme, you will want to consider the practical implications of your proposed route. Your group should look up the MSDSs or SDSs (see page 19) for all of the reagents and solvents required in your synthesis to evaluate potential hazards, as well as to get information about the reagents' physical properties and handling procedures. Your group may also be asked to consider the costs associated with your proposed synthesis. Pricing information can readily be found on the websites of chemical suppliers (see page 15).

An analysis of the route proposed in Figure 28.2 reveals several issues. The first step in the synthesis requires working with butadiene, which is a gas at room temperature and will require special handling techniques. An alternative source of butadiene is butadiene sulfone, which decomposes to butadiene and sulfur dioxide. Depending on the reaction conditions needed for the first step, this may be substituted for butadiene. Furthermore, PCl_3 is toxic, corrosive, and sensitive to water; anhydrous reaction conditions will be required (see Section 7.3).

[4] Chavez, M. R.; Ahao, P.; Zoltan, K.; Sherry, A. D. *Lett. Org. Chem.* **2004**, *1*, 194–200.

The second step requires toxic and environmentally hazardous CrO_3 and carcinogenic benzene; this is a cause for concern, and you should spend more time searching the literature for alternative oxidizing reagents. The book *Comprehensive Organic Transformations: A Guide to Functional Group Preparations,* by Richard Larock (Section 28.1), is a good source for other ways to carry out this allylic oxidation. This book has a **Transformation Index** which tabulates all of the transformations that appear in the book. The index has three columns. In the first column, labeled *To,* generic organic products of the transformation are listed. In the second column, labeled *From,* generic starting materials for the transformation are listed. The third column provides the relevant page numbers where the transformations are described. In the naming system of the Transformation Index, the proper entry for this example is: (To) 2-alkene-1-one, (From) alkene. The third column of this entry lists the pages where you can find reagents, including citations to the original journal articles, which have been used for oxidations similar to the one in the proposed synthesis. A perusal of the relevant pages leads to methods that utilize catalytic amounts of chromium(VI) or Se(IV) complexes in conjunction with more benign oxidants; in addition, there are several metal-free oxidants that accomplish similar transformations. These all would be more environmentally benign alternatives to the originally published procedure. Because this edition by Larock was published in 1999, you could perform a citation search on one or more of the listed journal articles to discover newer methods.

Summary

This example outlined a general approach to a complex synthesis—using *SciFinder*, articles from the primary literature, and an organic synthesis reference book—to plan the synthesis of a desired target molecule. A structure search using *SciFinder* is usually the first, most useful place to begin when planning a multistep synthesis. Once you obtain some likely routes, peruse the references identified in *Scifinder* and consult with your instructor to determine appropriate follow-up steps. These may involve searching for information from commercial suppliers, laboratory techniques books, and more general synthesis references. This planning will help you avoid pitfalls and dead ends in the laboratory and give you a much greater possibility of success.

INDEX

Note: Page numbers followed by f indicate figures; those followed by t indicate tables.

A

Ab initio molecular orbital methods, 115–116
Abbé refractometer, 208, 209f
Absolute uncertainty, 67
Absorption bands (peaks), 312–316, 312f
  absence of, 341
  bond vibrations and, 315–316
  combination, 315
  diagnostically useful, 338
  extra, 344
  intensity of, 316
  missing, 341, 343
  overtone, 314
  unexpected, 341–342
Abstracts, electronic, 502–503, 504–506
Accidents. *See* Laboratory safety
Accuracy, vs. precision, 67
ACD/CNMR Predictor, 427
Acetanilide, bromination of, 117–120
Acetone
  for drying glassware, 51
  as NMR solvent, 353, 354t
  as recrystallization solvent, 222t, 226t
Acetonitrile, in high-performance liquid
    chromatography, 290
Acetylene, chemical shifts of, 365
Acid anhydrides, IR spectra of, 333–334
Acid chlorides, IR spectra of, 328t, 333
Acid-base extraction, 147–149, 148f. *See also*
    Extraction
Acid-base reactions, solubility and, 131, 147–149
Acids, in resolution of racemic mixtures,
    241–242, 242f, 242t
Acros Organics database, 15, 16f, 37
Activated alumina, in solvent purification
    systems, 93
Adapters, 46f, 48f, 49f
  thermometer, 46f, 184, 184f
  tubing-needle, 97, 97f, 98f
Adsorbents
  in liquid chromatography, 270–272
    amount of, 274
    column volume of, 274–275, 274t
    polarity of, 270
    types of, 270–272
  in thin-layer chromatography, 256–257
Adsorption chromatography, 254
Advanced Chemical Development/NMR
    Predictor, 377
Air condensers, 48f, 49f, 87–88
Alcohol(s)
  IR spectra of, 328t, 330, 336, 340
  mass spectra of, 455–457, 456f
  oxidation to ketones
    case study of, 499–500
    in green chemistry, 30
Alcoholic sodium hydroxide, in base bath, 51
Aldehydes
  chemical shifts of, 365
  IR spectra of, 328t, 333, 336
Aldrich Catalog of Fine Chemicals, 38–39
Alkanes, IR spectra of, 327t
Alkenes
  bromination of, in green chemistry, 30–31
  ^{13}C chemical shifts of, 422–426
  C–H out-of-plane bending vibrations of,
      327–329, 329t
  chemical shifts of, 365
  IR spectra of, 327t, 336
  mass spectra of, 455, 455f
Alkyl carbons, ^{13}C chemical shifts of, 417–422
Alkyl protons, chemical shifts of, 366–370, 367t
Alkynes, IR spectra of, 327t
Allylic coupling, 381
α substituent
  ^{13}C chemical shift and, 417–418
  ^{1}H chemical shift and, 368
Alumina, in liquid chromatography, 272
Aluminum heating blocks, 75–77, 76f
Aluminum oxide, in thin-layer chromatography,
    257
AM1 method, 117
American Chemical Society
  CAS number and, 15, 505
  publications of, 502
Amides, IR spectra of, 328t, 330, 333, 335
Amines, IR spectra of, 328t, 330–331
1-Aminobutane, IR spectra of, 330, 331f
Anhydrides, IR spectra of, 328t, 333–334
Anhydrous conditions, 163–173
  drying agents for, 91–92, 163. *See also* Drying
      agents
p-Anisaldehyde, in thin-layer chromatography,
    262
Anisotropy, chemical shifts and, 365–366
Apparatus
  assembly of, 86–106
  glassware for. *See* Glassware
Aprons, 6
APT experiments, 427, 428f
APT spectra, 427, 428f
Aqueous extraction, 29, 143. *See also* Extraction

Aqueous waste, 21
Argon, for inert atmosphere reactions, 93, 94–95
Aromatic compounds
    C–H out-of-plane bending vibrations of, 327–329, 329t
    IR spectra of, 326, 326f, 328t, 329t, 336
Aromatic hydrocarbons
    ^{13}C chemical shifts of, 422–426
    mass spectra of, 454–455, 455f
Aromatic protons, chemical shifts of, 370–372
Asymmetric center, 240
Asymmetric stretching vibrations, 313, 313f, 314f
Atmospheric pressure, boiling point and, 174, 174t, 197, 197t, 198f
Atmospheric pressure chemical ionization (APCI), 445
Atom economy, 26–27
Attenuated total reflectance, in infrared spectrometry, 323–325, 323f, 324f
Automatic delivery pipet, 58f, 59
Axial cyclohexane conformers, equilibrium constants for, 113
Azeotropic distillation, 193–194, 193f, 193t, 194f

B
Background scan, 318, 318f
Balances, 52–54, 53f
Balloon assemblies, 97–98, 98f
Base(s)
    in acid-base extraction, 147–149, 148f. *See also* Extraction
    in acid-base reactions, solubility and, 131, 147–149
    in resolution of racemic mixtures, 241–242, 242f, 242t
Base bath, 50–51
Basis set, 116
Beakers, 45f
Beam splitter, 317–318, 317f
Beer-Lambert law, 466–467, 473, 475
Beilstein registry numbers, 505
Beilstein's Handbook of Organic Chemistry, 38
Bending vibrations, 313, 313f
    C–H out-of-plane, 327–329, 329t
Benzene, chemical shifts of, 366, 366f
Benzonitrile, IR spectrum of, 328t, 333, 336, 338
β substituent
    ^{13}C chemical shift and, 417–418
    ^{1}H chemical shift and, 368
Biochemical catalysis, 25–26
Blankets, fire, 12
Bleach, as chromium oxide substitute, 30
Boat conformation, 111
Boiling point, 174–180
    atmospheric pressure and, 174, 174t, 197, 197t
    definition of, 173
    determination of
        apparatus for, 175f

distillation in, 173, 176–205. *See also* Distillation
    microscale, 175–176, 175f
    miniscale, 175
    factors affecting, 174, 174t
    impurities and, 176
    vapor pressure and, 173
Boiling sticks, 73, 74
Boiling stones, 73–74
    in distillation, 204
    safety precautions for, 181
Bonds
    hydrogen, 127–129, 222
        NMR spectroscopy and, 397
    molecular vibrations and, 315–316
    order of, 315–316
Books. *See* Information sources
Boots-Hoechst-Celanese process, 28
Bounded zeroes, 66
Broadband decoupling, 409–411, 410f
Bromination
    of alkenes, in green chemistry, 30–31
    of benzene ring, energies of, 117–120
1-Bromopropane, mass spectrum of, 448–449, 448f
Bubblers, 95, 95f
    gas manifold with, 96–97, 96f
Büchner funnels, 45f, 137–138, 137f
Bumping, of liquids, 73
Bunsen burners, 81
    in capillary tube sealing, 220, 220f
    safety precautions for, 9–10
Burns. *See also* Laboratory safety
    chemical, 12–13
    management of, 13
    prevention of, 9, 74
    thermal, 9, 12–13
2-Butanone, mass spectrum of, 446–447, 457
Butene isomers, energies of, 114–115

C
^{13}C NMR spectra, 372–412, 408–412
^{13}C NMR spectroscopy, 408–441
    APT experiments in, 427, 428f
    broadband decoupling in, 409–411, 410f
    case study of, 429–431
    chemical shifts in, 412–427, 414t
        of alkenes, 422–426
        of alkyl carbons, 417–421, 418t
        alkyl substitution effects and, 412–413
        anisotropy and, 415–416
        of aromatic carbons, 422–426, 423t
        computer programs for, 426–427
        conjugation and, 415, 415t
        electronegativity and, 413–415
        hybridization and, 414–415
        quantitative estimation of, 417–427
        regions of, 413f
        ring structures and, 416–417

DEPT experiments in, 427–431, 428t, 429f
nuclear Overhauser enhancement in, 411
number of signals in, 411
overview of, 408
proton counting in, 427–428
sample preparation in, 408
solvent peaks and multiplicities in, 409
spin-spin splitting in, 409
symmetry in, 411
C–H out-of-plane bending vibrations, 327–329, 329t
Capillary columns, in gas chromatography, 294–250
Capillary tubes
  in boiling point determination, 176
  constricted, for thin-layer chromatography, 258, 258f
  sealing ends of, 220, 220f
CAplus, 505
Carbon dioxide
  sublimation of, 236
  supercritical, 24
Carbon NMR spectroscopy. See ^{13}C NMR spectroscopy
Carbonyl compounds, mass spectra of, 457–459, 458f
Carboxylic acids
  IR spectra of, 328t, 331, 331f, 333, 335
  in NMR spectroscopy, 354
CAS number, 15, 505
CAS Registry, 505
CASREACT, 505
Catalysts, in green chemistry, 25–26
Celite, 134, 136
Centrifugation, 140
Chair conformation, 111
Channeling, 286–287, 286f
ChemBioDraw Ultra, 426–427
ChemDraw Ultra, 426–427
Chemical Abstracts Service (CAS) number, 15, 505
Chemical burns, 12–13. See also Burns; Toxic exposures
Chemical fume hoods, 7–8, 7f
Chemical hygiene, 8
Chemical hygiene plan, 5
Chemical ionization, 445
Chemical names, 15, 15f
Chemical reactions. See also under Reaction
  anhydrous, 90–93, 91f
    drying agents for, 91–92, 163
  apparatus assembly for, 86–106
  designing, 485–497
    based on literature procedures, 488–494. See also Literature procedures
    case studies of, 497–500
    information sources for, 501–506. See also Information sources
    for multistep synthesis, 506–510

reaction time and, 496
  reagent addition and, 495–496
  scale modification and, 494–497
  workup and final purification of products and, 41–42, 492–494, 496–497
  in inert atmosphere conditions, 93–101. See also Inert atmosphere conditions
  noxious gas removal in, 103–106, 104f–106f
  reagent addition during, 89–90, 89f, 90f
  refluxing in, 87–90. See also Refluxing
    under inert conditions, 93–101
    reagent addition during, 89f, 90f
  setting up, 86–106
Chemical shifts, 359–377. See also ^{13}C NMR spectroscopy, chemical shifts in; nuclear magnetic resonance (^{1}H NMR) spectroscopy, chemical shifts in
Chemical splash goggles, 5–6
Chemical toxicology, 13–14, 17t. See also Toxic exposures
Chemicals, identifying, 15, 15f, 16f
Chemistry journals, 501–502
ChemNMR, 376–377, 426–427
Chiral center, 240
Chiral chromatography, 248
Chiral shift reagents, 249
Chirality, 240
3-Chloroethylbenzene, mass spectrum of, 449, 449f
Chloroform
  in ^{13}C NMR spectroscopy, 409
  in ^{1}H NMR spectroscopy, 353–354
  NMR spectrum of, 381, 381f
Chlorophyll, 469–470, 470f
Chromatography
  adsorption, 254
  chiral, 248
  flash, 270, 275–281
  gas, 254, 291–308
  liquid, 254, 270–291
  mobile phase in, 253, 254
  principles of, 253–254
  reverse-phase
    liquid, 272
    thin-layer, 257
  separation in, 254
  stationary phase in, 253, 254
  thin-layer, 254, 255–269
Chromatography funnels, 279–280, 279f
Chromium oxide, bleach as substitute for, 30
Chromophores, 467
Cinnamaldehyde, IR spectrum of, 335f, 336, 338
Cinnamyl alcohol, NMR spectrum of, 387–388, 388f
Claisen adapter, 46f, 48f, 49f, 90f, 200, 201f
  as drying tube, 92
Cleaning glassware, 50–51
Clothing, protective, 6
Color wheel, 470, 470f
Colored compounds, light absorption by, 469–470

Column chromatography. *See* Liquid
    chromatography
Column volume, 274–275, 274t
Combination bands, 315
Compendium of Organic Synthetic Methods, 503
Comprehensive Organic Name Reactions and
    Reagents, 503
Comprehensive Organic Transformations: A Guide
    to Functional Group Preparations, 503
Compressed gas cylinders, 94–95, 94f
Computational chemistry, 107–125
    ab initio molecular orbital methods in, 115–116
    basis set in, 116
    conformational searching in, 123
    definition of, 107
    density functional theory in, 120
    dihedral driver in, 123
    energy functional in, 120
    global minimum in, 122–123
    interpretation of results in, 123–124
    local minimum in, 122, 123
    method selection in, 121
    molecular dynamics simulation in, 123
    molecular mechanics method in, 109–115
    programs in, 67–68
    quantum mechanics methods in, 115–120
    rotamers and, 121
    Schrödinger wave equation in, 115
    semiempirical molecular orbital approach in,
      116–120
    sequential searching in, 123
    sources of confusion and pitfalls in, 121–124
    structure selection in, 121–122
Computer programs
    in computational chemistry, 67–68. *See also*
      Computational chemistry
    for NMR chemical shifts, 376–377
Condensers
    air, 48f, 49f, 87–88
    microscale, 48f, 49f
    miniscale, 46f
    water-jacket, 87, 88
    West type, 46f
Conical vials, 48f, 49
    volume markings on, 61
Conjugated compounds, UV/VIS spectroscopy
    and, 469
Conjugation, IR spectrum and, 334
Connectors, glassware, 48f, 49–50, 49f, 50f
Contact lenses, 5–6
Cooling baths, 85, 85f
Cotton wool filters, 133
Coupling constants, 378
    dihedral angle and, 383, 383f
    magnitude of, 382–383, 382t, 383f
Cow receiver, 201, 201f
Craig tube, for recrystallization, 232–234, 233f
CRC Handbook of Chemistry and Physics, 38, 39
Crystallization, 221

intermolecular forces in, 131
recrystallization and, 221–235. *See also*
    Recrystallization
Cuts. *See also* Laboratory safety
    prevention of, 8–9
Cuvettes, 474
Cyclohexane conformers
    energies of, 110–112
    equilibrium constants for, 113
Cyclopentane, IR spectrum of, 311, 312f

D

Dalton's law, 177
Databases, 504–506
    for identifying chemicals, 15, 15f, 16f
Dean-Stark trap, 194, 194f
Decantation, 140
Decomposition, melting point and, 219–220
Degassing, of liquids/solutions, 99–100, 99f
Degenerate energy levels, 349, 349f
Density functional theory, 120
DEPT experiments, 427–431, 428t, 429f
Design-based experiments, 2, 485–500, 490f,
    492f, 506–510, 507f
Desk equipment, 45, 45f
Desiccators, 92–93, 92f
Deuterated solvents
    in ^{13}C NMR spectroscopy, 409
    in ^{1}H NMR spectroscopy, 353–354, 354t
Deuterium oxide, in NMR spectroscopy, 354, 354t
Developing chamber, 260–261, 260f
Developing solvents, 255, 255f
    polarity of, 265–266, 266t
    problems with, 268
    rapid method for testing, 266–267, 266f
    selection of, 265–267, 266t, 268
Dewar flasks, 85, 85f
DFT (Density Functional Theory), 120
Diamagnetic shielding, 362–366
Diastereomers, 241–243. *See also* Enantiomer(s);
    Racemic mixtures
    preparation of, 241–243
Diastereotopic protons, in NMR spectroscopy,
    389–391
Dichloromethane
    alternatives to
      in aqueous extraction, 31
      in bromination of alkenes, 31
    in extraction, 145t
      vs. ethyl acetate, 29
Dictionary of Organic Compounds, 503
Diethyl ether
    as extraction solvent, 145t
    as recrystallization solvent, 222, 222t, 226t
Digital thermometers, 62–63
Dihedral angle, coupling constant and, 383, 383f
Dihedral driver, 123
Dimethyl sulfoxide-d_6, as NMR solvent, 354t
Diode-array spectrometers, 472, 472f

Diphenylethyne, IR spectrum of, 343, 344f
Dipole-dipole interactions, 128–129
Dipole-induced dipole interactions, 128–129
Dispensing pumps, 56–57, 57f
Dispersive spectrometers, 316–317
Disposable gloves, 6–7, 6t
Distillation, 176–205
  azeotropic, 193–194, 193f, 193t, 194f
  capillary tube in, 176
  composition of vapor above solution in, 177–178
  definition of, 173
  fractional, 179, 188–192, 189f–192f
    apparatus for, 190, 191f
    collecting fractions in, 191–192
    definition of, 179, 188
    fractionating columns in, 188–189, 189f
    microscale, 192, 192f
    miniscale, 190–192
    rate of distillation in, 191
    rate of heating in, 190
    sources of confusion and pitfalls in, 203–205
    when to use, 204
  intermolecular forces in, 131
  phase diagram for, 178, 178f
  in separation of mixtures, 176–180
  simple, 179–183
    apparatus for, 180–182, 181f, 183, 184–186, 184f
    microscale, 184–188, 184f–187f
    miniscale, 180–183, 181f, 184f
    short-path, 183, 184f
    sources of confusion and pitfalls in, 203–205
    vapor pressure vs. volume of distillate in, 179, 179f
    when to use, 204
  steam, 194–196, 196f
    when to use, 204
  temperature measurement in, 62–63
  temperature variation in, 204–205
  temperature-composition diagrams for, 178, 178f
  vacuum, 197–202
    apparatus for, 200–202, 201f
    definition of, 197
    microscale, 202–203
    miniscale, 200–203, 201f
    pressure monitoring during, 198f, 199f, 297–200
    procedure for, 202
    short-path, 201–202, 201f
    sources of confusion and pitfalls in, 203–205
    when to use, 204
  when to change receiving flasks in, 205
Distilling heads, 45f, 46f, 48f, 185, 185f, 186f, 192, 192f
Distribution coefficient (K), 144–145
Double-beam spectrometers, 471, 471f
Double-bond equivalents, 403
Doublets, in spin-spin coupling, 378
Dow-BASF process, for propylene oxidation, 26–28

Downfield shifts, 365
Draft shield, 52, 53f
Dropping funnels, 143f, 144f
Dry ice, melting of, 236
Dry ice baths, 85, 85f
Drying agents, 163–169
  amount of, 168–169
  anhydrous, 91–92, 163
  clumping of, 165–166, 169
  filtration of, 166–168, 167f
  guidelines for use, 165–166, 168–169
  properties of, 164t
  selection of, 164–165, 164t, 165t
  separation of, 166–168, 167f
  sources of confusion and pitfalls with, 168–169
  white liquid around, 169
Drying tubes, 45f, 48f, 90–92, 91f

E

Electrical fires. *See* Fires
Electron impact mass spectrometry, 442–444
Electrospray ionization, 445–446
Elsevier, publications of, 502
Elution solvents, 270, 272–273, 273t, 274t
  amount of, 274, 274t
  in flash chromatography, 279–280
  in high-performance liquid chromatography, 290
  problems with, 286–287
Emulsions, in extraction, 151–152
Enantiomer(s), 240–243
  definition of, 240
  $+/-$ isomers of, 240–241
  in racemic mixtures, 240–241. *See also* Racemic mixtures
  transformation into diastereoisomers, 241–243
Enantiomeric analysis
  chiral chromatography in, 248
  enantiospecific organic synthesis in, 243
  high-performance liquid chromatography in, 248
  modern methods of, 243–248
  NMR, 249
  polarimetry in, 243–248
  resolution in, 240–243, 242f
    with acids or bases, 241–242, 242f, 242t
  sources of confusion and pitfalls in, 250–251
Enantiomeric excess, 249–250
Enantiomeric ratio, 249
Enantiomerism, 240
Encyclopedia of Reagents for Organic Synthesis, 503
Energy, functional, 120
Energy, molecular, 109–115
Energy levels, degenerate, 349, 349f
Environmental considerations, 22–32. *See also* Green chemistry
Enzymatic resolution, of racemic mixtures, 242–243
Equatorial cyclohexane conformers, equilibrium constants for, 113

Erlenmeyer flasks, 45f, 49f
  volume markings on, 61
Error analysis, 64–72, 70t, 72t
Esters, IR spectra of, 328t, 333, 336, 337
Etching, of glassware, 51
Ethanol
  in bromination of alkenes, 31
  as recrystallization solvent, 222, 222t, 226t
Ether
  as extraction solvent, 145t
  as recrystallization solvent, 222, 222t, 226t
Ethyl acetate
  in aqueous extraction, 29
  as extraction solvent, 145t
    vs. dichloromethane, 29
  as recrystallization solvent, 222, 222t, 226t
Ethyl propanoate, NMR spectrum of, 352, 353f,
        369, 369f, 378–379
  spin-spin coupling and, 377–391
Ethyl *trans*-2-butenoate, NMR spectrum of, 381
  ^{13}C, 409–411
    APT, 427, 427f, 428f
    DEPT, 427–428, 428t, 429f
    2D COSY, 431–435
Ethylbenzene, mass spectrum of, 455, 455f
Ethylene, chemical shifts of, 336f, 365
Eutectic point, 213
Eutectic temperature, 212–213
Eutectics, 212–213
Evaporator, rotary, 171–173, 171f–172f
Exact numbers, 66
Explosions. *See also* Laboratory safety
  prevention of, 10–11
Extraction
  acid-base, solubility and, 147–149, 148f
  aqueous phase of, 29, 143
  definition of, 142
  determining container contents in, 162
  distillation in, 170
  distribution coefficient in, 144–145
  drying agents in, 163–169, 164t. *See also* Drying
        agents
    amount of, 168
    guidelines for use, 165–166
    selection of, 164–165, 165t
    separation of, 166–168, 167f, 168f
  emulsions in, 151–152
  evaporation in, 169–170, 171–173
  flowchart of, 146, 147f
  general procedure for, 143–147
  in green chemistry, 29
  intermolecular forces in, 131
  liquid-liquid, 142
  microscale, 155–161
    equipment for, 156–158, 156f, 157f
    with organic phase less dense than water,
        158–159, 159f, 160f
    with organic phase more dense than water,
        160–161, 161f

Pasteur pipet in, 157–158, 157f, 158f
      procedure for, 156–158
    miniscale, 152–155, 153f–154f
    mixture temperature in, 149–150
    with no visible separation phase, 162
    number of, 145–146
    organic phase of, 143, 161
    partition coefficient in, 144–145
    procedure for, 149–152
    product recovery in, 169–173
    salting out in, 151
    separatory funnel in, 152–154, 153f, 154f, 162
      venting of, 150, 150f
    single-extraction, 145–146
    solute distribution in, 144
    solute remaining after, 145–146
    solvents in, 142–143, 145–146, 145t
      density of, 143, 145t, 149, 150f, 154–155
      immiscible, 142
      properties of, 145t
      removal of, 169–173
    sources of confusion and pitfalls in, 161–162
    three-layer, 162
    washing in, 151
Eye protection, 5–6
Eye wash stations, 12–13

F
Feedstocks, renewable, 27–28
Fermi resonance, 314f, 315
Fiesers' Reagents for Organic Synthesis, 38–39, 503
Filter, quadrupole mass, 444, 444f
Filter aids, 134
Filter paper, 132–133, 133f, 133t
  clogged, 141
Filter-tip pipets, 60, 60f
Filtrate, 132, 141
Filtration, 132–142
  cloudy filtrate in, 141
  of drying agents, 166–167, 167f
  filtering media in, 132–134, 133f, 133t
  gravity, 134–136, 134f
    small-scale, 135–136
  intermolecular forces in, 131
  sources of confusion and pitfalls in, 140–141
  vacuum, 137–140, 137f–139f
    in recrystallization, 230, 230f, 232, 232f
Fingerprint region, of IR spectrum, 326, 326f, 336
Fire blankets, 12
Fires. *See also* Laboratory safety
  management of, 11–12
  prevention of, 9–10, 74
First aid kits, 13
Fishhooks, 454
Flame ionization detectors, 296–297, 297f, 300
Flammable waste, 21
Flash chromatography, 270, 275–281
  adsorbent removal in, 281
  apparatus in, 275, 276f, 277f

column elution in, 279–280, 280f
column preparation in, 275–278
  dry adsorbent method of, 277–278
  slurry method of, 278
compound recovery in, 281
sample preparation and application in, 278–282
Flasks, 45f
  desk equipment, 45f
  microscale, 48f, 49, 49f
  miniscale, 46f
Fluids, supercritical, 24
Fluorescence, in thin-layer chromatography, 261–262
Fluorinated mulling compounds, 321–322
Fluorolube mull, 322
Flushing, with inert gases, 98–99
Force fields, 109–110
Formaldehyde, chemical shifts of, 336f, 365
Four-diamond hazard labels, 18–19, 19f
Fourier transform spectrometers
  infrared, 317–319, 317f, 318f
  NMR
    in ^{13}C spectroscopy, 408
    in ^{1}H spectroscopy, 350–352, 351f, 352f
Fractional distillation, 179, 188–192, 189f–192f. *See also* Distillation, fractional
Fractionating columns, 188–189, 189f, 190f
Fragment ions, 453–459, 462
Free-induction decay, 350
Fritted glass disc, 138
FTIR spectrometers, 317–319, 317f, 318f
Fume hoods, 7–8, 7f, 103
Functional groups, IR spectrum of, 326f, 327–338, 327t–328t
Fundamental nitrogen rule, 447
Funnels, 45f, 49f
  Büchner, 45f, 137–138, 137f
  chromatography, 279–280, 279f
  dropping, 46f, 143, 143f
  for extractions, 143, 143f, 144f
  Hirsch, 45f, 49f, 137–139, 137f
  powder, 45f, 55, 55f
  pressure-equalizing, 101, 101f
  separatory, 46f, 143, 143f, 144f
    chromatography, 279–280, 279f
    in extraction, 149–150, 150f, 152–154, 153f
    venting of, 150, 150f
  in vacuum filtration, 137, 137f

G

γ substituent
  ^{13}C chemical shift and, 417–418
  ^{1}H chemical shift and, 368–370
Gas chromatograph–mass spectrometer (GC-MS), 444–445
Gas chromatography, 254, 291–308
  advantages and disadvantages of, 291–292
  capillary-column, 294f, 295, 301
  chromatogram in, 297–298, 307, 307f

extra peaks in, 304
flame ionization detectors in, 296–297, 297f, 300
injection technique in, 301–302, 301f, 303
instrumentation in, 293–294, 293f, 294f
interpretation of results in, 305–307, 306f, 307f, 307t
liquid-phase temperature range in, 295
mass spectrometry and, 305, 444–445
mobile phase in, 292
overview of, 292–293
packed-column, 295, 301
peak area determination in, 299, 305–306
peak enhancement in, 305
procedures for, 299–302, 301f
quantitative analysis in, 305–307
record keeping for, 302
recorders in, 297–298
response factors in, 306–307
retention time in, 298, 299f, 305
separation in, 292–293, 302, 392f
  problems with, 303–304
sources of confusion and pitfalls in, 303–304
spectroscopic methods in, 305
stationary phase in, 292, 294–296, 296t
syringes for, 301–302, 301f
thermal conductivity detectors in, 297, 297f, 300
trace impurities and, 304
vs. high-performance liquid chromatography, 287
Gas manifold, with bubblers, 96–97, 96f
Gas traps, 103–106, 104f–106f
GC. *See* Gas chromatography
Geminal coupling, 381
Geminal groups, 372
Glass fiber filters, 133
Glass wool filters, 133
Glasses, 5–6
Glassware, 42, 44–51. *See also specific types*
  cleaning and drying of, 50–51
  cuts from, 8–9
  desk equipment, 45, 45f
  etching of, 51
  greasing of, 46–47, 46f, 200
  icons for, 44, 44f
  microscale, 48–50, 48f–50f
  miniscale, 44, 45–47, 46f, 47f
  safe handling of, 8–9
  standard taper, 46f
  Williamson, 49–50, 49f, 50f
Global minimum, 123
Globally Harmonized System of pictograms, 16–18, 18f
Gloves, 6–7, 6t
Goggles, 5–6
Graduated cylinders, 45f, 56, 56f
Graduated pipets, 58–59, 58f. *See also* Pipets
  plastic transfer, 61, 61f
Gravity filtration, 134–136, 135f
  small-scale, 135–136

Greasing, joint, 46–47, 46f
  in vacuum distillation, 200
Green chemistry, 22–32
  in academic laboratories, 28–31
  atom economy in, 26–27
  Boots-Hoechst-Celanese process in, 28
  bromination of alkenes in, 30–31
  catalysts in, 25–26
  chemical syntheses in, 26, 29–31
  extraction in, 29
  in industrial processes, 24–28
  new developments in, 31
  oxidation of alcohols to ketones in, 30
  reaction efficiency in, 27
  renewable feedstocks in, 27–28
  solvents in, 24–25, 25t, 29
  supercritical carbon dioxide in, 24
  synthetic efficiency in, 28
  Twelve Principles of, 23–24, 24t
  water in, 24–25
Grubbs apparatus, 93
Grubbs test, 71, 72t
Guard column, 288
Guided-inquiry experiments, 485–487
Gyromagnetic ratio, 408

H

^{1}H nuclear magnetic resonance spectroscopy,
     310, 348–407. *See also* Nuclear magnetic
     resonance (^{1}H NMR) spectroscopy
Half-chair conformation, 112
Halogenated waste, 20
Handbooks, 38–39
Hazardous materials. *See* Laboratory safety;
     Toxic exposures
H-bonding, 127–129
Heat guns, 80, 80f
Heating blocks, 75–77, 76f
Heating devices, 10f, 74–84
  conventional, 74–81
  microwave reactors, 81–84, 82f, 83f
  safety precautions for, 9–10, 74
Heating mantles, 74–75, 75f
Hexane
  as extraction solvent, 145t
  as recrystallization solvent, 222, 222t, 226t
  solubility in, 130, 222
1-Hexene
  IR spectrum of, 332, 332f, 338
  mass spectrum of, 455, 455f
Hickman distilling head, 48f, 185, 186f, 192, 192f
High-performance liquid chromatography, 254,
     287–291
  apparatus for, 288–289, 288f
  detectors for, 289, 289f
  elution solvents for, 290
  in enantiomeric analysis, 248
  reverse-phase, 272, 288–289, 289f
  sample preparation in, 290–291

UV/VIS spectroscopy and, 466
  vs. gas chromatography, 287
High-resolution mass spectrometry, 450–451
Hirsch funnel, 45f, 49f, 137–139, 137f
Hit list, for mass spectral libraries, 451–453, 452f
Holdup volume, 183
Hoods, 7–8, 7f, 103
Hot plates, 10f, 75–77, 76f
  with magnetic stirring devices, 75–77, 78–79, 79f
  safety precautions for, 9–10, 10f
HPLC. *See* High-performance liquid
     chromatography
Hydrates, 163
Hydrogen bonding, 127–129, 222
  molecular vibrations and, 315–316
  NMR spectroscopy and, 397
Hydrogen bromide, in bromination of alkenes, 31
Hydrogen nuclear magnetic resonance
     spectroscopy, 310, 348–407. *See also*
     nuclear magnetic resonance (^{1}H NMR)
     spectroscopy
Hydrogen peroxide
  in bromination of alkenes, 31
  in propylene oxidation, 26–28
Hydrophobic effect, 130–131
HyperChem/HyperNMR, 377

I

Ice baths, 85, 85f
Immiscible solvents, 142
Implosions, 11
Indexes, electronic, 502–503, 504–506
Induced dipole-induced dipole interactions, 129
Inert atmosphere conditions, 93–101
  adding reagents in, 99, 101–103, 103f
  degassing liquids for, 99–100, 99f
  balloon assemblies for, 97–98, 98f
  flushing (purging) in, 98–99
  gases for, 93–95, 94f
  gas manifold for, 96–97, 96f
  liquid transfer in
    by funnel, 101
    by syringe, 101–103
    mineral oil bubbler for, 95, 95f
  refluxing under, 100–101, 100f
  safety precautions for, 94
Information sources, 36–39, 501–506
  chemistry journals, 501–502
  electronic abstracts and indexes, 502–503,
     504–506
  handbooks, 38–39
  literature procedures and, 494–497. *See also*
     Literature procedures
  online, 37–38, 502, 504–506
  for organic compounds, 36–39
  in planning multistep synthesis, 506–510
  reference books, 502–504
  searching methods for, 504–506
  textbooks, 504

for toxicology, 14–19
Infrared (IR) spectrometers, 316–319
  dispersive, 315–316
  Fourier transform, 317–319, 317f, 318f
  nuclear magnetic resonance, 349–350
Infrared (IR) spectroscopy, 310, 311–348
  attenuated total reflectance in, 323–325, 323f,
      324f
  background scan in, 318, 318f
  case study of, 340–341
  dispersive, 315–316
  Fourier transform spectrometer in, 317–319,
      317f, 318f
  instrumentation in, 316–319, 317f. *See also*
      Infrared (IR) spectrometers
  interpretation of results in, 325–339
    case study of, 340–341
    procedure for, 338–339
  IR spectrum in, 311–316. *See also* Infrared (IR)
      spectrum
  mulls in, 321–322
  potassium bromide pellets in, 322–323
  sample cards in, 323
  sample (solution) cells in, 320, 320f
  sample preparation in, 319–325
    cast films in, 321
    for liquids, 319–320, 320f
    neat sample in, 320
    for solids, 321–323
    thin films in, 320
  solvents in, 320–321
  sources of confusion and pitfalls in, 341–344
  stretching vibrations in, 327t–328t, 330–338
  transmittance in, 316
Infrared (IR) spectrum, 311–316
  absorption bands (peaks) in, 312–316, 312f
    absence of, 341
    bond vibrations and, 315–316
    broadened, 342
    combination, 315
    diagnostically useful, 338
    extra, 344
    flattened, 342
    indistinct, 342
    intensity of, 316
    missing, 341, 343
    overtone, 314, 314f
    unexpected, 341, 343
  of alcohols, 328t, 330, 336
  of aldehydes, 328t, 333, 336
  of alkenes, 327t, 329t, 336
  of amides, 328t, 330, 333, 335
  of amines, 328t, 330
  of 1-aminobutane, 330, 331f
  of aromatic compounds, 329t, 336
  atomic mass and, 316
  of benzonitrile, 328t, 333, 336, 338
  bond order and, 315–316
  C–C stretch in, 327t–328t

C–H out-of-plane bending vibrations and,
    327–329, 329t
C–H stretch in, 327t–328t, 332–333, 335
C–O stretch in, 328t, 336–337
C=C stretch in, 328t, 333f, 336
C=O stretch in, 328t, 333–336
of carboxylic acids, 328t, 331, 331f, 333, 335
C≡C stretch in, 328t, 333
of cinnamaldehyde, 335, 335f, 338
C≡N stretch in, 328t, 333
C≡O stretch in, 328t, 333
complexity of, 314–315
conjugation and, 334
of diphenylethyne, 343, 344f
of esters, 328t, 333, 336, 337
Fermi resonance and, 314f, 315
of functional groups, 326, 326f, 327t–328t,
    330–338
of 1-hexene, 332, 332f, 338
interpretation of, 325–339
of methyl acetate, 336, 336f, 337
of 4-methyl-3-penten-2-one, 334, 335f, 336, 338
of 4-methylpentan-2-one, 334
middle range of, 311, 312f
molecular vibrations and, 311–316, 313f, 314f
    bending, 313, 313f
    stretching, 313, 313f
N–H stretch in, 328t, 330–331, 335
of nitro compounds, 328t, 337
of 3-nitrotoluene, 337f, 338
NO$_2$ stretch in, 328t, 337, 337f
of Nujol, 342, 343f
O–H stretch in, 328t, 330–331, 331f
of phenylacetylene, 332, 332f, 336
of propanoic acid, 331, 331f
of 2-propanol, 330, 330f
regions of, 326
    aromatic, 326, 326f, 328t, 337
    fingerprint, 326, 326f, 336
    functional group, 326, 326f, 327t–328t
ring strain and, 334
sloping baseline in, 342
sp hybridization and, 332–333
stretching vibrations in, 327t–328t, 330–338
unexpected peaks in, 341
wavenumber in, 313
Injuries, 8–9. *See also* Burns; Laboratory safety;
    Toxic exposures
  management of, 12–13
  prevention of, 8–9
Immiscible solvents, 142
Inquiry-driven experiments, 485–487
Integrated spectroscopy problems, 476–484
Interferogram, 318
Intermolecular forces, 127–131
  in crystallization, 131
  dipole-dipole interactions, 127–128
  dipole-induced dipole interactions, 128–129
  in distillation, 131

Intermolecular forces (*Contd.*)
  in extraction, 131
  in filtration, 131
  hydrogen bonds, 127–128
  molecular vibrations and, 315–316
  NMR spectroscopy and, 397
  in separation and purification, 131
  solubility and, 130–131
  van der Waals forces, 130
Iodine visualization, in thin-layer
      chromatography, 263
IR spectroscopy. *See* Infrared (IR) spectroscopy
(+)-Isomer, 240–241
(−)-Isomer, 240–241
Isomers
  of enantiomers, 240–241
  optical, resolution of, 240–243
Isotopes
  exact masses of, 450–451, 450t
  relative abundance of, 447–450, 448t
IUPAC naming system, 15

J

Jacketed condenser, 48f, 87, 88
Joint clips, 46f
Joint greasing, 46–47, 46f
  in vacuum distillation, 200
Journals, 501–502

K

K (distribution coefficient), 144–145
Karplus curve, 383, 383f
Keck clips, 181–182, 184f
Ketones
  alcohol oxidation to
    case study of, 499–500
    in green chemistry, 30
  IR spectra of, 328t
  mass spectra of, 457–459, 458f

L

Lab coats, 6
Labels
  safety, 18–19, 18f, 19f
  waste container, 20
Laboratory glassware, 42, 44–51. *See also*
      Glassware
Laboratory jacks, 85–86, 86f
Laboratory notebooks, 32–35
Laboratory safety, 3–22. *See also under* Safety
  accident management, 11–13
  accident prevention, 9–11
  chemical exposures, 5–8, 12–14
  cuts and burns, 8–9, 12–13
  fires, explosions, and implosions, 9–12, 74
  general rules, 4–5
  glassware, 44, 44f
  information sources, 14–19
  microwave reactors, 84

personal protective equipment, 5–7, 7t
  safety equipment, 12–13
  toxic exposures, 5–8, 12–13
  waste management, 5, 20–21
Latex gloves, 6–7, 7t
LC. *See* Liquid chromatography
LD_{50}, 14, 17t
Le Bel, Joseph, 240
Leading zeroes, 65
Library research. *See* Information sources
Light
  infrared. *See* Infrared (IR) spectrum
  plane-polarized, 244
Like-dissolves-like rule, 221
Limiting reagents, 36
Liquid(s). *See also* Reagents; Solvent(s)
  boiling point of, 174–180
  bumping of, 73
  degassing of, 99–100, 99f
  extraction of, 142–163. *See also* Extraction
  superheated, 73
  transfer of
    with pipet, 61, 61f
    with syringe, 59, 90, 90f, 101–103, 103f
  vapor pressure of, 173
  volume of, measurement of, 55–61, 56f–58f,
      60f, 61f
Liquid chromatography, 254, 270–291
  adsorbents in, 270–272
    amount of, 274
    column volume of, 274, 274t
    polarity of, 270
  apparatus for
    in flash chromatography, 275–277, 276f, 277f
    in high-performance liquid
        chromatography, 288, 288f, 289f
    microscale, 281–285, 282f, 284f
    miniscale, 275, 276f, 277f
  channeling in, 286–287, 286f
  columns in
    dimensions of, 273–275, 274t
    in flash chromatography, 275–278, 276f, 277f
    guard, 288
    in high-performance liquid
        chromatography, 288, 288f
    microscale, 281–285, 282f, 284f
    overloading, 285
    preparation of, 275–278, 281–285
    problems with, 285–287
  definition of, 270
  diffuse bands in, 287
  elution solvents in, 270, 272–273, 273t, 274t
    amounts of, 274–275, 274t
    in flash chromatography, 279–280
    in high-performance liquid
        chromatography, 290
    problems with, 285, 286–287
  flash, 270, 275–281. *See also* Flash
      chromatography

high-performance, 254, 287–291, 288f, 289f. *See also* High-performance liquid chromatography
microscale, 281–285, 282f, 284f
  Williamson, 283–285
miniscale, 275–281. *See also* Flash chromatography
mobile phase in, 270, 271f
nonhorizontal bands in, 286
overview of, 270
separation in, 270, 320f
sources of confusion and pitfalls in, 285–287
stationary phase in, 270, 271f
steps in, 285
tailing in, 287
Liquid chromatography–mass spectrometry (LC-MS), 445–446
Liquid-liquid extraction, 142
Literature of organic chemistry, 501–506. *See also* Information sources
Literature procedures, 494–497
  case studies of
    oxidation of secondary alcohol to ketone, 499–500
    solvatochromic dye synthesis, 497–499
  interpretation and implementation of, 488–494
    analyzing reaction in, 490
    monitoring reaction progress and, 491–492
    reaction conditions and, 491
    reagent handling and transfer and, 491
    safety concerns and, 491
    steps in, 488
    understanding reaction workup and, 492–494
  modifying scale of, 494–497
Local diamagnetic shielding, 363
Local minimum, 122, 123
London dispersion forces, 129

M

Magnetic resonance, 349–350, 352f
Magnetic sector mass analyzer, 443–444, 443f
Magnetic spin vane, 48f
Magnetic stirring devices, 74
  with hot plates, 75–77, 78–79, 79f
  for microwave reactors, 84
MALDI (matrix-assisted laser desorption/ionization), 445
Manometers, in vacuum distillation, 197–199, 198f
Mass, tare, 52–53
Mass spectrometers, 442–446, 442f–444f
Mass spectrometry, 310, 442–462
  base peak in, 446
  case study of, 459–461
  data presentation in, 446
  electron impact, 442–444
  fragment ions in, 453–459, 462
  fundamental nitrogen rule in, 447
  gas chromatography and, 305, 444–445
  high-resolution, 450–451

instrumentation for, 442–446, 442f–444f
  isotopes in
    exact masses of, 450–451, 450t
    relative abundance of, 447–450, 448t
  with liquid chromatography, 445–446
  M+1 and M+2 peaks in, 447–449
  mass spectral libraries in, 451–453, 452f, 453f
  mass-to-charge (m/z) ratio in, 441, 442–444
  molecular ion in, 442–444, 446–450, 461–462
  radical cations in, 442
  Rule of Thirteen in, 446–447
  sources of confusion and pitfalls in, 461–462
Mass spectrum
  base peak in, 446
  of 1-bromopropane, 448–449, 448f
  of 2-butanone, 446–447, 457
  of 3-chloroethylbenzene, 449, 449f
  of ethylbenzene, 455, 455f
  of 1-hexene, 455, 455f
  libraries of, 451–453, 452f, 453f
  M+1 and M+2 peaks in, 447–449
  of methyl nonanoate, 457–458, 458f
  of 2-methyl-2-butanol, 456, 456f
  of orange oil, 444–445, 452–453, 452f, 453f
Mass-to-charge (m/z) ratio, 441, 442–444
Material Safety Data Sheets (MSDSs), 19
Matrix-assisted laser desorption/ionization (MALDI), 445
McLafferty rearrangement, 458
McLeod gauge, 199–200, 199f
Measurement techniques, 52–73
  for volume, 55–61
  for weight, 52–54, 53f
Measurement uncertainty, 64–72
  absolute, 67
  accuracy vs. precision and, 67
  estimation of, 69–71
  human error and, 64
  outliers and, 71, 72t
  propagation of, 67–69
  random errors and, 65
  rejection of data and, 71, 72t
  relative, 67
  rounding and, 66–67
  significant figures and, 65–67
  standard deviation and, 69–71, 70t
  systematic errors and, 64–65
  vs. exact numbers, 66
Measurements, recording of, 34–35
MEDLINE, 505
Melting point apparatus, 62
  analog, 214–215, 215f
  digital, 213, 214f
Melting point/range, 211–220
  definition of, 211
  determination of, 215–218
    decomposition in, 219–220
    documentation of, 217
    for mixtures, 218

Melting point/range (*Contd.*)
   procedure for, 215–218
   sample heating in, 216–217
   sample preparation in, 215–216
   sources of confusion and pitfalls in, 219–220
   sublimation and, 219
   thermometer calibration in, 216
  eutectic point and, 213
  factors affecting, 212–213
  heating rate and, 219
  identifying compounds with, 218
  impurities and, 212, 213
  lower limit of, 213
  melting behavior and, 212–213
  of mixtures, 212–213, 212f, 218
  reporting of, 217
  sublimation and, 219, 236
  theory of, 212–213
The Merck Index: An Encyclopedia of Chemicals,
   Drugs, and Biologicals, 38, 39
Mercury thermometers, 62
Methanol
  in high-performance liquid chromatography,
   290
  NMR spectrum of, 397–398, 397f
  as recrystallization solvent, 222, 222t, 226t
Methyl acetate, IR spectrum of, 336, 336f
Methyl nonanoate, mass spectrum of, 457–458,
   458f
2-Methyl-1-butanol, NMR spectrum of, 389–391
2-Methyl-2-butanol, mass spectrum of, 456, 456f
4-Methyl-3-penten-2-one, IR spectrum of, 334,
   335f, 336, 338
4-Methylpentan-2-one, IR spectrum of, 334, 334f
Michelson interferometer, 318, 318f
Micropipets, for thin-layer chromatography, 258
Micropore filters, 133–134
Microscale glassware. *See also* Glassware
  standard taper, 48–49, 48f, 49f
  Williamson, 49–50, 49f, 50f
Microwave heating, in green chemistry, 31
Microwave reactors, 81–84, 82f, 83f
Microwave-assisted organic synthesis (MAOS),
   81–84, 82f, 83f
Mid infrared spectrum, 311, 312f
Mineral oil baths, 79–80
Mineral oil bubblers, 95, 95f
  gas manifold with, 96–97, 96f
Miniscale standard taper glassware, 45–47,
   46f–47. *See also* Glassware
Miscibility, 128, 222t
Mixtures, separation of, distillation in, 176–180.
   See also Distillation
MNDO method, 117
Molar absorptivity, 466–468, 475
Molar extinction coefficient, 466, 468, 468t
Mole fraction, 177–178
Molecular energy, 109–115
Molecular ion (M^+), 442–444, 446–450, 461–462

Molecular mechanics method, 109–115
Molecular modeling, 107. *See also* Computational
   chemistry
Molecular orbital methods
  ab initio, 115–116
  semiempirical, 116–120
Molecular Orbital Package (MOPAC), 117
Molecular rotation, 248
Molecular sieves, 93
Monomode microwave reactors, 82–83, 82f
MOPAC, 117
MS. *See* Mass spectrometry
Mulls, in infrared spectroscopy, 321–322
Multimode microwave reactors, 82–83, 82f
Multistep synthesis planning, 506–510

N
N+1 rule, 379–381
Name Reactions and Reagents in Organic Synthesis,
   504
Nanometer, 465
Nature, 502
Nitriles, IR spectra of, 328t
Nitro compounds, IR spectra of, 328t, 337
Nitrogen, for inert atmosphere reactions, 93, 94–95
3-Nitrotoluene, IR spectrum of, 337f, 338
NMR. *See* Nuclear magnetic resonance
Nonaqueous waste, 21
Notebooks, laboratory, 32–35
Nuclear energy levels, 349, 349f
Nuclear magnetic resonance (NMR) analysis,
   enantiomeric, 249
Nuclear magnetic resonance (^{1}H NMR)
   spectroscopy, 310, 348–407
  case studies of, 398–404
  chemical shifts in, 359–377
   of alkyl protons, 366–370, 367t
   anisotropy and, 365–366
   of aromatic protons, 370–372
   computer programs for, 376–377
   definition of, 359
   deshielding and, 363–366, 376
   diamagnetic shielding and, 362–366
   downfield, 365
   electronegativity effects on, 364–365
   estimation of, 366–377
   geminal groups and, 372
   hindered rotation and, 376
   measurement of, 359–360
   molecular structure and, 359–360
   reference point for, 359–360
   regions of, 361, 361f
   ring structures and, 366, 376
   significance of, 359
   spectrometer frequency and, 360
   substituent effects in, 368
   units of, 359–361
   upfield, 365
   of vinyl protons, 372–374, 372t

chiral shift reagents for, 249
data acquisition problems in, 392
data organization in, 399–400
degree of saturation and, 399
diastereotopic protons in, 389–391
double-bond equivalents in, 399, 403
doublet of quartets in, 353–354
doublets in, 353–354, 378
extra signals in, sources of, 395, 395t
Fourier transform, 350–352, 351f
free-induction decay in, 350
high-field, 352
hydrogen bonding and, 397
index of unsaturation in, 399
instrumentation for, 350–352
integration in, 358–359
magnetic field homogeneity and, 392–393
mixtures in, 394–396
nuclear energy levels and, 349, 349f
nuclear spin and, 349–350, 349f
obtaining NMR spectrum in, 356
overview of, 348–350
proton counting in, 358–359
proton exchange and, 397–398, 397f
quartets in, 378, 380
reference calibration in, 354–355
sample preparation for, 355–356, 357, 392
sample recovery in, 356
second-order effects in, 387–389
shielding/deshielding in, 362–366
signature patterns in, 403
singlets in, 379
solvents in, 353–354
    NMR signals of, 395, 395t
    reference compounds in, 354–355
    selection of, 354t
sources of confusion and pitfalls in,
        392–393
spectra interpretation in, 391–398
spinning sidebands in, 393
spin-spin coupling in, 377–391
    allylic (four-bond), 381
    coupling constants and, 378, 382–383,
        382t, 383f
    dihedral angle and, 383, 383f
    doublet signal in, 378
    geminal (two-bond), 381
    Karplus curve for, 383, 383f
    long-range, in aromatic rings, 381
    multiple, 383–387
    N+1 rule and, 379–381
    one-bond, 381
    Pascal's triangle and, 380–381, 380f
    signal splitting and, 380, 380f
    splitting tree in, 379, 380f
    vicinal (three-bond), 377–381, 383
standard reference substances for, 354–355
strong coupling in, 393
theoretical basis for, 349–350

triplets in, 379
tubes in, 355
    cleaning of, 356–357
    collar for, 356, 356f
    filling of, 355–356, 356f
two-dimensional correlated, 431–435
virtual coupling in, 393–394
vs. ^{13}C NMR spectroscopy, 408
Nuclear magnetic resonance (^{13}C NMR)
        spectroscopy, 408–441. See also ^{13}C NMR
        spectroscopy
Nuclear magnetic resonance spectrum
    ^{13}C, 408–412
    of chloroform, 381f
    of cinnamyl alcohol, 387–389, 388f
    DEPT, 427–428, 428t, 429f
    of ethyl propanoate, 352, 353f, 358, 358f, 369,
        369f, 378–379, 381
    of ethyl trans-2-butenoate, 409–411
    of methanol, 397–398, 397f
    of 2-methyl-1-butanol, 388f, 389–391
    of tert-butyl acetate, 360–361
    of 1, 1, 2-tribromo-2-phenylethane, 378
    2D COSY, 431–435
Nuclear Overhauser enhancement (NOE), 411
Nuclear spin, 349–350, 349f
Nujol mulls, 321–322, 342, 343f

O

Oil baths, 79–80
Oiling out, in recrystallization, 235
Online resources, 37–38, 504–506. See also
        Information sources
Open tubular columns
    support-coated, 295
    wall-coated, 295
Optical activity
    chirality and, 240
    definition of, 240
    enantiomeric excess and, 249–250
    measurement of, 243–248. See also Polarimetry
        high-performance liquid chromatography
        in, 248
    NMR analysis in, 249
Optical isomers, resolution of, 240–243
Optical purity, determination of, 249–250
Organic chemistry, literature of, 501–506. See also
        Information sources
Organic compounds
    chirality of, 240
    information sources for, 36–39. See also
        Information sources
    pure, 221
    toxicity of, 13–14. See also Toxic exposures
Organic extraction, 143. See also Extraction
Organic reactions. See Chemical reactions
Organic Reactions, 503
Organic solvents, 25, 25t. See also Solvent(s)
Organic Syntheses, 503

Out-of-plane bending vibrations, 327–329, 329t
Overtone bands, 314
Oxidation
    of alcohols to ketones
        case study of, 499–500
        in green chemistry, 30
    propylene
        atom economy in, 26–27
        with hydrogen peroxide, 26–28
Oxone, 30

P
Partition coefficient (K), 144–145
Pascal's triangle, spin-spin coupling and,
        380–381, 380f
Pasteur pipets, 59–61, 60f
    in extraction, 156f, 157–158, 157f, 159f
    in filtration, 135–136, 136f
    preparation of, 136
    safety precautions for, 135
Peaks. See Absorption bands (peaks)
Pentane, as extraction solvent, 145t
Percent yield, 35
    calculation of, 35–36
Personal protective equipment, 5–7, 7t
Petroleum ether
    as extraction solvent, 145t
    as recrystallization solvent, 222, 222t, 226t
Phase diagram, for distillation, 178, 178f
Phenylacetylene, IR spectrum of, 332, 332f, 336
Phosphomolybdic acid, in thin-layer
        chromatography, 262
Pictograms, Globally Harmonized System for,
        16–18, 18f
Pipets
    automatic delivery, 58f, 59
    dripping from, 60–61
    graduated, 58–59, 58f, 61f
    Pasteur. See Pasteur pipets
    plastic transfer, 61, 61f
    pumps, 58, 58f
Planck's Law, 467
Plane-polarized light, 244
Plastic transfer pipets, 61, 61f
PM3 model, 117
PM6 model, 117
Poisons. See Toxic exposures
Polarimeter(s), 243–244
    automatic digital, 244, 244f, 246
    manual, 243–244, 246–247, 247f
Polarimeter tubes, 245–246, 245f
Polarimetry, 243–248
    alternatives to, 248–249
    analyzing results in, 247–248
    enantiomeric excess and, 249–250
    instrumentation for, 243–244, 244f, 246–247, 247f
    molecular rotation and, 248
    monochromatic light in, 245
    sample preparation for, 245

    specific rotation and, 247
    techniques in, 245–247
Polarity, of solvents, 222, 222t
Polarizability, 129
Potassium bromide pellets, in infrared
        spectrometry, 322–323
Powder funnel, 45f, 55, 55f
Precipitate, 132
Precision, vs. accuracy, 67
Prelaboratory information sources, 36–39
Pressure
    atmospheric, boiling point and, 174, 174t, 197,
        197t, 198f
    vapor. See Vapor pressure
Pressure-equalizing funnels, 101, 101f
Propanoic acid, IR spectrum of, 331, 331f
2-Propanol, IR spectrum of, 330, 330f
Propylene oxidation
    atom economy in, 26–27
    with hydrogen peroxide, 26–28
Protective attire, 6
Protective Groups in Organic Synthesis, 504
Proton(s)
    chemical shifts of. See Nuclear magnetic
        resonance (^{1}H NMR) spectroscopy,
        chemical shifts in
    diastereotopic, in NMR spectroscopy, 389–391
Proton counting
    in ^{13}C NMR spectroscopy, 427–428
    in ^{1}H NMR spectroscopy, 358–359
Proton exchange, in NMR spectroscopy, 397–398,
        397f
Pumps, dispensing, 56–57, 57f
Pure organic compounds, 221
Purging, with inert gases, 98–99
Purification. See Separation and purification

Q
Quadrupole mass filter, 444, 444f
Quantum mechanics methods, 115–120
Quartets, in spin-spin coupling, 378

R
Racemic mixtures
    enantiomers and, 240–243. See also
        Enantiomer(s); Enantiomeric analysis
    preparing diastereomers from, 241–243
    resolution of
        acid/base, 242–242, 242f, 242t
        enzymatic, 242–243
        vs. enantiospecific organic synthesis, 243
Radiation, infrared. See Infrared (IR) spectrum
Radical cations, in mass spectrometry, 442
Raoult's law, 178
Reaction. See Chemical reactions
Reaction apparatus
    assembly of, 86–106
    glassware for. See Glassware
Reaction efficiency, 27

Reaction tubes, 49f
  volume markings on, 61
Reaction vials, 48f
  volume markings on, 61
Reagents. *See also* Liquid(s)
  addition during reaction, 89–90, 89f, 90f
  chiral shift, 249
  extraction of, 142–163. *See also* Extraction
  limiting, 36
  measurement of, 52–54, 55–57
  solid
    transfer of, 54–55, 54f, 55f
    weighing of, 52–54, 53f
  transfer of, 54–61, 61f
    with pipet, 61, 61f
    with syringe, 59, 90, 90f, 101–103, 103f
  visualization, in thin-layer chromatography,
    262–263
  volume of, measurement of, 55–61
  weighing of, 54
Reagents for Organic Synthesis, 38–39
Reaxys, 505
Recrystallization, 221–235
  Craig tube for, 232–234, 233f
  crystal formation in, 221
    promotion of, 234–235
  definition of, 221
  ensuring dry crystals in, 225
  failure of, 234–235
  gravity filtration for, 134–136
  insoluble impurities in, 225
  maximum product recovery in, 225
  oiling out in, 235
  procedure for, 223–225
    microscale, 231–234, 232f
    miniscale, 228–231, 228f–230f
  scale of, 224–225
  seed crystals in, 224
  solvents for, 221–223, 222t, 226t, 234–235
    amounts of, 234–235
    boiling point of, 222
    miscible, 226
    paired, 226–227, 226t
    polarity of, 222
    properties of, 221–222, 222t
    selection of, 223, 225–227
    solubility and, 221–223, 223f
  sources of confusion and pitfalls in, 234–235
  theory, 221
Reference books, 502–504
Refluxing, 87–90
  under anhydrous conditions, 90–93, 90f, 91f
  under inert conditions, 100–101, 100f. *See also*
    Inert atmosphere conditions
  methods and apparatus for, 87–89, 88f
  reagent addition during, 89–90, 89f, 90f
Refractive index, 206–208
  definition of, 206
  measurement of

  instrumentation for, 208, 208f, 209f
  procedure for, 208–210
  temperature correction in, 210
  temperature and, 207
  wavelength and, 207
Refractometer, 208, 208f, 209f
Refractometer detectors, in high-performance
    liquid chromatography, 289–290
Refractometry, 206–211
  sources of confusion and pitfalls in, 211
Relative uncertainty, 67
Relaxation, 349
Renewable feedstocks, 27–28
Research, library. *See* Information sources
Response factors, 306–307
Reverse-phase chromatography
  liquid, 272
    high-performance, 288–289
  thin-layer, 257
R_f values, in thin-layer chromatography,
    263–264, 263f, 268
Rheostats, for heating mantles, 75
Ring strain, IR spectrum and, 334
Ring structures, chemical shifts and, 366, 376,
    416–417
RM1 method, 117
Rotamers, 121
Rotary evaporator, 171–173, 171f–172f
Round-bottomed flasks, 45f, 46f, 48f, 49
Rounding, 66–67
Royal Society of Chemistry, publications of, 502
Rule of Thirteen, 446–447

S

Safety Data Sheets (SDSs), 19
Safety equipment, 12–13
Safety glasses/goggles, 5–6
Safety information, sources of, 14–19
Safety labels, 18–19, 18f, 19f
Safety measures, 3–22. *See also* Laboratory safety
Safety showers, 12
Salting out, 151
Sand baths, 77, 77f
Satellite accumulation areas for chemical waste, 20
Schrödinger wave equation, 115
Science, 502
SciFinder, 505, 506–510
Scoopula, 45f
Semiempirical molecular orbital (MO) approach,
    116–120
Separation and purification
  centrifugation in, 140
  crystallization/recrystallization in, 221–235.
    See also Recrystallization
  decantation in, 140
  distillation in, 176–205. *See also* Distillation
  extraction in, 142–163. *See also* Extraction
  filtration in, 132–142. *See also* Filtration
  intermolecular forces in, 131

Separatory funnels, 46f, 143, 143f, 144f
  chromatography, 279–280, 279f
  in extraction, 149–150, 150f, 152–154, 153f
  venting of, 150, 150f
Shimming, 356
Short-path distillation, 183, 184f
Showers, safety, 12
SI (similarity index), 452–453
Sigma-Aldrich database/website, 15, 15f, 37–38,
    37f, 38f
Significant figures, 34, 65–67
  in calculations, 66
  rounding and, 66–67
  rules for, 65–66
Silica gel
  in liquid chromatography, 272, 273t, 274t.
      See also Liquid chromatography,
      adsorbents in
  in thin-layer chromatography, 256–257, 265
Silicone oil baths, 79–80
Similarity index (SI), 452–453
Simple distillation. See Distillation, simple
Single-mode microwave reactors, 82–83, 82f
Singlets, in spin-spin coupling, 379
Sodium 2,2-dimethyl-2-silapentane-5-sulfonate
    (DSS), in NMR spectroscopy, 355
Sodium chloride disks, in infrared spectrometry,
    320, 320f
Sodium D line, 207, 245
Software. See also Computational chemistry
  for NMR chemical shifts, 376–377
Solid waste, 21
Solubility, 221–223, 223f
  acid-base chemistry and, 131, 147–149
  intermolecular forces and, 130–131
Solutes, solubility of, 222, 222t
Solvatochromic dye, synthesis of, 497–499
Solvent(s). See also Liquid(s)
  azeotropes from, 193–194, 193f, 193t, 194f
  in ^{13}C NMR spectroscopy, 409
  density of, 143, 149, 155
  deuterated, 353–354, 354t
  developing, 225f, 255, 255f, 265–267, 265f,
      266t, 268
    polarity of, 265–266, 266t
    problems with, 268
    rapid method for testing, 266–267, 266f
    selection of, 265–267, 266t, 268
  dielectric constant for, 222t
  elution, 270, 272–273, 273t, 274t
    amount of, 274, 274t
    in flash chromatography, 279–280
    in high-performance liquid
        chromatography, 290
    problems with, 286–287
  extraction, 142, 145t
    density of, 143, 145t, 155
    in green chemistry, 29
    properties of, 145t

in green chemistry, 24–25, 25t, 29
in high-performance liquid chromatography,
    290
immiscible (insoluble), 142
in infrared spectroscopy, 320–321
in liquid chromatography, 270, 272–273, 273t,
    274t
miscible, 226
in NMR spectroscopy, 353–354, 354t
  reference compounds in, 354–355
  selection of, 354t
  signals of, 395, 395t
organic, 25, 25t
paired, 226–227, 226t
polarity of, 222, 222t
properties of, 221–222, 222t
purification systems for, 93
recrystallization, 221–223, 222t, 226t. See also
      Recrystallization, solvents for
solubility and, 130–131, 221–223, 223f
supercritical carbon dioxide as, 24
in UV/VIS spectrometry, 473–474, 474t, 475
water as, 24–25
  in recrystallization, 222t
Solvent front, 268
sp hybridization, IR spectra and, 332–333
Sparging, 99–100, 99f
Spatula, 45f
Specific rotation, 247
Spectrometers
  double-beam, 471, 471f
  Fourier transform
    infrared, 317–319, 318f, 352f
    NMR
      in ^{13}C spectroscopy, 408
      in ^{1}H spectroscopy, 350–352, 351f
  infrared, 316–319
    dispersive, 315–316
    Fourier transform, 317–319, 317f, 318f
    NMR, 350–352
  mass, 442–446
  UV/VIS, 310, 471–474, 471f, 472f
    operation of, 473–474
Spectroscopy, 309–484
  in gas chromatography, 305
  infrared, 310, 311–348. See also Infrared (IR)
      spectroscopy
  integrated approach to, 476–482
  nuclear magnetic resonance, 310, 348–407
  overview of, 309–310
  ultraviolet and visible light, 310, 465–484.
      See also UV/VIS spectroscopy
  vibrational, 311
Spin, nuclear, 349–350, 349f
Spinning sidebands, 393
Spin-spin coupling, 377–391. See also Nuclear
      magnetic resonance (^{1}H NMR)
      spectroscopy, spin-spin coupling in
Spin-spin splitting, in ^{13}C NMR spectroscopy, 409

Splash goggles, 5–6
Splitting trees, 379, 380f
Standard deviation, measurement uncertainty and, 69–71, 70t
Standard taper glassware, 44–47. *See also* Glassware
  microscale, 44, 48–49, 48f
  miniscale, 44, 45–47, 46f, 47f
Steam baths, 78, 78f
Steam distillation, 194–196, 196f
Stereocenters, 240, 242
Steric compression, 414
Steric energy, molecular, 109–115
Stir/hot plates, 75–77, 78–79, 79f
  safety precautions for, 9–10, 10f
Stirring devices. *See* Magnetic stirring devices
Stoppers, 46f
Strain energy, molecular, 109–115
Stretching vibrations, 313, 313f
  in infrared spectrum, 327t–328t, 330–338
Sublimation, 236–239
  advantages of, 236–237
  apparatus for, 237–238, 237f, 238f
  definition of, 236
  melting point and, 219, 236
  microscale, 238–239
  procedure for, 238–239
  sources of confusion and pitfalls in, 239
Substituent effects, in NMR spectroscopy, 417–418
Supercritical carbon dioxide, as solvent, 24
Supercritical fluids, 24
Superheated liquids, 73
Support-coated open tubular columns, 295
Symmetric stretching vibrations, 313, 313f
Syringes
  for column injection, 301–302, 301f
  for reagent transfer, 59, 90, 90f, 101–103, 103f

T

Tailing, in liquid chromatography, 287
Tare mass, 52–53
Tared watch glass, 232
Temperature, eutectic, 212–213
Temperature probes, 62–63, 62f
Temperature-composition diagram, for distillation, 178, 178f
Tetramethylsilane, in NMR spectroscopy, 355, 359–360
Textbooks, 504
Theoretical plates, 189
Theoretical yield, 35–36
Thermal conductivity detectors, 297, 297f, 300
Thermometers, 45f, 62–63
  adapters for, 46f, 184, 184f
  calibration of, 63, 63f, 63t
  digital, 62–63
  mercury, 62
  nonmercury, 62

  in water baths, 78–79, 79f
Thieme Chemistry, publications of, 502
Thin-layer chromatography (TLC), 254, 255–269
  adsorbents in, 256–257
  analysis of results in, 263–264
  developing chamber in, 260–261, 260f
  developing solvents in, 225f, 255, 265–267, 265f, 266t, 268
  developing the chromatogram in, 255
  fluorescent indicators in, 261–262
  known standards in, 259–260
  mobile phase in, 255
  overview of, 255–256
  plates for, 257
    backing for, 257
    development of, 260–261, 260f
    spotting of, 258, 259f, 267–268, 269
    trimming of, 257
  principles of, 255–256
  quantitative information from, 269
  reverse-phase, 257
  R_f values in, 263–264, 263f, 268
  safety precautions for, 262
  sample application in, 257–260, 259f
  solvent front in, 256, 268
  sources of confusion and pitfalls in, 267–268
  steps in, 264–265
  in synthetic organic chemistry, 267
  visualization methods in, 256, 261–263, 262f, 268
Three-necked flasks, 46f
TLC. *See* Thin-layer chromatography
Toluene
  as recrystallization solvent, 222t
  UV spectrum of, 469, 469f
Total energy, molecular, 109–115
Toxic exposures, 5–8, 13–14
  acute, 13
  chronic, 13
  information sources for, 14–19
  LD_{50} and, 13–14, 17t
  management of, 12–13
  physical effects of, 14
  prevention of, 5–8
  severity of, 13–14
  terminology for, 17t
  testing and reporting for, 14
Toxic metal waste, 21
Trailing zeroes, 66
Transfer pipets, 61, 61f
Transferring methods
  for liquids
    with pipet, 61, 61f
    with syringe, 59, 90, 90f, 101–103, 103f
  for solids, 54–55, 54f, 55f
Transmittance, 316
Traps, gas, 103–106, 104f–106f
Triplets, in spin-spin coupling, 379
Twist-boat conformation, 111–112

Two-dimensional correlated spectroscopy, 431–435
  heteronuclear (C,H)-correlated, 434–435, 434f
  homonuclear (H,H)-correlated, 432–433, 433f
  long-range coupling and, 435

U
Ultraviolet detectors, in high-performance liquid chromatography, 289–290
Ultraviolet spectroscopy. *See* UV/VIS spectroscopy
Ultraviolet visualization, in thin-layer chromatography, 261–262, 262f
Uncertainties. *See* Measurement Uncertainties
Upfield shifts, 365
UV transparent cells, 474
UV/VIS spectrometers, 471–474
  calibration of, 473
  diode-array, 472, 472f
  dispersive, 471–472, 471f
  operation of, 473–474
UV/VIS spectroscopy, 311–348, 465–484
  applications of, 466
  Beer-Lambert law and, 466–467, 473, 475
  colored compounds and, 469–470
  conjugated compounds and, 469
  cuvettes in, 474
  electronic excitation and, 466–470
  electronic transitions and, 467–468, 467f, 468t
  instrumentation for, 471–474
  molar extinction coefficient and, 466, 468, 468t
  overview of, 465–466
  procedure for, 472–474
  samples for
    dust particles in, 475
    preparation of, 472–474
  solvents for, 473–474, 474t, 475
  sources of confusion and pitfalls in, 474–475
  UV transparent cells in, 474

V
Vacuum adapter, 46f
Vacuum distillation, 197–202. *See also* Distillation, vacuum
Vacuum filtration, 137–140, 137f–139f
  in recrystallization, 230, 230f, 232, 232f
Vacuum gauge, 200, 200f
Vacuum manifold, 200, 200f
Van der Waals forces, 130
Vanillin, in thin-layer chromatography, 262
Van't Hoff, Jacobus, 240
Vapor pressure
  calculation of, 176–177
  definition of, 173
  in distillation, 173
Vapor pressure–mole fraction relationship, 176–177, 177f
Vapor pressure–temperature relationship, 174f
  boiling point and, 174, 174f
Vapors, noxious, removal of, 103–106, 104f–106f
Variable transformers, for heating mantles, 75
Vials, reaction (conical), 48f
  volume markings on, 61
Vibrational spectroscopy, 311. *See also* Infrared (IR) spectroscopy
Vicinal coupling, 377–381
  coupling constant for, 382–383, 383f
Vinyl protons, chemical shifts of, 372–374, 372t
Virtual coupling, 393–394
Visible light spectroscopy (VIS), 310, 465–484. *See also* UV/VIS spectroscopy
Volume, measurement of, 55–61, 56f–58f, 60f

W
Wall-coated open tubular columns, 295
Washing, in extraction, 151
Waste management, 5, 20–21
Watch glass, tared, 232
Water. *See also* Liquid(s)
  codistillation with, 194–196, 196f
  hydrogen bonding in, 127–128
  polarity of, 129
  solubility in, 130–131
  as solvent, 24–25
    in recrystallization, 222t
Water aspirators, 104, 104f, 105, 105f
Water baths, 78–79
Water waste, 21
Water-jacket condenser, 48f, 87, 88
Wavenumber, 313
Web of Science, 505–506
Websites, 504–506. *See also* Information sources
  for organic compounds, 37–38
Weighing methods
  for liquids, 54
  for solids, 52–54, 53f
Weighing paper, 54–55, 54f, 55f
West condenser, 46f
Whatman filter paper, 132, 133t
Wiley Online Library, 502
Williamson microscale glassware, 49–50, 49f, 50f. *See also* Glassware
Williamson reaction tube, in boiling point determination, 175, 175f
World Wide Web, 506

Y
Yield, percent, 35–36

Additive parameters for predicting NMR chemical shifts of aromatic protons in CDCl$_3$

	Base value	7.36 ppm[a]	
Group	ortho	meta	para
—CH$_3$	–0.18	–0.11	–0.21
—CH(CH$_3$)$_2$	–0.14	–0.08	–0.20
—CH$_2$Cl	0.02	–0.01	–0.04
—CH=CH$_2$	0.04	–0.04	–0.12
—CH=CHAr	0.14	–0.02	–0.11
—CH=CHCO$_2$H	0.19	0.04	0.05
—CH=CH(C=O)Ar	0.28	0.06	0.05
—Ar	0.23	0.07	–0.02
—(C=O)H	0.53	0.18	0.28
—(C=O)R	0.60	0.10	0.20
—(C=O)Ar	0.45	0.12	0.23
—(C=O)CH=CHAr	0.67	0.14	0.21
—(C=O)OCH$_3$	0.68	0.08	0.19
—(C=O)OCH$_2$CH$_3$	0.69	0.06	0.17
—(C=O)OH	0.77	0.11	0.25
—(C=O)Cl	0.76	0.16	0.33
—(C=O)NH$_2$	0.46	0.09	0.17
—C≡N	0.29	0.12	0.25
—F	–0.32	–0.05	–0.25
—Cl	–0.02	–0.07	–0.13
—Br	0.13	–0.13	–0.08
—OH	–0.53	–0.14	–0.43
—OR	–0.45	–0.07	–0.41
—OAr	–0.36	–0.04	–0.28
—O(C=O)R	–0.27	0.02	–0.13
—O(C=O)Ar	–0.14	0.07	–0.09
—NH$_2$	–0.71	–0.22	–0.62
—N(CH$_3$)$_2$	–0.68	–0.15	–0.73
—NH(C=O)R	0.14	–0.07	–0.27
—NO$_2$	0.87	0.20	0.35

a. Base value is the measured chemical shift of benzene in CDCl$_3$ (1% solution).

Selected data on common acid and base solutions

Compound	Molarity	Density (g/mL)	% by weight
Acetic acid (glacial)	17	1.05	100
Ammonia (concentrated)	15.3	0.90	28.4
Hydrobromic acid (concentrated)	8.9	1.49	48
Hydrochloric acid (concentrated)	12	1.18	37
Nitric acid (concentrated)	16	1.42	71
Phosphoric acid (concentrated)	14.7	1.70	85
Sodium hydroxide	6	1.22	20
Sulfuric acid (concentrated)	18	1.84	95–98

Additive parameters for predicting NMR chemical shifts of alkyl protons in CDCl$_3$[a]

Base values

Methyl	0.9 ppm
Methylene	1.2 ppm
Methine	1.5 ppm

Group (Y)	Alpha (α) substituent	Beta (β) substituent	Gamma (γ) substituent
	H—C—Y	H—C—C—Y	H—C—C—C—Y
—R	0.0	0.0	0.0
—C=C	0.8	0.2	0.1
—C=C-Ar[b]	0.9	0.1	0.0
—C=C(C=O)OR	1.0	0.3	0.1
—C≡C-R	0.9	0.3	0.1
—C≡C-Ar	1.2	0.4	0.2
—Ar	1.4	0.4	0.1
—(C=O)OH	1.1	0.3	0.1
—(C=O)OR	1.1	0.3	0.1
—(C=O)H	1.1	0.4	0.1
—(C=O)R	1.2	0.3	0.0
—(C=O)Ar	1.7	0.3	0.1
—(C=O)NH$_2$	1.0	0.3	0.1
—(C=O)Cl	1.8	0.4	0.1
—C≡N	1.1	0.4	0.2
—Br	2.1	0.7	0.2
—Cl	2.2	0.5	0.2
—OH	2.3	0.3	0.1
—OR	2.1	0.3	0.1
—OAr	2.8	0.5	0.3
—O(C=O)R	2.8	0.5	0.1
—O(C=O)Ar	3.1	0.5	0.2
—NH$_2$	1.5	0.2	0.1
—NH(C=O)R	2.1	0.3	0.1
—NH(C=O)Ar	2.3	0.4	0.1

a. There may be differences of 0.1−0.5 ppm in the chemical shift values calculated from this table and those measured from individual spectra.
b. Ar = aromatic group.

Table 22.5 on page 372 contains additive parameters for predicting NMR chemical shifts of vinyl protons.